Stephan Pauleit • Adrien Coly
Sandra Fohlmeister • Paolo Gasparini
Gertrud Jørgensen • Sigrun Kabisch
Wilbard J. Kombe • Sarah Lindley
Ingo Simonis • Kumelachew Yeshitela
Editors

Urban Vulnerability and Climate Change in Africa

A Multidisciplinary Approach

Editors
Stephan Pauleit
Chair for Strategic Landscape Planning and Management
Technical University of Munich (TUM)
Freising, Germany

Sandra Fohlmeister
Chair for Strategic Landscape Planning and Management
Technical University of Munich (TUM)
Freising, Germany

Gertrud Jørgensen
Department of Geosciences and Natural Resource Management
University of Copenhagen
Copenhagen, Denmark

Wilbard J. Kombe
Institute of Human Settlements Studies
Ardhi University
Dar es Salaam, Tanzania

Ingo Simonis
School of Mathematics, Statistics and Computer Science
University of Kwazulu Natal
Durban, South Africa

Adrien Coly
Department of Geography
University of Gaston Berger
Saint Louis, Senegal

Paolo Gasparini
University of Naples Federico II
AMRA S.c.a.r.l.
Naples, Italy

Sigrun Kabisch
Department of Urban and Environmental Sociology
Helmholtz Centre for Environmental Research - UFZ
Leipzig, Germany

Sarah Lindley
School of Environment, Education and Development
University of Manchester
Manchester, UK

Kumelachew Yeshitela
Ethiopian Institute of Architecture, Building Construction and City Development (EiABC)
Addis Ababa University
Addis Ababa, Ethiopia

ISSN 1876-0899 ISSN 1876-0880 (electronic)
Future City
ISBN 978-3-319-03984-8 (hardcover) ISBN 978-3-319-03982-4 (eBook)
DOI 10.1007/978-3-319-03982-4

Library of Congress Control Number: 2015933631

Springer Cham Heidelberg New York Dordrecht London

Cover illustration: The aftermath of the flooding of April 2014 in Dar es Salaam, Tanzania. Roads and businesses are submerged under water, making access impossible to homes, schools and workplaces. Local newspapers reported the flood as one of the most destructive events the city has experienced in years due to the extent of the damage both to public infrastructure and private property.

Photo: Eric Schaechter; commissioned by the Helmholtz Centre for Environmental Research-UFZ ©2014 UFZ

English language review: Dr. Cynthia Skelhorn, Qatar Green Building Council, Doha, Qatar

Printed on acid-free paper

Springer International Publishing AG Switzerland is part of Springer Science+Business Media (www.springer.com)

Preface

Africa, a continent of diversity and contrasts, is often perceived through news stories of political unrest and humanitarian crisis. Documentaries on the wealth of natural resources and wildlife might also be part of the general image. Yet, many other and very different facets of this continent exist; one of these, frequently overlooked by media coverage, is its enormous drive towards urbanisation. Fuelled by strong overall population growth and migration from the rural areas towards cities, African urban areas are expanding at an incredible pace. The city of Dar es Salaam, for instance, which is one of the case study cities treated in this book, is predicted to double its population from over four million currently to more than eight million in the coming 15 years. While the potential of urbanisation for furthering economic development and social progress is increasingly recognised by African leaders, it also poses enormous challenges – rapid urban population growth goes hand-in-hand with informal settlements, which usually lack basic infrastructure. Poverty is rampant. To successfully address these challenges of urban growth alone would be more than sufficient to absorb all of the efforts of policy-makers and researchers alike for the next decades.

Africa is also the continent where climate change will be particularly severe. Increasing frequencies and intensities of heat waves, droughts, sea level rise and storm surges, rainstorms and landslides are predicted to hit the continent, and are doing so already today. Urban areas are especially vulnerable as they are often poorly planned and built in zones that are exposed to natural hazards. Consequently, climate related disasters are claiming an ever-increasing toll on human lives and threatening the development of Africa's urbanising societies.

And yet, despite all of these challenges, urbanisation is the future of the African continent due to the developmental potential it offers. The task ahead is thus to find ways for more sustainable and climate resilient urban areas. At present, an understanding of how urbanisation and climate change interact in Africa is still scarce. A lack of regional and local data on climate change makes it difficult to translate potential changes into hazards. Knowledge on the vulnerability of urban society and its ecosystems, built environment and critical infrastructure, such as roads and hospitals, is widely absent. Considering these factors, what action is due? As the high degree of informality indicates, the capacity for planning and management of urban areas is very low in Africa. How can governance of urban areas be strengthened and civil society be supported to build climate resilience?

CLUVA – '**CL**imate Change and **U**rban **V**ulnerability in **A**frica' was a research project funded by the European Union within its 7th Framework Programme between 2010 and 2013 to address these questions. For the first time, a systematic study of African cities' vulnerability to climate change was applied in an inter- and trans-disciplinary research project. The main target was to develop an integrated approach to the assessment of urban vulnerability to climate change and to identify suitable strategies that may enhance the climate resilience of African cities. Not least, the project aimed to transfer new knowledge and tools to practice and increase the research capacity in this field in Africa.

The consortium consisted of six African and seven European partner organisations, led by AMRA, the Research Center for Analysis and Monitoring of Environmental Risks, located in Naples, Italy. The project team comprised expertise from a range of disciplines, including climate change modelling, hazard and risk assessment, the study of vulnerability, urban planning and governance analysis. Researchers from the following organisations were involved: Addis Ababa University, Ethiopia • AMRA Scarl, Italy (coordinator) • Ardhi University, Tanzania • Centro Euro-Mediterraneo per i Cambiamenti Climatici Scarl, Italy • Council for Scientific and Industrial Research, South Africa • Helmholtz-Zentrum für Umweltforschung UFZ, Germany • Københavns Universitet, Denmark • Norsk Institutt for By- Og Regionforskning, Norway • Technische Universität München, Germany • Université de Ouagadougou, Burkina Faso • Université de Yaoundé I, Cameroon • Université Gaston Berger de Saint Louis, Senegal • University of Manchester, UK.

In CLUVA, we concentrated on Sub-Saharan cities because of the particular need for a better evidence base in this huge part of Africa. Five case study cities were selected to represent the different geographic settings, processes of urbanisation and climate change impacts to be expected. These were:

- Saint Louis, a comparatively small town in Senegal, which is designated as a world heritage site due to its history as a trading town and its richness in colonial architecture. Located on islands of the Senegal river and a small strip of land between the river and the Atlantic ocean, it is mainly threatened by sea level rise and increased flooding from the Senegal river.
- Ouagadougou, the capital of Burkina Faso in the Sahel zone. Perhaps surprisingly, pluvial flooding puts the city most at risk but drought events and the consequent migration of rural people to the city are also a major threat.
- Douala in Cameroon, the large city at the mouth of the Wouri river where riverine and pluvial flooding are already now posing great problems which will be exacerbated by sea level rise.
- Addis Ababa, the capital of Ethiopia, and Dar es Salaam, Tanzania, where again, riverine flooding is the major environmental risk. Dar es Salaam may also be affected by a rising sea level and tropical storms.

This book provides a synthesis of the most relevant results of the CLUVA project. Its ambition is to cover the entire approach developed and applied in CLUVA. In addition, it contains contributions by renowned scholars in urban planning, social and political science, which helped us to substantially widen, and sometimes also challenged, our own insights from the CLUVA research. For instance, while the focus of CLUVA was on adaptation to climate change, Susan Parnell convincingly argues for the need to link the predominant adaptation and anti-poverty agendas with planning for mitigation in her invited contribution to the book. It is recognised that such an approach will require a fundamental reform and strengthening of urban planning in African cities.

Certainly, the book cannot give definite answers to the questions stated above; it intends, however, to offer the reader new perspectives on the challenges of urbanisation and climate change and to illustrate entry points for further action, both from local governments as well as the international community. The editors of this book and the researchers that were intensely involved in the CLUVA project believe that we made some useful advances in:

- Developing and applying novel approaches and tools to climate modelling which resulted in improved climate change data, hazard, vulnerability and risk assessments, as well as scenario modelling to support urban planning;
- Generating and disseminating new knowledge on the entire chain of climate change science for adaptation of Sub-Saharan cities, from producing detailed information of climate change scenarios to specific assessments of how social vulnerabilities of local communities interact with the physical vulnerability of houses in informal settlements and critical infrastructures. For the first time, the project produced information on the ecosystem services provided by the cities' green and blue spaces that may inform green infrastructure planning for adaptation. Governance and urban planning studies gave clear insights into current barriers to adaptation but also identified approaches that may work for building climate resilience. The project resulted in 45 deliverables and a number of scientific, policy-orientated and popular publications, conference presentations and posters (see www.cluva.eu);
- Building capacity by establishing new networks between researchers from Africa and Europe and between researchers and local stakeholders which we hope will persist and can even be strengthened in the future. At the time of writing, consortium partners are already involved in new projects that build on the CLUVA results. We are particularly proud of the extensive number of students that were able to do their PhD research within the CLUVA project. Several PhD workshops were organised, and Master students have also made valuable contributions to the project. Hopefully, this next generation will boost the science on climate change and urbanisation in Africa. Beyond strengthening scientific capacity, the project further co-produced a set of teaching modules on Master's level in the framework of UN Habitat's Cities and Climate Change Academy (CCCA) that involves a number of universities worldwide.

These achievements provide us with optimism that CLUVA has been able to make a significant impact on the science and practice of climate change adaptation in Africa. However, it is up to the reader of this book and those stakeholders that were collaborating with us during the project to decide whether we were able to fulfil our ambitions.

The CLUVA project was funded as a collaborative project by the European Union's 7th Framework Programme (Grant agreement no: 265137). We are thankful to our project officers from Directorate-General for Research and Innovation for having been greatly supportive of our project.

Working in a large consortium from two continents has been challenging, but in the end, a highly rewarding enterprise. Not least, establishing and maintaining an effective communication required a great effort from all of the partners. The project would not have come to a successful ending without a highly dedicated coordinator. We are deeply grateful to Guy Weets for safely guiding us through all of the project's stages and never losing his good temper when confronted with the difficult task to keep ends together. He was supported by an excellent project management team at AMRA Scarl, Italy, among which Angela Di Ruocco and Alfonso Rossi Filangieri need to be commended. The members of CLUVA's Scientific Advisory Board played an important role in supporting the project with constructive comments and advice. These were: Prof. Jochen Zschau, University of Potsdam, Claude Ngomsi and Bernhard Barth, both UN-Habitat.

We would further like to thank all the people associated with our case study cities – whether politicians, administrators, members from NGOs and development aid organisations or residents – for their great help, dedication and kindness by which they supported the project through hosting and attending workshops, meetings and conferences, being available for interviews, providing us with data, and more. Our partners Bernhard Barth and Fernando Cabrera (UN-Habitat), Vilma Hossini (UN University) and David Dodman (IIED) need to be especially mentioned for their willingness to join forces in the frame of the Cities and Climate Change Academy (CCCA). We are also very grateful to the African Union for hosting our final conference in Addis Ababa.

Sincere thanks go to Ben Wisner, Mark Pelling, Adolfo Mascarenhas, Ailsa Holloway, Babacar Ndong, Papa Faye, Jesse Ribot, David Simon and Susan Parnell who greatly enriched this book with their invited contributions.

Cynthia Skelhorn merits a particular thank you for her outstanding language editing efforts which went far beyond improving English writing but also included a careful review of the manuscript's content.

Besides, we would like to address our gratitude towards all internal and external reviewers of the book's contributions for their highly valuable and important comments on the chapters' early versions.

Finally, we would like to thank Catherine Cotton and Ria Kanters for accepting the book proposal initially for the Future City series, and Nel van der Werf and Izabela Witkowska from Springer Publishers for their extended patience and support in producing this book.

Freising - Munich, Germany
16.02.2015

Stephan Pauleit
Chair of the Editorial Board

Contents

Chapter 1
Urbanisation and Climate Change in Africa: Setting the Scene

Angela Di Ruocco, Paolo Gasparini, and Guy Weets

Abstract African urban population is developing with a growth rate of 3.2 % per year on average, the highest in the world. Almost all Africa's cities with one million inhabitants or more are currently located in areas exposed to at least one natural hazard. Inevitably, natural disasters in African urban areas are more likely to occur as expanding cities place an increasing number of people in the path of extreme natural events, often those in vulnerable accommodation, reliant on poor infrastructure with little resilience to impacts. According to the IPCC, Africa is one of the most vulnerable continents to climate change so climate drivers are likely to aggravate this situation still further. In spite of this knowledge, East, West and Central Africa are among the regions of the world that are least covered by climate change studies. The "Climate Change and Urban Vulnerability in Africa (CLUVA)" project was conceived to address this issue for selected African cities (Addis Ababa in Ethiopia, Dar es Salaam in Tanzania, Douala in Cameroon, Saint Louis in Senegal and Ouagadougou in Burkina Faso). CLUVA downscaled climate change projections to a resolution of 8 km for the case study cities and focused on the assessment of five main climate change affected hazards: floods, droughts, desertification, heat waves and sea level rise. Innovative methodologies have been developed in a multi-disciplinary approach, both for climate change vulnerability assessment, and for the definition of new risk mitigation and adaptation strategies, aiming to provide planners and policy makers with tools for the development of more climate resilient cities.

Keywords Africa • Urbanisation • Climate change • Hazard • Multi-risk • Vulnerability

A. Di Ruocco (✉) • G. Weets (✉)
AMRA – Center for the Analysis and Monitoring of Environmental Risk, Via Nuova Agnano 11, 80125 Naples, Italy
e-mail: angela.diruocco@amracenter.com; guy.weets@gmail.com

P. Gasparini (✉)
University of Naples Federico II, AMRA S.c.a.r.l., Naples, Italy
e-mail: paolo.gasparini@amracenter.com; paolo.gasparini@na.infn.it

S. Pauleit et al. (eds.), *Urban Vulnerability and Climate Change in Africa*, Future City 4,
DOI 10.1007/978-3-319-03982-4_1

Introduction

The evaluation of disaster risk in urban areas is becoming increasingly important. Most of the world's population and means of production are concentrated in these areas, with a consequent rapid growth in the number of deaths and the amount of devastation caused by disasters. In particular, nearly three quarters of the world's urban population and most of its largest cities are now in low- and middle-income nations. Since 1950, there has been a sevenfold increase in the urban population of low- and middle-income nations and a much-increased concentration of people and economic activities in low-lying coastal zones and other areas particularly exposed to flooding and extreme weather events. The United Nations Population Division suggests that almost all of the world's population growth up to 2025 and beyond will be in urban areas in low- and middle-income nations. How this very large and rapidly growing urban population is served and governed today has a major implication for development and for reducing disaster risk in the near future (Dodman et al. 2009).

Despite many international and local initiatives on disaster risk management and advances in scientific knowledge, the social and economic impacts of natural disasters on emerging economies and developing countries are growing. This is due to fragile economies that are unable to absorb the shocks caused by natural hazards affecting an increasing vulnerability on the exposed population, and further, as a result of the rapid growth of urban population, weak institutions and rampant conflicts (Pelling and Wisner 2009). Most of these countries are currently struggling to implement an effective risk reduction strategy, and climate change has the potential to rapidly exacerbate this situation, particularly across the African continent. In fact, among developing countries, those in Africa could be the most affected by climate change. Climate change threatens Africa's cities and rapidly urbanising coasts where about 40 % of Africa's population – some 400 million people – lives. According to projections of the United Nations, by around 2030, this figure is expected to grow to approximately 50 % of Africa's projected total population of around 1.5 billion. Since 2010, the urban population growth rate in Africa has been 3.2 % per year on average. This is the highest in the world, resulting in more urban areas with bigger populations, as well as an expansion of the existing ones (UNDESA 2012; UN-Habitat 2014). Fifty cities in Africa currently have populations of more than a million, and by 2015, nine cities are expected to be added to this group.

In many urban areas, rates of economic growth and infrastructure development have lagged behind urbanisation rates, resulting in high levels of unemployment, inadequate standards of housing and services, and negative impacts on human health and development. Environmental disasters and conflicts have also caused many people to flee rural areas and to seek refuge in urban centres. Weather-related disasters are causing increasing damage to water supplies, which are already scarce in many places. Other critical infrastructure, such as energy, transport, and telecommunications – already struggling to keep pace with urbanisation – may face additional threats from climate change related hazards. These vulnerabilities add to the significant sustainability challenges faced by African cities, including urban sprawl, population growth, pollution and the loss of biodiversity. The vulnerability of African cities is considered to be influenced not only by changing biophysical conditions, but also by highly dynamic social, economic, political, institutional and technological processes. Thus, urban and land use planners, managers and researchers within the African context need reliable forecasts of the local impacts of climate change and better equipment for strengthening the coping capacities of urban communities. Improved knowledge of the direct and indirect impacts of climate change and the spatial and temporal scales over which these are felt will ultimately benefit urban communities more widely, both within and beyond the developing world.

Urbanisation in Africa and Disaster Risk

Population Trends

In 2009 Africa's total population for the first time exceeded one billion. Whereas it took 27 years for the continent's population to double from 500 million to one billion people, the next 500 million will be added to the population in less than 20 years. Around 2027, Africa's demographic growth is expected to slow down so that it will likely take 24 years to add the next 500 million, reaching the two billion mark around 2050 (UN-Habitat 2010). According to recent projections of the United Nations (UNDESA 2012; UN-Habitat 2014) until 2050, the population will continue to grow in each of the five African geographical regions (Eastern, Middle, Northern, Southern and Western Africa). Western and Eastern Africa will remain the most populated areas, as shown in Fig. 1.1.

While the global average of urbanised population passed the 50 % mark in 2010, Africa is still predominantly rural with a current percentage of urban population around 40 %, but this percentage has grown continuously since 1950. According to projections produced by the United Nations, the urban population trend is expected to continue to rise until around 2050, both for the African continent as a whole and for each of the five African geographical regions. Figure 1.2 shows these rising trends in urban population based on data provided by the United Nations (UNDESA 2012). According to these data, by around 2030, some 50 % of Africa's total population will have become urban dwellers, and by the early 2040s, African cities will be home to nearly one billion people, a number that is equivalent to the continent's entire population in 2009.

If the African urban population continues to grow as expected, as early as 2015 African cities with populations of at least one million will outnumber those of Europe or North America (Table 1.1). Figures 1.3a, b show the location of large cities with at least one million people in the world in 1950 and 2025. It can be seen that by 2025, the largest African cities are expected to be located mainly in the Sub-Saharan region and the largest agglomerations will be in coastal areas.

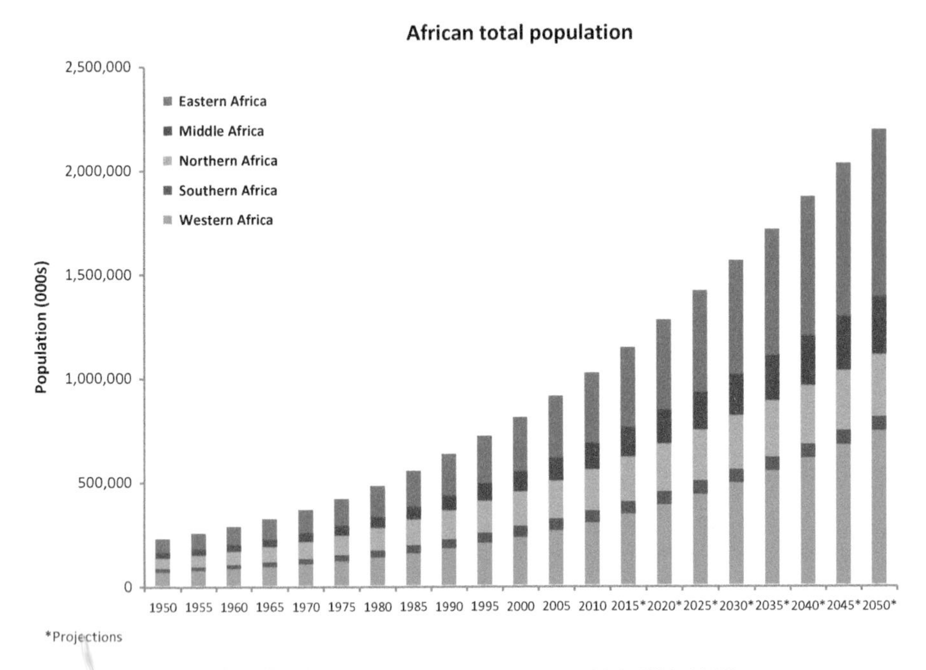

Fig. 1.1 African total population trend until 2050 (Based on: UNDESA 2012)

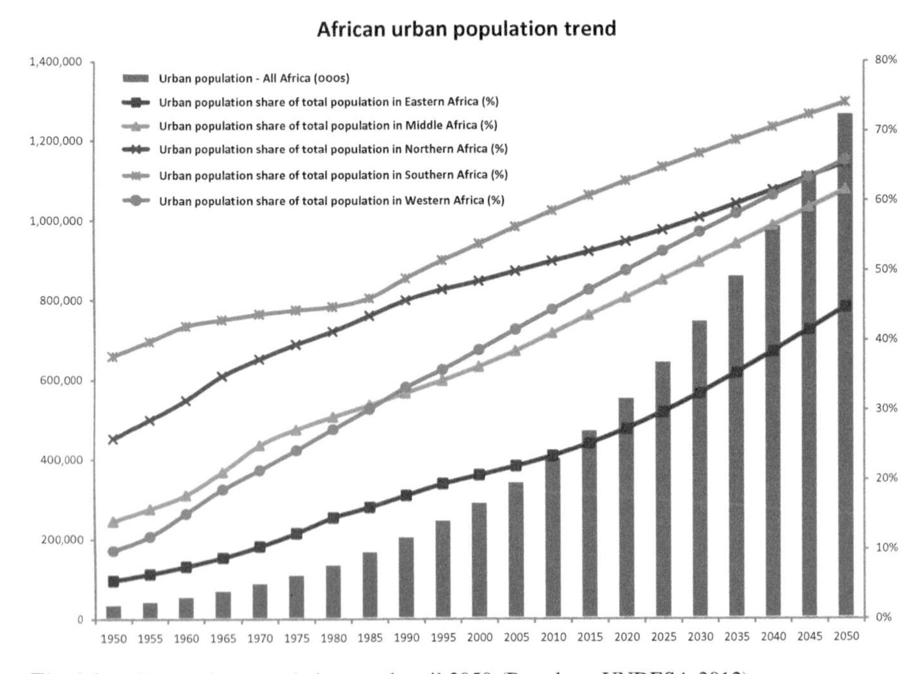

Fig. 1.2 African urban population trend until 2050 (Based on: UNDESA 2012)

Table 1.1 Distribution of large cities with at least one million people by major geographic area, 1950–2025

Region	1950	1955	1960	1965	1970	1975	1980	1985	1990	1995	2000	2005	2010	2015	2020	2025
Africa	2	3	3	6	8	9	15	20	24	28	37	46	50	59	76	93
Asia	26	34	40	48	54	61	73	88	113	137	168	188	226	276	315	359
Europe	23	23	28	34	36	38	45	47	49	50	53	53	54	55	60	63
Latin America and the Caribbean	8	8	11	14	17	20	27	31	39	43	50	54	63	69	76	85
Northern America	14	16	22	24	28	31	33	34	36	39	41	45	50	53	56	62
Oceania	2	2	2	2	2	2	3	4	5	5	6	6	6	6	6	6

Based on: UNDESA (2012)

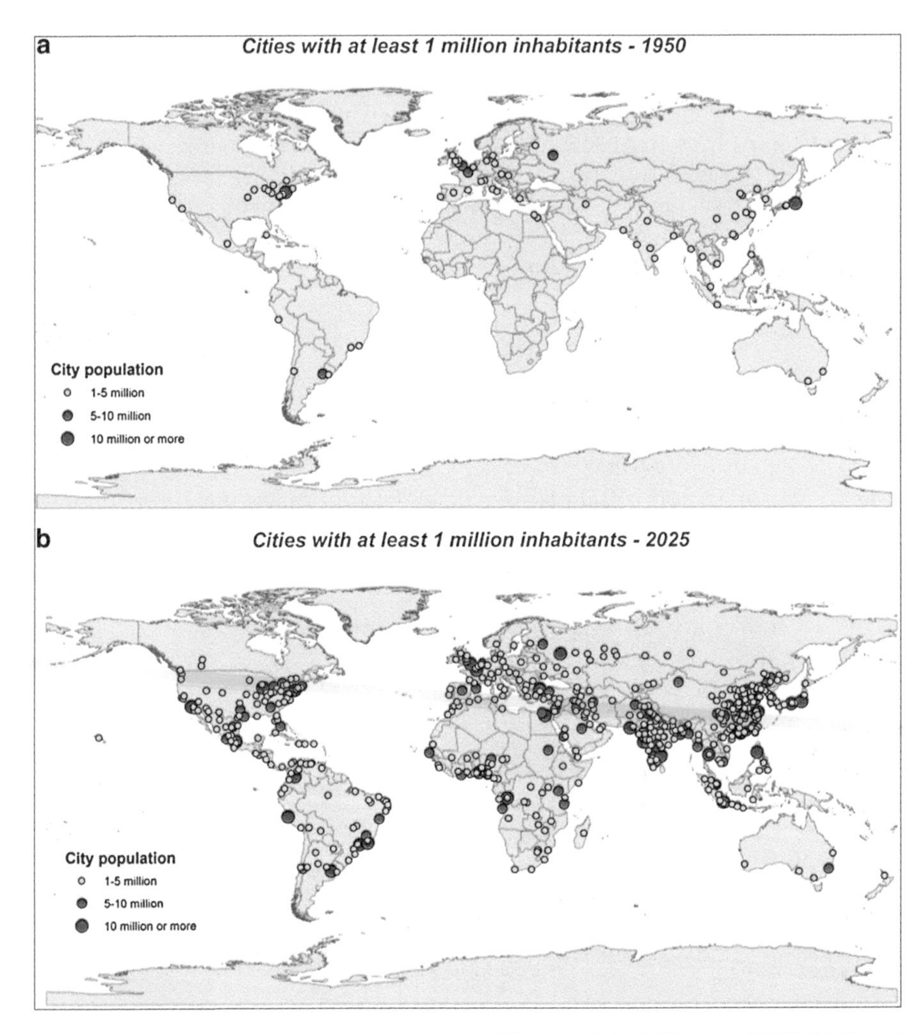

Fig. 1.3 Location of large cities with at least one million people in 1950 (**a**) and 2025 (**b**) (Based on: UNDESA 2012)

Exposure and Vulnerability to Natural Disasters

Almost all of Africa's cities with one million inhabitants or more are currently located in areas exposed to one or more natural hazards (cyclones, droughts, earthquakes, floods, landslides, volcanoes) (Fig. 1.4). Only 10 % of these cities are not threatened by any of these natural hazards (UNDESA 2012). This means that a considerable number of people and economic activities are potentially exposed to natural disasters. Looking ahead, this number could climb further due to the expected increasing proportion of the population in urban areas.

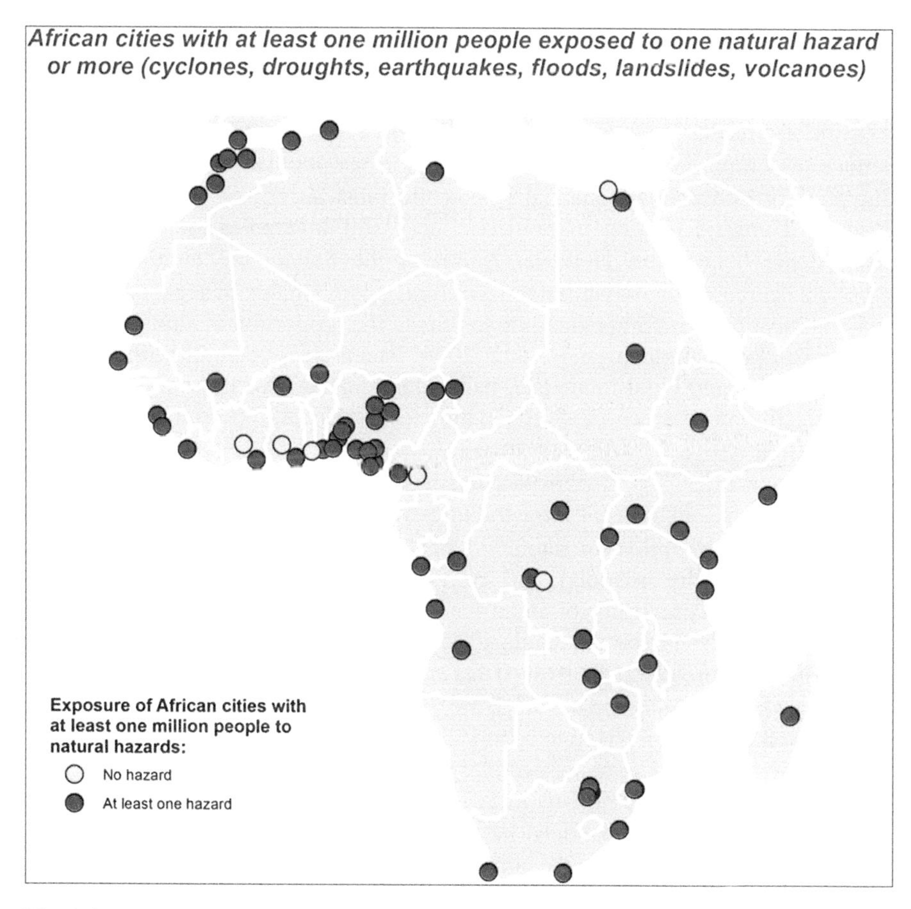

Fig. 1.4 African cities with at least one million inhabitants currently exposed to one or more natural hazards (Based on: UNDESA 2012)

In addition to the growing exposure to natural disasters, a significant proportion of the African urban population is ill served by protective infrastructure and services. In fact, a key characteristic of current African urbanisation is that, unlike many cities in Asia and Latin America, urbanisation in Africa appears to have become decoupled from economic development (Cohen 2006). The vast majority of African cities are economically marginalised in the new global economy and lack direct foreign investment; most of them are run by administrations that understandably struggle to provide basic infrastructure and essential services given that their physical growth completely outstrips the macro-economic performance required to support and sustain it (Cohen 2006). The 'solution' to this institutional gap has been for much urban expansion to occur outside of the official and legal frameworks of building codes, land-use regulations, and land transactions. A substantive part of the housing stock continues to be created and modified informally and illegally (usually on land that is occupied, sub-divided or developed illegally in the first place) (Dodman et al. 2009).

In recent years, Africa as a whole has shown that informal settlements can be effectively reduced as 24 million African slum dwellers saw their living conditions improve during the first decade of the twenty-first century. Progress has been uneven across the continent, though. Northern Africa collectively managed to reduce the proportion of slum dwellers in its urban population from 20 to 13 %. Egypt, Morocco and Tunisia have been the most successful countries in this respect. Tunisia has been particularly successful in eradicating slums within its national boundaries. However, south of the Sahara the number of slum dwellers decreased by only 5 % (or 17 million), with Ghana, Senegal and Uganda making most progress and managing to reduce the proportion of slum dwellers in their urban populations by more than 20 %. Much work remains with regard to urban poverty and slum incidence, particularly in view of the rapid growth of African cities, most of which results in the proliferation of informal settlements in Sub-Saharan Africa (UN-Habitat 2010).

Population expansion in African cities has created and will continue to create serious challenges in terms of infrastructure. Much service provision is private and informal (for water, often for sanitation, healthcare and solid waste management, sometimes even for schools) and of poor quality or inadequate coverage. In addition, a high proportion of urban economic activity and of livelihoods derived from this is outside the formal, regulated economy (Dodman et al. 2009). The levels of infrastructure provision in urban Africa are the lowest across the world's regions: average water and sanitation coverage are 89 % and 69 %, respectively; electricity 69 %; paved roads 28 %; fixed telephone lines 4 %; mobile telephones and internet connectivity 57 % and 10 %, respectively (UN-Habitat 2012).

The rapid population growth in Africa is also inducing a unique effect on land cover. Cities are expanding far beyond formal administrative boundaries, largely propelled by the use of the automobile and land speculation. A large number of cities are characterised by land-consuming suburban sprawling patterns that often extend over large areas. Cities are becoming endless expanses, with high degrees of fragmentation of the urban fabric. At the periphery, residential neighbourhoods are often characterised by semi-continuous low-density developments which, along with under-used spaces and fragmented built-up areas in the intermediate city-rings, are contributing to dramatic reductions in residential densities (UN-Habitat 2012). This in turn puts pressure on urban ecosystems and the urban ecosystem services upon which much of the population ultimately depends for health and wellbeing (Douglas 2012).

The haphazard creation and spread of informal settlements and unregulated enterprises, the discontinuous expansion of urban zones and the limited provision of associated basic services are all indicators that African cities are developing in a problematic manner, with little or no official control. Inevitably, natural disasters are likely to happen more often as these dysfunctional cities place an increasing number of people in the path of extreme natural events of all forms (Fig. 1.4). Those living in vulnerable accommodation, reliant on absent or sub-standard infrastructure and with little inherent resilience to impacts are likely to be the main victims and to suffer the most hardship as a result (IMechE 2013).

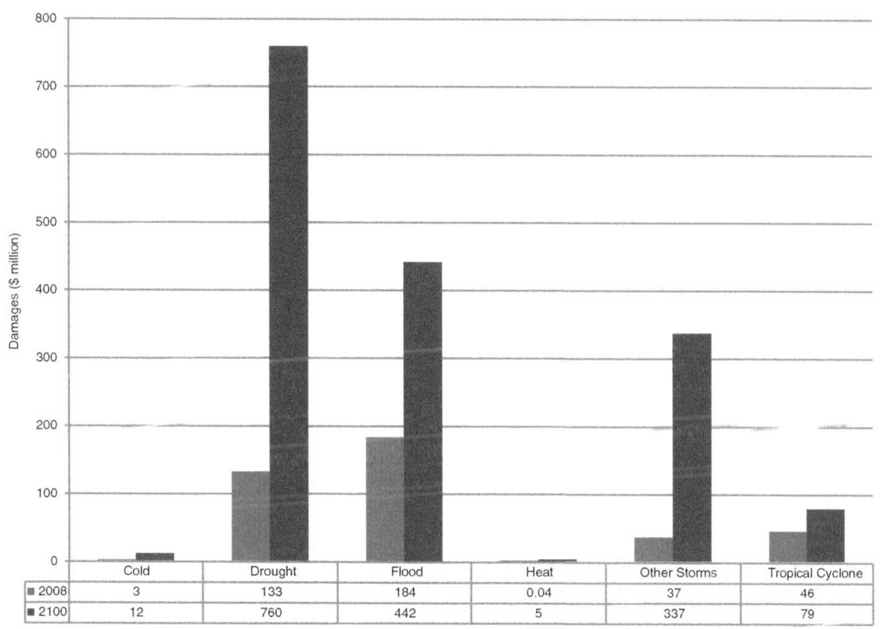

	Cold	Drought	Flood	Heat	Other Storms	Tropical Cyclone
2008	3	133	184	0.04	37	46
2100	12	760	442	5	337	79

Fig. 1.5 Current (2008) and projected damages (2100) from extreme events without climate change in Africa (Based on: Mendelsohn and Saher 2010)

The estimation of natural disaster damages in Africa remains extremely challenging as data are often poorly reported or simply not available. Based on the available data for Africa, during the period 1990–2008 climate-related hazards ("extreme events") resulted in an average of USD 400 million a year in damages (Mendelsohn and Saher 2010). Of these damages, floods account for 45 % and droughts for 33 %.

Without conscious change in adaptation policies to extreme events, baseline damages, not including the effects of climate change, are expected to quadruple to USD 1.6 billion a year from economic and population growth alone. Floods and droughts are expected to continue to be the main sources, but damages from droughts are also anticipated to become more prominent (Fig. 1.5).

Much of the natural disaster risk associated with large African cities is the result of major sections of the rapidly growing population being forced to settle on sites which are inadequate for this purpose. Even if most of the large African cities which exist today were initially formed on safe sites, their very large population expansion and the failure to make provision for infrastructure and other risk-reduction measures are now pushing parts of their population onto dangerous sites (Dodman et al. 2009). In addition, when lower income groups have difficulties in affording or accessing land for housing, they are more likely to resort to 'fragile' sites, i.e. those which are more likely to expose them to environmental hazards.

Box 1.1: Small Urban Centres in Africa
A considerable proportion of the population of Africa live in tens of thousands of small urban centres and in hundreds of thousands of large villages that have several thousand inhabitants each, which could be considered as small urban centres in their own right (Dodman et al. 2009). In terms of major disasters, it is likely that risk is lower in most of a country's small urban centres compared to its larger ones due to a lower potential for inhabitants to be living on dangerous sites. However, most small urban centres (and large villages) in low- and middle-income nations have a lack of infrastructure and limited services. The weakness and lack of investment capacity within their governments would suggest that large sections of their population face extensive risks (i.e. the risk of premature death, injury and impoverishment from all events whose impact is too small to be classified as major disasters). Small urban centres that begin to grow very rapidly may also face a large increase in extensive risk as the backlogs in infrastructure and services increase (Satterthwaite 2006; Dodman et al. 2009). Chapter 5 provides an overview of the challenges of climate change in small African cities and a discussion about the ways in which small urban centres provide opportunities to tackle these challenges more effectively.

These figures could be further aggravated by the additional effects of climate change. In fact, the impacts of climate change are likely not only to increase the frequency and/or the intensity of events that could cause disasters, but also to accentuate their uneven social consequences (Dodman et al. 2009). Further research to improve understanding of climate change impacts and their wider consequences in urban Africa is therefore urgently required.

The CLUVA Project: Assessing the Additional Effects of Climate Change in Urban Africa

Although Africa is considered a continent particularly vulnerable to climate change, its real impact, especially on a local scale, is still poorly understood. This is clearly shown by several reports summarising the present situation, such as those of the International Panel of Climate Change (IPCC) (IPCC 2007a, b), the "Background Paper of the Workshop on Impacts, Vulnerability and Adaptation In Africa" (UNFCCC 2006), and several documents produced by the World Bank (e.g. World Bank 2011). Predictions of climate change impacts on Africa in the twenty-first century are commonly based on the outputs from Global Circulation Models (GCMs) with low spatial resolution and which fail to represent two potentially important drivers of African climate variability, namely the El Niño/Southern Oscillation and land cover change (Hulme et al. 2001). The task of developing reliable predictions of future climate change in Africa is further complicated by the lack of accurate baseline data on the current climate and by the intricacies of present and past climate variations in space and time (DFID 2004a, b). In fact, many African regions are becoming recognised as having climates that, on intra-seasonal to decadal timescales, are among the most variable in the world. Changes in vegetation, hydrology and dust transfer from land surfaces to the atmosphere all modify large-scale atmospheric properties in the region and have the potential to impact climate variability. In spite of this, all the existing models consistently indicate that the spatial distribution patterns, frequency and intensity of weather-induced hazards are being changed significantly by climate change on a continental scale (IPCC 2007b).

A qualitative representation of the impact of expected climate change on four hazards (floods, droughts, sea level rise and cyclones) is shown in Fig. 1.6.

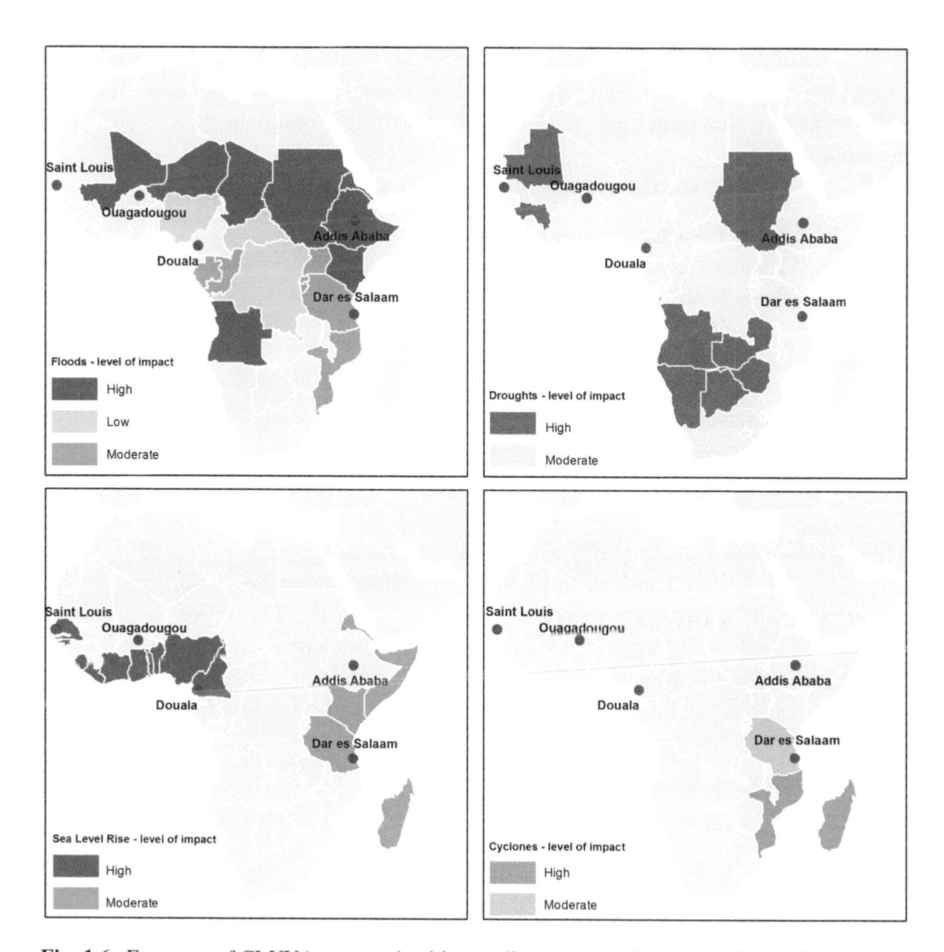

Fig. 1.6 Exposure of CLUVA case study cities to climate change hazards (Adapted from: World Bank 2009)

However, for impact studies, particularly in areas characterised by a diverse and heterogeneous land surface, quantitative fine-scale climate information is needed.

The main objective of the European Commission's 7th Framework Program project CLUVA (CLimate change and Urban Vulnerability in Africa)[1] has been to assess the impact of climate change on the urban vulnerability of five African case study cities and to develop innovative tools and methodologies through which they can be more effectively managed. The selection of the cities was based on three criteria: the cities must represent the most typical climates of Sub-Saharan Africa, face significant climate-related hazards and have a local university with the capacity and willingness to effectively contribute to the research. This was important because the CLUVA project also aimed to enhance research capacity within local academic institutions, and therefore, to support more autonomous research activity into the analysis of climate change and its impacts over the longer term. Finally, the project emphasised the importance of working with local stakeholders so that much of the research activity was carried out in close collaboration with practitioners. In turn, this helped to consolidate and further develop academic-practitioner networks and support the sort of knowledge exchange activities that help to ensure that academic research has maximum practical benefit for local decision-makers.

According to these criteria, the following cities were selected:

- Addis Ababa, Ethiopia
- Dar es Salaam, Tanzania
- Douala, Cameroon
- Ouagadougou, Burkina Faso
- Saint Louis, Senegal.

The CLUVA Case Study Cities

The CLUVA cities, located in East and West Africa, encompass coastal, estuary, inland, and highland characteristics and feature different weather conditions such as tropical dry, tropical humid and Sub-Saharan climate. The cities range from medium to large (i.e. from hundreds of thousands to millions inhabitants) and are confronted not only with increasing weather related hazards, but also with the pressure of a contrasting population mix (modern and traditional) (Jean-Baptiste et al. 2011).[2]

---

[1] Seventh Framework Programme, Grant agreement no. 265137: "CLimate change and Urban Vulnerability in Africa", 2010–2013, www.cluva.eu

[2] Cities were involved in research activities and selected for in-depth research to different extent, depending on their interest/disposition/suitability; for further information on case study selection, please see the individual chapters of this volume.

Addis Ababa is the capital and by far the largest city of Ethiopia. Hosting 30 % of the urban population of Ethiopia, Addis Ababa is one of the fastest growing cities on the continent. Its population has nearly doubled every decade. Its population of four million people in 2008 is set to increase to 12 million by 2024 (UN-Habitat 2008a). The capital covers an area of about 540 km^2 from which 290 km^2 is built-up (Jean-Baptiste et al. 2011). The city is a self-governing chartered city with its own city council. It is divided into 10 Sub-Cities. Among them, Kolfe Keranio, located west of the city, has the highest number of inhabitants in contrast to Akaki Kaliti in the South with the lowest population (Melesse 2005). These Sub-Cities are further divided into a total of 116 Weredas/Kebeles, which are the lowest level of city administration (Jean-Baptiste et al. 2011). Despite a diversified city economy, a low level of income persists and progress is uneven across different social groups. This leads in turn to restricted access to health and education (Maccioni and Zebenigus 2010). According to the data provided by UN-Habitat, 69 % of all employment in Addis Ababa is informal and 69 % of households are located in slum areas (UN-Habitat 2010). Addis Ababa has a pronounced rainfall peak during the boreal summer (June to August) and exhibits a rainfall minimum during the boreal winter (December to February). The city has a temperate climate due to its high-altitude location in the subtropics. Average monthly temperatures vary between 10 and 20 °C, and are the lowest during summer, due to the prevailing cloud cover experienced during this season (CSIR and CMCC 2013). Drought and heat waves are the most common climate related hazards threatening Addis Ababa, but flooding is also increasingly affecting several areas of the city (Giugni et al. 2011; Jean-Baptiste et al. 2011).

Dar es Salaam (DSM) is the largest city in Tanzania with an estimated population of 3.4 million and an annual population growth of 4.1 % (Jean-Baptiste et al. 2011). DSM is the fastest growing region among 26 others in Tanzania and ranked amongst the ten fastest growing cities worldwide. The population is expected to exceed 4.5 million in 2020 (Jean-Baptiste et al. 2011; UN-Habitat 2008a, b). The city is located on the East African coast and has a wider city-region covering almost 1,400 km^2. The region is headed by the Dar es Salaam Regional Commissioner whilst the city is managed by the Dar es Salaam City Council. The area is sub-divided into three autonomous municipal councils or districts: Kinondoni (531 km^2) to the North, Ilala (210 km^2) in the Centre and Temeke (652 km^2) to the South. Each council is divided into 11 sub-divisions and each of these is further segmented into 73 Wards (Jean-Baptiste et al. 2011; UN-Habitat 2009). Sub Wards, locally known as Mtaas, are DSM's lowest administrative level (Jean-Baptiste et al. 2011). Compared to other Tanzanian regions, Dar es Salaam hosts the highest (65.1 %) percentage of working-age people 15–64 years (URT 2006). Urban unemployment, however, continues to persist due to the differential between DSM's annual migration rate (10 %) and its annual economic growth (4 %). The literacy rate has increased steadily over the last decades and Kiswahili is the most dominant language (Jean-Baptiste et al. 2011). The present day climate of Dar es Salaam is characterised by the strong seasonal rainfall cycle, with the "long rains" from March to May, and the "short rains" from November to January. These rainfall maxima are induced by meridional displacements of the Inter-tropical Convergence Zone. It experiences peak temperatures during the austral summer from December to February, due to the peak in solar radiation (CSIR and CMCC 2013). The city is particularly susceptible to climate threats like sea level rise and coastal erosion, drought, heat waves and water scarcity, strong winds and flooding (Jean-Baptiste et al. 2011; Dodman et al. 2011).

Douala is the economic capital and the largest city of Cameroon with a population of about 2.1 million people in 2011. Douala's population is equivalent to around 20 % of Cameroon's urban population and 11 % of the country's total population. It has an annual growth rate of 5 %, more than double the national average of 2.3 % (Jean-Baptiste et al. 2011). The city is divided into six communes, each with its own headquarters: Douala 1 (Bonandjo), Douala 2 (Newbell), Douala 3 (Logbaba), Douala 4 (Bonassama), Douala 5 (Kotto), Douala 6 (Monako). The first five communes are urban areas while the sixth is a rural zone. The city is led by a community council of 37 members and two government representatives (Jean-Baptiste et al. 2011). Douala is a major port and industrial centre (Kemajou et al. 2008), but there is also a significant urban agricultural activity within the metropolitan area. Children's access to primary school is universal (SITRASS/SSATP 2004). Self-medication and recourse to traditional practitioners are frequent because of the difficulties in reaching healthcare facilities (Jean-Baptiste et al. 2011). Douala experiences a wet, tropical monsoonal climate, with the average total annual rainfall exceeding 3,000 mm. Rainfall peaks during the boreal summer and autumn (June–November) when the monsoonal circulation dynamics are best developed. Average monthly temperatures are generally in the range from 25 to 30 °C throughout the year (CSIR and CMCC 2013). The main climate related hazards affecting the city of Douala are flooding, induced by the Wouri river system, and heat waves (Tonyé et al. 2012).

Ouagadougou is the capital of the Republic of Burkina Faso. It extends over 520 km^2, of which 217.5 km^2 are urbanised. 70 % of the industrial activities of the country are concentrated in the capital which hosts a population of 1.5 million inhabitants. In 2020, the capital is expected to reach 3.4 million inhabitants, making it one of the most rapidly growing cities in the region. Ouagadougou faces several urban challenges; among them is poverty, with 50 % of the population living in poor conditions. Women are particularly vulnerable with less access to education and employment, and they are also more exposed due to restrictions on their access to land (Jean-Baptiste et al. 2011). Ouagadougou is made up of five districts, 30 sectors and 17 villages. Since 1995, a council of 90 members elected for a 5-year mandate has governed the city (Jean-Baptiste et al. 2011). The Mayor of the city is the executive leader of the municipal authorities. The Ouagadougou region has a pronounced rainfall maximum during the boreal summer (June to August), and experiences very dry winters (December to February). The summer rainfall maximum occurs in relation to the West African monsoon. The city has a tropical climate with average monthly temperatures ranging between 20 and 30 °C (CSIR and CMCC 2013). Drought, flooding, heat waves and desertification represent the most critical climate related hazards for Ouagadougou (Tonyé et al. 2012).

The city of ***Saint Louis*** is an archipelago located on low-lying islands encompassing the Langue de Barbarie spit, Ndar Island and the Sor district along the east–west axis (Diagne 2007). The city is surrounded by low-lying floodplains and marshes while sitting on the edge of the Sahel. As a result, Saint Louis experiences heat wave periods and droughts throughout the year and flooding during the rainy season. Tidal flooding is also increasingly affecting Saint Louis due to rising sea levels. The main economic activity is fishing. Its population was about 200,000 inhabitants in 2011 and is expected to grow with a rate of 2.4 %. Urban growth, poverty and natural hazards constitute the main problems for the socio-economic stability in the city (Jean-Baptiste et al. 2011). Saint Louis is sub-divided into 20 districts, each of which is made up of 22 quarters or neighbourhoods. 33 % of households are affected by unemployment and/or low incomes and the majority of people affected by flooding in the city are the very poorest. Guet ndar district is one of the most populated zones of West Africa with 15 people per room (Jean-Baptiste et al. 2011). Saint Louis exhibits a very dry climate with an average annual rainfall of less than 300 mm. The bulk of the city's rainfall falls during the boreal summer (June to August). During summer the average monthly temperatures are around 25 °C, with the winters being cooler (CSIR and CMCC 2013).

The CLUVA Approach

CLUVA comprised a partnership of interacting researchers with expertise in several different fields from a range of African and European institutions dealing with climate modelling, natural and social risks, urban planning and governance. The heterogeneity of the available expertise enabled the project team to apply a strong set of multi-scale and multi-disciplinary quantitative and probabilistic methods to the five African case study cities. Furthermore, the cooperation between African and European partners over a period of three years and the outcomes of the project have significantly advanced the research capacity in Africa with the potential for long lasting effects.

Planning and implementing the CLUVA project was challenging in many respects. It required assembling a strong multi-disciplinary team; establishing a sound, functioning network between East African, West African and European research professionals; coping with language barriers; and finding appropriate ways to communicate and exchange data, documents and other research materials. The project involved working out probabilistic models of hazard and vulnerability with very limited data. It necessitated dealing with the fact that the effects of climate change are very complex and interwoven, as local human factors act to intensify the consequences of pure climatic processes. The project team had to work out the uncertainties of the proposed scenarios and communicate them. Finally, it had to propose feasible low cost risk reduction strategies, tools and actions at infrastructural, social and governance levels and present them in a way decision-makers could understand.

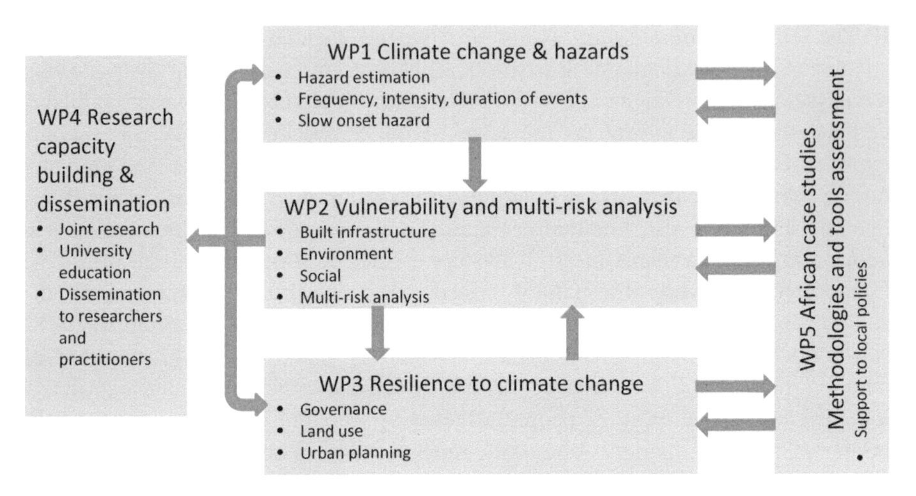

Fig. 1.7 The structure of the CLUVA project and its five Work Packages (WP)

The research activities of the CLUVA project were carried out through five interconnected Work Packages (WP) (Fig. 1.7), four of them responsible for the core research and one for its implementation:

- The ***'climate driven hazard' group (WP 1)*** worked on high resolution (8 km and 1 km) regional climate change projections up to the year 2050 for an area surrounding the five cities. This enabled the probability of extreme meteorological events to be assessed including the frequency and intensity of temperature and precipitation over the five case study cities.
- The ***'vulnerability and multi-risk assessment' group (WP 2)*** worked on the multiple dimensions of vulnerability (physical, social and ecological) and on the development of a probabilistic multi-risk framework in order to consider the various combinations of hazards and vulnerabilities in the African urban context. The developed framework includes the evaluation of physical damages and social context conditions representing indirect losses.
- The ***'innovative land use and governance strategies development' group (WP 3)*** worked through two strands of activity. The first involved analysing the case study cities' individual governance structures and working on measures that can be implemented locally. The second combined both vulnerability and land use indicators to identify areas and communities which are considered to be especially vulnerable and at risk.
- The ***'capacity building' group (WP 4)*** made a significant investment in research capacity building by: (1) offering young researchers and PhD students associated with the project a bespoke set of training workshops; (2) closely integrating them in the research teams and work programmes; and (3) joining forces with other donors (UN-Habitat, UN University) on developing Master's level climate change related modules and fostering their integration into the case study cities' university curricula.

- The ***city implementation group (WP 5)***: in addition to the four main research groups, an implementation group was developed for each city. The group's three objectives were: (1) to manage the interaction between the project team and local stakeholders; (2) to adapt the methods developed in CLUVA to specific local conditions; and (3) to help identify other specific risks considered important by the local stakeholders and not covered by the main research groups. This group was also responsible for producing reports for each city and supporting stakeholders over the longer term, i.e. once the CLUVA project had formally finished.

Climate Change and Natural Hazards in Africa

Climate Change Projections

CLUVA downscaled three IPCC scenarios to a resolution of 8 km for the case study cities, unprecedented for the African continent. The IPCC scenarios are neither scientifically controlled predictions nor forecasts. They are projections computed assuming possible trends in key driving forces, such as the rate of technological change, prices and so on. In fact, in order to assess how the climate is going to evolve in the future, it is necessary to have an idea of the atmospheric concentrations of greenhouse gases in the years to come, and their emissions from natural as well as man-made sources. For this purpose, IPCC has defined a set of 'emission scenarios', describing future releases (until 2100) of greenhouse gases, aerosols, and other pollutants into the atmosphere. The possibility that any single emission path will occur as described in a single scenario is highly unlikely. Thus, multiple instances are needed to provide a clear view on the potential range of reasonable scenarios.

Due to the requirements of policy-makers and decision-makers, the need for climate change information on the regional-to-local scale is one of the central issues within the global change debate. Until 2007, future climate projections were made through 40 scenarios according to the Special Report on Emissions Scenarios (SRES, IPCC 2000). Each one of these represents a different demographic, social, economic, technological, and environmental development trajectory. Afterwards, a new set of scenarios, the Representative Concentration Pathways (RCP), was developed. Distinguishable by their hypothetical 2100 radiative forcing level, they constitute a new way for providing inputs to climate models (Moss et al. 2010; IPCC 2013). Basically, they consider possible changes to each of the components (particularly atmospheric composition) known to influence the balance between incoming and outgoing radiation, and therefore, climate. So far, four RCP scenarios exist and each assumes a different level of radiative forcing by the year 2100: 3, 4.5, 6 and 8.5 W/m^2. In the CLUVA project, the IPCC emission scenarios considered are the A2, at the higher end of the SRES emission scenarios, and the 4.5 and 8.5 of the RCP set, i.e. one average and one extreme scenario.

Main Climate Dependent Hazards in Africa

Climate change is expected to create short- to long-term impacts on African urban areas, with increasing frequency and severity of extreme weather events, such as hurricanes, storm surges, floods and heat waves. Furthermore, semi-permanent or permanent effects such as sea level rise, falling groundwater tables or increased desertification are also likely to happen (UN-Habitat 2010).

The particular combination of the expected climate change impacts in the five CLUVA case study cities varies with their latitude, region, and among coastal and inland areas. In particular, existing research indicates that the coastal cities and those located on lagoons, estuaries, deltas or the mouths of large rivers (i.e. Saint Louis, Dar es Salaam and Douala) are likely to experience sea-level rises, increased flooding, more frequent heat waves and drought episodes. Inland and highland cities (e.g. Ouagadougou and Addis Ababa) will probably become more exposed to flooding as well as to more frequent heat waves and droughts. The cities of Saint Louis and Ouagadougou sitting at the edge of Sahel could also suffer an increasing desertification problem.

Thus, the CLUVA research activity focused on the assessment of the main hazards expected to be enhanced by climate change in the selected CLUVA cities, i.e. floods, droughts, desertification, heat waves and sea level rise. Chapter 2 of this book provides – for each CLUVA city – a detailed description of the CLUVA climate change scenarios in the five case study cities and the impacts on the previously mentioned natural hazards, whose main features are briefly described in the following paragraphs.

Floods

Flooding is one of the major natural hazards, which disrupts the prosperity, safety and amenity of human settlements. The term *flood* refers to a flow of water over areas that are habitually dry (Jonkman and Kelman 2005). A main source of floods is rainfall, which can develop into riverine or flash flooding when the volume of water exceeds the capacity of watercourses (Jha et al. 2012). Flood hazard is strongly dependent on the patterns and intensity of rainfall. It is generally assessed through the evaluation of its impact parameters, such as water level and velocity, and its associated probability of occurrence.

Flooding in urban areas is not only related to heavy rainfall and extreme climatic events; it is also related to changes in the built environment itself. Urbanisation restricts the infiltration of floodwaters by covering large parts of the ground with roofs, roads and pavements. Thus, natural channels are obstructed and, due to the construction of drains, water moves to rivers and streams more rapidly than it did under natural conditions. Large-scale urbanisation and population increases have led to large numbers of people, especially the poor, settling and living in floodplains in and around urban areas. As people crowd into such areas, human impacts on

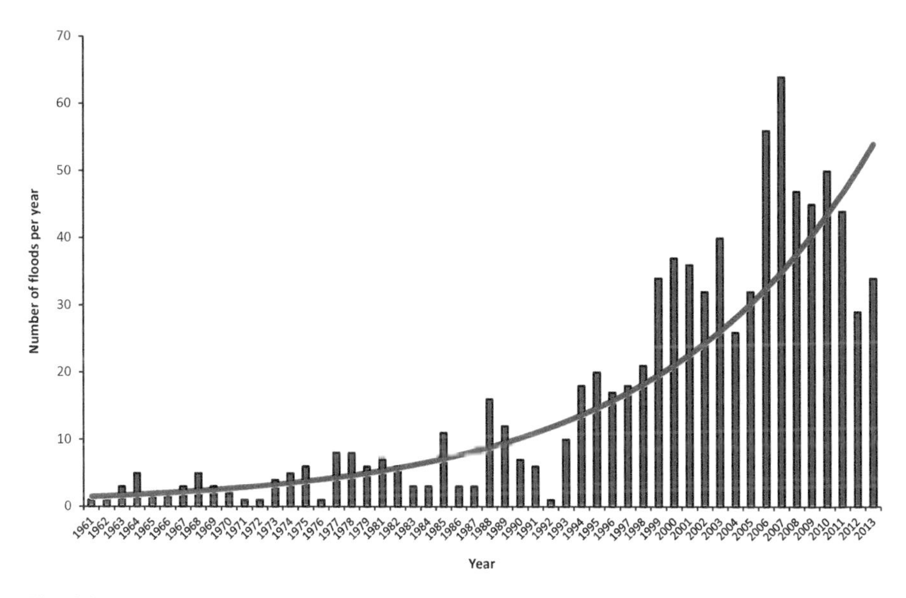

Fig. 1.8 Number of reported floods per year in Africa during the past half century (Based on: EM-DAT 2014)

urban land surfaces and drainage intensify. Even moderate storms can – under such conditions – produce quite high flows in rivers because of surface runoff from hard surfaces and drains (Douglas et al. 2008).

The effects of climate change on the pattern of flooding in Africa are superimposed on these human drivers of local land surface changes. In fact, climate change appears to be altering the frequency of flooding in Africa, in particular of rare large floods. Prolonged heavy rains may increase in volume and occurrence. Many African urban areas have experienced extreme flooding since 1995, as also reported by the CRED Dataset (EM-DAT 2014), showing an increasing trend in the number of reported flood-related disasters in Africa over the past 20 years (Fig. 1.8). In particular, flooding is a common threat for the five CLUVA cities, as also confirmed by several examples of flooding that occurred in African urban areas in the last two decades. In fact, heavy rains and cyclones in February and March 2000 in Mozambique led to the worst flooding in 50 years and brought widespread devastation to the capital city, Maputo, as well as to the city of Matola. In 2002, heavy rains caused by unusually high temperatures over the Indian Ocean killed more than 112 people in East Africa. Floods and mudslides forced tens of thousands of people to leave their homes in Rwanda, Kenya, Burundi, Tanzania and Uganda. In August 2006, in Addis Ababa, floods killed more than 100 people and destroyed homes in eastern Ethiopia after heavy rains caused a river to overflow (Douglas et al. 2008). In 2011, after unusual intense rains, the city of Dar es Salaam was subjected to massive floods that killed 20 people (BBC News 2011).

Damage has also occurred in West Africa. In fact, since 1995 floods have tended to cause increasing damage in Ghana, particularly in coastal areas. The cities of

Accra and Kumasi have been particularly severely affected, with many forced to leave their homes (Douglas et al. 2008). In August 2000, three days of intense rains have left the city of Douala (Cameroon) facing its worst flooding on record. The floods affected a third of the homes and caused the residents to take refuge on roofs of high buildings and trees (BBC News 2000). In September 2009, torrential rains and flooding affected 600,000 people in 16 West African nations. The worst hit countries were Burkina Faso, Senegal, Ghana and Niger (Di Baldassarre et al. 2010). In Burkina Faso, one of the most affected countries, an estimated 150,000 people were left homeless, mostly in the capital Ouagadougou which was hit by the worst flooding event in the last 90 years (BBC News 2009). In July and August 2012 more than 1.5 million people were affected by heavy rains that fell on West and Central Africa. Niger, Chad, Senegal and Nigeria together accounted for more than 90 % of identified affected people. In Senegal, since the beginning of the rainy season in July, torrential rains caused local flooding in several areas, including Saint Louis (OCHA 2012).

Droughts

Although 'drought' has several definitions, the central element in these definitions is *water deficit*. In general, drought is defined as an extended period – a season, a year, or several years – of deficient rainfall relative to the statistical multi-year average for a region (NASA Earth Observatory 2000). This deficiency results in a water shortage for some activity, group, or environmental sector. A more in-depth definition of drought includes four sub-definitions, including meteorological, hydrological, agricultural and socio-economic drought (UNECA 2008).

The underlying cause of most droughts can be related to changing weather patterns manifested through the excessive build-up of heat on the Earth's surface, meteorological changes which result in a reduction of rainfall, and reduced cloud cover, all of which result in higher evaporation rates. The resulting drought effects are exacerbated by human activities, such as deforestation, overgrazing and poor cropping methods, which reduce water retention of the soil, and improper soil conservation techniques, which lead to soil degradation. Urbanisation also enhances the longer-term impacts of droughts because many who are forced to leave rural areas never return. The post-drought population displacement leads to a lack of labour in the countryside and adds further stress on food supplies to urban centres (Glantz 1987).

Statistics per continent show that the highest number of reported droughts was registered for Africa (Table 1.2). In particular, all five CLUVA case study cities, with the exception of Douala, are susceptible to increasing drought episodes. In fact, the African countries experiencing a higher frequency of occurrence of drought episodes are those concentrated in the Sahelian region, the Horn of Africa and in Eastern Africa (Fig. 1.9). Current climate scenarios predict that the driest regions of the world will become even drier, signalling a risk of persistence of drought in many parts of Africa (arid, semi-arid and dry sub-humid areas), which will therefore bear bigger and sustained negative impacts (UNECA 2008).

Table 1.2 Summarised table of drought disasters sorted by continent from 1900 to 2013

	# of events	# of deaths	Total affected pers.	Damage (000 USD)
Africa	293	847,143	366,825,799	2,984,593
Americas	135	77	69,835,746	50,471,139
Asia	153	9,663,389	1,712,816,029	34,251,865
Europe	42	1,200,002	15,488,769	25,481,309
Oceania	22	660	8,034,019	11,526,000
World	645	11,711,271	2,173,000,362	124,714,906

Based on: EM-DAT (2014)

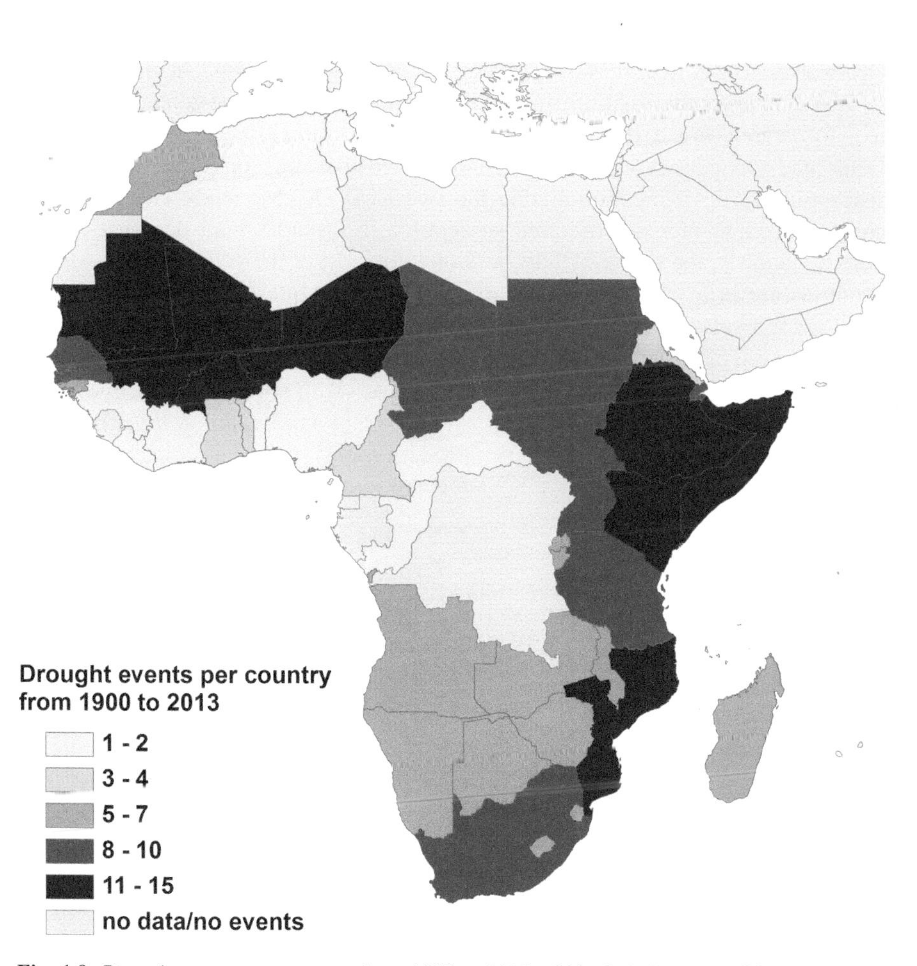

Fig. 1.9 Drought events per country from 1900 to 2013 within Sub-Saharan Africa (Based on: EM-DAT 2014)

Desertification

Desertification, as defined by the International Convention on Desertification, is the degradation of the land in arid, semi-arid and sub-humid dry areas caused by climate change and human activities (UNCCD 1994). It is accompanied by a reduction in the natural potential of the land and depletion in surface and groundwater resources. Above all, it has negative repercussions on the living conditions and the economic development of the people affected. Even though the cycles of drought years and climatic changes can contribute to the advance of desertification, it is mainly caused by changes in the ways humans use natural resources, mainly by over-grazing, land clearance, the over-cropping of cultivated and woodland areas and more generally using land in a way that is inappropriate for the local conditions (Koohafkan 1996).

It is estimated that two-thirds of Africa's land resources are already degraded to some degree (Fig. 1.10). Land degradation particularly affects the cities of Saint Louis and Ouagadougou, and it is these cities that are consequently the most susceptible to desertification among the five CLUVA case study cities. Climate change is set to increase the area susceptible to drought, land degradation and desertification in the region (UNECA 2008). As most of the economies of African countries are primarily agro-based, most of the desertification problems in rural areas are a result of poverty related agricultural practices and other inappropriate land use systems. Deforestation, especially to meet energy needs and expand agricultural land, is another serious direct cause of desertification in the region (UNECA 2008).

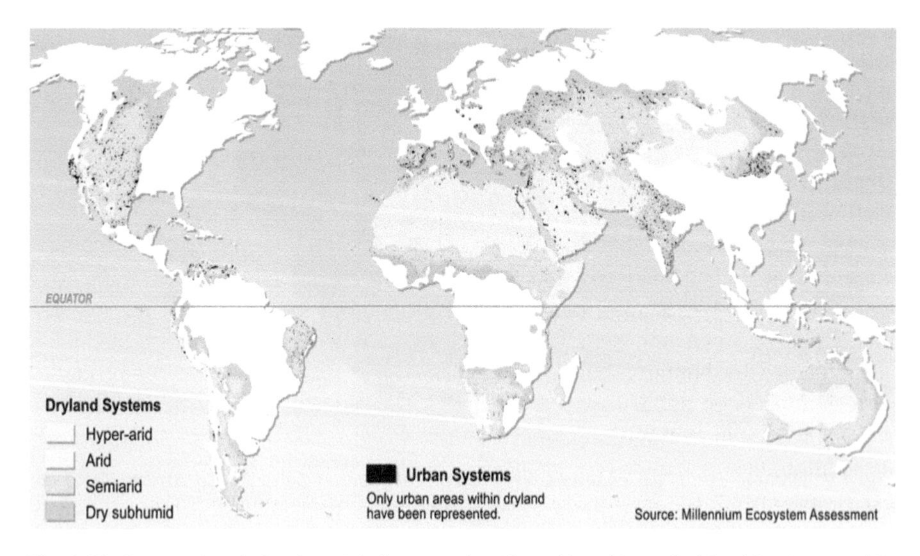

Fig. 1.10 Present-day drylands and their categories (dry subhumid, semiarid, arid or hyper-arid), based on Aridity Index values (Source: Millennium Ecosystem Assessment 2005)

The above direct causes of desertification are driven by a complex set of underlying factors including the high levels of poverty and conflicting land use patterns in the region, high population growth rates, and climate change. Without suitable alternatives, poor people are forced to exploit land resources, including fragile lands, for survival (food production, medicine, fuel, fodder, building materials and household items). Given that most drylands in Africa are poverty hotspots as well, the risk of desertification is high in many of these areas, as the poor inevitably become both the victims and agents of environmental damage and desertification. Consequently, the effects of desertification extend beyond the affected dryland areas. Indeed, as the level of vulnerability due to the combined impacts of desertification and socio-economic susceptibility increases, the greater the probability of human migration becomes (UNECA 2008).

Desertification is displacing large populations of people and is forcing them to leave their homes and lands in search of better livelihoods. In 2008 it was estimated that 135 million people – the combined populations of France and Germany – could be displaced by desertification. The problem appears to be most severe in Sub-Saharan Africa, the Sahel and the Horn of Africa. Some 60 million were estimated to eventually move from the desertified areas of Sub-Saharan Africa towards Northern Africa and Europe by the year 2020 (UNECA 2008). It is already reported that in the past 20 years, nearly half of the total male population in Mali has migrated at least once to neighbouring African countries (96 %) or to Europe (2.7 %). In Burkina Faso, desertification can be identified as the cause of 60 % of the growth of the main urban centres. In Kenya one of the consequences of desertification is a constant flow of rural poor to Nairobi. The population of Nairobi has grown by 800 % from 350,000 in 1963 to 2,818,000 in 2005 (UNECA 2008). Migration will exert stress on the poorly developed and limited public infrastructure in urban areas and may exacerbate conflicts already witnessed in the region as a result of scarcity of grazing land and water.

Heat Waves

The World Meteorological Organization and the World Health Organization have not yet issued a standard definition of the term ‘heat wave’ even though there is a consensus on two qualitative requirements for an event to be called a heat wave: high temperatures and extended duration. These characteristics are the result of the interaction between atmospheric, oceanic and land surface processes frequently accompanied by humid conditions and low precipitation. Heat waves can induce serious impacts on public health causing heat stress, temporary changes in lifestyle and an increase in mortality and morbidity, especially for the most vulnerable groups. In addition, heat waves have a strong impact from a social and economic point of view, increasing forest fires, losses in agricultural resources and ecosystems, and inducing a strain on infrastructure (like power generation, water supply, transportation, etc.) (Kuglitsch et al. 2010).

Both empirical and numerical studies have reached a consensus that warming over the African continent is a reality (Fontaine et al. 2013). However, the recent

Working Group I contribution to the IPCC Fifth Assessment Report (WGI AR5) indicates that there is only a low to medium confidence in observed trends in daily temperature extremes across Africa due to insufficient published evidence (IPCC 2013). What evidence exists, points to an increase in heat waves and warm spells in North and Southern Africa (Perkins and Alexander 2012; Donat et al. 2014; Kruger and Sekele 2013) but little is known about other areas or the specific case in urban areas.

Increasing heat wave phenomena could affect all CLUVA case study cities, but it is expected that impacts will be particularly marked in inland African cities. It is these cities which are likely to face higher ambient temperatures and more frequent heat waves as well as stronger Urban Heat Island (UHI) phenomena, whereby air and surface temperatures are elevated in relation to surrounding rural areas (see Chap. 4). The most affected cities will be those already experiencing heat stress and related problems during the summer season, as well as those in the Sahel, on or near the boundary between the desert and the bush, such as Kano and Ouagadougou (UN-Habitat 2010).

Sea Level Rise

Global mean projections indicate that global warming is leading to a rise in sea levels in the coming decades (UN-Habitat 2008c). To assess how much and how fast the global mean sea level will rise, scientists have come up with two fundamentally different approaches. The first approach is based on physical models, which aim to describe quantitatively the physical processes that contribute to global sea level rise, i.e. thermal expansion of the ocean water by warming, mass addition coming primarily from melting land ice and depth change of the global ocean basins by movements of the Earth's crust. The second approach is based on models, which utilise the link between observed sea level rise and observed global temperature changes in the past in order to predict the future (Rahmstorf 2012).

In order to forecast sea-level change at a particular coastal location, the sum of global, regional and local trends related to changing ocean and land levels must be established. Indeed, local sea level change can deviate from the global mean for a number of reasons, including the effect of local wind and/or vertical land movements, related to tectonic processes or anthropogenic causes (e.g. oil extraction) (Rahmstorf 2012).

Over the period 1901 to 2010, global mean sea level rose by 0.19 m, and for the most conservative climate change scenario the expected sea level rise by the year 2100 ranges from 0.28 to 0.61 m (IPCC 2013). The projected rise in sea levels could result in catastrophic flooding affecting coastal cities. Thirteen of the world's 20 megacities are situated along coastlines; therefore, any rise in sea levels is potentially devastating for millions of urban dwellers and the global economy (UN-Habitat 2008c).

The Low Elevation Coastal Zone (LECZ) – the continuous area along coastlines that is less than 10 m above sea level – represents 2 % of the world's land area but in 2000 contained 10 % of its total population and 13 % of its urban population (McGranahan et al. 2007; UN-Habitat 2008c; Dodman et al. 2009).

Table 1.3 Urban population at risk from sea level rise

	% of LECZ urban to total urban	% of urban in LECZ
Africa total	11.5 %	58.2 %
Northern Africa	17.6 %	50.6 %
Sub-Saharan Africa	8.7 %	57.6 %
Average developing countries	14.4 %	54.2 %

Source: UN-Habitat (2008c)

In Northern Africa, 18 % of the urban population lives in the low elevation coastal zone, while in Sub-Saharan Africa, the figure is 9 % of the total urban population (Table 1.3). Coastal towns are by far the most developed of Africa's urban areas and, by implication, the concentration of residential, industrial, commercial, agricultural, educational and military facilities in coastal zones is high. Nearly 60 % of Africa's total population living in low elevation coastal zones is urban, representing 11.5 % of the region's total urban population (UN-Habitat 2008c). The major coastal cities in Africa that could be severely affected by the impact of rising sea levels include Abidjan, Accra, Alexandria, Algiers, Cape Town, Casablanca, Dakar, Dar es Salaam, Djibouti, Durban, Freetown, Lagos, Libreville, Lome, Luanda, Maputo, Mombasa, Port Louis, Saint Louis and Tunis (UN-Habitat 2008c).

Although the proportion and number of urban dwellers in coastal African cities is relatively small compared to Asian cities, African cities will be among those most adversely affected by sea level rise, as they are poorly equipped to cope with its impacts. Many cities in Africa and other less-developed regions do not have the infrastructure to withstand extreme weather conditions. Lack of adequate drainage, embankments and preparedness could therefore lead to devastating consequences in coastal African cities (UN-Habitat 2008c). Furthermore, some features of urban development increase the risk of flooding. Water drains more rapidly from built-over land, increasing peak flows and flood risks, particularly if the built drainage system is not adapted accordingly. Developers have drained wetlands, sometimes reducing malaria prevalence or opening up valuable land for urban development, but also removing a buffer against tidal floods (McGranahan et al. 2007). Particularly in delta regions, land compaction, subsidence due to groundwater withdrawal and reductions in the rate of sediment deposition (due to water regulation) can lead, in effect, to sea level rise, increasing the risk of tidal flooding (as well as creating various other problems) (Ericson et al. 2006).

Modelling Vulnerability of African Cities to Climate Related Hazards

Vulnerability to climate change induced hazards and its rapid dynamics are the most important issues for assessing risk and resilience in African cities. Although understanding the current and future patterns of hazards in African urban areas is important, it is vulnerability which is the major factor determining how hazards translate into loss of life, livelihoods and human wellbeing. Indeed, in the African context vulnerability acts as a multiplication factor meaning that even hazards of mild intensity can become disasters of disproportionate magnitude. The factors affecting vulnerability to climate hazards are complex, multi-faceted, inter-related and subject to rapid change. While this makes its study challenging, it is none-theless vitally important for assessing risk and resilience in African cities.

Therefore, a key requirement for developing strategies to improve local resilience in African cities is to understand vulnerability and consider its multifaceted aspects, including those related to the ***physical***, ***social*** and ***ecological*** characteristics of urban areas. Opportunities for adaptation associated with each aspect also need to be explored in order to inform and properly contextualise the process of developing innovative risk mitigation strategies. Therefore, a key element of CLUVA was to explore these vulnerabilities further by assessing their characteristics and proposing a suite of adaptation measures that might be adopted in response.

Physical Vulnerability

Physical vulnerability relates to all physical components of a system. In the African urban areas, lifelines (e.g. sewage systems and roadways) and informal settlements (e.g. adobe houses, houses made of cement, bricks, mud and wood houses) are the most vulnerable components of the physical system. Informal settlements contain some of the urban structures that are most physically vulnerable to damage from climate change hazards. The vulnerability level is variable, depending on which materials are used, how the buildings are constructed and where the buildings are located (Davis 1984; Kouakou and Morel 2009). Sewer systems and other lifelines are mostly affected by riverine floods, flash floods and landslides. The water retention capacity of urban catchments is often drastically reduced due to an increased amount of impervious surface areas. Consequently, the drainage capacity of sewer systems is frequently inadequate, leading to urban flooding. The serviceability of the road network is a key factor in order to provide assistance to those in need and to carry equipment and materials for search and rescue activities. Natural hazards amplified by climate change can adversely affect the safety, operations, and maintenance of transportation infrastructure.

CLUVA has investigated methods to assess the flood vulnerability of the built environment in African urban areas. In particular, Chap. 3 provides an overview of the activities and the findings related to the vulnerability to flooding of the urban built environment in the context of the CLUVA case study cities. This includes the identification of urban flood hotspots in relation to the built structures and lifelines of three African case study cities (Addis Ababa, Dar es Salaam and Ouagadougou) and a probabilistic methodology to perform micro-scale (at district level) evaluation of building vulnerability and risk to flooding for the example of Dar es Salaam.

Vulnerability of Urban Ecosystems

Although vulnerable to the effects of climate change, urban ecosystems are associated with a range of benefits for human health and wellbeing and therefore have a potential for building resilience. Urban green infrastructure can provide important provisioning services (e.g. food, wood) and also regulating services which mitigate the negative effects of urbanisation (e.g. flood water retention, improved infiltration and UHI mitigation) (Millennium Ecosystem Assessment 2005). Regulating ecosystem services are therefore particularly relevant for meeting the challenges of planning for future climate variability (Anderson et al. 2006; Lafortezza et al. 2009; Bartens et al. 2009). However, this positive potential of urban ecosystems may not be realised because of the pressures that green structures face now and in the future due to uncontrolled city development and, importantly, to climate change itself. It is therefore important that research is undertaken to improve understanding of all aspects of urban ecosystem services in African cities.

The analysis, quantification and mapping of important ecosystem services of the urban green structure have been addressed in CLUVA by using Urban Morphology Types (UMTs), i.e. "integrating spatial units linking human activities and natural processes" (Gill et al. 2008: 211).

The identification of important ecosystem services associated with urban green infrastructure and the assessment of their vulnerability not only to climate change, but also to urban development, have been addressed by CLUVA in the five case study cities. Details of these activities are provided in Chap. 4 where the adopted methods as well as the main research results are described. It also includes the evaluation of prospects for urban green structure as a measure for adapting African cities to climate change. This evaluation was based on urban spatial scenarios developed for the two cities of Dar es Salaam and Addis Ababa. Specific information on the methods adopted to model these scenarios is provided in Chap. 8.

Social Vulnerability

A wide range of drivers influences the vulnerability of people living in African urban areas. This encompasses demographic change; increasing population pressure by climate refugees; poverty; new formation of formal and informal settlements; and climate change, to name just a few. Each of these drivers significantly influences the vulnerability and resilience of households being exposed to various hazards. The concept of social vulnerability is useful for better understanding the above drivers and pressures as it attempts to explain, "what it is about the condition of people that enables a hazard to become a disaster" (Tapsell et al. 2005: 3). Therefore, it helps to elucidate the relevant characteristics of social entities "in terms of their capacity to anticipate, cope with, resist and recover from the impacts of a natural hazard" (Blaikie et al. 1994: 9). This understanding not only considers exposure, but more importantly, the resources and capacities for being prepared for and coping with hazards (Timmerman 1981; Chambers 1989). This concept is particularly relevant in the African context where not only rapid urbanisation is occurring, but also the consequences of climate change are likely to have far-reaching consequences (Kiunsi and Lupala 2009; Diagne and Ndiaye 2009).

The analysis of social vulnerability has been addressed by CLUVA in selected African cities exploring the exposure and the susceptibility of specific neighbourhoods and their residents to the consequences of natural hazards. This analysis has included also the assessment of the coping capacity and the adaptation level of these communities to the impacts of natural hazards (Jean-Baptiste et al. 2011). Chapter 6 covers all relevant aspects of this analysis, focusing in particular on the assessment of social vulnerability of households and communities in flood prone urban areas.

An Integrated Multi-risk Framework Considering Climate Change Impacts on African Cities

This chapter has already explained the complex and multi-faceted challenges that many African cities are facing. Researchers have responded by developing an integrated multi-risk framework specifically designed to handle the connections between non-stationary climate-related hazards, physical vulnerabilities and social fragilities.

In the common practice, risk evaluation related to different sources is generally done through independent analyses, adopting disparate procedures and spatial-temporal resolutions. Such a strategy of risk evaluation has some evident major drawbacks as, for example, it is difficult (if not impossible) to compare the risk of different origins, and the implicit assumption of independence of the risk sources leads to the neglect of possible interactions among threats and/or cascade effects (Marzocchi et al. 2012).

A multi-hazard perspective implies considering different independent hazard sources or assessing possible interactions between them as well as cascading effects. Conversely, a multi-vulnerability perspective implies considering the response of the different kinds of exposed elements (such as the built environment, green elements representing urban ecosystems, or the social context) to the effects of the different hazards considered as well as dealing with complex interactions and cascading phenomena in the anthropogenic system. The concept of multi-risk brings these ideas together and refers to a complex variety of combinations of risk (i.e. various combinations of hazards and various combinations of vulnerabilities) (Garcia-Aristizabal et al. 2013). Accordingly, a multi-risk framework enables an assessment to be made of the impact of a combination of hazardous events on the essential infrastructure and people in urban areas and an exploration of how the associated risk changes over time. It involves two levels of analysis: a first level analysis, in which an evaluation of potential physical damages is performed (e.g. for buildings, infrastructure and urban ecosystems); and a second-level analysis, in which a set of social context conditions essential for indirect losses is considered (Garcia-Aristizabal et al. 2013).

A quantitative risk and multi-risk assessment system should also be probabilistic in order to consider uncertainties (Marzocchi et al. 2009, 2012; Gasparini et al. 2014). In fact, this is a fundamental requirement for robust risk-based decision making because it encourages decision-makers to appreciate the extent of uncertainties associated with their decisions.

This analysis can provide different outputs of interest for risk-based decision-making. On the one hand, quantitative results in terms of loss exceedance curves and expected annual losses (or consequences) can be used for comparing and ranking the risks, and assessing the effects of different risk mitigation options. On the other hand, a qualitative output using indicator-based analyses may provide complementary information that highlights the influence of particular social context conditions.

An example of application of this multi-risk analysis has been performed by CLUVA in a pilot area in Dar es Salaam in order to demonstrate how multi-risk analysis can support the risk-based decision-making. Relevant information on the multi-risk approach adopted by CLUVA as well as the results of its application to the Tanzanian city are provided in Chap. 7.

Governance Strategies and Policies

African cities clearly need to become more resilient towards climate change. But even in developed countries, adaptation to climate change is a new task for the cities, and, although both administration and the political level are increasingly aware of the need, no routine or commonly agreed practices have been developed yet (Jørgensen et al. 2014). However, the most feasible way for a city to adapt to climate change seems to be the integration of climate change into existing urban policies, instead of considering it as a "stand alone task" (Bicknell et al. 2009; Jørgensen et al. 2014).

Basic measures need to be taken in planning for urban adaptation to climate change, including: (i) improved land use management and higher building standards; (ii) upgrading of urban drainage and storm water management; and (iii) enhanced environmental planning with focus on the urban green infrastructure and its ecosystem services.

Such measures, apparently simple, require knowledge, adequate planning and implementation instruments, and economic power. As climate change adaptation is a complex and cross-cutting problem, it needs a combined multi-level and cross-sectoral approach. Making cities more resilient also requires different types of knowledge. Cities should prioritise high resolution terrain data and vulnerability mapping. Relevant information should be compiled in one place.

The highly informal urban development in African cities clearly raises challenges in relation to such measures. Indeed African cities are not in the lead when integrated climate change strategies are concerned, and they suffer from informal settlements, which are often located in hazard-prone areas and characterised by above average poverty and limited capacity to cope with disasters at household and neighbourhood levels (Sjaastad and Cousins 2009). This adds further difficulties to making and implementing climate change adaptation strategies.

CLUVA has investigated how disaster risk reduction and climate change adaptation are integrated into existing urban planning and governance systems in selected case study cities, developing an applicable method for making such an analysis. Furthermore, innovative land use and governance strategies to enhance resilience of urban areas towards climate change evolved from this research. Chapters 8, 9, 10, and 11 focus on these activities, addressing the issues of strategic urban planning, governance strategies and the need to deal both with adaptation and mitigation, aiming to reduce vulnerability and improve the coping capacity and resilience to climate change of African cities.

Conclusion

The urban vulnerability of developing countries, particularly those located in Sub-Saharan Africa, is "dynamically complex", being related to a combination of rapidly changing socio-economic, physical and environmental factors coupled with increasing urbanisation. In this context, even small hazardous events can cause serious impacts in the urban environment because the high vulnerability magnifies their effects. Consequently, climate change is likely to play a role as an aggravating factor in an already highly vulnerable situation.

In order to develop strategies and take action for reducing the vulnerability of African cities to climate related hazards and to promote the resilience of their local communities, the CLUVA project has followed a unique approach to address different research issues described in detail in the following chapters.

First, the combined pressures of urbanisation and climate change on the African continent and the impacts these will have on cities are presented (Chap. 2). Then, the vulnerability of three main elements of the urban system is explored: built environment and infrastructure, urban ecosystems and people (Chaps. 3, 4, and 6). Rich material from five case studies is provided to discuss in-depth the factors that make these elements vulnerable to climate change and outline ways for increasing their adaptive capacity. The section is concluded by a chapter that presents a novel approach of multi-risk modeling as an aggregate measure for risk assessment of African cities (Chap. 7).

The next section is dedicated to the role of urban spatial scenario design modelling, governance and urban planning for climate change adaptation, which is discussed from diverse perspectives (Chaps. 8, 9, and 10).

Furthermore, new insights into ongoing investigations on small cities (Chap. 5) and on strategies to foster transformative climate mitigation and adaptation in the African city (Chap. 11) are presented.

A final chapter synthesises the different dimensions of the CLUVA project to draw overall conclusions, and invites the interested reader to share a perspective view on how African cities can begin to cope with urbanisation and climate change (Chap. 12).

References

Anderson BG, Rutherford ID, Western AW (2006) An analysis of the influence of riparian vegetation on the propagation of flood waves. Environ Model Softw 21:1290–1296

Bartens J, Day SD, Harris JR, Dove JE, Wynn TM (2009) Transpiration and root development of urban trees in structural soil stormwater reservoirs. J Environ Manag 44(4):646–657

BBC News (2000) Floods hit Cameroon city. http://news.bbc.co.uk/2/hi/africa/865819.stm. August 4, 2000. Accessed 28 Aug 2014

BBC News (2009) UN warns on West Africa floods. http://news.bbc.co.uk/2/hi/africa/8239552.stm. September 5, 2009. Accessed 28 Aug 2014

BBC News (2011) Tanzania floods: heavy rains inundate Dar es Salaam. http://www.bbc.co.uk/news/world-africa-16299734. December 22, 2011. Accessed 28 Aug 2014

Bicknell J, Dodman D, Satterthwaite D (eds) (2009) Adapting cities to climate change: understanding and addressing the development challenges. Earthscan, London

Blaikie P, Cannon T, Davis I, Wisner B (1994) At risk: natural hazards, people's vulnerability, and disaster. Routledge, London/New York

Chambers R (1989) Editorial introduction: vulnerability, coping and policy. IDS Bull-I Dev Stud 20(2):1–7

Cohen B (2006) Urbanisation in developing countries: current trends, future projections, and key challenges for sustainability. Technol Soc 28(2006):63–80

CSIR, CMCC (2013) CLUVA deliverable D1.7 climate maps and statistical indices for selected cities. CLUVA-EC FP7 project

Davis I (1984) Adapting to hazards: some observations on the relationship of the extreme climatic forces of flooding and high winds to the maintenance and planning of settlements for low-income families within developing countries. Energy Build 7(3):195–203

DFID (2004a) Key sheet 10, climate change in Africa. Global and Local Environment Team, Policy Division, Department for International Development, London, UK

DFID (2004b) Key sheet 7 adaptation to climate change: the right information can help the poor to cope. Global and Local Environment Team, Policy Division, Department for International Development, London, UK

Di Baldassarre G, Montanari A, Lins H, Koutsoyiannis D, Brandimarte L, Blöschl G (2010) Flood fatalities in Africa: from diagnosis to mitigation. Geophys Res Lett 37:L22402. doi:10.1029/2010GL045467

Diagne K (2007) Governance and natural disasters: addressing flooding in Saint Louis, Senegal. Int Inst Environ Dev (IIED) 19(2):552–562

Diagne K, Ndiaye A (2009) History, governance and the millennium development goals: flood risk reduction in Saint-Louis, Senegal. In: Pelling M, Wisner B (eds) Disaster risk reduction: cases from urban Africa. Earthscan, London, pp 147–167

Dodman D, Hardoy J, Satterthwaite D (2009) Urban development and intensive and extensive risk. International Institute for Environment and Development (IIED), London, UK, and IIED-America Latina, Buenos Aires, Argentina. Contribution to the global assessment report on disaster risk reduction 2009

Dodman D, Kibona E, Kiluma L (2011) Tomorrow is too late: responding to social and climate vulnerability in Dar es Salaam, Tanzania. Case study prepared for the global report on human settlements 2011. UN-Habitat, Nairobi. Available at: http://mirror.unhabitat.org/downloads/docs/GRHS2011/GRHS2011CaseStudyChapter06DaresSalaam.pdf. Accessed 1 Feb 2014

Donat MG, Peterson TC, Brunet M, King AD, Almazroui M, Kolli RK, Boucherf D, Al-Mulla AY, Nour AY, Aly AA, Ali Nada TA, Semawi MM, Al Dashti HA, Salhab TG, El Fadli KI, Muftah MK, Eida SD, Badi W, Driouech F, El Rhaz K, Abubaker MJY, Ghulam AS, Erayah AS, Mansour MB, Alabdouli WO, Al Dhanhani JS, Al Shekaili MN (2014) Changes in extreme temperature and precipitation in the Arab region: long-term trends and variability related to ENSO and NAO. Int J Climatol 34(3):581–592

Douglas I (2012) Urban ecology and urban ecosystems: understanding the links to human health and wellbeing. Curr Opin Environ Sustain 4:385–392

Douglas I, Alam K, Maghenda M, Mcdonnell Y, Mclean I, Campbell J (2008) Unjust waters: climate change, flooding and the urban poor in Africa. Environ Urban 20(1):187–205. doi:10.1177/0956247808089156

EM-DAT (2014) The OFDA/CRED international disaster database. www.emdat.be. Université catholique de Louvain, Brussels, Belgium. Accessed 1 May 2014

Ericson JP, Vorosmarty CJ, Dingman SL, Ward LG, Meybeck M (2006) Effective sea-level rise and deltas: causes of change and human dimension implications. Glob Planet Chang 50:63–82

Fontaine B, Janicot S, Monerie PA (2013) Recent changes in air temperature, heat waves occurrences, and atmospheric circulation in Northern Africa. J Geophys Res-Atmos 118(15):8536–8552

Garcia-Aristizabal A, Marzocchi W, Ambara G, Uhinga G (2013) CLUVA deliverable D2.14 reports and map on multi-risk Bayesian scenarios on one selected city (Dar Es Salaam, Tanzania). CLUVA-EC FP7 project

Gasparini P, Di Ruocco A, Russo R (2014) Natural hazards impacting on future cities. In: Gasparini P, Manfredi G, Asprone D (eds) Resilience and sustainability in relation to natural disasters: a challenge for future cities, SpringerBriefs in Earth sciences. Springer, Cham/Heidelberg/New York/Dordrecht/London, pp 67–76

Gill SE, Handley JF, Ennos AR, Pauleit S, Theuray N, Lindley SJ (2008) Characterising the urban environment of UK cities and towns: a template for landscape planning in a changing climate. Landsc Urban Plan 87:210–222

Giugni M, Adamo P, Capuano P, De Paola F, Di Ruocco A, Giordano S, Iavazzo P, Sellerino M, Terracciano S, Topa ME (2011) CLUVA deliverable D1.2 hazard scenarios for test cities using available data. CLUVA-EC FP7 project

Glantz MH (1987) Drought and economic development in Sub-Saharan Africa. In: Glantz MH (ed) Drought and hunger in Africa: denying famine a future. Cambridge University Press, Cambridge

Hulme M, Doherty R, Ngara T, New M, Lister D (2001) African climate change: 1900–2100. Clim Res 17:145–168

IMechE (Institution of Mechanical Engineers) (2013) Natural disasters saving lives today, building resilience for tomorrow. Institution of Mechanical Engineers, London

IPCC (2000) Special report on emission scenarios. Cambridge University Press, Cambridge

IPCC (2007a) Climate change 2007: the physical science basis. In: Solomon S, Qin D, Manning M, Chen Z, Marquis M, Tignor KBM, Miller HBM (eds) Contribution of Working Group I to the fourth assessment report of the Intergovernmental Panel on Climate Change. Cambridge University Press, Cambridge/New York, 996 pp

IPCC (2007b) Climate change 2007: impacts, adaptation and vulnerability. In: Parry ML, Canziani OF, Palutikof JP, van der Linden PJ, Hanson CE (eds) Contribution of Working Group II to the fourth assessment report of the Intergovernmental Panel on Climate Change. Cambridge University Press, Cambridge/New York

IPCC (2013) Climate change 2013: the physical science basis. In: Stocker TF, Qin D, Plattner G-K, Tignor M, Allen SK, Boschung J, Nauels A, Xia Y, Bex V, Midgley PM (eds) Contribution of Working Group I to the fifth assessment report of the Intergovernmental Panel on Climate Change. Cambridge University Press, Cambridge/New York

Jean-Baptiste N, Kuhlicke C, Kunath A, Kabisch S (2011) CLUVA deliverable D2.11 review and evaluation of existing vulnerability indicators in order to obtain an appropriate set of indicators for assessing climate related vulnerability. CLUVA-EC FP7 project

Jha AK, Bloch R, Lamond J (2012) Cities and flooding. A guide to integrated urban flood risk management for the 21st century. The World Bank, Washington, DC

Jonkman SN, Kelman I (2005) An analysis of the causes and circumstances of flood disaster deaths. Disasters 29(1):75–97

Jørgensen G, Herslund LB, Lund DH, Workneh A, Kombe W, Gueye S (2014) Climate change adaptation in urban planning in African cities: the CLUVA project. In: Gasparini P, Manfredi G, Asprone D (eds) Resilience and sustainability in relation to natural disasters: a challenge for future cities, SpringerBriefs in Earth sciences. Springer, Cham/Heidelberg/New York/Dordrecht/London, pp 25–37

Kemajou A, Bergossi O, Tatietse TT, Diboma BS (2008) Is industrial development incompatible with constraints of industrial ecology in Cameroon? Int Sci J Altern Energy Ecol 6(62):194–203

Koohafkan AP (1996) Desertification, drought and their consequences. Available at: http://www.fao.org/docrep/x5317e/x5317e01.htm. Accessed 28 Aug 2014

Kouakou CH, Morel JC (2009) Strength and elasto-plastic properties of non-industrial building materials manufactured with clay as a natural binder. Appl Clay Sci 44(1):27–34. doi:10.1016/j.clay.2008.12.019

Kiunsi RB, Lupala J (2009) Building disaster-resilient communities: Dar es Salaam, Tanzania. In: Pelling M, Wisner B (eds) Disaster risk reduction – cases from urban Africa. Earthscan, London

Kruger A, Sekele S (2013) Trends in extreme temperature indices in South Africa: 1962–2009. Int J Climatol 33:661–676

Kuglitsch FG, Toreti A, Xoplaki E, Della-Marta PM, Zerefos CS, Turkes M, Luterbacher J (2010) Heat wave changes in the eastern Mediterranean since 1960. Geophys Res Lett 37:L04802. doi:10.1029/2009GL041841

Lafortezza R, Carrus G, Sanesi G, Davies C (2009) Benefits and well-being perceived by people visiting green spaces in periods of heat stress. Urban For Urban Green 8(2):97–108

Maccioni L, Zebenigus D (2010) Educity. In: Angélil M, Hebel D (eds) Cities of change: Addis Ababa. Transformation strategies for urban territories in the 21st century. Birkhäuser Verlag, Basel/Boston/Berlin, pp 66–75

Marzocchi W, Mastellone ML, Di Ruocco A, Novelli P, Romeo E, Gasparini P (2009) Principles of multi-risk assessment. Office for Official Publications of the European Communities, Luxembourg

Marzocchi W, Garcia-Aristizabal A, Gasparini P, Mastellone ML, Di Ruocco A (2012) Basic principles of multi-risk assessment: a case study in Italy. Nat Hazards 62(2):551–573

McGranahan G, Balk D, Anderson B (2007) The rising tide: assessing the risks of climate change and human settlements in low elevation coastal zones. Environ Urban 19(1):17–37. doi:10.1177/0956247807076960

Melesse M (2005) City expansion, squatter settlements and policy implications in Addis Ababa: the case of Kolfe Keranio Sub-City. Working papers on population and land use change in central Ethiopia, no. 2. Acta Geographica, Trondheim

Mendelsohn R, Saher G (2010) The global impact of climate change on extreme events. Background paper for the report "Natural Hazards, UnNatural Disasters. The Economics of Effective Prevention" (2010). World Bank, Washington, DC

Millennium Ecosystem Assessment (2005) Ecosystems and human well-being: desertification synthesis. World Resources Institute, Washington, DC

Moss RH, Edmonds JA, Hibbard KA, Manning MR, Rose SK, van Vuuren DP, Carter TR, Emori S, Kainuma M, Kram T, Meehl GA, Mitchell JFB, Nakicenovic N, Riahi K, Smith SJ, Stouffer RJ, Thomson AM, Weyant JP, Wilbanks TJ (2010) The next generation of scenarios for climate change research and assessment. Nature 463(7282):747–756. doi:10.1038/nature08823

NASA Earth Observatory (2000) Drought: the creeping disaster. http://earthobservatory.nasa.gov/Features/DroughtFacts/. Accessed 28 Aug 2014

OCHA (2012) West and Central Africa. Overview: impact of floods. Report of the Office for the Coordination of Humanitarian Affairs, 15 Sept 2012. UN Office for the Coordination of Humanitarian Affairs, Geneva, Switzerland

Pelling M, Wisner B (eds) (2009) Disaster risk reduction: cases from urban Africa. Earthscan, London

Perkins SE, Alexander LV (2012) On the measurement of heat waves. J Clim 26:4500–4517

Rahmstorf S (2012) Modeling sea level rise. Nat Educ Knowl 3(10):4

Satterthwaite D (2006) Outside the large cities; the demographic importance of small urban centres and large villages in Africa, Asia and Latin America. Human settlements discussion paper. Urban Change-3, IIED, London

SITRASS/SSATP (2004) Poverty and urban mobility in Douala. Final report. SSATP report no 09/04/Dla. Africa Region, World Bank, Washington, DC

Sjaastad E, Cousins B (2009) Formalisation of land rights in the South: an overview. Land Use Policy 26(1):1–9

Tapsell SM, Tunstall SM, Green C, Fernandez A (2005) Social indicator set. FLOODsite report T11-07-01. Flood Hazard Research Centre, Enfield

Timmerman P (1981) Vulnerability, resilience and the collapse of society. Institute for Environmental Studies, University of Toronto, Toronto

Tonyé E, Yango J, Tsalefac M, Ngosso AB, Nguimbis J, Ngohe Ekam PS, Moudiki C, Mgaba MP, Tamo T, Pancha Moluhcluva PT, Giugni M, Capuano P, Topa ME, Kassenga G, Yeshitela K, Coly A, Toure H (2012) Deliverable D5.2 report on climate related hazards in the selected cities. CLUVA-EC FP7 project

UNCCD (United Nations Convention to Combat Desertification) (1994) Elaboration of an international convention to combat desertification in countries experiencing serious drought and/or desertification, particularly in Africa. U.N. Doc. A/AC. 241/27, 33 I.L.M. 1328, United Nations, New York, USA

UNDESA (United Nations, Department of Economic and Social Affairs, Population Division) (2012) World urbanization prospects: the 2011 revision. CD-ROM edition. United Nations, New York. Available online: http://esa.un.org/unup/CD-ROM/Urban-rural-Population.htm. Accessed 1 Feb 2014

UNECA (United Nations Economic Commission for Africa) (2008) Africa review report on drought and desertification. UNECA (United Nations Economic Commission for Africa), Addis Ababa

UNFCCC (United Nations Framework Convention on Climate Change) (2006) Background paper on impacts, vulnerability and adaptation to climate change in Africa for the African workshop on adaptation implementation of decision 1/CP10 of the UNFCCC Convention, Accra, 21–23 Sept 2006

UN-Habitat (2008a) Ethiopia: Addis Ababa urban profile. United Nations Human Settlements Programme, Nairobi

UN-Habitat (2008b) The state of African cities 2008. A framework for addressing urban challenges in Africa. United Nations Human Settlements Programme, Nairobi

UN-Habitat (2008c) State of the world's cities 2008/2009. Harmonious cities. United Nations Human Settlements Programme, Nairobi

UN-Habitat (2009) Tanzania: Dar es Salaam city profile. United Nations Human Settlements Programme, Nairobi

UN-Habitat (2010) The state of African cities 2010. Governance, inequality and urban land markets. United Nations Human Settlements Programme, Nairobi

UN-Habitat (2012) State of the world's cities report 2012/2013: prosperity of cities. United Nations Human Settlements Programme, Nairobi

UN-Habitat (2014) The state of African cities 2014. Re-imagining sustainable urban transitions. United Nations Human Settlements Programme, Nairobi

URT (2006) Tanzania census 2002, Analytical report. National Bureau of Statistics, Ministry of Planning, Economy and Empowerment, Dar es Salaam, Tanzania

World Bank (2009) Making development climate resilient: a world bank strategy for sub-Saharan Africa. Report no. 46947-AFR. Sustainable Development Department, Washington, DC

World Bank (2011) Sustainable land management for climate change mitigation and adaptation. Environment and Development English, Paperback, Washington, DC, 104 p

Chapter 2
The Impacts of Climate Change on African Cities

Maurizio Giugni, Ingo Simonis, Edoardo Bucchignani, Paolo Capuano, Francesco De Paola, Francois Engelbrecht, Paola Mercogliano, and Maria Elena Topa

Abstract Changes in the frequency and intensity of extreme events have significant impacts and are one of the most serious challenges faced by society in coping with a changing climate.

During the last ten years, floods have caused more damage than in the previous 30 years; for example, a flood event of 22 December 2011 in Dar es Salaam led to 20 deaths, considerable damage and loss of livelihoods. Moreover, higher temperatures have led to a high rate of evaporation and very dry conditions in some areas

---

M. Giugni (✉) • F. De Paola
Department of Civil, Architectural and Environmental Engineering DICEA, Università di Napoli Federico II, Via Claudio 21, 80125 Naples, Italy

AMRA - Center for the Analysis and Monitoring of Environmental Risk, Via Nuova Agnano 11, 80125 Naples, Italy
e-mail: giugni@unina.it; depaola@unina.it

I. Simonis
School of Mathematics, Statistics and Computer Science, University of Kwazulu Natal, University Rd, Westville, 3630 Durban, South Africa
e-mail: ingo.simonis@geospatialresearch.de

E. Bucchignani • P. Mercogliano
Centro Euro-Mediterraneo sui Cambiamenti Climatici (C.M.C.C.), Via Maiorise, 81043 Capua, CE, Italy

Centro Italiano Ricerche Aerospaziali (C.I.R.A.), Via Maiorise, 81043 Capua, CE, Italy
e-mail: e.bucchignani@cira.it; p.mercogliano@cira.it

P. Capuano
Dipartimento di Fisica "E.R. Caianiello", Università di Salerno, 84084 Fisciano, SA, Italy
e-mail: pcapuano@unisa.it

F. Engelbrecht
Climate Studies, Modelling and Environmental Health, CSIR Natural Resources and the Environment, Pretoria, South Africa
e-mail: FEngelbrecht@csir.co.za

M.E. Topa
AMRA - Center for the Analysis and Monitoring of Environmental Risk, Via Nuova Agnano 11, 80125 Naples, Italy
e-mail: maria_elena_83@hotmail.it

S. Pauleit et al. (eds.), *Urban Vulnerability and Climate Change in Africa*, Future City 4,
DOI 10.1007/978-3-319-03982-4_2

of the world. More generally, there is evidence of a global increase in the occurrence of severe weather events.

Africa is one of the most vulnerable continents to climate change and variability, a situation aggravated by a low adaptive capacity. Adaptation policies and actions will be most effective if based on the best knowledge concerning present and future climate.

This chapter aims to highlight and quantify the impact of climate change on the African cities Addis Ababa (Ethiopia), Dar es Salaam (Tanzania), Douala (Cameroon), Ouagadougou (Burkina Faso) and Saint Louis (Senegal), through the analysis of observed data and model projected changes in temperature and rainfall, synthesising the results of the CLUVA project (Climate Change and Urban Vulnerability in Africa). The climate projections were performed following emissions scenarios prepared by the Intergovernmental Panel on Climate Change's Fourth and Fifth Assessment Reports, using a suite of global climate models and downscaling techniques.

The aim of the project is to evaluate the impact of climate change on the urban scale and, consequently, the possible works of adaptation. For this purpose, innovative downscaling techniques of climate projections were developed in order to achieve results at the urban scale.

Keywords Climate projections • Rainfall and temperature variability • Hazards • African cities

Introduction to Climate Change

The shortwave radiation coming from the sun is the main driver of the Earth's climate system. The Earth's albedo is about 30 %, implying that 30 % of the incoming solar energy is reflected back to space by objects in the atmosphere (e.g. clouds) and by the Earth's surface. That is, 70 % of the solar energy that enters the atmosphere is absorbed by the oceans, continents, and the atmosphere itself. This energy is later re-emitted as infrared radiation and transferred by sensible and latent heat fluxes. In fact, the energy balance on the planet is maintained by the longwave radiation that is constantly emitted back to space. It is due to certain gases – commonly referred to as greenhouse gases, which absorb considerable amounts of the outgoing longwave radiation, that the atmosphere maintains an average surface temperature of about 14 °C. Without these gases, the temperature at the surface of Earth would be about −16 °C. The warming effect of the greenhouse gases is called the 'natural greenhouse effect,' with the main greenhouse gases being carbon dioxide (CO_2) and water vapour.

The CO_2 increase in the atmosphere caused mainly by the burning of fossil fuels and deforestation, in combination with an associated increase in water vapour, has increased the capacity of the atmosphere to absorb heat. Compared to the pre-industrialisation era, the current CO_2 concentration is about 40 % higher.

The 'Fourth Assessment Report' (AR4) of the Intergovernmental Panel on Climate Change (IPCC), published in 2007, conveys a clear message: the phenomenon referred to as 'global warming' may be linked directly to the anthropogenic induced increase in CO_2 and water vapour in the atmosphere, which results in an 'enhanced greenhouse effect' (IPCC 2007). The enhanced greenhouse effect is responsible for the temperature increases that have been recorded worldwide since the industrial revolution. Systematic increases in the average global surface temperature have been recorded over the last century, but they have been particularly strong (and alarming) over the last few decades. Still, the amount of CO_2 released into the atmosphere is continuing to grow. If current trends in emissions persist, the amount of CO_2 in the atmosphere will have doubled by approximately the year 2050 (from the pre-industrial value of 280 ppm, to about 560 ppm). An important objective of the United Nations Framework Convention on Climate Change (UNFCCC) is to keep the increase in global surface temperatures below 2 °C (compared to the pre-industrial global temperature) – it is thought that 'dangerous climate change' may be avoided by keeping the global rise in surface temperature below this threshold level. It is estimated that CO_2 concentrations will need to stabilise at about 450 ppm for this target to be achieved (IPCC 2001, 2012).

The first law of thermodynamics is useful to understand the principles and drivers of climate change. According to this law, the increased ability of the atmosphere to absorb energy leads either to the increased internal energy level of the atmosphere (global warming), or an increased ability of the atmosphere to perform work. The warming is experienced as the worldwide rise in temperatures recorded over the last century. Some of this excess of energy in the atmosphere is effectively used in it to induce shifts in climate regimes, as current research has shown. Both the tropical and extra-tropical regimes are expanding polewards, and a general increase in the frequency of severe weather events has also been recorded over recent decades. These effects are commonly referred to as anthropogenic induced 'climate change'.

The second law of thermodynamics states that a generally warmer atmosphere is capable of carrying more water vapour, and may therefore be expected to be more moist. However, this effect should not be expected to lead to homogeneous rainfall increases across the planet. Despite the atmosphere's increased potential to carry water within a warmer climate, other effects such as shifting climate regimes or changing circulation patterns may override this effect, and in fact lead to drier conditions in some areas. For example, an area located in today's extra-tropical climate regime may in future become part of the poleward-expanding, subtropical high-pressure belt, the latter an area known for its relatively dry conditions.

In summary, quantitative physical models are required to provide meaningful projections for the various climatic regions on Earth. Such projections can be obtained using Global Climate Models (GCMs). A GCM is a mathematical model of the general circulation of a planetary atmosphere or ocean, taking into account thermodynamic terms for various energy sources (radiation, latent heat) as well as the laws of momentum, mass and energy conservation. Although based on physical laws, the projections of these models are subject to a number of uncertainties, since no model can fully represent the complexity of the Earth system. For more information on climate change in general, Eggleton (2014) is a great point to start. McGuffie and Stocker are good entry points into climate change modeling (Stocker 2011; McGuffie and Henderson-Sellers 2014).

Downscaling of Global Models

GCMs are a powerful tool to perform climate projections at global scale, but they are generally unsuitable to provide climate data for impact studies as they are characterised by coarse resolutions of 1–5° in latitude or longitude. This is insufficient to analyse many important phenomena that occur at spatial scales of few tens of kilometres. In addition, local topographic peculiarities such as rain shadows or wind tunnel effects are often not resolved by global models with resolutions of about 250 × 250 km per grid cell. Thus, in order to analyse local and regional scales, the global climate models have to be downscaled to resolutions between 1 × 1 km and 80 × 80 km.

Two approaches are available for downscaling of global climate information: *dynamical* (Giorgi and Hewitson 2001) and *statistical* techniques (Schmidli et al. 2007). In the first case, process-based physical dynamics of the regional climate system are explicitly solved. In the second case, a statistical relationship between local climate variables (predictands) and large scale variables (predictors) is established.

One of the most effective tools for providing high-resolution climate analysis through downscaling is represented by Regional Climate Models (RCMs), which are able to provide an accurate description of climate variability on a local scale. Moreover, RCMs show the capability to provide a detailed description of climate extremes that are often more important than mean values.

In order to capture these effects, downscaling techniques are applied to the global models, resulting in projections with resolutions of 50 × 50 km to 8 × 8 km per grid cell or even finer. As the grid cells become smaller, the computational costs (in terms of elapsed time) become higher, which is why very high resolution experiments with grid cells around 1 × 1 km may currently be conducted only for small areas.

In principle, two downscaling techniques may be distinguished: dynamical downscaling, using high-resolution, numerical regional climate models; and statistical downscaling, based on statistical relationships between large-scale predictor variables and regional ones. Each technique has advantages and disadvantages. Dynamical downscaling simulates climate mechanisms without any a priori assumptions about relationships between the current and the future climate. Though being state of the science, they come with the drawbacks of being expensive in terms of computing resources and professional expertise, may be sensitive to uncertain parameterisations and may propagate biases from the GCM to the regional scale. Statistical downscaling is much cheaper to perform from a computational perspective and may correct for biases of the GCM, but assumes that the relationships between large-scale and local climate are rather constant. It also requires substantial amounts of observed data and does not capture climate mechanisms. We applied both techniques within the CLUVA project and a multi-model ensemble of simulations of present-day and future climate has been made available for each of the cities. Such an ensemble of projections is useful to describe the range of uncertainty associated with the projected future climate for each of the CLUVA case study cities (for a full introduction on the CLUVA project and its case study cities, see also Chap. 1).

Dynamical Downscaling

In the context of the CLUVA project, dynamical downscaling has been performed using two regional models: conformal-cubic atmospheric model (CCAM, CSIR); and COnsortium for Small-scale MOdelling-Climate Local Model (COSMO-CLM, CMCC).

Dynamical downscaling with CCAM was conducted for an ensemble of six global projections of future climate change, as obtained from Coupled Global Climate Models (CGCMs) that contributed to AR4 of the IPCC. The downscaling was performed for the period 1961–2050. Through the process of multiple nudging, the projections have been downscaled to a resolution of 60–80 km over Africa, and subsequently to a horizontal resolution of about 8 km over domains that include the cities under consideration (Addis Ababa, Dar es Salaam, Ouagadougou, Saint Louis and Douala). CCAM is a variable-resolution atmospheric GCM. When applied in stretched-grid mode, it functions as a regional climate model. The model is highly suitable for the purpose of regional climate modelling, due to its computational efficiency and variable-resolution formulation that allows great flexibility in downscaling GCM data, through the application of a multiple nudging technique (McGregor 2005). The model's ability to simulate present-day climate characteristics has been investigated over Africa (Engelbrecht et al. 2009, 2011) and for various other climate regions (Nunez and McGregor 2007; Lal et al. 2008)

A three-phase, multiple downscaling strategy was followed to obtain the 8 km resolution simulations. The first phase in the downscaling procedure was forcing the model with the bias-corrected sea-surface temperature (SST) and sea-ice output of six different CGCMs used in AR4 of the IPCC, for the period 1961–2050. All the projections are for the A2 emission scenario of the Special Report on Emission Scenarios (SRES). CCAM was first integrated globally at quasi-uniform resolution (of about 200 km in the horizontal), forcing the model with the SSTs and sea-ice of each host model, and with CO_2, sulphate and ozone forcing consistent with the A2 scenario. In a second phase of the downscaling, the model was integrated in stretched-grid mode over Africa, with the modestly-stretched grid providing a resolution of about 60–80 km. The higher-resolution simulations were nudged within the quasi-uniform 200 km simulations, through the application of a digital filter using a 4,000 km length scale. The filter was applied at six-hourly intervals and from 900 hPa upwards. The final step in the downscaling procedure involved performing the 8 km resolution simulations over five different domains (each domain surrounding one of the CLUVA cities). These simulations were nudged within the 60–80 km resolution simulations. Model output for the 200 km resolution ensemble members is available at six-hourly intervals, whilst output for the 60–80 km and the 8 km resolution simulations is available at hourly intervals for selected variables. All data is available at http://www.cluva.eu. The ensemble of six global projections included:

- csiro: Version 3.5 of the Coupled Global Climate Model (CGCM) by the Commonwealth Scientific and Industrial Research Organisation in Australia (CSIRO)
- ukmo: The third version of the United Kingdom's Met Office Hadley Centre Coupled Ocean-atmosphere Global Climate Model
- gdfl20: Version 2.0 of the Coupled Global Climate Model of the Geophysical Fluid Dynamics Laboratory (GFDL) of the National Oceanic and Atmospheric Administration in USA
- gfdl21: Version 2.1 of the Coupled Global Climate Model of the Geophysical Fluid Dynamics Laboratory (GFDL) of the National Oceanic and Atmospheric Administration in USA
- mpi: The fifth generation climate model developed by the Max Planck Institute in Germany
- miroc: The MIROC3.2-medres Model for Interdisciplinary Research on Climate 3.2, medium resolution version, of the Japanese Agency for Marine-Earth Science and Technology.

The dynamical downscaling with COSMO-CLM was conducted for two global climate change projections, as obtained from the coupled global climate model CMCC-CM using the RCP4.5 and RCP8.5 radiative forcing scenarios developed in the framework of the 5th Coupled Model Intercomparison Project (CMIP5) (http://cmip-pcmdi.llnl.gov/cmip5/). RCP4.5 and RCP8.5 identify two different concentration pathways, which approximately result in a radiative forcing of, respectively, 4.5 Wm^{-2} and 8.5 Wm^{-2}, at year 2100, relative to pre-industrial conditions. RCP4.5 is a stabilisation scenario and assumes that climate policies, in this instance the introduction of a set of global greenhouse gas emissions prices, are invoked to achieve the goal of limiting emissions and radiative forcing. RCP8.5 is a high emissions scenario, projecting high population growth, slow per capita income growth and little convergence by high and low countries, representing the 90th percentile of the reference emissions range. COSMO-CLM (Rockel et al. 2008) was used at a spatial resolution of about 8 km. Three different domains have been defined to perform the climate simulations: West domain (including Ouagadougou, Douala and Saint Louis), Lower East domain (including Dar es Salaam), and Upper East domain (including Addis Ababa). These simulations have been performed considering the time period 1950–2050. Model output is available at six-hourly intervals.

Statistical Downscaling

The downscaling of models can be performed also through statistical or stochastic techniques. Statistical approaches assume that the relationship between large scale climate variables (e.g. grid box rainfall and pressure) and the actual rainfall measured at one particular rain gauge will always be equal (Schmidli et al. 2007). So, if that relationship is known for current climate, the GCM projections of future climate can be used to predict how the rainfall, measured at that rain gauge, will change in the future. Further, stochastic ensembles of small-scale predictions from the output of atmospheric models can be performed using nonlinearly transformed spectral models (Rebora et al. 2006); the precipitation fields generated by stochastic procedures are consistent meteorological forecasts, in particular the total rainfall volume is conserved.

A similar approach called *hybrid downscaling* (Maraun et al. 2010) has been used to further increase the spatial-temporal resolution of COSMO-CLM and to remove the model output bias. For the hybrid downscaling, the output of the regional model (dynamical downscaling) is used as the input of the statistical downscaling. In particular, in the framework of the CLUVA project, the stochastic downscaling procedure RainFARM (Rainfall Filtered Auto Regressive model) has been used to downscale the COSMO-CLM precipitation fields. RainFARM (Rebora et al. 2006) belongs to the family of algorithms called *metagaussian models*. RainFARM uses simple statistical properties of large-scale meteorological predictors, e.g. the shape of the power spectrum, and generates small-scale rainfall fields propagating this information to a smaller scale. It is able to preserve the total amount of precipitation projected by the large-scale model and also its linear correlation structure, both in space and time, as well as the position of the larger rainfall structures. The output of COSMO-CLM has been downscaled to a spatial resolution of about 1–2 km for small areas, including the five cities of interest.

Results for Africa

The projected climate futures obtained with COSMO-CLM at about 8 km spatial resolution have been analysed by comparing the simulated values of temperature (2 m above ground, referred to as 2 m temperature in this chapter) and precipitation, averaged over 2021–2050 with analogue averaged values over the recent past period 1971–2000, subdivided by seasons (boreal winter (DJF) and boreal summer (JJA)). For the West Domain the temperature analysis for RCP4.5 reports a general positive change of about +1.5 °C in DJF, while in JJA the projected increase ranges between +1 °C and +2 °C. Results for RCP8.5 (Fig. 2.1) have a similar distribution. The precipitation changes in this domain are appropriate for analysis only in JJA, considering that very little rainfall occurs in DJF. For both scenarios the percentage variations in precipitation in JJA mainly fall within the range of −20 % and +20 %.

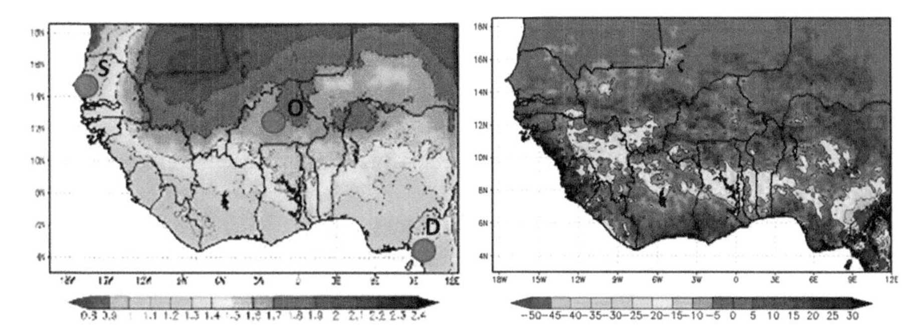

Fig. 2.1 West domain – seasonal average 2 m temperature change (*left*) and percentage change of seasonal precipitation (*right*) for the time period 2021–2050 (RCP8.5 scenario) with respect to 1971–2000, JJA (locations of the case cities are highlighted: Saint Louis (*S*), Ouagadougou (*O*), Douala (*D*))

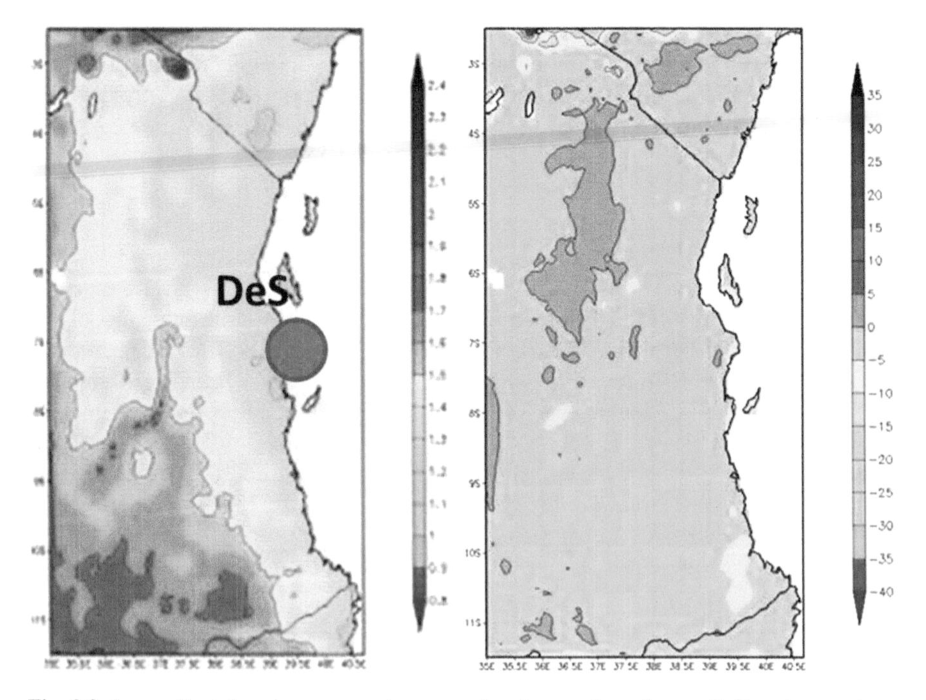

Fig. 2.2 Lower East domain – seasonal average 2 m temperature change (*left*) and percentage change of seasonal precipitation (*right*) for the time period 2021–2050 (RCP8.5 scenario) with respect to 1971–2000, JJA (locations of case cities are highlighted: Dar es Salaam (*DeS*))

In the Lower East domain, the temperature changes for RCP4.5 generally vary from 1 to 1.5 °C, with the lower values along the coast. There are no substantial differences between DJF and JJA (just slightly higher minimums in the second case). RCP8.5 changes are only slightly higher than the other scenario; again, values rise by moving inland with a range of 1.2–1.6 °C in DJF and 1.4–1.7 °C in JJA (Fig. 2.2).

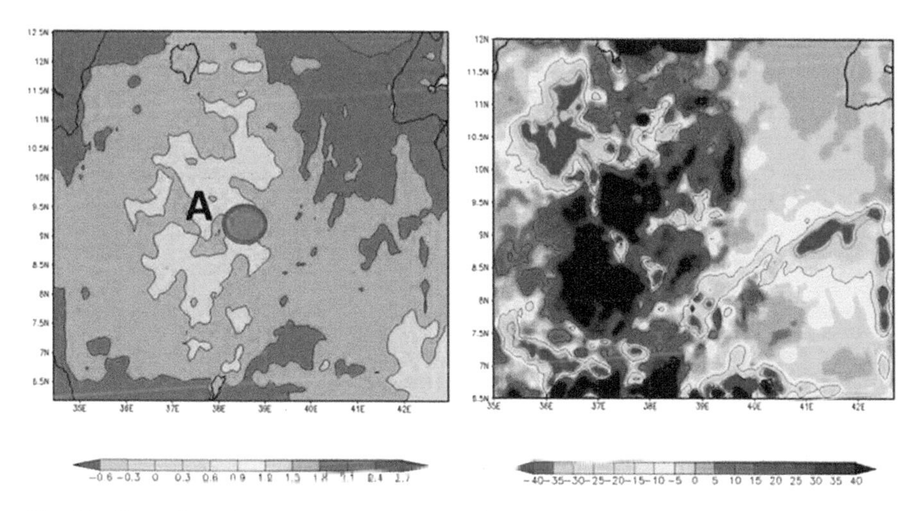

Fig. 2.3 Upper East domain – seasonal average 2 m temperature change (*left*) and percentage change of seasonal precipitation (*right*) for the time period 2021–2050 (RCP8.5 scenario) with respect to 1971–2000, JJA (locations of case study cities are highlighted: Addis Ababa (*A*))

Considering precipitation, a significant change in DJF is found (scenario RCP4.5), with percentage reductions in the internal area as large as 40 %, while the trend is inverted approaching the coastal line (up to +35 %). The RCP8.5 model behaves similarly, but it features less strong changes. No significant values are registered in JJA (+/−5 % for both scenarios).

In the Upper East Domain, the projected temperature changes range generally between 0.9 and 1.5 °C, with an average of about 1.2 °C for the DJF season, while in JJA the projected range of change varies between 1.5 and 1.8 °C. RCP4.5 and RCP8.5 behave similarly (Fig. 2.3), but the latter scenario is characterised by a relatively small increase in values. In DJF, RCP4.5 precipitation is projected to change hitting peaks of +40 % in high orography zones, whereas they rapidly decrease to 0 in other areas; in JJA, the most considerable changes are located in the western half of the domain, ranging from −40 to +40 %. The RCP8.5 scenario features smaller changes in DJF (up to +30 %), while results are roughly the same in JJA.

The CCAM projected changes in average annual temperatures (°C) and annual rainfall totals (percentage change) are shown in Fig. 2.4, for regions surrounding each of the five CLUVA cities, for the period 2041–2050 relative to 1961–1990. For each region, the ensemble averages of the projected changes are shown. The uncertainties associated with the projections are illustrated by means of Tables 2.3 and 2.4. For the Dar es Salaam region, a general increase in precipitation is projected by the ensemble average and most ensemble members, across the domain. This is consistent with most GCMs projecting East Africa becoming wetter under climate change (Christensen et al. 2007). Temperature increases of approximately 2 °C are projected for the interior parts of the domain, with somewhat smaller changes projected along the coastal areas, due to the moderating effects of

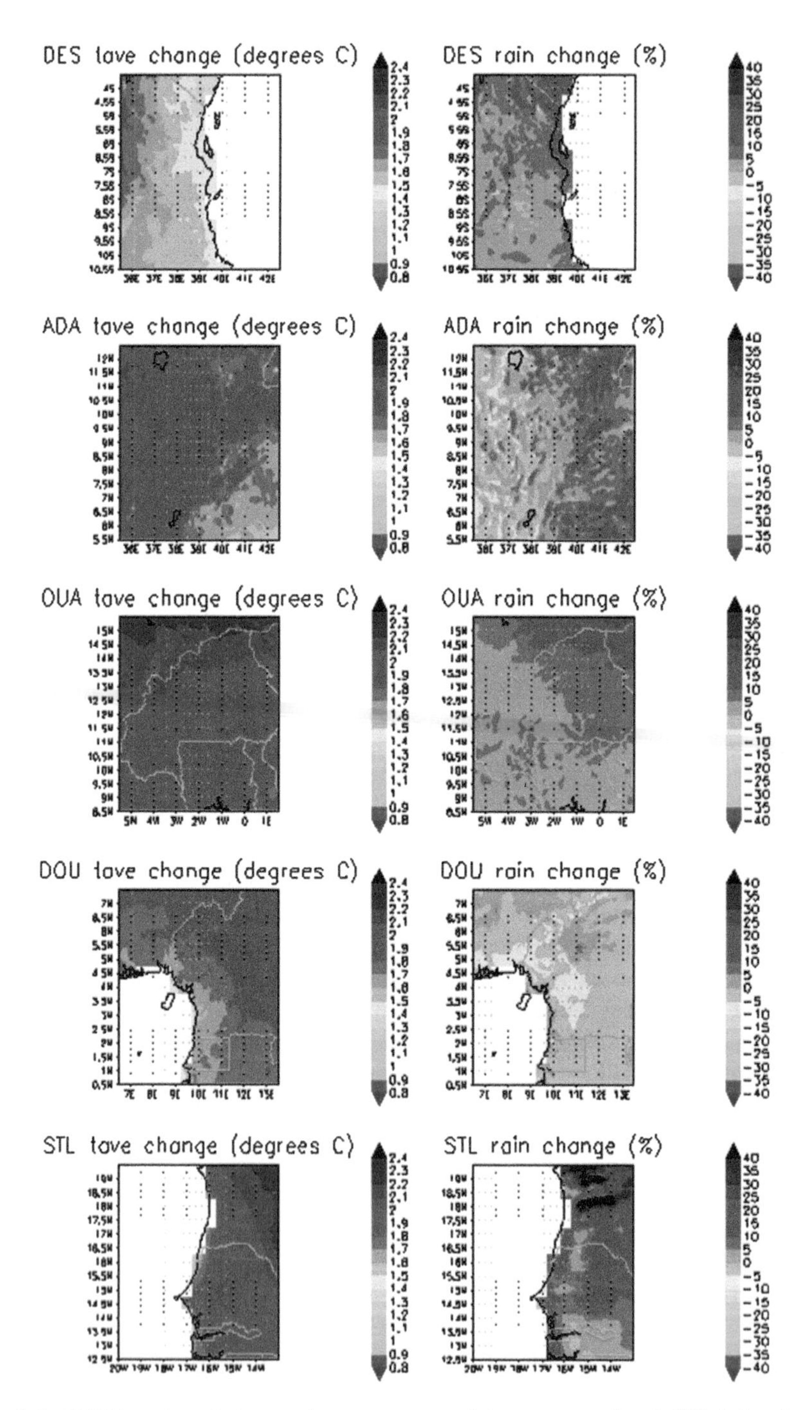

Fig. 2.4 CCAM projected changes in average annual temperatures (tave) (°C, *left columns*) and annual rainfall totals (mm, *right columns*) over the five CLUVA city regions, for the period 2041–2050 relative to 1961–1990, per city, the ensemble average of the projected changes are shown

Table 2.1 Lower and upper bounds of the different CMCC scenario (RCP 8.5 and 4.5) projected changes in seasonal rainfall (expressed as a percentage change) over the different CLUVA domains

	RCP4.5		RCP8.5	
Domain	DJF	JJA	DJF	JJA
West	0	−20 %–+20 %	0	−20 %–+20 %
Lower East	−40 %–0	0	−30 %–0	0
Upper East	0–+40 %	−40 %–+40 %	0–+30 %	−40 %–+40 %

The changes shown are for the period 2041–2050 relative to the period 1961–1990

Table 2.2 Lower and upper bounds of the ensemble of CCAM projected changes in seasonal and annual average temperatures (°C) over the different CLUVA domains

	DJF		JJA		Annual	
Domain	Lower bound (°C)	Upper bound (°C)	Lower bound (°C)	Upper bound (°C)	Lower bound (°C)	Upper bound (°C)
ADA	+1.5	+2.4	+1.4	+2.7	+1.5	+2.3
DES	+1.2	+1.9	+1.3	+2.2	+1.3	+1.9
OUA	+1.9	+2.6	+1.1	+2.4	+1.6	+2.5
DOU	+1.8	+2.3	+1.4	+2.3	+1.5	+2.2
STL	+1.6	+2.3	+1.6	+2.2	+1.7	+2.3

The changes shown are for the period 2041–2050 relative to the period 1961–1990

the ocean. Rainfall increases are likewise projected over the Addis Ababa region by the ensemble average, except over the far western part of the domain, where rainfall decreases are projected. Most ensemble members indeed project a corresponding signal of rainfall increases over Addis Ababa; however, a minority of projections indicate that rainfall decreases are also plausible. Temperature increases in the order of 2 °C are projected, with larger increases plausible over the north-western parts of the domain. Relatively large temperature increases of about 2.5 °C are projected over the northern parts of the Ouagadougou domain. The region is also projected to become generally wetter, by the ensemble average and most members. Temperature increases in the order of 2 °C are projected over the Douala domain by most ensemble members, whilst the region is also projected to become generally drier. Wetter futures are also plausible, however, as indicated by a minority of ensemble members. The Saint Louis region is projected to become generally wetter, with temperature increases as large as 2.5 °C projected over the eastern parts. The projected rainfall increases are linked to the projected northward displacement of the Inter Tropical Convergence Zone (ITCZ) and the expansion of the tropical belt under climate change.

The range of projected changes across the CSIR and CMCC simulations for each of the different CLUVA domains considered in the modelling performed are displayed in Tables 2.3 and 2.2 for average temperatures and Tables 2.1 and 2.4 for annual rainfall.

Table 2.3 Lower and upper bounds of the different CMCC scenario (RCP 8.5 and 4.5) projected changes in seasonal average temperatures (°C) over the different CLUVA domains

	RCP4.5		RCP8.5	
Domain	DJF	JJA	DJF	JJA
West	+1.5°	+1°–+2°	+1.5°	+1°–+2°
Lower East	+1°–+1.5°	+1°–+1.5°	+1.1°–+1.6°	+1.1°–+1.6°
Upper East	+0.9°–+1.5°	+1.5°–+1.8°	+1°–+1.6°	+1.6°–+1.9°

The changes shown are for the period 2041–2050 relative to the period 1961–1990

Table 2.4 Lower and upper bounds of the ensemble of CCAM projected changes in seasonal and annual rainfall totals (expressed as a percentage change) over the different CLUVA domains

	DJF		JJA		Annual	
Domain	Lower bound (%)	Upper bound (%)	Lower bound (%)	Upper bound (%)	Lower bound (%)	Upper bound (%)
ADA	−34.2	+78.2	+39.0	+70.1	−4.0	+9.4
DES	−0.1	+9.6	−71.9	−61.3	−0.5	+11.2
OUA	−34.5	+109.5	+56.6	+99.3	−10.0	+5.5
DOU	−19.5	+23.8	+14.4	+52.7	−3.4	+15.3
STL	−40.1	+36.4	+41.2	+198.4	−29.3	+46.4

The changes shown are for the period 2041–2050 relative to the period 1961–1990

Overview of the Use of Historical and Projected Data

The climate projections were performed using different models and radiative forcing scenarios. In order to assess the impact of climate change on natural hazards, a stepwise approach has been carried out based on the following activities:

- Calculation of indices and application of extreme-value theory to better understand historical data and to allow proper interpretation of climate projection data;
- Statistical approaches for assessing changes between historical and projected data;
- Assessment of observed climate change.

All projections have been compared against available historical data from the dataset that has been obtained from the Royal Netherlands Meteorological Institute (www.climexp.knmi.nl). This dataset is maintained by the United States National Oceanic and Atmospheric Administration (NOAA) National Climatic Data Center (NCDC). NCDC/NOAA collects daily station data from around the world. The data is from two main sources: the weather services and the GTS (Global Telecommunication System, the system used to exchange data to make weather forecasts). Where necessary, the data has been complemented by additional datasets from www.tutiempo.net, which provides data mostly obtained by airport administrations.

Table 2.5 Location, elevation (m above sea level) and recording period of the meteorological stations

	Longitude	Latitude	Altitude [masl]	Available data range
Addis Ababa	38.8	9.03	2,355	1964–2010
Dar es Salaam	39.2	−6.86	55	1958–2010
Douala	9.73	4	10	1976–2010
Ouagadougou	−1.51	12.35	316	1973–2010
Saint Louis	−16.45	16.05	4	1973–2010

Together, both data sets provide some of the best freely available data, though reliability cannot be fully evaluated. The available data is described in Table 2.5.

The following Tables 2.6 and 2.7 provide an overview of historical data and projected data. Table 2.6 shows average values, evaluated in the different range dataset for historical data and climate projections, cumulative annual rainfall, while Table 2.7 shows mean annual temperatures.

In the following sections, different methodologies developed for the analysis of different hazards are described. The hazard analysis has been performed for all CLUVA case study cities. Examples of hazard analysis are reported and discussed, based on the historical data and on some of models and scenarios of climate projections provided. The discussion focuses on those that highlighted the most significant results.

Assessment of Climate Related Hazards

This section provides an overview of the methodology used to assess current and future climate-related hazards in the five CLUVA case study cities. Results for all cities are presented in the following paragraphs (Fig. 2.5).

Rainfall

The rainfall probability curve, also called the IDF (Intensity – Duration – Frequency) curve is a tool, which characterises an area's rainfall pattern. By analysing past rainfall events, statistics about rainfall reoccurrence can be determined for various standard return periods. This, in turn, provides an essential input into flood modelling exercises through which many wider impact studies rely. The return period of an event of a given magnitude may be defined as the average recurrence interval between events, which are greater than or equal to a specified magnitude. Rainfall intensity i is the average rainfall depth h that falls per time increment so that $i = h/t$.

Table 2.6 Mean annual rainfall (mm) observed and calculated using different climate change scenarios

Mean annual rainfall [mm]									
		8 km							
		CMCC COSMO		CSIR CCAM					
	Observed	RCP 8.5	RCP 4.5	A2	A2	A2	A2	A2	A2
	historical data	CMCC CM	CMCC CM	csiro	ukmo	mpi	miroc	gfdl20	gfdl21
Addis Ababa (1964–2010 historical data; 2011–2050 climate projections)	732	991	978	1,060	1,093	1,097	1,132	1,112	1,048
Dar es Salaam (1958–2010 historical data; 2011–2050 climate projections)	1,114	1,052	1,068	1,239	1,142	1,234	1,160	1,137	1,152
Douala (1976–2010 historical data; 2011–2050 climate projections)	2,435	3,193	3,285	2,287	2,225	2,884	2,352	2,431	2,727
Ouagadougou (1973–2010 historical data; 2011–2050 climate projections)	636	761	735	780	808	897	882	870	870
Saint Louis (1973–2010 historical data; 2011–2050 climate projections)	283	352	314	324	334	497	437	453	381

Table 2.7 Mean annual temperature (degree Celsius) observed and calculated using different climate change scenarios

Mean annual temperature [°C]									
		8 km							
		CMCC – CCLM		CSIR CCAM – A2					
	Observed historical data	RCP 8.5	RCP 4.5	csiro	ukmo	mpi	miroc	gfdl20	gfdl21
Addis Ababa (1964–2010 historical data; 2011–2050 climate projections)	16.6	17.7	17.6	18.3	17.7	17.5	17.9	17.8	17.9
Dar es Salaam (1958–2010 historical data; 2011–2050 climate projections)	24.9	26.7	26.7	26.8	26.6	26.5	27.1	26.8	26.8
Douala (1976–2010 historical data; 2011–2050 climate projections)	26.4	27.4	29.3	27.8	27.2	27.1	27.6	27.3	27.4
Ouagadougou (1973–2010 historical data; 2011–2050 climate projections)	28.7	29.8	29.8	30.1	29.6	29.3	29.5	29.7	29.6
Saint Louis (1973–2010 historical data; 2011–2050 climate projections)	25.5	26.5	26.3	27.8	27.5	27.8	27.7	28.0	28.0

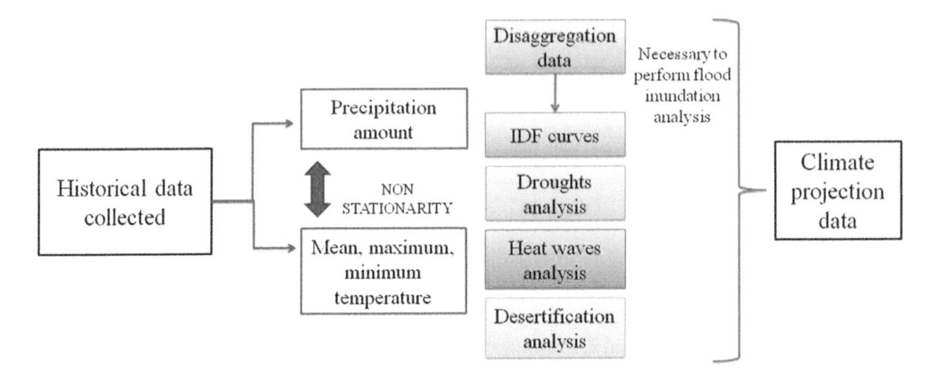

Fig. 2.5 Framework of the performed analysis

The IDF analysis starts by gathering a time series record of different observed rainfall durations. It then involves extracting data on the annual extremes associated with each rainfall duration within the time series. Finally a probability distribution is generated for the whole time series in order to estimate rainfall characteristics. Since the historical record only contained daily rainfall data it was necessary to generate a synthetic sequence of rainfall with the same statistical properties as those in the observed record in order to define the appropriate extreme values for shorter rainfall durations (i.e. 10 min, 30 min, 1 h, 3 h, 6 h, 12 h) (Giugni et al. 2012).

In order to calculate the extreme values in smaller time windows, the daily rainfall can be successively disaggregated using two models:

- a cascade-based disaggregation model;
- a short-time intensity disaggregation method.

Assuming that daily rainfall is derived from a marked Poisson process, i.e. rainfall lag and depths are drawn from exponential Probability Density Functions (PDFs), it is possible to use a simple stochastic model of daily rainfall that describes the occurrence of rainfall as a compound Poisson (De Paola et al. 2013a, b).

In a *cascade-based disaggregation model* (Güntner et al. 2001), precipitation data of daily resolution are converted into 12-h, 6-h, or 3-h duration values, based on the principles of multiplicative cascade processes.

The *short-time intensity disaggregation model* (Connolly et al. 1998) is used to develop three fine-resolution time intervals, equal to 1 h, 30 min and 10 min.

In more detail, for each case study city, the available daily rainfall data, for all years of observations, were disaggregated into seven durations (10 min, 30 min, 1 h, 3 h, 6 h, 12 h and 24 h). In particular, using the cascade-based model, the values for the 3, 6 and 12 h were obtained. In order to evaluate durations less than 3 h (10 min, 30 min and 1 h), the short time intensity disaggregation model was used.

For the application of the short time intensity model, a freely downloaded software was used named *Cra.Clima.Rain* (http://www.sipeaa.it/ASP/ASP2/Rain.asp). Once the maximum values for the seven considered durations were obtained, these values were fitted by the classical Gumbel function (Maximum Extreme Value Type 1) evaluating the IDF curves, expressed in the form $\mu(d) = a_{\mu}d^{n}$ and calculating K_T, growth rate, values for different return periods (De Risi et al. 2013).

The illustrated procedure assisted in overcoming the limited number of available data (only one meteorological station for each case study and only daily data). The same procedure was also used for the projected data; in more detail, the projected 24-h data were disaggregated, as shown before, and then, using the Gumbel function the IDF curves were evaluated in order to take climate change into account.

Temperature and Heat Waves

Temperature is associated with several types of extreme events, such as heat waves and cold spells. Both have impacts on human health, the physical environment, ecosystems, or energy consumption (Christensen et al. 2007; Meehl et al. 2007; Trenberth et al. 2007).

Heat waves generally refer to periods of exceptionally warm temperatures, but no universal technical definition exists (Robinson 2001). Heat waves cause heat stress and temporary changes in lifestyle and may have adverse health consequences for the affected population, producing an increase in mortality and morbidity especially for those most vulnerable to excess heat such as those with heightened biophysical susceptibility due to age or ill-health.

Different approaches are commonly adopted to define the meteorological criteria for assessing when a heat wave takes place. Each method differs in its definition of a threshold as follows: (1) a threshold is based on the statistical comparison with historical meteorological baselines; for example, the definition is based on exceeding threshold values defined by identifying the observed highest values in time series data in a specific area; (2) a threshold is derived from statistical analyses of the relationship between weather indices and mortality; or (3) a threshold is derived from bio-meteorological studies of human comfort under conditions of high temperatures and high humidity. Most countries use absolute thresholds of air temperature (mean, maximum or minimum) or a combination of air temperature and a measure of humidity. For example, the National Weather Service of the USA recognises a heat wave when the daily maximum Heat Index (HI) exceeds 40.6 °C and the night-time minimum exceeds 26.7 °C, for two consecutive days, while in Greece a heat wave is recognised when the daily maximum temperature exceeds 38 °C, for three consecutive days. Here, we define a heat wave as a period in which the maximum temperatures are over the 90th percentile of the monthly/daily distribution for at least three days, fixing a minimum for T_{max} to exclude months with cooler temperatures. The 90th percentile has been evaluated over a climatologic base period (1961–1990) or shorter periods in cases of data unavailability. The choice of a simple approach is also justified by the consideration that the accuracy of forecasts is relatively high. This avoids the increase in uncertainty that occurs as the number of input variables increases.

The use of percentiles allows comparison of results from different locations characterised by different climate features, focusing on the same part of the temperature distribution. These indices only describe moderately extreme events. In fact, the choice of 90th percentile corresponds to a return period of 36 days per year, and therefore, does not indicate rare extreme events, i.e. the most extreme events with the greatest impacts for human health, urban infrastructure and the natural environment.

Droughts

Drought is a deficiency of precipitation from expected or "normal" amounts that, when extended over a season or longer time period, is insufficient to meet demands. Common indicators of droughts include meteorological variables, such as precipitation and evaporation, as well as hydrological variables, such as stream flow, groundwater, reservoir, or lake levels, and soil moisture (Rossi and Cancelliere 2003). Here, we focus on the Standardized Precipitation Index (SPI) and the Method of Run. The main advantages of the SPI is its standardised nature, which makes it particularly suited for comparing drought conditions among different time periods and regions with different climatic conditions (McKee et al. 1993, 1995). Positive SPI values indicate greater than average precipitation and negative values indicate less than average precipitation. Table 2.8 reports the drought classification based on SPI values adopted by the

Table 2.8 SPI classification (McKee et al. 1995)

SPI	Class
$\geq$2.0	Extremely wet
From 1.5 to 1.99	Very wet
From 1.0 to 1.49	Moderately wet
From −0.99 to 0.99	Near normal
From −1.0 to −1.49	Moderately dry
From −1.5 to −1.99	Very dry
−2 and less	Extremely dry

US National Drought Mitigation Center, which can be used in the African context as well.

The Method of Run analysis has been proposed as a method for identifying drought periods and for evaluating the statistical properties of droughts. According to this method a drought period coincides with a *negative run*, defined as a consecutive number of intervals where a selected hydrological variable remains below a chosen threshold.

Then, a run is defined as a sequence of intervals characterised by deviations with the same sign, preceded and followed by at least one interval with deviation of the opposite sign. Negative deviations are usually referred to as deficit, whereas the corresponding run is termed as negative run or drought (Rossi and Cancelliere 2003).

Desertification

Desertification, as defined in the International Convention on Desertification, is the degradation of the land in arid, semi-arid and sub-humid dry areas caused by climatic changes and human activities. It is accompanied by a reduction in the natural growing potential of the land and depletion of surface and groundwater resources. Above all, it has negative repercussions on the living conditions and the economic development of the people affected. Desertification not only occurs in natural deserts, but can also take place on land which is prone to desertification processes (UNCCD 2002).

A number of methodologies have been used to assess desertification. In the CLUVA project, the methodological approach for the assessment and mapping of desertification dynamics is based on the general methodology described by Kosmas et al. (1999). This methodology allows the mapping of Environmentally Sensitive Areas (ESAs) to desertification through the evaluation of an Environmentally Sensitive Area Index (ESAI).

The key indicators for mapping ESAs to desertification, which can be used at regional or national level, can be divided into four broad categories defining the qualities of soil, climate, vegetation and management (stress indicators). Each index results from a set of parameters, shown in Fig. 2.6, obtained by databases and thematic cartography at different scales.

Scores were assigned to each parameter, ranging from 1, referring to the best conditions, to 2, for the worst conditions, according to their relevance for the processes of desertification. Classes and scores were redefined considering soil, vegetation and climate specific characteristics for Sub-Saharan West Africa.

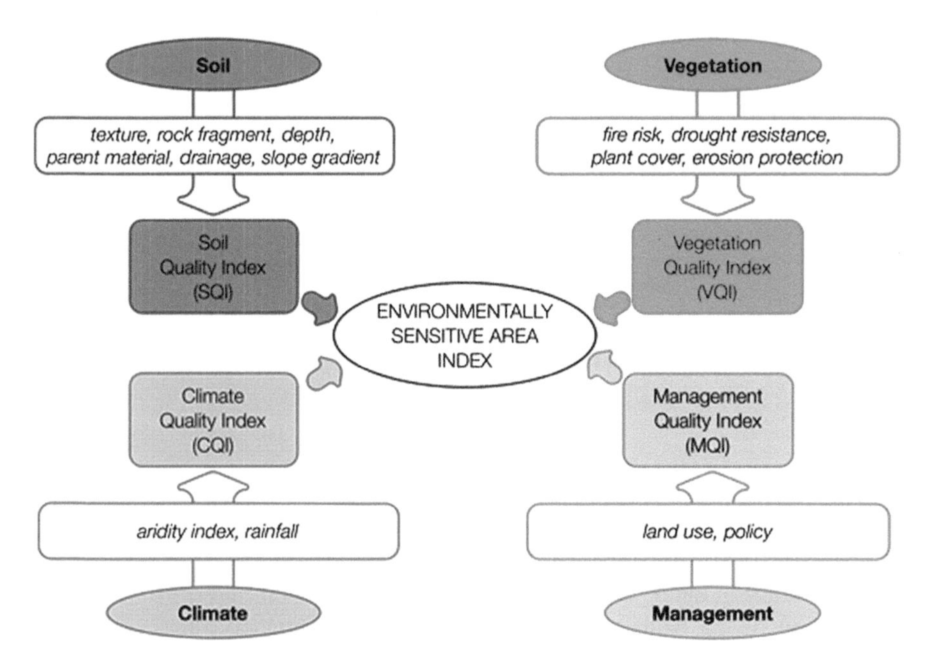

Fig. 2.6 Flowchart of the Environmentally Sensitive Area Index (ESAI)

Soil Quality Index (SQI) factors have been rearranged taking into account the specific geopedological properties of study areas. Based on the collected data, a new class (moderately drained soils) has been added to the study due to the absence of well drained soils and in order to better discriminate soils at different drainage capacity.

The adaptation of the Vegetation Quality Index (VQI) to local conditions was made starting from classification of the dominant natural vegetation and agricultural crops in Sub-Saharan West Africa and, subsequently, assigning the scores in order to match the specific relevance of the four factors.

For the Climate Quality Index (CQI), data collected from twelve meteorological stations distributed in West Africa have been used for redefining the classes and assigning the scores of two factors, Rainfall and Aridity Index, evaluated by the method of Thornthwaite (Singh and Xu 2001).

The ESAI results from the geometric average of these four factors and was ordered into eight classes of sensitivity degree, from "non-affected" to "critical."

Finally, all local data, geographically managed and processed within a GIS, allowed the generation of maps identifying Environmentally Sensitive Areas (ESAs). This process found two case study cities, Ouagadougou and Saint Louis, to be more sensitive to desertification.

Main Results and Comparisons with Historical Data

Addis Ababa

Rainfall Analysis and Evaluation of Intensity/Duration/Frequency (IDF) Curves

The IDF curves show an increase in frequency of extreme events, whereas the intensity of the events decreases. This means that – across all climate projections – we see a higher number of high precipitation events per year, but the mean intensity is lower compared to events experienced in the past. Figure 2.7 illustrates IDF curves for historical data (blue) and two climate change projection data sets, RCP8.5 (red, time period 2010–2050; purple, time period 2040–2050) and Ukmo (green, time period 2010–2050; cyan, time period 2040–2050). With growing duration (x-axis), historic events produced considerably more rain (y-axis) than projected events. The comparison of the mean values for the two time periods (2010–2050 vs. 2040–2050) shows rather similar curve shapes, indicating only slight changes within the next 40 years. In terms of frequency, as shown by the curves of growing factor variation (K_T) in Fig. 2.7b, the effects of climate change involve a rise of frequency of extreme events. In fact, keeping constant K_T, the corresponding return period value is reduced when considering the climate projections.

Heat Waves

For Addis Ababa, the duration of heat waves and number of hot days are strongly correlated. The duration of heat wave episodes shows highly increased mean values across all projections. Figure 2.8 illustrates this development. Whereas heat waves with a duration of six days have been observed in the past (green column), 14, 15, or even 22 days are projected by the various climate change models. The frequency distribution f(E) shows a similar picture. Comparing five 20-year periods, a clear increase in both duration and intensity can be observed, illustrated using the RCP4.5 scenario in Fig. 2.8.

Drought Analysis

Based on the monthly average rainfall, the SPI was evaluated, with reference to a 24-month aggregation scale, both for the historical data and for the projection data. The results are shown in Table 2.9. It is interesting to note that the minimum value of SPI highlights extremely dry conditions (referring to the US SPI classification),

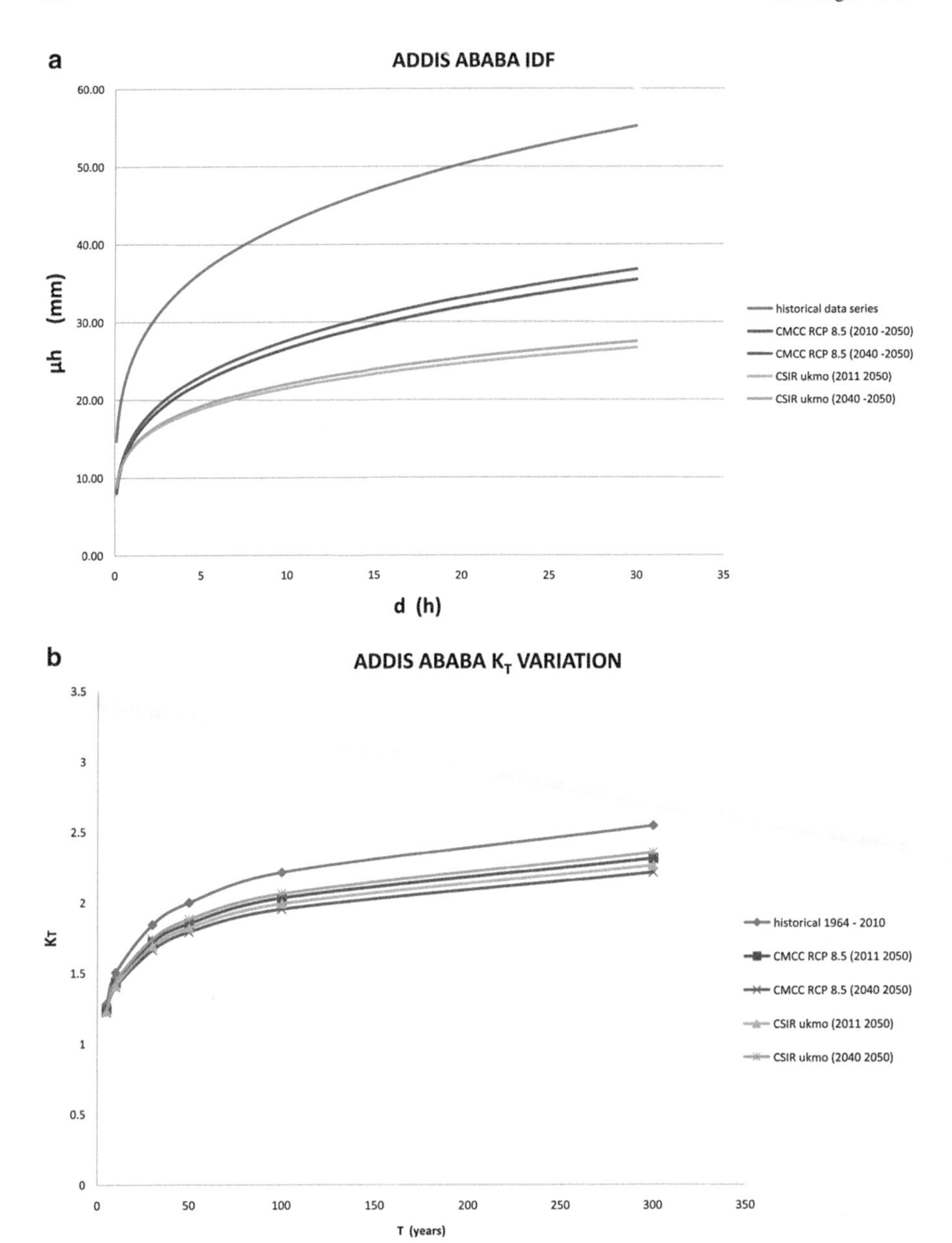

Fig. 2.7 IDF curves of historical data (*blue*) compared to projected data of two different models and scenarios, in terms of (**a**) intensity and (**b**) frequency

both based on historical data and on climate projections. The different projections show a very uneven picture. Whereas the RCP scenarios project considerably drier conditions, the A2 model projections show less severe minimum values and a trend towards wetter conditions.

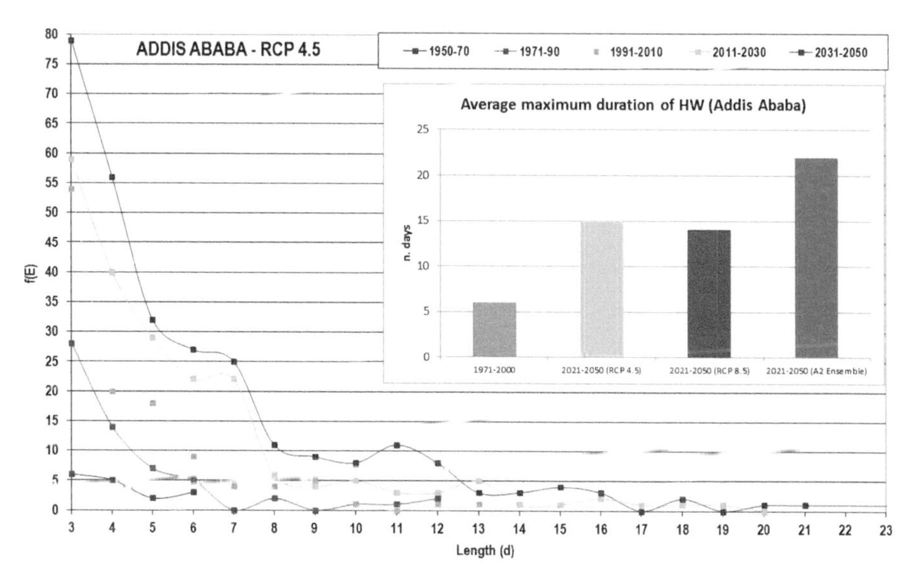

Fig. 2.8 Frequency distribution (f(E)) of the maximum duration of heat waves for different periods and projection models

The data collected for Addis Ababa were also analysed by the *Method of Run*. The method was applied using mean annual rainfall as the threshold. The following aspects have been evaluated:

- Duration: number of consecutive intervals in which the variable takes values less than the chosen threshold;
- Cumulative deficit: the sum of negative differences in a drought event;
- Intensity: ratio between cumulative deficit and duration;
- Return period associated to the event.

The results of Method of Run are shown, for brevity, only with reference to the historical data and different projection scenarios. In Fig. 2.9a, only historical data were considered, with a threshold equal to the mean annual rainfall, 732.4 mm. In Fig. 2.9b, the results are shown considering the projection data with reference to the scenario RCP 4.5 with a threshold equal to 978.6 mm. When considering the climate projections, the mean annual rainfall is increasing, but the return period is decreasing and duration of drought events is increasing. Thus, the Method of Run stresses that, although the mean annual rainfall is increasing in the climate projections, the duration of dry periods also increases.

Table 2.9 Minimum, mean and maximum values for 24 months aggregation time scale SPI application

SPI values									
	Historical data	RCP 8.5	RCP 4.5	A2 ukmo	A2 mpi	A2 miroc	A2 gdf 21	A2 gdf 20	A2 csiro
Min	−3.26	−4.58	−4.62	−2.51	−2.30	−3.37	−2.24	−3.30	−2.48
Mean	0.03	0.10	0.08	0.00	0.00	0.00	0.00	0.00	0.00
Max	1.77	1.52	1.62	2.12	2.40	2.20	1.76	2.50	2.45

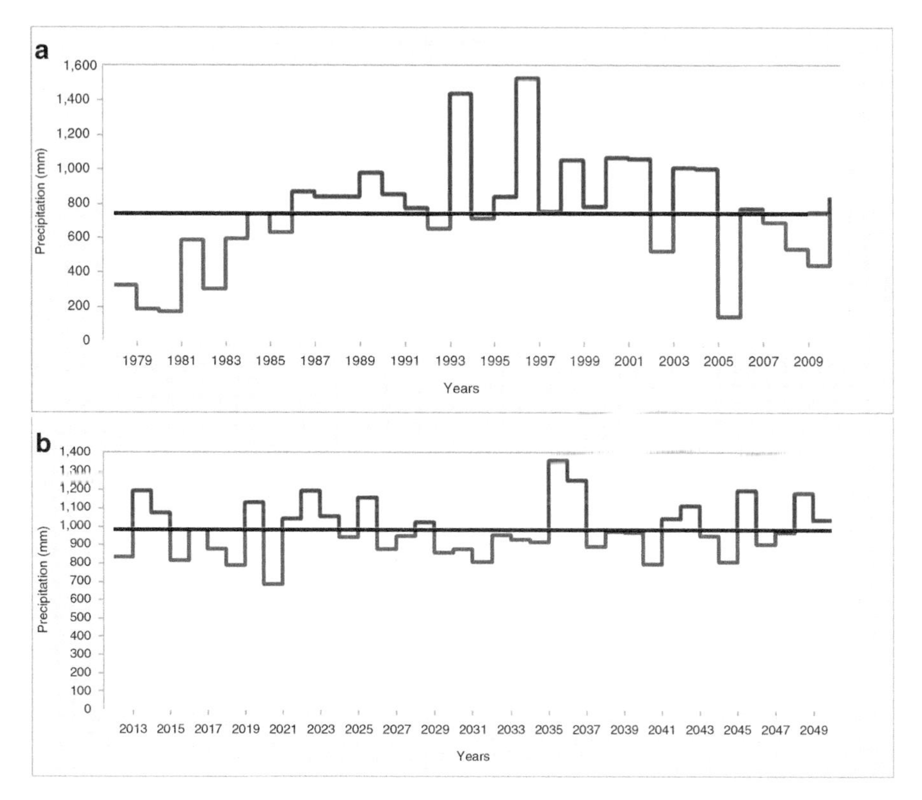

Fig. 2.9 Application of Method of Run using data projections (**a**) historical data and (**b**) scenario RCP 4.5

Dar es Salaam

Rainfall Analysis and Evaluation of Intensity/Duration/Frequency (IDF) Curves

From the analysis of the IDF curves, no significant difference occurred between the two RCP scenarios. The intensity of events decreases (Fig. 2.10). Additionally, both scenarios show an increase in frequency of severe rain events.

Drought Analysis

The results of the application of SPI for a 24-month aggregation scale are shown in Table 2.10. Considering the SPI classification schema, the conditions are rather dry. The minimum values remain rather stable for both RCP climate change projections, whereas RCP8.5 indicates an increase of the maximum value.

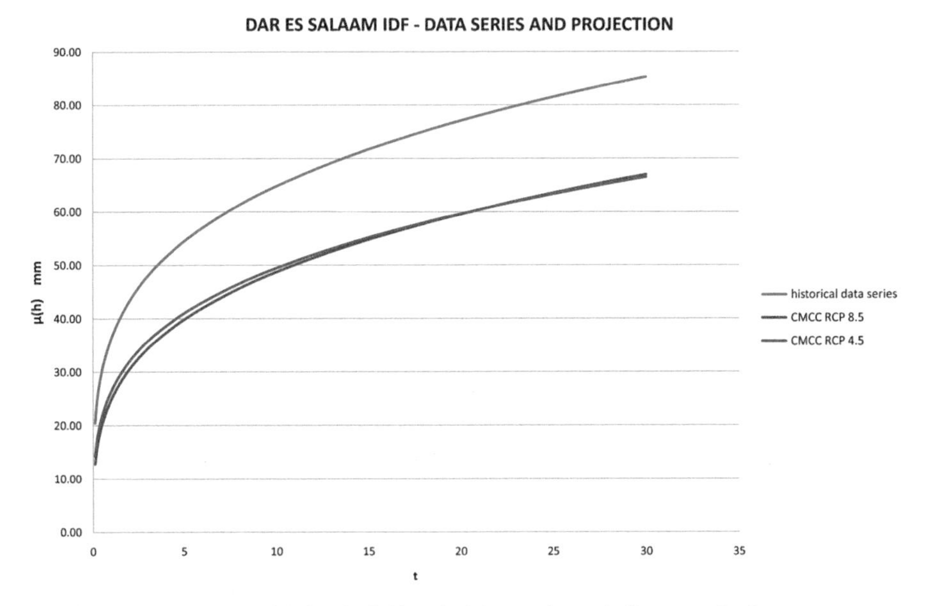

Fig. 2.10 IDF curves considering both historical data series and climate projections

Table 2.10 Minimum, mean and maximum values for 24-month aggregation scale SPI application

SPI values			
	Historical data	RCP 8.5	RCP 4.5
Min	−2.32	−2.337	−2.193
Mean	0.00	−0.003	−0.001
Max	2.31	3.169	2.536

The data collected for Dar es Salaam were also analysed by the *Method of Run*. Through climate projections, the threshold for applications, i.e. the mean annual rainfall, remains practically the same, with no significant variations of SPI compared to the evaluation using historical data.

Heat Waves

In the future, as shown in Fig. 2.11, a clear increase in the average duration of heat waves is projected and the number of events with maximum length lasting 5 days could increase from 3 to 24 (33 for RCP 8.5) over 100 years (from 1950–1970 to 2030–2050).

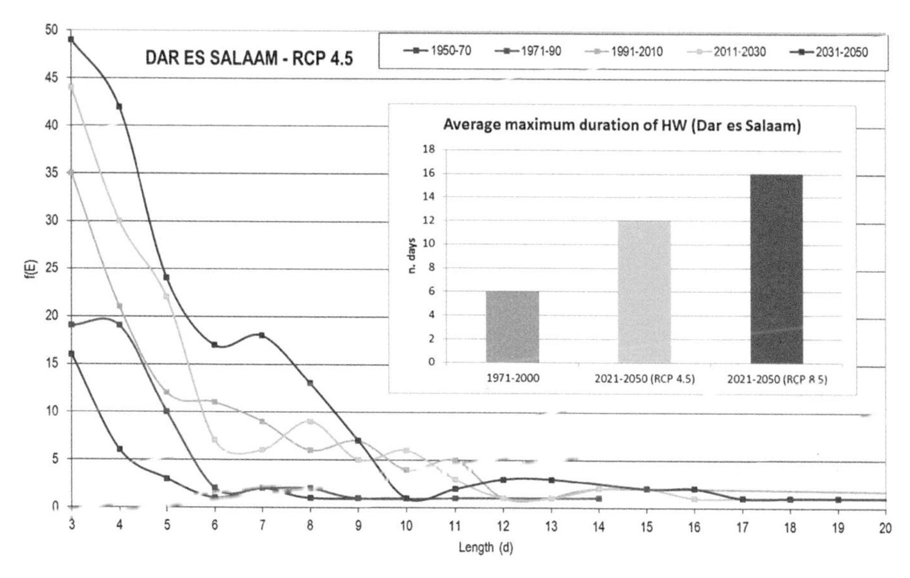

Fig. 2.11 Frequency distribution (f(E)) of maximum length heat waves for different independent periods

Douala

Rainfall Analysis and Evaluation of Intensity/Duration/Frequency (IDF) Curves

From the analysis of the IDF curves, no significant difference occurred between the two RCP scenarios, 8.5 and 4.5. The intensity of events decreases, as Fig. 2.12 illustrates. Additionally, both scenarios show an increase in frequency of severe rain events.

Drought Analysis

The SPI analysis shows a clear tendency towards drier conditions. The historical minimum value of −3.55 drops to −5.2/−5.25 in both scenarios, whereas the maximum value (Table 2.11) remains more stable, with both projections providing even values of 2.13 and 2.03, respectively, compared to 2.73 calculated from the historical data.

The data collected for Douala were also analysed by the *Method of Run*. Through climate projections, an increase in the number of drought events was highlighted.

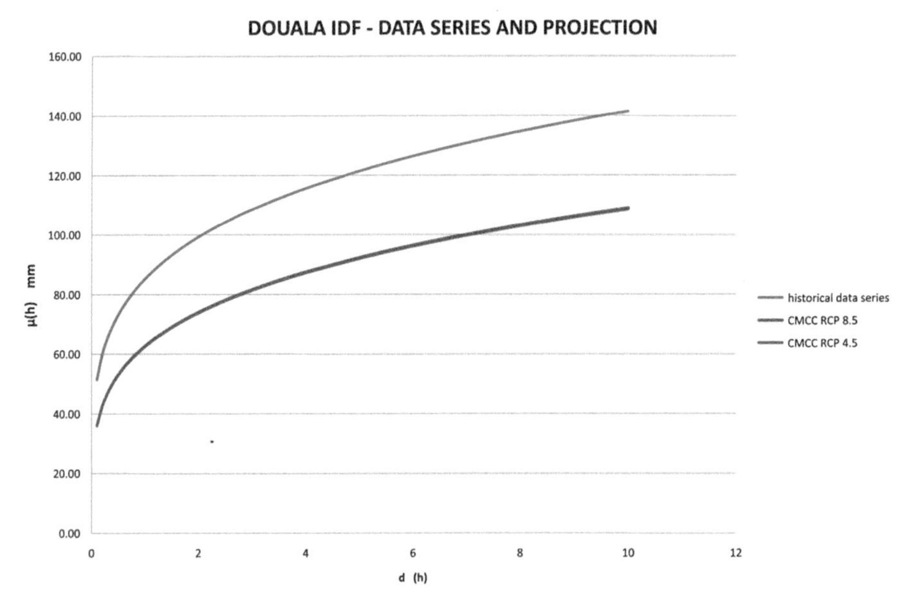

Fig. 2.12 IDF curves considering both historical data series and climate projections

Table 2.11 Minimum, mean and maximum values for 24-month aggregation time scale SPI application

SPI values			
	Historical data	RCP 8.5	RCP 4.5
Min	−3.55	−5.25	−5.20
Mean	0.00	0.15	0.16
Max	2.73	2.13	2.03

Heat Waves

The maximum duration of heat wave shows a mean value increasing from 3 to 15 days (RCP4.5) or 17 days (RCP8.5), respectively, illustrated in Fig. 2.13. Thus, Douala shows the highest increase in heat wave duration compared to the other cities. Splitting the 100 years period (1950–2050) into five consecutive intervals of 20 years each, a general increase in heat wave duration and intensity is observed, with the longest heat waves lasting more than 20 days.

Ouagadougou

Rainfall Analysis and Evaluation of Intensity/Duration/Frequency (IDF) Curves

The IDF curves (Fig. 2.14) analysis for Ouagadougou does not show the general decrease in intensity as observed at other cities. All three curves – historical data, RCP4.5, and RCP8.5 – show rather similar shapes.

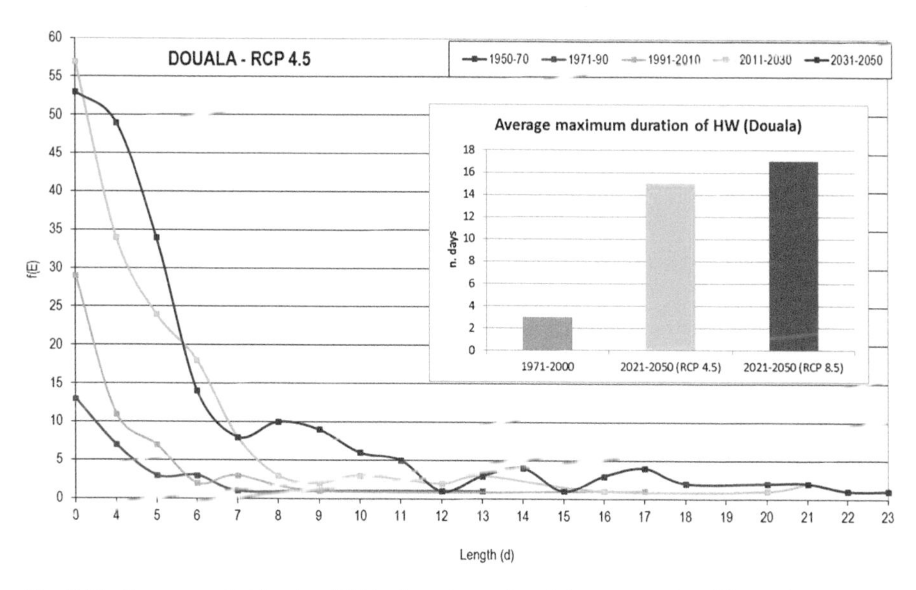

Fig. 2.13 Frequency distribution (f(E)) of maximum length heat waves for different independent periods

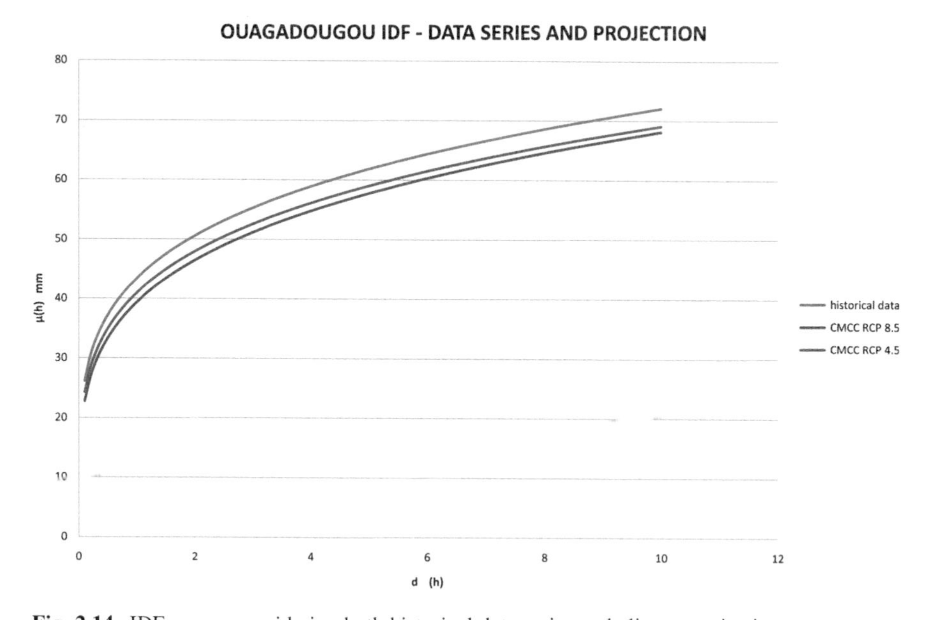

Fig. 2.14 IDF curves considering both historical data series and climate projections

Drought Analysis

The results of SPI reported in Table 2.12, referring to a 24-month time aggregation scale, show a strong increase in dry conditions, as the value drops from historical −2.71 to −4.28. Both projection scenarios, RCP 8.5 and 4.5, produce similar

Table 2.12 Minimum, mean and maximum values for 24-month aggregation time scale SPI application

SPI values			
	Historical data	RCP 8.5	RCP 4.5
Min	−2.71	−4.28	−4.28
Mean	−0.002	0.01	0.01
Max	2.53	2.50	2.35

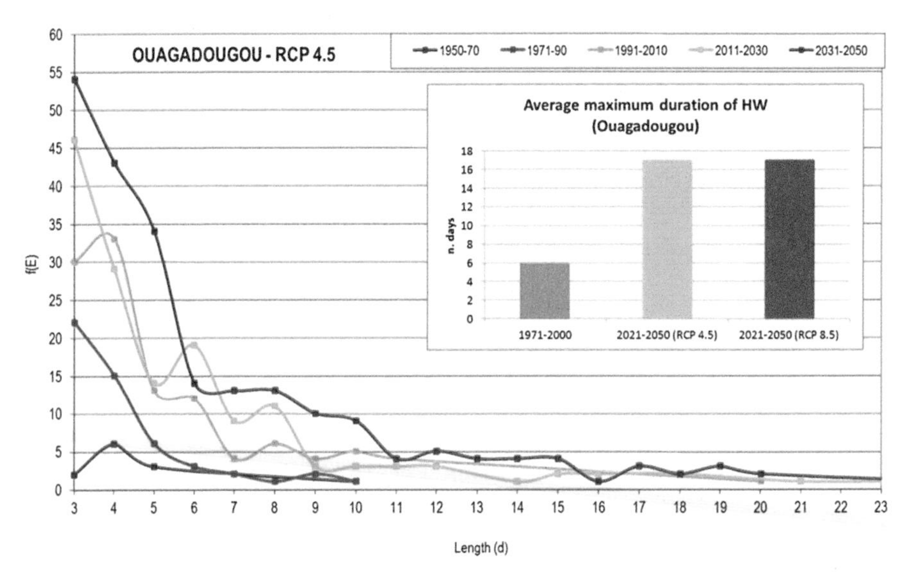

Fig. 2.15 Frequency distribution (f(E)) of maximum length heat waves for different independent periods

results. The wet conditions remain rather stable, with a historical value of 2.53, compared to 2.50 and 2.35 for RCP8.5 and RCP4.5, respectively.

The data collected for Ouagadougou were also analysed by the *Method of Run* and the results show an increase in the number of drought events, from six events, considering only historical data, to 10 events, when considering climate projections (Giugni et al. 2013).

Heat Waves

For Ouagadougou, as shown in Fig. 2.15, the length of heat wave episodes shows a mean value increasing from 6 to 17 (RCP4.5 and RCP8.5) days, while the number of events with maximum length lasting 5 days increases from 3 to 34 (37 for RCP8.5) over 100 years (from 1950–1970 to 2030–2050).

Table 2.13 Type and scale of the sources used for calculation of soil (SQI), vegetation (VQI), and management (MQI) quality index

	Burkina Faso
SQI	Carte pédologique de reconnaissance de la République de Haute-Volta (by O.R.S.T.O.M., 1968 – 1/500,000) (Michel 1967)
	Milieu naturels du Burkina Faso by I.R.A.T. – 1/1,000,000 (Guillobez 1985)
	Ressources en sols by O.R.S.T.O.M. – 1/500,000 (Boulet and Alboucq 1976)
	Carte géologique de reconnaissance de la République de Haute-Volta by D.G.M. – 1/1,000,000 (D.G.M. 1976)
VQI	GLOBCOVER by ESA/UNEP/FAO/JRC/IGBP/GOFC-GOLD (Bontemps et al. 2009)
MQI	GLOBCOVER by ESA/UNEP/FAO/JRC/IGBP/GOFC-GOLD (Bontemps et al. 2009)
	Ministère des Ressources Animales, Burkina Faso

Desertification

The study area includes the urban and peri-urban areas of Ouagadougou (about 2,800 km^2), with a population of around 1,950,000 in 2009. It is highly urbanised and is the most densely populated area of the country. Table 2.13 shows the type and scale of the sources used for calculation of soil (SQI), vegetation (VQI), and management (MQI) quality index.

Results show that the whole study area is affected by sensitivity to desertification, with a dominance of *moderate* quality soils (WRB 2006).

The study area is characterised by *moderate* and *low* vegetation quality. *Moderate* vegetation quality is attributed to rain-fed cropland usually characterised by a lower fire risk and a good plant cover.

The study area is characterised by *moderate* and *low* quality of management. The enforcement of policy for environmental protection is *very low* and restricted to a small part of the study area. The only protected area is the Urban Park of Ouagadougou, also known as Fôret Classée du Barrage de Ouagadougou, which is an ancient sacred forest of 260 ha. The heavy demographic pressure led to an overexploitation of natural resources resulting in the degradation of the environment and a loss of biological diversity.

The projection data (i.e. daily rainfall, temperature) entered in the ESAI model for the parameters, allowed a revised evaluation of Climate Quality Index CQI, and consequently, a revised assessment of climate change impact on sensitivity to desertification in the study area. The sub-indices Aridity Index (AI) and Mean Annual Rainfall have been evaluated with reference to data collected from 1973 to 2010 and then by considering data produced through the CLUVA climate projections.

The comparison between the different maps (Fig. 2.16), identifying the Environmentally Sensitive Areas (ESAs) produced on the basis of the historical climatic data and projection data and, in particular, the CQI, highlights how the increase in rainfall produces a reduction of classes for both the sub-indices which constitute the CQI; therefore, the future situation appears less critical.

Further, no marked differences are found in the results between the two RCP scenarios.

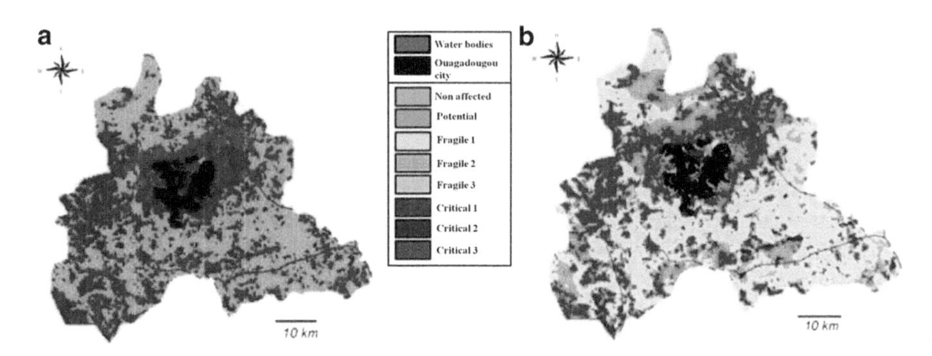

Fig. 2.16 ESAs maps evaluated using (**a**) historical data and (**b**) RCP climate projections

Saint Louis

Rainfall Analysis and Evaluation of Intensity/Duration/Frequency (IDF) Curves

The analysis of the curves (Fig. 2.17) does not show any significant difference between the two climate change scenarios. In contrast to other cities, Saint Louis shows a significant increase in both intensity as well as frequency of severe rain events.

Drought Analysis

The results of SPI reported in the following Table 2.14, referring to a 24-month aggregation scale, show that the conditions will not change substantially compared to the past.

The application of *Method of Run* in Saint Louis shows that there is an increase of mean annual rainfall, so threshold for applications is increasing and consequently the drought events are decreased.

Heat Waves

The length of heat wave episodes shows a mean value increasing from 5 to 9 (RCP4.5 and RCP8.5) days, thus Saint Louis shows the lowest increase in the heat wave length, with respect to the other cities. The frequency distribution (f(E)) plot of hot days duration (Fig. 2.18) for five separate 20-year periods, shows the temporal change of heat wave characteristics. This distribution has become longer tailed with time. For example, the number of events with maximum length lasting 5 days could increase from 9 to 18 (21 for RCP8.5) over 80 years (from 1971–1990 to 2030–2050).

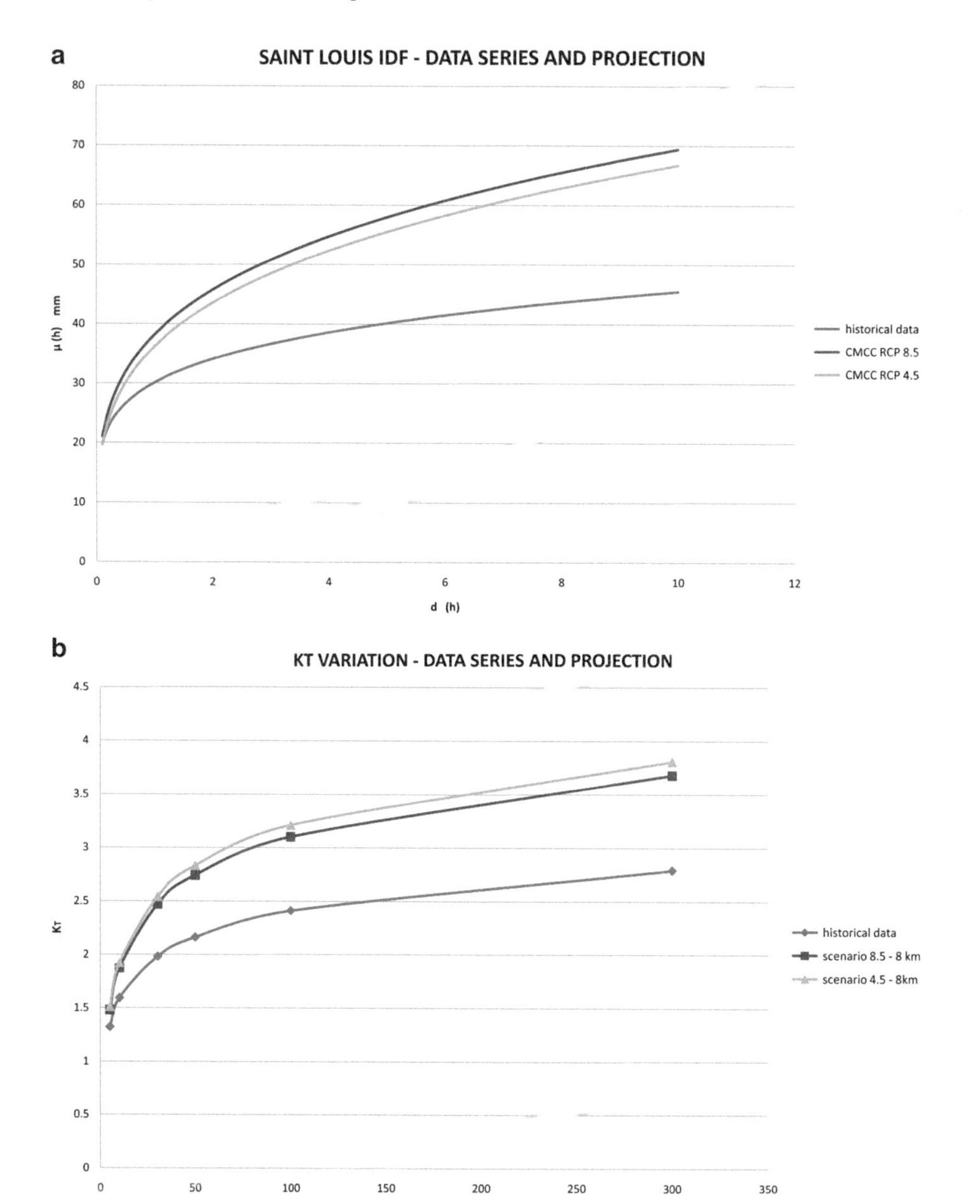

Fig. 2.17 IDF curves illustrating the growing intensity of (**a**) severe rain events and (**b**) the variation of growing factor K_T

Desertification

Urban and peri-urban area of Saint Louis can be divided into three main regions: coastal area, agricultural expansion region and the Senegal River valley.

Table 2.14 Minimum, mean and maximum values for 24-month aggregation time scale SPI application

SPI values			
	Historical data	RCP 8.5	RCP 4.5
Min	−2.77	−2.578	−2.04
Mean	0.01	−0.032	−0.058
Max	2.37	2.88	3.09

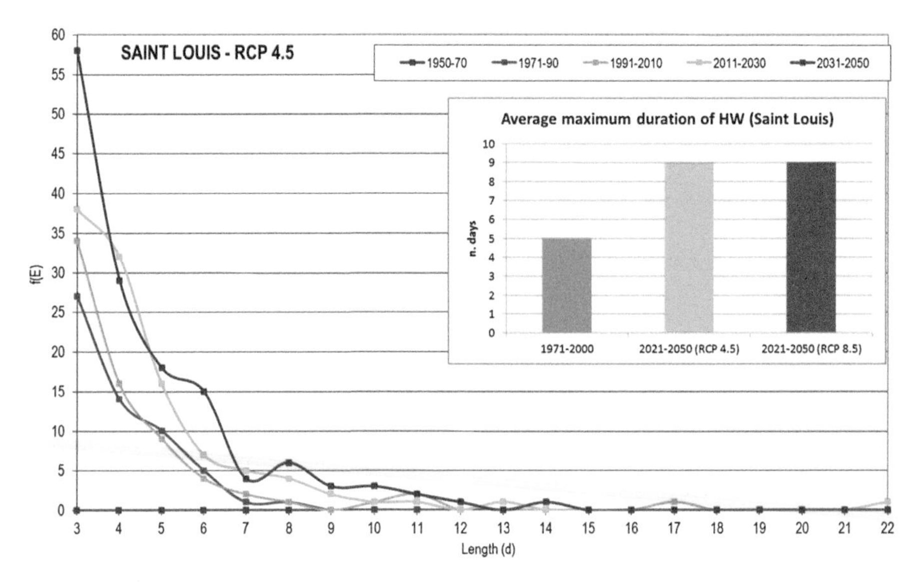

Fig. 2.18 Frequency distribution (f(E)) of maximum length heat waves for different independent periods

Table 2.15 Type and scale of the sources used for calculation of soil (SQI), vegetation (VQI), and management (MQI) quality index

	Senegal
SQI	Carte pédologique du Sénégal by O.R.S.T.O.M. – 1/1,000,000 (Maignein 1965)
	Carte geologique du Senegal by BRGM – 1/200,000 (Michel 1967)
VQI	GLOBCOVER by ESA/UNEP/FAO/JRC/IGBP/GOFC-GOLD (Bontemps et al. 2009)
	Tappan et al. (2004)
MQI	LADA: Land Degradation Assessment in Drylands (http://www.fao.org/nr/lada/index.php)
	Tappan et al. (2004)

Table 2.15 shows the type and scale of the sources used for calculation of soil (SQI), vegetation (VQI), and management (MQI) quality index.

The southern part of the study area results to be characterised by *high* quality of soils, while the coastal zone shows *moderate* soil quality. The worst levels of soil quality mainly occur in the northern part of the study area, especially near the city of Saint Louis.

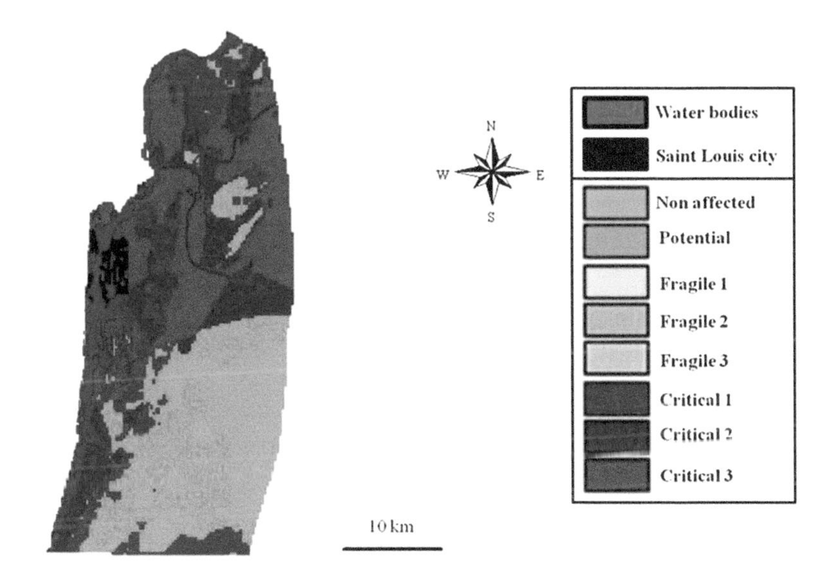

Fig. 2.19 ESAs index map for Saint Louis

The study area is characterised by *low* vegetation quality, with critical condition for most of the area.

The climate quality of the study area, evaluated by using both historical data and climate projections, can be defined as *low*, so the map in Fig. 2.19, representing the spatial distribution of the Index of Environmentally Sensitive Areas (ESAI), suggests that the northern part of Saint Louis area is critically sensitive to desertification, mainly due to the overexploitation of natural resources.

Conclusion

This chapter provided an overview of the impacts of climate change based on different projections and downscalings, produced for the time period 2010–2050. The projections target different locations, including major African cities such as Saint Louis, Ouagadougou, Douala, Addis Ababa and Dar es Salaam. The results have been shown and discussed for various climate change indicators, including changes in temperature and precipitation regimes and subsequent effects such as desertification. The diversity of the African climate, high rainfall variability and a very sparse observation network make the calibration of projections very difficult at sub regional levels. Therefore, all results need to be considered with care. CLUVA addressed this aspect using and comparing a collection of different global models and downscaling techniques and successfully established the sensitivity to climate change in Africa on a new spatial scale and temporal resolution.

The analysis of the climate projection data shows a general increase in mean annual precipitation and temperature for all the case study cities. Regarding rainfall trends and in particular IDF curves, the analysis of the growing factor K_T highlights that the effects of climate change in all the case study cities involve a rise of frequency of extreme events. Moreover, the analysis of rainfall extreme events underlines a decrease in intensity, with the exception of Saint Louis. However, assuming a 'Business As Usual' scenario of population growth, runoff of rainwater in the African cities is expected to increase due to the decreased permeability that occurs in urban environments. Therefore, the hydraulic risk is expected to increase in the future. The rainfall pattern also negatively influences drought events despite increasing mean annual rainfall. The analysis, based on SPI and Method of Run, highlights extremely dry conditions in the future.

For all studied cities, temperature is expected to increase by at least 1 °C. This, of course, is reflected in the analysis of heat waves, which stresses an increase in frequency of hot days and duration of heat waves. The future warming causes a further increase in parameters associated with heat waves. The strong correlation between duration of heat waves and number of hot days shows that temperature rise could mainly generate longer average durations, increasing the impacts of heat waves on society. In fact, the capacity of adaptation can be reduced with prolonged exposure to high temperature and humidity. The expected persistence of long-lived heat waves, those lasting at least 1.5–2 weeks, is clearly longer with respect to the reference climatic period (1961–1990). Moreover, during the 100 years analysed, short-lived but intense waves are more than doubled.

It is also interesting to note that desertification processes, in the case study cities Ouagadougou and Saint Louis, are influenced mainly by the environmental impact of population growth (i.e. overexploitation of natural resources), rather than by climatic change.

Finally, the results show that climate changes, although producing a severe increase in heat waves, are not expected to produce dramatic increases of flood, drought and desertification hazards in the next few decades in the selected cities. Nevertheless, vulnerability and exposure may increase significantly due to human factors, so that the overall risk is very likely to increase during the next decades. For example, the uncontrolled growth of urban areas as a result of population growth and a lack of city planning may result in the reduction of green areas. In consequence, unpredictable increases of both urban flash floods and heat waves are likely to occur. Often, uncontrolled urbanisation also reduces both amount and extent of areas available for flood management and increases the number of houses and businesses located in flood prone areas.

Ongoing urbanisation is also strongly linked to increasing temperatures through the formation of "heat islands" – areas with higher temperatures due to radiant heat from buildings, infrastructures and other human activities. Climate change is projected to exacerbate these problems.

In summary, the crucial issue is to develop strategies that are able to reduce the vulnerability of African cities to climate-related hazards, in order to promote the resilience of local communities. This requires a joint approach, combining scientific research, technical knowledge, dialogue and partnership with a variety of government levels and institutions. These strategies should also take into account that climate change projections continuously change. Consequently, the assessment of climate change effects should be regularly updated.

References

Bontemps S, Defourny P, Bogaert E, Arino O, Kalogirou V, Perez JR (2009) Products description and validation report, GLOBCOVER 2009

Boulet R, Alboucq G (1976) Carte pédologique de reconnaissance de la République de Haute-Volta, Agronomiques déduites de la carte Pédologique. ORSTOM, Ministère de la Coopération, Paris

Christensen JH, Hewitson B, Busuioc A, Chen X, Gao I, Held R, Jones R, Kolli W, Kwon R, Laprise V, Magaña Rueda L, Mearns L, Menéndez C, Räisänen J, Rinke A, Sarr A, Whetton P (2007) Regional climate projections. In: Solomon S, Qin D, Manning M, Chen Z, Marquis M, Averyt KB, Tignor M, Miller HL (eds) Climate change 2007: the physical science basis. Contribution of Working Group I to the fourth assessment report of the Intergovernmental Panel on Climate Change. Cambridge University Press, Cambridge/New York, pp 847–940

Connolly RD, Schirmer J, Dunn PK (1998) A daily rainfall disaggregation model. Agric For Meteorol 92:105–117

De Paola F, Giugni M, Garcia A, Bucchignani E (2013a) Stationary vs. non-stationary of extreme rainfall in Dar es Salaam (Tanzania). IAHR Congress Tsinghua University Press, Beijing

De Paola F, Giugni M, Topa ME (2013b) Probability density function (Pdf) of daily rainfall heights by superstatistics of hydro-climatic fluctuations for African test cities. Wulfenia 20(5):106–126

De Risi R, Jalayer F, De Paola F, Iervolino I, Giugni M, Topa ME, Mbuya E, Kyessi A, Manfredi G, Gasparini P (2013) Flood risk assessment for informal settlements. Nat Hazards 69(1):1003–1032. doi:10.1007/s11069-013-0749-0

Direction Générale de la Météorologie (DGM) du Burkina (1976) Carte géologique de reconnaissance de la République de Haute-Volta 1/1000000

Eggleton T (2014) A short introduction to climate change. Cambridge University Press, Cambridge

Engelbrecht FA, McGregor J, Engelbrecht C (2009) Dynamics of the Conformal-Cubic Atmospheric Model projected climate-change signal over southern Africa. Int J Climatol 29:1013–1033

Engelbrecht FA, Landman WA, Engelbrecht C, Landman S, Bopape M, Roux B, McGregor JL, Thatcher M (2011) Multi-scale climate modeling over Southern Africa using a variable-resolution global model. Water SA 37:647–658

Giorgi F, Hewitson B (2001) Regional climate information—evaluation and projections. In: Houghton JT, Ding Y, Griggs DJ, Noguer M, van der Linden PJ, Dai X, Maskell K, Johnson CA (eds) Climate change 2001: the scientific basis, contribution of working group I to the third assessment report of the Intergovernmental Panel on Climate Change. Cambridge University Press, Cambridge/New York, pp 583–638

Giugni M, De Paola F, Topa ME (2012) Deliverable: hazard scenarios for test cities using available data. CLUVA CLimate change and Urban Vulnerability in Africa – Seventh Framework Programme Environment (FP7-ENV-2010)

Giugni M, Capuano P, De Paola F, Di Ruocco A, Topa ME (2013) Deliverable: probabilistic hazard and multihazard scenarios in GIS environment for the relevant hazards in selected cities. CLUVA CLimate change and Urban Vulnerability in Africa – Seventh Framework Programme Environment (FP7-ENV-2010)

Guillobez S (1985) Milieu naturels du Burkina Faso 1/1000000. IRAT Institut de Recherches Agronomiques Tropicales et des Cultures Vivriéres, Montpellier

Güntner A, Jonas Olsson J, Calver A, Gannon B (2001) Cascade-based disaggregation of continuous rainfall time series: the influence of climate. Hydrol Earth Syst Sci 5:145–164

IPCC (2001) Third assessment report: climate change, the scientific basis. In: Houghton JT, Ding Y, Griggs DJ, Noguer M, van der Linden PJ, Dai X, Maskell K, Johnson CA (eds) Cambridge University Press, Cambridge/New York

IPCC (2007) Climate change 2007: synthesis report. Contribution of Working Groups I, II and III to the fourth assessment report of the Intergovernmental Panel on Climate Change. In: Pachauri RK, Reisinger A (eds) IPCC, Geneva

IPCC (2012) Managing the risks of extreme events and disasters to advance climate change adaptation. Cambridge University Press, Cambridge/New York

Kosmas C, Kirkby M, Geeson N (1999) The Medalus project Mediterranean desertification and land use. European Commission Project ENV4 CT 95 0119

Lal M, McGregor JL, Nguyen KC (2008) Very high-resolution climate simulation over Fiji using a global variable-resolution model. Clim Dyn 30(2–3):293–305. doi:10.1007/s00382-007-0287-0

Maignein R (1965) Carte pédologique du Sénégal à l'échelle de 1:1,000,000, Service cartographique., O.R.S.T.O.M. (Agency France), Centre de Dakar, Paris

Maraun D, Wetterhall F, Ireson AM, Chandler RE, Kendon E, Widmann M, Brienen S, Rust HW, Sauter T, Themessl M, Venema V, Chun KP, Goodess C, Jones R, Onof M, Vrac C, ThieleEich I (2010) Precipitation downscaling under climate change: recent developments to bridge the gap between dynamical downscaling models and the end user. Rev Geophys 48:RG3003. doi:10.1029/2009RG000314

McGregor JL (2005) C-CAM geometric aspects and dynamical formulation. CSIRO atmospheric research technical paper, vol 70, CSIRO Atmospheric Research, Aspendale, VIC, Australia

McGuffie K, Henderson-Sellers A (2014) The climate modelling primer. Wiley-Blackwell, New York

McKee TB, Doesken NJ, Kleist J (1993) The relationship of drought frequency and duration to time scales. In: Eighth conference on applied climatology, 17–22 January, Anaheim, CA, pp 179–184

McKee TB, Doesken NJ, Kleist J (1995) Drought monitoring with multiple time scales. In: Preprints of 9th AMS conference on applied climatology, 15–20 January, Dallas, TX, pp 233–236

Meehl GA, Stocker TF, Collins WD, Friedlingstein P, Gaye AT, Gregory JM, Kitoh A, Knutti R, Murphy JM, Noda A, Raper SCB, Watterson IG, Weaver AJ, Zhao ZC (2007) Global climate projections. In: Solomon S, Qin D, Manning M, Chen Z, Marquis M, Averyt KB, Tignor M, Miller HL (eds) Climate change 2007: the physical science basis. Contribution of Working Group I to the fourth assessment report of the Intergovernmental Panel on Climate Change. Cambridge University Press, Cambridge/New York, pp 747–845

Michel P (1967) Carte geologique du Senegal au 1:200,000. In: Michel P Les bassins des fleuves Senegal et Gambie: etude geomorpholgique, France. Bureau de recherches geologiques et minieres (BRGM) Senegal, Direction des mines et de la geologie, Paris, France

Nunez M, McGregor JL (2007) Modelling future water environments of Tasmania, Australia. Clim Res 34(1):25–37. doi:10.3354/cr034025

Rebora N, Ferraris L, Von Hardenberg J (2006) Rainfall downscaling and flood forecasting: a case study in the Mediterranean area. Nat Hazards Earth Syst 6:611–619

Robinson PJ (2001) On the definition of a heat wave. J Appl Meteorol 40:762–775

Rockel B, Will A, Hense A (2008) The regional climate model COSMO-CLM (CCLM). Meteorol Z 17:347–348

Rossi G, Cancelliere A (2003) At-site and regional drought identification by redim model. In: Rossi G, Cancelliere A, Pereira LS, Oweis T, Shatanawi M, Zairi A (eds) Tools for drought mitigation in Mediterranean regions, vol 44, Water science and technology library. Springer Science + Business Media, Dordrecht, pp 37–54

Schmidli J, Goodess M, Frei C, Haylock MR, Hundecha Y, Ribalaygua J, Schmith T (2007) Statistical and dynamical downscaling of precipitation: an evaluation and comparison of scenarios for the European Alps. J Geophys Res 112:D04105. doi:10.1029/2005JD007026

Singh VP, Xu C (2001) Evaluation and generalization of temperature-based methods for calculating evaporation. Hydrol Process 15:305–319

Stocker T (2011) Introduction to climate modelling. In: Hutter K (ed) Advances in geophysical and environmental mechanics and mathematics. Springer, Heidelberg

Tappan GG, Sall M, Wood EC, Cushing M (2004) Ecoregions and land cover trends in Senegal. J Arid Environ 59:427–462

Trenberth KE, Jones PD, Ambenje P, Bojariu R, Easterling D, Klein Tank A, Parker D, Rahimzadeh F, Renwick JA, Rusticucci M, Solden B, Zhai P (2007) Observations: surface and atmospheric climate change. In: Solomon S, Qin D, Manning M, Chen Z, Marquis M, Averyt KB, Tignor M, Miller HL (eds) Climate change 2007: the physical science basis. Contribution of Working Group I to the fourth assessment report of the Intergovernmental Panel on Climate Change. Cambridge University Press, Cambridge/New York, pp 235–336

UNCCD (2002) United Nations convention to combat desertification in those countries experiencing serious drought and/or desertification, particularly in Africa: text with annexes, and a preface by the UNCCD Secretariat. United Nations Convention to Combat Desertification Secretariat, Bonn

WRB World reference base for soil resources (2006) 2nd. World soil 103. FAO, Rome

Chapter 3
Vulnerability of Built Environment to Flooding in African Cities

Fatemeh Jalayer, Raffaele De Risi, Alphonce Kyessi, Elinorata Mbuya, and Nebyou Yonas

Abstract Urban built structures and lifelines in African cities are particularly vulnerable to extreme weather-related events such as flooding. This chapter provides an overview of the activities and the findings related to the vulnerability to flooding of the urban built environment in the context of the CLUVA project. First, the urban hotspots to flooding for the built structures and the lifelines are identified for three African case study cities. In the next step, a probabilistic methodology is employed in order to perform micro-scale evaluation of building vulnerability and risk to flooding for the case study city of Dar es Salaam based on historical rainfall data. This methodology is developed specifically for vulnerability assessments based on incomplete knowledge and relies on various data-gathering techniques such as orthophoto boundary recognition, field surveys and laboratory tests. The micro-scale evaluation of the building vulnerability is also performed using rainfall data for a projected 1-year interval in 2050. The results in terms of economic loss and number of people affected are discussed and compared to the evaluation performed based on historical rainfall data. In this comparison, the negative effect

F. Jalayer (✉)
Center for the Analysis and Monitoring of Environmental Risk (AMRA), Scarl, Via Nuova Agnano 11, 80125 Naples, Italy

Department of Structures for Engineering and Architecture, University of Naples Federico II, Via Claudio 21, 80125 Naples, Italy
e-mail: fatemeh.jalayer@unina.it

R. De Risi
Department of Structures for Engineering and Architecture, University of Naples Federico II, Via Claudio 21, 80125 Naples, Italy
e-mail: raffaele.derisi@unina.it

A. Kyessi • E. Mbuya
Institute of Human Settlements Studies, Ardhi University, P.O. Box 35124, Dar es Salaam, Tanzania
e-mail: akyessi@gmail.com; mbuyataz@yahoo.com

N. Yonas
Ethiopian Institute of Architecture, Building Construction and City Development, Addis Ababa University, P.O. Box 518, Addis Ababa, Ethiopia
e-mail: nebyou_yonas@yahoo.com

S. Pauleit et al. (eds.), *Urban Vulnerability and Climate Change in Africa*, Future City 4,
DOI 10.1007/978-3-319-03982-4_3

of urbanisation on flooding risk is emphasised. The findings presented in this chapter can be translated into strategic adaptive measures for urban structures and lifelines to flooding.

Keywords Africa • Flood-prone • Physical vulnerability • Fragility • Risk assessment

Introduction

Africa is probably the continent most vulnerable to climate change and its adverse effects, despite being the continent least responsible for global greenhouse gas emissions. The adverse effects of climate change in Africa may include reduced agricultural production, worsening food security, the increased incidence of flooding and drought, spreading disease and an increased risk of conflict over scarce land and water resources. Therefore, it is clear that immediate action is needed in both reducing the global carbon emissions and adapting to the adverse effects of climate change. Hence, climate change and its associated impacts on the natural, physical and social systems calls for the integration of a forward-looking perspective into decision-making processes to ensure that climate change adaptation strategies are fully addressed. Evidence is increasing about correlation between the phenomenon of climate change and extreme weather-related events. As evidence of the link between climate change and extreme events, one can cite the increases in: frequency of heat waves; temperature maxima; likelihood of flash floods and droughts; and hurricane intensity. Additional evidence includes changes in rainfall patterns and intensity, and sea-level rise (CRED 2012; also see Chap. 2). The weather-related extreme events are transformed into natural disasters when they affect vulnerable areas. Therefore, assessment and prediction of the adverse effects of climate change and extreme weather-related events, and in particular, identifying vulnerable areas, are undoubtedly important steps in an integrated climate change adaptation strategy.

Africa has the highest rate of urbanisation around the world. However, the infrastructure development and economic growth in urban areas lag behind the rapidly-growing urbanisation phenomenon. In simple terms, the cities develop too quickly and the time margin between growth and development is not large enough to properly absorb new inhabitants (see Chap. 1 for a more detailed discussion). This leads to high levels of unemployment, inadequate standards of housing and services, and impacts on human health and development. These are amongst the reasons why the urbanised areas are potentially vulnerable to weather-related extreme events. One of the most significant consequences of the rapid urbanisation process is the phenomenon of informal settlements, also known as squatter settlements, shanty towns, and slums. Although they differ slightly in their definitions, these terms all refer to generally poor standards of living. General estimates indicate that about 60 % of the urban population in Africa lives in informal settlements and shanty towns (UN-Habitat 2009). Table 3.1 reports the proportion of the urban population living in slums in Africa in the last 20 years (UN-Habitat 2010).

Table 3.1 Proportion of the urban population in Sub-Saharan Africa living in slums (UN-Habitat 2010)

Year	1990	1995	2000	2005	2007	2010
Percentage of urban population living in slums in Sub-Saharan Africa	70.0 %	67.6 %	65.0 %	63.0 %	62.4 %	61.7 %

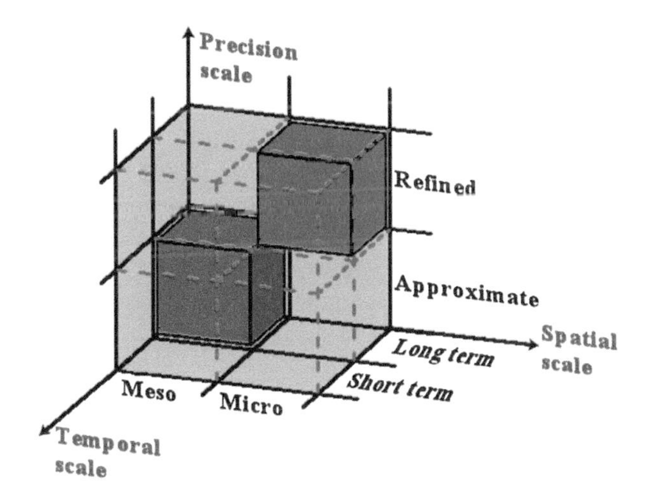

Fig. 3.1 Methods used in the CLUVA project for risk and vulnerability assessment: their spatio-temporal scale versus level of precision (De Risi 2013)

This chapter provides an overview of the activities carried out in the context of the CLUVA project on the vulnerability of the built environment to flooding. Figure 3.1 provides a schematic view of the precision and the spatio-temporal scale of the methods adopted herein for vulnerability assessment to flooding. As it can be observed in the schematic diagram in Fig. 3.1, the adopted methods in this chapter are applicable in two spatial scales, namely meso- and micro-scale.

While the meso-scale level is more suitable for the purpose of city-level decision-making, the micro-scale level is more relevant for considerations at the neighbourhood level. The *meso-scale* risk assessment methods discussed herein are approximate procedures adopted in order to efficiently identify the *urban flooding hotspots* for the built environment. Urban flooding hotspots can be defined as zones where high flooding hazard and high concentration of exposure coincide. The urban hotspots can be identified for various types of built environment, such as residential areas, road networks, sewage system, etc. In this chapter, urban flooding risk hotspots are identified for residential areas and major urban roads for the two cities of Addis Ababa (Ethiopia) and Dar es Salaam (Tanzania). Moreover, potentially flood-prone areas are delineated for the city of Ouagadougou (Burkina Faso).

Once the urban hotspots are identified, more accurate and refined methods, referred to as *micro-scale* risk assessment methods, can be adopted in order to evaluate the flooding risk in a small spatial extent. This chapter presents a probability-based methodology for evaluating the flooding risk for the informal settlements for the case study city of Dar es Salaam. Informal settlements are chosen because they are particularly vulnerable to flooding and because they are usually constructed in the zones susceptible to flooding. Additionally, as demonstrated in Table 3.1, a large percentage of the urban population in Sub-Saharan Africa lives in this type of housing (assuming that all houses in informal settlements are of a poor quality, which is not always the case). Furthermore, a preliminary study on the impact of the climate projections on micro-scale flood risk assessment is presented for the city of Dar es Salaam. As depicted in the schematic diagram in Fig. 3.1 and can be reasonably expected, the precision and the spatial scale of the adopted methods coincide. That is, the more refined risk assessment methods are suitable for micro-scale studies and vice versa. As far as the temporal scale is concerned, this chapter deals with long-term risk assessment. In other words, the findings of this work are most suitable for strategic urban adaptation planning.

Meso-scale Methods for Urban Flooding Hotspot Identification

We have employed various GIS (Geographic Information System)-based datasets for identifying the urban flooding hotspots for residential buildings and urban corridors (i.e. major roads). This is done by overlaying a map of potentially flood-prone areas (identified by the Topographic Wetness Index, TWI) and a map of Urban Morphology Types (UMT) classified as residential or as urban corridors. The TWI (Qin et al. 2011; see Box 3.1) allows for the delineation of a portion of a hydrographic basin potentially exposed to flood inundation by identifying all the areas characterised by a topographic index that exceeds a given threshold. The UMTs (see Chap. 4; Pauleit and Duhme 2000; Gill et al. 2008; Cavan et al. 2012) form the foundation of a classification scheme which brings together facets of urban form and function. The distinction of UMTs at a meso-scale makes a suitable basis for the spatial analysis of cities.

The TWI threshold value depends on the resolution of the Digital Elevation Model (DEM), topology of the hydrographic basin and the constructed infrastructure (Manfreda et al. 2008, 2011). Assuming that the TWI map for the entire urban territory can be characterised by a single threshold value, one can calibrate this threshold based on both the results of detailed delineation of the inundation profile for selected zones and/or information about the areal extent of previous historical flooding events. In particular, the fact that the historical flooding extent can be used for the calibration of the TWI threshold is especially useful in cases where one needs to use the TWI map for the delineation of urban risk hotspots – eliminating the need for performing detailed hydraulic analysis. In fact, assuming that the spatial extent of the flood-prone areas is not significantly affected by the return period of the flooding event (see e.g. De Risi and Jalayer 2013), the calibration based on historical flooding extent is done without considering the return period and/or intensity of the specific flooding event. In this study, the TWI threshold is calibrated based on the calculated inundation profiles for various return periods (the cases of Dar es Salaam and Addis Ababa) and historical flooding spatial extent (the case of Ouagadougou) for selected zones within the basin through Bayesian parameter estimation. The urban flooding risk hotspots are delineated in the GIS environment by overlaying the map of TWI and the UMT units classified as residential and major urban roads for various percentiles of the TWI threshold (see Box 3.2). Additional information related to exposure such as population density and demographic information can be integrated by overlaying geo-spatial datasets created based on census data. Differences in exposure characteristics are then assessed for a range of different residential types including, for example, between condominium/multi-storey, single storey stone/concrete and areas predominantly associated with informal construction.

Delineation of Flood-Prone Areas Based on Historical Information: The Case of Ouagadougou

On September 1, 2009, an unprecedented deluge of rain hit the capital city of Ouagadougou and resulted in widespread damage (destruction of buildings and infrastructure). More than 25 cm of rainfall in 12 h turned the streets of Ouagadougou into fast-flowing rivers. The infrastructure was severely affected as the floods cut off electricity, fresh water and fuel supplies. The number of people severely affected by the flooding was estimated to vary between 109,000 and 150,000 (Vega 2011). In the probabilistic framework described in this chapter, the areal extent of this flooding event is used in order to delineate the flood-prone areas for Ouagadougou. The TWI was calculated in the GIS framework and based on the DEM of the city (scale 1:30,000, vertical resolution: 30 m).

The TWI threshold for Ouagadougou has been calibrated based on the inundated area of the 2009 flooding event. Based on the information available at the internet site MapAction (2011), it was possible to geo-reference this image and generate a spatial dataset. As it can be seen from Fig. 3.2b, two inundated areas have been identified for the 2009 flooding event. In the following, the smaller area is called Area 1 and the larger area is referred to as Area 2. It is possible to observe in Fig. 3.2b that the *TWI* map with a threshold equal to the maximum likelihood *TWI* threshold τ_{ML} matches the spatial extent of the areas inundated by the 2009 flooding event. Moreover, the map of the flood-prone area up-scaled for the entire city of Ouagadougou is shown in Fig. 3.2a.

Box 3.1: The TWI and the Calibration of the TWI Threshold

The topographic wetness index (TWI, Qin et al. 2011), which is used as a proxy for delineating the potentially flood-prone areas in the meso-scale, allows for the delineation of a portion of a hydrographic basin that is potentially exposed to flood inundation. This is achieved by identifying all the areas characterised by a topographic index that exceeds a given threshold. The TWI has a purely topographic interpretation (basically measures the capability to accumulate water) and is quite straightforward to calculate in the GIS environment for very large areal extents.

One of the most important components of the TWI map is the TWI threshold because the potentially flood-prone areas are identified as those with TWI greater than the threshold. In this study, the TWI threshold is updated through a probabilistic method based on available information such as inundation profiles for various return periods or historical flooding data. In calculating the threshold, the first important step is to delineate the spatial window W that contains the additional information that will be used. Then, the threshold is chosen as the value that maximises the probability of the correct identification of the flood-prone areas based on the available additional information within window W.

The probability of correct identification for a given threshold value can be calculated as the sum of two probability terms (1) the probability that the TWI map and the additional information both indicate that the area within W is flood-prone; and (2) the probability that the TWI map and the additional information both indicate that the area within window W is not flood-prone. These probability terms are estimated as the ratio of areal extents. The estimation of probability terms as a ratio of areal extents renders the procedure particularly suitable for GIS-based implementation. Finally, the TWI threshold, calibrated based on information available within W, is used in order to create the map of potentially flood-prone areas at the meso-scale.

Urban Flooding Risk Hot-Spot Identification: The Case of Dar es Salaam

The topographic wetness index for the city of Dar es Salaam is based on the Digital Elevation Model (DEM) of the city presented in Giugni et al. (2012) (Year: 2008, scale 1:10,000, vertical resolution: 10 m, reduced to 2 m through spatial interpolation). In order to calibrate the TWI threshold for Dar es Salaam, the inundation profiles for various return periods have been calculated for Suna located in between Ilala and Kinondoni districts. Suna results as flood-prone based on past flooding experiences. The inundation profile has been calculated by bi-dimensional simulation of flood volume propagation using the software FLO2D (using historical rainfall records, the DEM, and the calculation of the hydrograph based on the

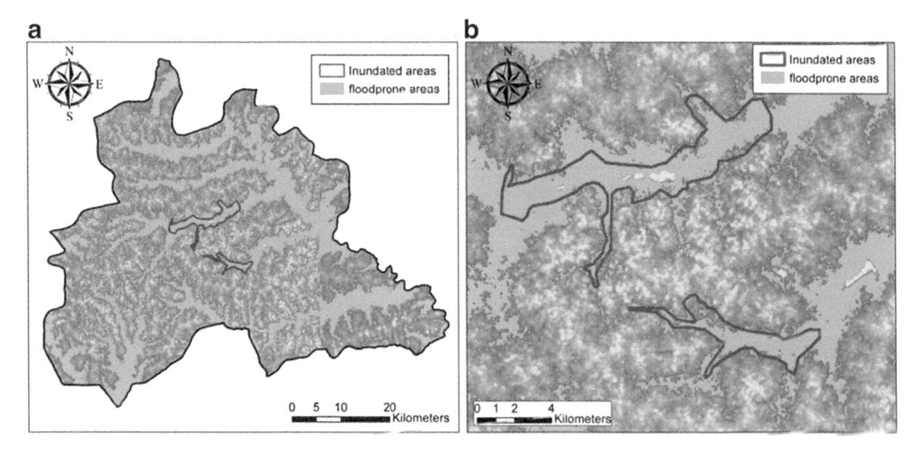

Fig. 3.2 (**a**) Up-scaling the flood-prone areas for the entire city of Ouagadougou, (**b**) Matching of the flood-prone area and the inundation extent based on the 2009 flooding event for the spatial window of interest in Ouagadougou (De Risi and Jalayer 2013)

curve number method) assuming a simulation time of 45 h. The outcome of the flood propagation is reported in terms of maximum flow depth and velocity with reference to six considered return periods ($T_R = 2$, 10, 30, 50, 100 and 300 years; see Chap. 2 and De Risi et al. 2013a for more information). The UMTs for Dar es Salaam are classified and delineated based on aerial photos acquired in December 2008 from the Ministry of Lands, Housing and Human Settlement Development (Cavan et al. 2012). The urban risk hotspots are delineated by overlaying the flood-prone areas (TWI), the UMT's classified as residential or major road corridors and the population density dataset based on the 2006 City Census (NBS 2006).

The Residential UMT

The spatial units characterised as residential cover about 47 % of the entire city surface, in which about 60 % of the population is concentrated. This category is further divided into five sub-categories: (a) condominium and multi storey buildings (covering 0.2 % of the city surface and containing around 0.5 % of the population); (b) villa and single storey stone/concrete buildings (covering 7.8 % of the city surface and containing around 17 % of the population); (c) mud and wood construction (covering 0.4 % of the city surface and containing 2 % of the population); (d) scattered settlement (covering 25 % of the city surface and containing 5.3 % of the population); and (e) mixed residential (covering 13 % of the city and containing 36 % of the population).

Figure 3.3 illustrates the delineated urban hotspots (the coloured zones), obtained by overlaying the UMT and the TWI datasets, for different estimates of TWI threshold (see Boxes 3.1 and 3.2 for further details). Furthermore, the information on population density obtained is integrated in order to estimate the number

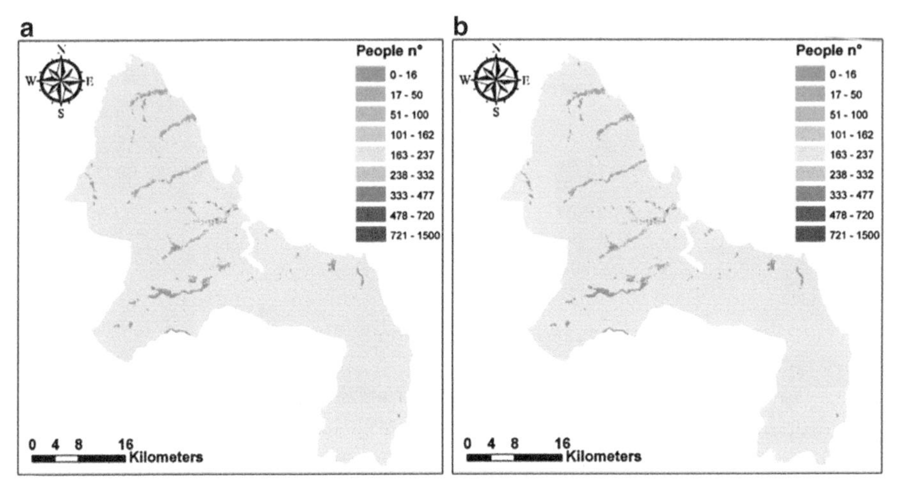

Fig. 3.3 Urban flooding risk hotspots in Dar es Salaam delineated for (**a**) maximum likelihood estimate of the TWI threshold; and (**b**) 50th percentile estimate of the TWI threshold for residential areas (De Risi and Jalayer 2013)

Box 3.2: Alternative Estimates of the TWI Threshold

In this chapter we have adopted three different statistics of the TWI threshold, namely, 16th and 50th percentiles and the maximum likelihood estimate. The 16th percentile threshold denoted by τ_{16} indicates that with 16 % probability the TWI threshold τ is going to be less than τ_{16}. In a similar manner, the 50th percentile threshold denoted by τ_{50} indicates that with 50 % probability the TWI threshold τ is going to be less than τ_{50}. The maximum likelihood estimate denoted by τ_{ML} is the most probable value of the TWI threshold.

Table 3.2 Exposure to flooding risk in terms of the percentage of residential area and the percentage of people living in the residential areas affected by flooding (De Risi and Jalayer 2013)

	τ_{ML}	τ_{16}	τ_{50}
% of residential area	3.5 %	23.1 %	3.5 %
% of people in residential area	6.4 %	37.8 %	6.5 %
% of city population	3.9 %	22.9 %	3.9 %

of affected people by flooding for different statistics of the TWI threshold. Table 3.2 shows the percentage of residential areas affected by flooding (the areal extent of hotspots illustrated in Fig. 3.3 normalised by total residential area) and the percentage of people living in the residential area affected (estimated population normalised by total population in the residential area), for alternative estimates of the TWI threshold.

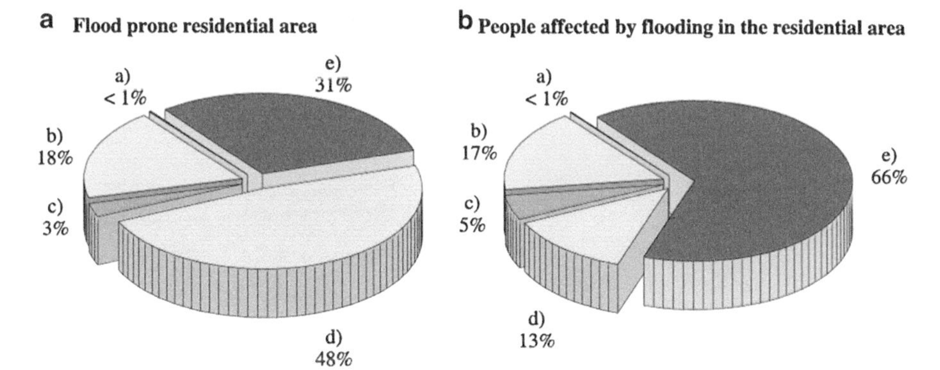

Fig. 3.4 Breakdown of the residential flooding risk hot-spots in terms of (**a**) residential types; and (**b**) their population (De Risi and Jalayer 2013). Abbreviations for the building types are described in the text

It can be noted that the percentages of people affected by flooding in the residential areas of Dar es Salaam in correspondence to τ_{50} and τ_{16}[1] varies between 6.5 and 37.8 % of the total Dar es Salaam population living in the residential areas and between 3.9 and 22.9 % of the total population of the city.

The pie chart shown in Fig. 3.4 illustrates the percentage break-down of the areal extent of residential flood hotspots (corresponding to τ_{ML} equal to 19.53) in terms of different residential sub-classes; namely, (a) condominium/multi-storey; (b) single storey stone/concrete; (c) mud/wood construction; (d) scattered settlement; and (e) mixed residential (see Cavan et al. 2012 for a description of the residential sub-classes). It can be observed that around 66 % of the population in the flood-prone residential areas lives in residential types labelled as *mixed*. This residential type constitutes around 31 % of the flood-prone residential buildings. On the other hand, around half of the residential buildings identified as flood-prone belong to the scattered settlements category. The inhabitants of the scattered settlements constitute around 13 % of the total population in the residential areas affected by flooding.

The Major Road Corridors UMT

The UMT class *major road corridors* covers about 0.5 % of the whole city surface (see Fig. 3.5a). The urban road corridor hotspots (the red zones) delineated for the maximum likelihood estimate of TWI threshold are shown in Fig. 3.5b.

[1] Note that we have not included the 84th percentile for the TWI threshold. This is done because the areal extent of the flood-prone areas corresponding to the 84th percentile of the TWI threshold is going to be much smaller than those corresponding to 50th and 16th percentiles.

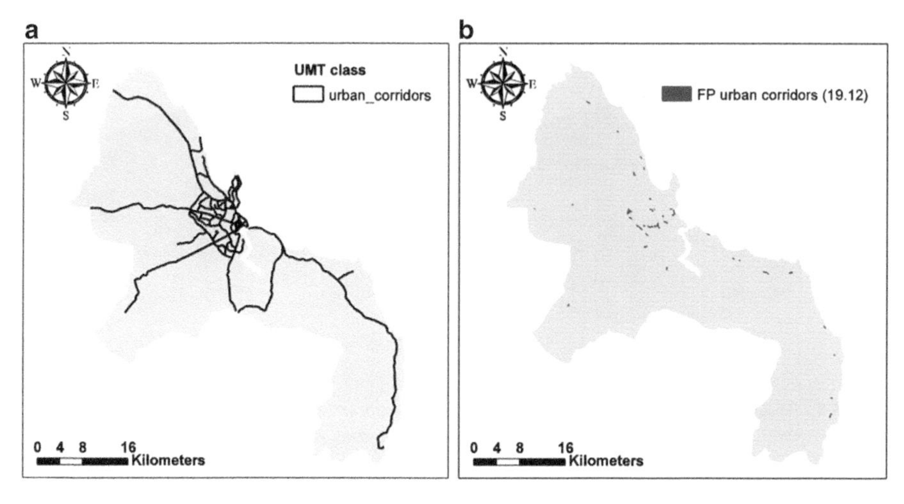

Fig. 3.5 (**a**) The map of urban corridors (major urban roads) in Dar es Salaam; (**b**) The urban corridors' risk hotspots for flooding delineated for maximum likelihood estimate of the TWI threshold (De Risi and Jalayer 2013)

Table 3.3 Flood risk exposure expressed in terms of the percentage of the urban corridors potentially exposed to flooding (De Risi and Jalayer 2013)

	τ_{ML}	τ_{16}	τ_{50}
% of urban corridors area	3.8 %	29.8 %	3.9 %

Table 3.3 reports the percentage of roads affected by flooding (estimated as the areal extent of urban corridor hotspots illustrated in Fig. 3.5b normalised by the total area of urban corridor in Fig. 3.5a for alternative estimates of the TWI threshold (i.e. τ_{ML}, τ_{16} and τ_{50})). It can be noted that the percentages of major urban roads affected by flooding in Dar es Salaam corresponding to τ_{50} and τ_{16} varies between 3.9 and 29.8 % of the total areal extent of urban roads in the city.

Urban Flooding Risk Hot-Spot Identification: The Case of Addis Ababa

In the following, the methodology described earlier in this chapter is implemented in order to identify urban flooding hotspots in relation to residential buildings and major urban corridors in the city of Addis Ababa. Table 3.4 shows the percentage of residential areas affected by flooding (the areal extent of hotspots obtained, illustrated in Fig. 3.6, normalised by total residential area) and the percentage of

Table 3.4 Exposure to flood risk in terms of the percentage of residential hotspots (normalised to the areal extent of the residential units) and people affected by flooding (normalised by the population living in the residential areas) (Jalayer et al. 2014)

	τ_{ML}	τ_{16}	τ_{50}
% of residential area	4.6 %	22.2 %	3.1 %
% of population in residential area	6.0 %	26.9 %	3.7 %

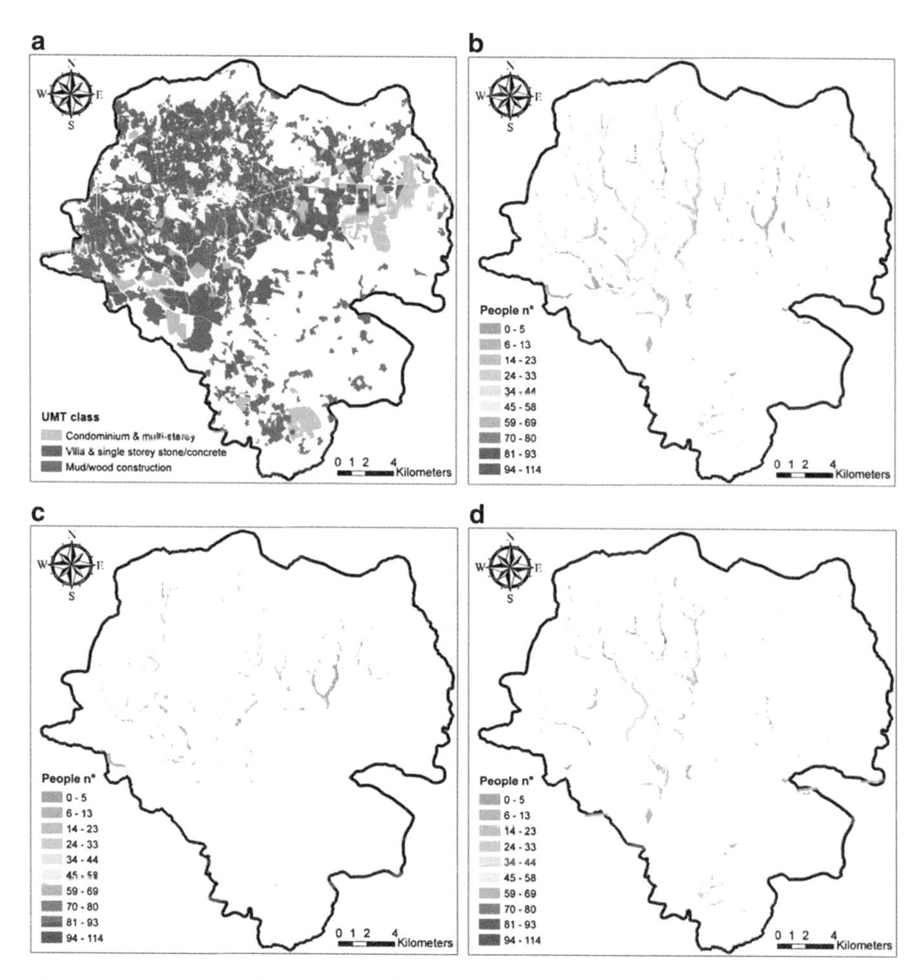

Fig. 3.6 The urban residential flood risk hotspots for Addis Ababa: (**a**) map of the residential areas in the city; (**b**) the residential flooding hotspots delineated by τ_{ML}; (**c**) the flooding hotspots for villa & single storey (τ_{ML}); (**d**) the flooding hotspots for mud and wood construction (τ_{ML}) (Jalayer et al. 2014)

Table 3.5 Exposure to flood risk in terms of the estimated percentage of residential population affected by flooding (normalised by the entire city population, Jalayer et al. 2014)

	τ_{ML}	τ_{16}	τ_{50}
% of city population	2.6 %	12.5 %	1.7 %

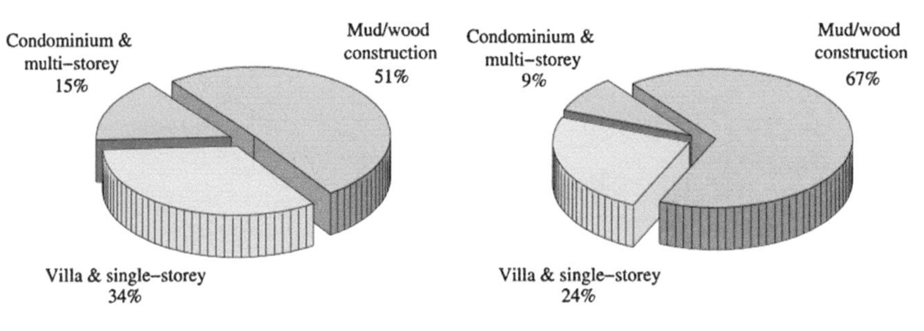

Fig. 3.7 Breakdown of the residential flooding hotspots in Addis Ababa in terms of (**a**) residential types; and (**b**) their population (Jalayer et al. 2014)

people living in the residential areas affected by flooding (estimated population living in residential hotspots normalised by total population in the residential area), for alternative estimates of the TWI threshold. It can be observed that the percentages of the residential population affected by floods associated with τ_{50} and τ_{16} range from 3.7 to 26.9 %. Meanwhile, the percentages of the residential flooding hotspots area associated with τ_{50} and τ_{16} range from 3.1 to 22.2 %.

Table 3.5 shows that the percentages of the total Addis Ababa population affected by floods associated with τ_{50} and τ_{16} range from 1.7 to 12.5 %.

The pie chart shown in Fig. 3.7 illustrates the percentage breakdown of the areal extent of the residential flood hotspots (corresponding to τ_{ML} equal to 17.05) in terms of different residential sub-classes; namely, (a) condominium/multi-storey; (b) single storey stone/concrete; and (c) mud/wood construction (see Cavan et al. 2012 for a description of the residential sub-classes). It can be observed that 67 % of the population in the flood-prone residential areas lives in mud and wood constructions (constituting 51 % of the total residential area). These constructions are particularly vulnerable to damage from flooding.

In-situ verification (Fig. 3.8) illustrates the points where observations were made during a recent field trip (October 2012). The figure is overlaid with the map of the residential flooding risk hotspots reported in Fig. 3.6b (for the τ_{ML} estimate). Figure 3.8a shows a one-storey house with visual signs of water on the walls; Fig. 3.8b shows a bridge destroyed due to flooding in a central part of the city; Fig. 3.8c shows the owner of a typical mud and wood house indicating with his hands the level of flooding experienced in the past; various objects reminiscent

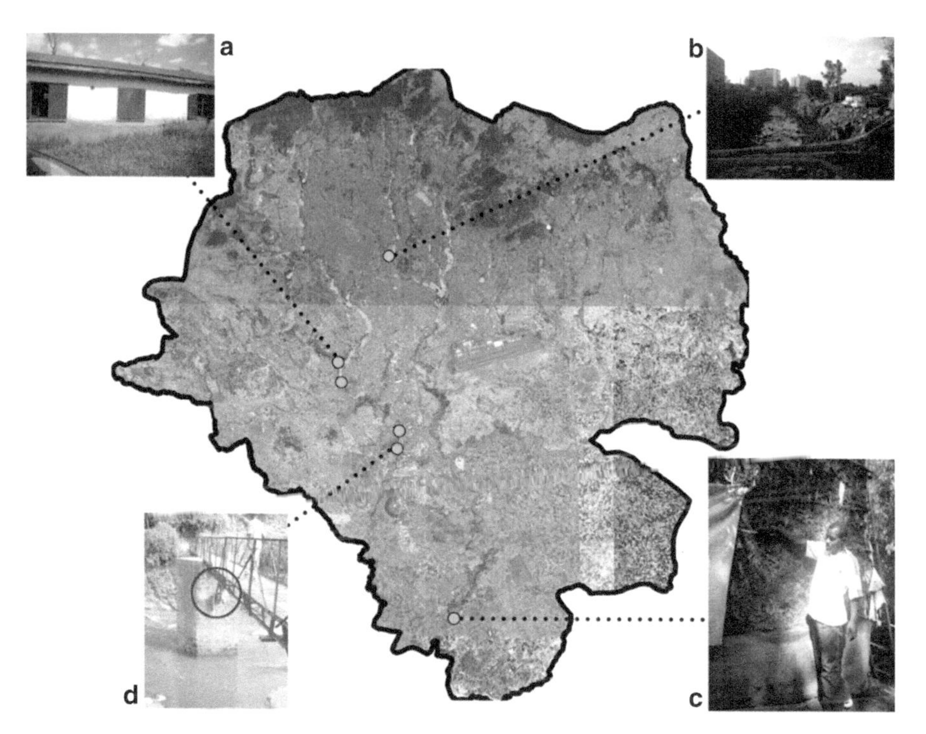

Fig. 3.8 In-situ observations for various locations on the map of the residential flooding risk hotspots for Addis Ababa (Pictures: F. Jalayer and N. Yonas 2012) (Jalayer et al. 2014). See text for detailed description

of previous inundations can be observed on the pedestrian bridge in Fig. 3.8d. These observations can be viewed as additional field evidence in support of the information provided by the TWI map.

The Major Road Corridors

The UMT class *major road corridors* covers about 3.4 % of the whole city surface. Below in Fig. 3.9, the delineated urban road corridor hotspots (the red zones) are shown for the maximum likelihood estimate of the TWI threshold.

Table 3.6 reports the percentage of the roads affected by flooding (estimated as the areal extent of flooding hotspots illustrated in Fig. 3.9 normalised by the total area of urban corridor) for alternative estimates of the TWI threshold.

This information can be quite useful for identifying the major roads that are most likely to be blocked in case of flooding. Furthermore, intersecting the flooding hotspots for major roads with the map of points of principal interest for the city (e.g. fire stations, hospitals, police stations, schools) would be useful as the basis of a preliminary connectivity study for the road network of Addis Ababa in case of flooding.

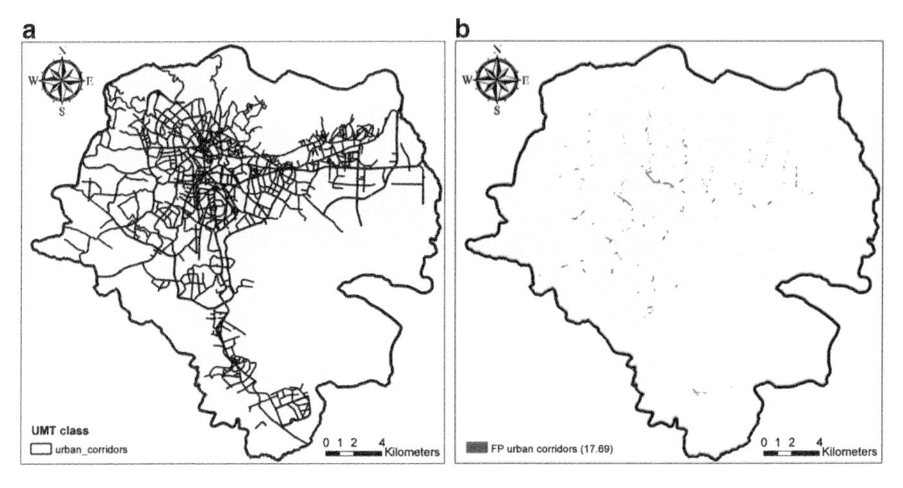

Fig. 3.9 (**a**) Urban corridor morphology type for Addis Ababa; (**b**) Urban corridor flood hotspots delineated for the maximum likelihood TWI threshold (Jalayer et al. 2014)

Table 3.6 Exposure assessment as a percentage of urban corridors (Jalayer et al. 2014)

	τ_{ML}	τ_{16}	τ_{50}
% of urban corridors area	5.2 %	23.4 %	3.4 %

Box 3.3: The Meso-scale Methodology and Its Use for Urban Planning

Once the Meso-Scale Methodology is developed as a software package, it can be used by urban planners and other practitioners to develop their own maps of the urban hotspots. The maps obtained based on the proposed procedure provide a quick screening tool to the urban planner in order to efficiently identify the zones that need his/her immediate or long-term actions. These actions can include for example adoption of more refined small-scale risk assessment procedures and undertaking various prevention strategies. The prevention strategies range from planning for structures that help in mitigating the flood risk, to relocation policies (if advisable), territory restriction measures and actions that aim to increase public awareness.

Micro-scale Flood Risk Assessment for Informal Settlements: The Case of Dar es Salaam

Once the flood risk hotspots are identified through a meso-scale risk evaluation procedure (see Box 3.3), more accurate and refined calculations can be adopted in order to evaluate the flooding risk at the micro-scale. In our case, a scale of 1:2,000 has been adopted for micro scale risk assessment for the city of Dar es Salaam. As mentioned in the introduction, informal settlements have been chosen as the basis for the study both because they are particularly vulnerable to flooding and because they are usually constructed in the zones susceptible to flooding. Additionally, as demonstrated in Table 3.1, a large percentage of the urban population in Sub-Saharan Africa lives in this type of housing. Flood risk assessment has been performed based on a modular probabilistic methodology that is described in the next section.

General Probability-Based Framework

The adopted methodology for micro-scale flood risk assessment of the informal settlements, can be summarised in a single equation (Eq. (3.1)), where λ_{LS} denotes the risk expressed as the mean annual rate of exceedance of a given limit state (*LS*). The limit state refers to a threshold (e.g. critical water height $h_{f,c}$ or critical velocity $v_{f,c}$) for a structure, beyond which, it no longer fulfils a specified functionality. $\lambda(h_f)$ denotes the mean annual rate of exceedance of a given flood height h_f at a given point in the considered area, known also as the *flood hazard*. $P(LS/h_f)$ denotes the flood fragility for the limit state *LS* expressed in terms of the conditional probability of exceeding the limit state threshold given the flood height h_f, known also as the *flood fragility* for the portfolio of buildings.

$$\lambda_{LS} = \int_{h_f} P\left(LS \middle| h_f\right) \cdot \left|d\lambda\left(h_f\right)\right| \cdot dh_f \tag{3.1}$$

The risk λ_{LS} is calculated in terms of the mean annual frequency of exceeding the limit state *LS* for each node of the lattice covering the zone of interest by integrating fragility $P(LS/h_f)$ and the (absolute value of) hazard increment $|d\lambda(h_f)|$ over all possible values of flooding height. The mean annual frequency of exceeding the limit state λ_{LS} is later transformed into the annual probability of exceeding the limit state assuming a homogenous Poisson Process as a model for occurrence of limit-state-inducing events (see e.g. De Risi et al. 2013a). In this work, the limit state threshold is specified based on the flood height. That is, the limit state threshold can be defined as the critical water height, beyond which the structure exceeds the limit state in question. It is helpful in this context to consider the critical water height as a proxy for the structural capacity for the specified limit state.

It should be noted that Eq. (3.1) manages to divide the flood risk assessment procedure in two main modules, as follows: (1) the hazard assessment module which leads to the calculation of the mean annual frequency $\lambda(h_f)$ of exceeding a given flooding height h_f; and (2) the vulnerability assessment module which leads to the calculation of the flooding fragility curve in terms of the probability of exceeding specified limit state $P(LS/h_f)$. Figure 3.10 shows a flow chart of the proposed

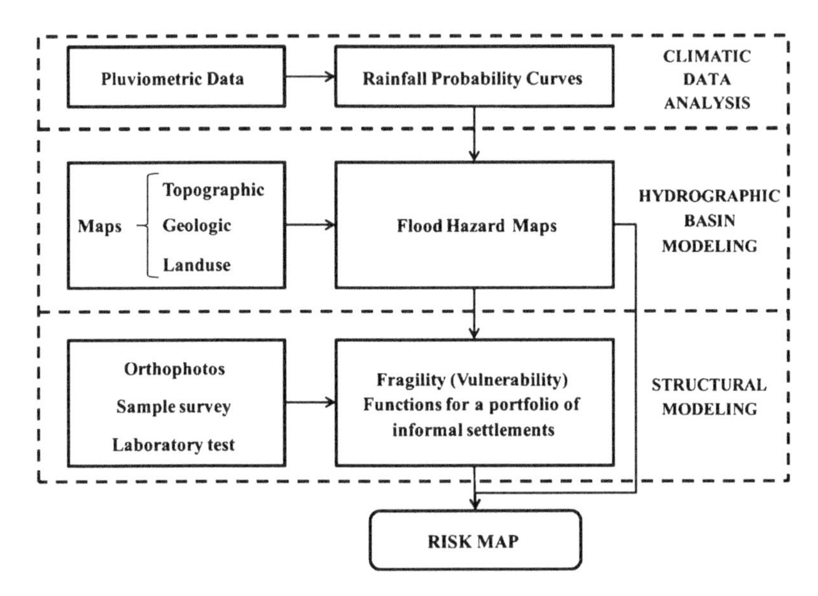

Fig. 3.10 Flow-chart representation of the modular methodology used for micro-scale flood risk assessment for informal settlements (De Risi et al. 2013a)

procedure. As it can be seen, this flow chart is constructed in a modular manner. In the climatic data analysis module, historical rainfall data and/or climate projections are transformed into rainfall probability curves (see Chap. 2 for more detail). This information, together with detailed topography of the area, geology maps and land-use or UMT maps, is then used in the hydrographic basin modelling module, in order to evaluate the basin hydrograph and to develop the flood hazard maps (i.e. inundation scenarios for various return periods). In the structural modelling module, the vulnerability of the portfolio of informal settlements is evaluated in terms of fragility functions for a specific limit state, based on orthophotos of the area and sample in-situ building surveys. Finally, the flood risk map is obtained by point-wise integration of the flood hazard map and the fragility functions for the informal settlements.

As shown, the proposed methodology integrates climatic data analysis, hydrographic basin modelling and structural fragility modelling in order to generate the risk map for the zone of interest. In the following sections, the above-mentioned steps are described in detail. It should be emphasised that the meso-scale methods described earlier in this chapter do not rely on Eq. (3.1) for calculating the flood risk. More specifically, the flood hazard assessment in the meso-scale methods is performed in an approximate manner distinguishing only between the potentially flood-prone areas and not potentially flood-prone areas. Another point that differentiates the meso-scale method from the micro-scale method is that the former does not rely on a quantitative estimate of physical vulnerability of buildings but uses the building typology as an indicator of vulnerability.

Analysing Meteorological Data and Predicting the Climate Trends

Flood hazard is expressed herein as the annual probability of exceeding a specific flood height (h_f). In the first place, the rainfall probability curve or the rainfall Intensity-Duration-Frequency (*IDF*) curve is extracted from the historical data or from down-scaling of regional climate projections (see also Chap. 2). The *IDF* curve is employed next as the input of a catchment rainfall-runoff model in order to determine the hydrograph and the peak discharge. This information, together with detailed geo-morphological maps of the zone and a DEM, is used as input into the hydraulic model in order to calculate the inundation profile. The results can be presented in terms of the annual rate/probability of exceeding a specific value of flooding height and flooding velocity for the nodes of a lattice covering the zone of interest.

Historical Rainfall Data

Riverine flooding events are strictly connected to rainfall patterns. Therefore, rainfall data time-series are essential pieces of information for determining the total flood discharge. They can be obtained as pluviometric time-series from governmental organisations and/or internet sources (e.g. www.tutiempo.net and www.knmi.nl). It is desirable that the pluviometric data are available as precipitation extremes (maxima) recorded over a range of time intervals. The rainfall maxima recorded for different intervals are used in order to construct the rainfall IDF curve (described later, see also Chap. 2 for a more detailed description). The rainfall time-series can also be used to evaluate the antecedent soil moisture condition (AMC). In hydrological modelling, antecedent moisture condition is usually associated with the pre-storm soil moisture deficit. This latter has a significant effect on the amount of rainfall drained by the river network, and finally, on the flooding potential of a rainstorm.

Climate Projections

Future climate patterns may lead to adverse effects on the frequency and/or intensity of extreme weather-related events such as floods. In compliance with the Intergovernmental Panel on Climate Change (IPCC) scenarios (IPCC 2007), climate projections are generally evaluated using a General Circulation Model (GCM). A GCM is a mathematical model that simulates the general atmospheric and oceanic circulation using a specific formulation of the Navier-Stokes equations, discretised with spatial resolutions in the order of around 100 km. Ideally, GCMs can be used to produce long-term simulations suitable for catastrophe modelling. However, for a realistic simulation of precipitation patterns, representing the vertical structure of the atmosphere as well as the effect of the terrain on atmospheric circulation, a model must have a resolution less than 100 km. This is not practical within GCMs because the calculation time increases exponentially. Therefore, using a GCM for direct simulation of local precipitation is not feasible. Moreover, the GCMs are based on simplified microphysics and may not provide a solid representation of precipitation in mountainous areas (Bellucci et al. 2012). The application of a Regional Climate Model (RCM) with horizontal spatial resolution of about 10 km can be useful for the description of the climate variability at a local scale. A RCM (see Bellucci et al. 2012) depends on the definition of boundary conditions that can be obtained based on the results of a GCM. Finally, through statistical downscaling it is possible to obtain climatological data for finer spatial resolutions, on the order of 1 km. This provides the precipitation data necessary for comprehensive flood modelling. The climate projections used in this work are provided by the CMCC (Centro Euro-Mediterraneo sui Cambiamenti Climatici). They have been obtained by following the IPCC 20C3M protocol for the twentieth Century, and using RCP4.5 and RCP8.5 radiative forcing emission scenarios developed in the framework of the 5th Coupled Model Intercomparison Project (WCRP 2008).

CMCC has performed a set of climate simulations over the time period 1950–2050 with the global model CMCC-MED (Gualdi et al. 2013) (spatial resolution 80 km) in the context of CLUVA (see Chap. 2 for more information). The initial conditions have been obtained from an equilibrium state reached by integrating the model for 200 years with constant greenhouse gas (GHG) concentrations corresponding to the 1950s. Once the climate model reaches equilibrium with the prescribed constant radiative forcing (GHG and aerosol concentrations), simulations have been conducted by increasing the GHG and aerosol concentrations in line with observed data. These simulations have been downscaled to a spatial resolution of 8 km with the non-hydrostatic RCM COSMO-CLM (Rockel et al. 2008), developed considering the spatial extent covered by the urban area of interest. The results were further down-scaled to a spatial resolution of 1 km by using a stochastic downscaling technique (Rebora et al. 2006) in order to render them suitable for modelling of precipitation patterns. In this work, the climate projections based on the RCP8.5 scenario and downscaled to 1 km resolution are taken into account. This may be considered a worst-case scenario in terms of gas emissions and temperature increase, although readers are reminded that all scenarios are equally plausible (Chap. 2).

Rainfall Curve: Historical Data and Climate Projection

Studying historical meteorological data and analysing the past and future climate patterns arguably constitute the first step in developing a probabilistic inundation model. Its output is usually expressed in terms of rainfall scenarios for various return periods (T_R), also known as the rainfall curves or the Intensity-Duration-Frequency (*IDF*) curves. The *IDF* curves are normally used, in lieu of sufficient data for direct probabilistic discharge modelling, in order to evaluate the peak discharge. In particular, the *IDF* curves present the probability of a given rainfall intensity and duration expected to occur in a particular location. The rainfall curve is calculated herein based on both historical data and a specific climate projection scenario. The historical rainfall data (H) span from 1958 to 2010 and the climate projection (CC) is provided from 1958 to 2050 (with the 1958–2010 period identical for both time-series). The historical rainfall data is obtained from a single meteorological station located in the DSM International Airport at 55 m altitude from the sea level, 6°86′ latitude and 39°20′ longitude. For consistency, the projection data are calculated for the same point. Rainfall height h_r is calculated as the maximum rainfall depth in millimetres calculated in a time-window of duration d. The annual rainfall extremes are then calculated for various time window durations (see Box 3.4). Figure 3.11a, b show the two rainfall height maxima time-series used herein for $d = 24$ h.

Box 3.4: Rainfall Curve Based on Incomplete Historical Records (Chap. 2)

The maximum annual rainfall data for a specific duration are not always available. In such cases, available data could be disaggregated to the desired durations. This involves generating synthetic sequences of rainfall for smaller time windows (e.g. 10 min, 30 min, 1 h, 3 h, 6 h, 12 h), with statistical properties equal to that of the observed daily rainfall. Herein, two alternative downscaling techniques are used in order to generate maximum rainfall values for the desired time windows. The short-time intensity disaggregation method (Connolly et al. 1998) has been used for simulation of smaller time windows (i.e. 10 min, 30 min, 1 h) and the random cascade-based disaggregation method (Olsson 1998; Güntner et al. 2001) has been used for larger time windows (i.e. 3 h, 6 h and 12 h).

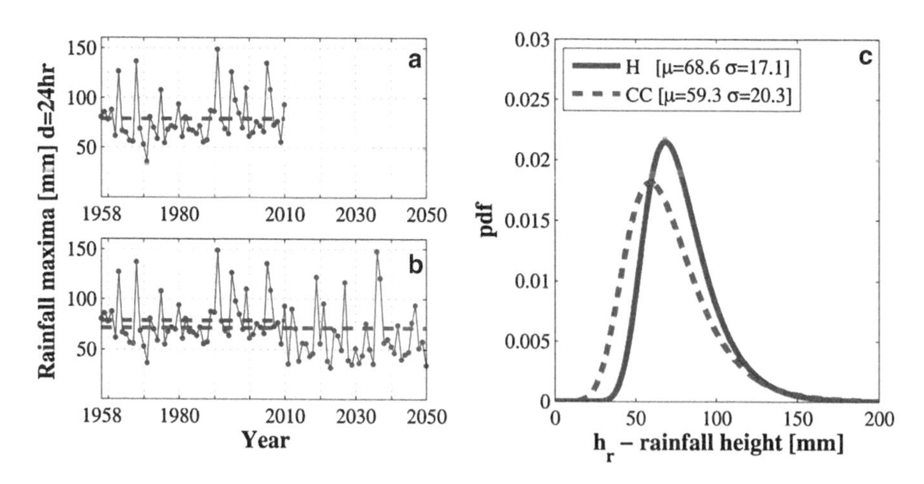

Fig. 3.11 The annual rainfall height maxima (**a**) H and (**b**) CC for a duration of 24 h (DSM, Airport Station) (**c**) density function for the maximum rainfall height for a duration of 24 h ((Jalayer et al. 2013b) © 2013 Taylor and Francis Group, London UK, used with permission)

The first time-series depicted in Fig. 3.11a is based on historical data only (hereafter referred to as H, 1958–2010). The second time-series shown in Fig. 3.11b, and hereafter referred to as CC, consists of H (1958–2010) plus climate projections for 2011–2050. A Gumbel extreme-value distribution is used to describe the annual rainfall height maxima in both cases CC and H (Fig. 3.11c). It is worth noting that considering the climate projections leads to a reduction of 14 % in the mean rainfall height.

The Results for DSM The *IDF* curve obtained based on historical data (H) is characterised by the following relationship (De Risi et al. 2013a):

$$h_r(d, T_R) = K_{T_R} \cdot 36.44 \cdot d^{0.25} \tag{3.2}$$

where $h_r(d,T_R)$ is the maximum annual rainfall height measured over a time-window of duration d, corresponding to a return period of T_R; K_{TR} is the growing factor and is a function of the coefficient of variation for the corresponding Gumbel probability distribution. The projected *IDF* curve (CC) is obtained through the same procedure, based on the climate change projection for scenario RCP8.5 with a spatial resolution of 80 km, spatially downscaled to a resolution of 1 km. It is characterised by the following relationship:

$$h_r(d, T_R) = K_{T_R} \cdot 31.70 \cdot d^{0.26} \tag{3.3}$$

Table 3.7 reports the values of the growing factor K_{TR} for various return periods based on the historical data (H) and climate projections (CC).

The rainfall curves for CC and H for two different return periods are plotted in Fig. 3.12a. It is possible to observe that for the city of Dar es Salaam, this climate scenario leads to a decrease in terms of rainfall intensity. In fact, the IDF curves that take into account the climate projection (the dotted lines in Fig. 3.12a) are lower than those evaluated based on historical data series (the solid lines in Fig. 3.12a). Nevertheless, although the flooding intensity considering the projections decreases with respect to the historical data, the growing factor demonstrates a slight increase (Table 3.7). Recalling that the growing factor is a function of the coefficient of variation for the extreme value distribution, one can deduct that having a higher coefficient of variation (with constant mean or central value) leads to higher probability for extreme rainfall events (i.e. in the tail of the distribution).

Table 3.7 Growing factors for the different return periods (Jalayer et al. 2013b)

T_R	2Ys	10Ys	30Ys	50Ys	100Ys	300Ys
H	0.95	1.42	1.70	1.83	2.01	2.29
CC	0.94	1.50	1.84	2.00	2.21	2.41

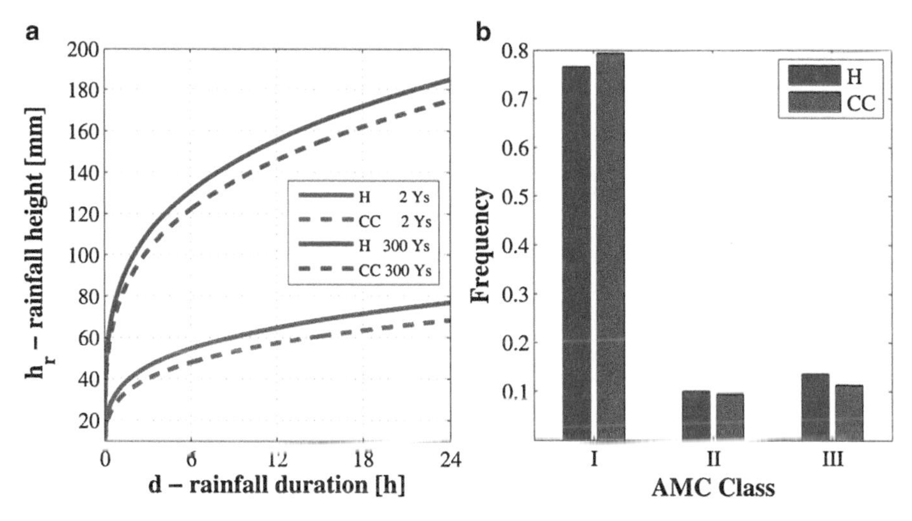

Fig. 3.12 (**a**) IDF curves related to return period of 2 years and 300 years for CC and H; (**b**) The histogram of the AMC classes for the growing season for various AMC classes ((Jalayer et al. 2013b) © 2013 Taylor and Francis Group, London UK, used with permission)

Antecedent Moisture Conditions

As mentioned before, the Antecedent Moisture Condition (AMC) is the relative moisture of the pervious soil surfaces prior to the rainfall event and reflects the level of soil moisture in a five day interval preceding the rainfall extreme event. Antecedent moisture is considered to be low when there has been little preceding rainfall and high when there has been considerable preceding rainfall prior to the extreme event. Determination of antecedent soil moisture content and classification into the antecedent moisture classes AMC I, AMC II and AMC III, representing dry, average and wet conditions, is an essential requirement for the application of the curve number procedure described next. The rainfall time-series (both CC and H) are both post-processed in order to obtain the histogram of AMC classes calculated for the data series available. Figure 3.12b illustrates such a histogram calculated based on both H and CC. No significant change in AMC can be observed between the two time-series. In this case, we considered watersheds to be AMC II, which is essentially an average moisture condition, even if there is a slightly higher likelihood for class III (Fig. 3.12b).

The Hazard Curves for the Case Study Area (Suna)

The case study neighbourhood for micro-scale flood risk assessment is chosen as a part of the Suna Sub Ward (Fig. 3.13) in the Kinondoni District in Dar es Salaam. Suna, located on the Western bank of the Msimbazi river with an extension of about 50 ha, is a historically flood-prone area. As it can be depicted from Fig. 3.13, Suna is located in a residential flooding hotspot delineated through the meso-scale

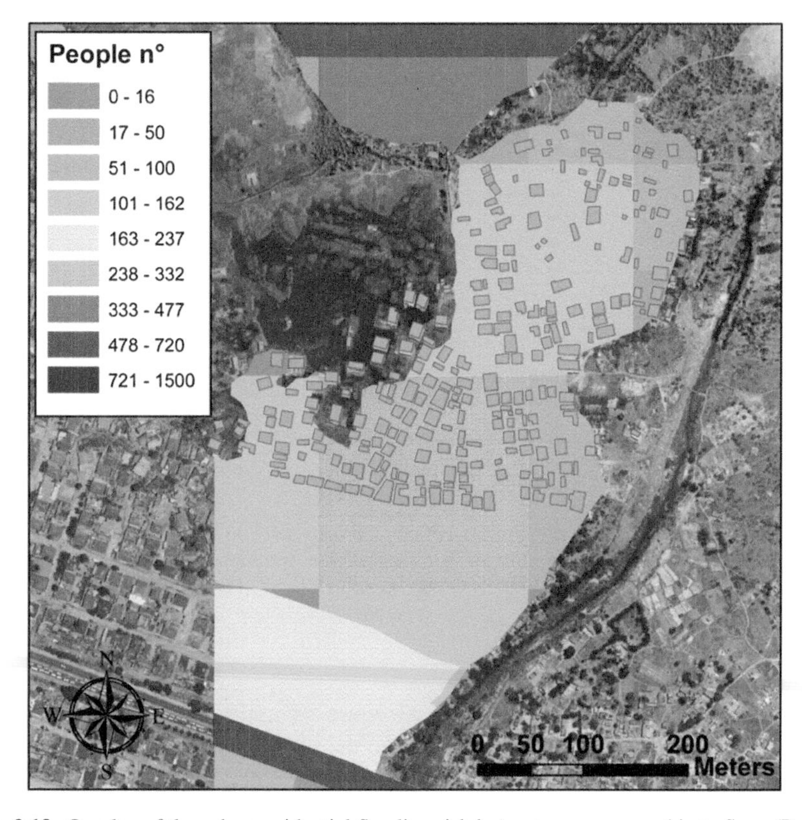

Fig. 3.13 Overlay of the urban residential flooding risk hotspots map zoomed in to Suna (Dar es Salaam) and a group of houses identified as flood-prone based on household interviews (highlighted on an orthophoto) (De Risi and Jalayer 2013)

method described earlier in this chapter. The Msimbazi river flows across the city of Dar es Salaam from the higher areas of Kisarawe in the coastal region and discharges into the Indian Ocean.

The case study area drains water from three different catchment areas (of about 250 km^2). The characteristics of the three catchments identified, the land-use, and geological maps are described in detail in De Risi et al. (2013a). The peak flow in the three catchments is estimated by employing the Curve Number (CN) method (Mockus 1972), with reference to six different return periods (e.g. 2, 10, 30, 50, 100 and 300 years) based on both historical data (H) and climate projections (CC). The CN is representative of the catchment runoff capacity. If the terrain is moderately permeable and only a small portion of the water flows as run-off (e.g. as in the case of bare land and green areas), a CN corresponding to an AMC II (CN2) is usually used. If there is a high degree of urbanisation (e.g. paved roads and high density of construction), the overflow can easily reach the main channel, a higher value of CN may be assigned (e.g. the CN corresponding to AMC III). In the following, the present condition is represented by CN2 (corresponding to AMC II, which may underestimate run-off in some areas). Meanwhile, CN3 (corresponding to

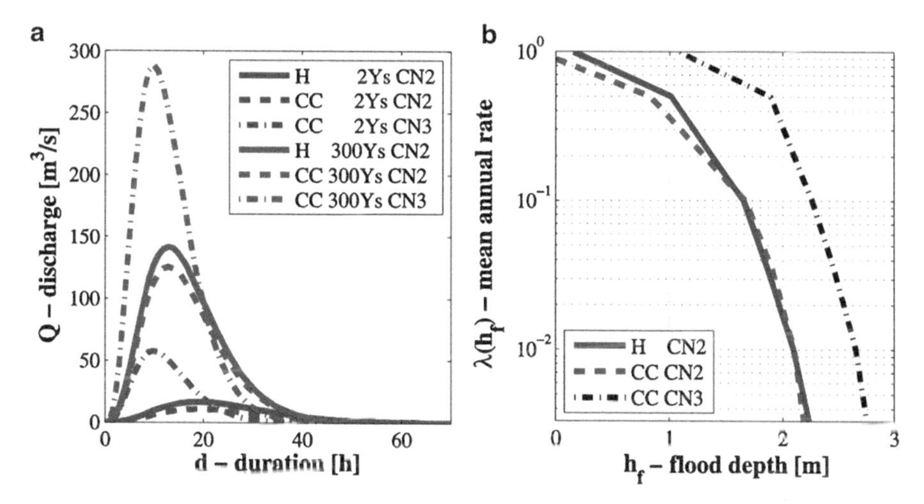

Fig. 3.14 (**a**) Hydrographs evaluated for catchment 1 ($T_R = 2$ and 300 years) with and without CC effects, for different CN classes; (**b**) The mean annual rate of exceedance of a specific flood height (the median curves calculated over the entire area of interest) ((Jalayer et al. 2013b) © 2013 Taylor and Francis, used with permission)

AMC III) is used to describe the future territorial situation, based on a hypothetical urban expansion in the catchment area. The inflow hydrographs (flooding discharge at the catchment's closing point) for catchment 1, the largest of the three catchments, corresponding to two return periods (2 and 300 years) and for two Curve Numbers (CN2 and CN3) are illustrated in Fig. 3.14a. It is possible to observe that for the same CN, the CC hydrographs are lower than those based on historical data (H). Changing the CN class, there is a substantial increase in the discharge with respect to the historical data of about 3.5 times for $T_R = 2$ years and 2 times for $T_R = 300$ year.

The hydrographs are used next to obtain the inundation profile in terms of flood height and velocity for the nodes within the lattice covering the area of interest for each return period T_R. The software FLO-2D (O'Brien et al. 1993; FLO-2D 2004) was used for a bi-dimensional simulation of the flooding volume propagation (based on the calculated hydrographs and a DEM) assuming a 45 h simulation time (i.e. the total duration of the hydrograph). The detailed results for the specific case study in terms of inundation maps are shown in De Risi et al. (2013a). Starting from the inundation profile, using the procedure illustrated in De Risi et al. (2013b), it is possible to obtain the hazard curves. These curves can be extracted for the centroid of each structure within the portfolio. The hazard curves, plotting the mean annual rate of exceeding various flooding heights (i.e. inverse of the return period), can be obtained based on CC and H. As a central statistic of the hazard curves for the portfolio of structures, the median hazard curve is calculated for each of the following three cases, namely, H CN2, CC CN2 and CC CN3 (Fig. 3.14b). It is possible to observe that, for the same CN, the median hazard curves obtained based on historical data are very similar to those obtained considering the climate projections. The opposite is also true, considering the CN3, the hazard values obtained considering the climate projections are substantially larger (approximately one order of magnitude) with respect to the hazard curve calculated based on historical data only.

Portfolio Vulnerability Assessment

In this work, a novel simulation-based and analytic methodology has been adopted for flood vulnerability assessment (De Risi et al. 2013a, b). This methodology employs the Bayesian parameter estimation for calculating the structural fragility for a class of structures, by characterisation of building-to-building variability and other sources of uncertainty based on a limited number of in-situ field surveys. In this method, the various sources of uncertainty are treated in a unified manner by employing the standard Monte Carlo Simulation. The portfolio of structures considered in this work is a group of informal settlements located in Suna Sub Ward, Dar es Salaam (Fig. 3.13). The following flooding actions are considered: hydrostatic pressure, hydrodynamic pressure and accidental debris impact. The informal buildings located in this neighbourhood reveal similar characteristics. For instance, they are all one-storey buildings, use cement-stabilised earth bricks as wall material, and have a roof system made up of corrugated iron sheets and wooden beams. Various building-specific field surveys (see De Risi et al. 2013a for more details) conducted by the CLUVA team in this neighbourhood reveal that the houses in this neighbourhood have similar geometrical patterns (the so-called Mozambique-style housing). As common adapting strategies, a significant portion of the inhabitants tend to build a barrier in front of the door or to build the house on a raised foundation (platform). The windows and doors are generally not water-tight.

The Fragility Assessment for the Class

The fragility curves derived herein correspond to the Collapse limit state (CO), defined as the critical flooding height in which the most vulnerable section of the most vulnerable wall in the building is going to exceed the allowable stress requirements. The critical water height for structural collapse is calculated by employing structural analysis taking into account the various sources of uncertainties in geometry, material properties and construction details. For a prescribed limit state, the simulation procedure leads to a set of different realisations of the critical water height reflecting the building-to-building variability in construction details and lack of information about material properties. These critical water height values are implemented inside an efficient and Bayesian procedure referred to as the robust fragility method (more detailed information about this procedure can be found in De Risi et al. 2013a, b and Jalayer et al. 2013a). The outcomes of this procedure are reported in terms of various prescribed percentiles for the plausible fragility curves. Figure 3.15a illustrates the 16th, 50th and 84th percentile fragility curves obtained for the Collapse limit state related to H with CN2. Note that the interval between the 16th and 84th percentile can be considered as a proxy

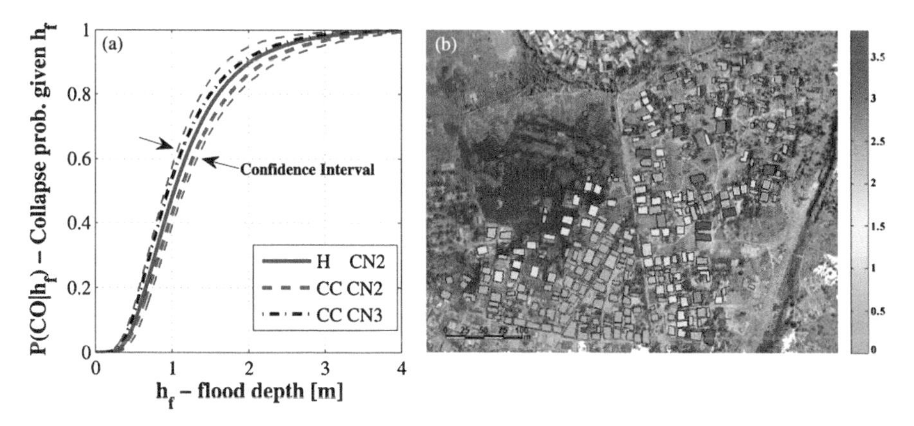

Fig. 3.15 (**a**) The fragility curves for the CO limit state for the class of informal building made up of cement-stabilised bricks in Suna neighbourhood (based on 20 simulations); (**b**) The household risk map: number of people affected (Collapse limit state only) ((Jalayer et al. 2013b) © 2013 Taylor and Francis Group, London UK, used with permission)

for plus/minus one standard deviation confidence interval. The figure also illustrates the 50th percentile fragility curves for CC CN2 and CC CN3. In fact, the fragility curves based on historical data only and the climate projections for different CN classes are not identical. This can be explained by the fact that velocity-dependent flood action such as hydro-dynamic pressure and accidental debris impact are taken into account. However, as can be observed, the CC CN2 and CC CN3 fragility curves are contained within the confidence interval for H CN2.

Flood Risk Assessment

The flooding risk λ_{ls} expressed as the mean annual rate of exceeding a given limit state *ls* can be calculated from Eq. (3.1). The annual probability of exceeding a limit state *P*(*ls*), assuming a homogeneous Poisson process model with rate λ_{ls}, can be calculated as:

$$P(ls) = 1 - \exp(-\lambda_{ls}) \tag{3.4}$$

Risk can be quantified by calculating the total expected loss or the expected number of people affected for the portfolio of buildings.

The expected repair costs (per building or per unit residential area), *E[R]*, can be calculated as a function of the limit state probabilities and by defining the damage state *i* as *i* and *i* + *1*:

$$E[R] = \sum_{i=1}^{N_{ls}} [P(ls_i) - P(ls_{i+1})] \cdot R_i \tag{3.5}$$

where N_{ls} is the number limit states that are used in the problem in order to discretise the structural damage; R_i is the repair cost corresponding to damage state i; and $P(ls_{Nls+1}) = 0$.

The expected number of people affected by flooding can also be estimated as a function of the limit state probabilities from Eq. (3.5), replacing R_i by the population density (per house or per unit residential area).

Figure 3.15b illustrates the expected value of the number of people endangered by flooding per year and per household,[2] taking into account only the collapse limit state probabilities in Eq. (3.5) (based on only historical data, CN2 class). Table 3.8 reports the total annual expected replacement costs (collapse limit state only) normalised to the total replacement cost for the entire portfolio, considering both the historical data H and the climate change projections CC, based on two different land-cover scenarios (CN2 and CN3).

It can be observed that CC and the CN3 class (un-controlled urbanisation) might lead to an increase of about 30 % in the expected annual replacement costs. The whole procedure for the assessment of risk and vulnerability to flooding in the micro-scale for a class of structures is implemented in a graphical user interface called Visual Vulnerability and Risk (VISK, Flooding; see Box 3.5).

Table 3.8 Expected annual repair costs normalised to the total replacement cost for the entire portfolio (considering only the Collapse (CO) limit state)

	$F - \sigma$	F	$F + \sigma$
H CN2	36 %	33 %	30 %
CC CN2	30 %	28 %	25 %
CC CN3	66 %	63 %	60 %

Box 3.5: Flood Risk Assessment Software (VISK): Visual Vulnerability and Risk

Once the procedure described herein is developed as a software package, it can be used for performing refined flood risk assessment in a micro-scale (e.g. scale 1:2,000). This spatial scale allows for identifying the housing units at risk and obtaining a more accurate estimate of the number of people potentially vulnerable to flooding. In fact, in the context of CLUVA, a software has been developed for micro-scale flood risk assessment.

(continued)

[2] Note that the differences in risk between various households are only due to differences in the hydraulic profile between different points on the map and do not reflect variations in the quality of construction between various houses. This is because the same fragility curve is used in order to characterise the vulnerability to flooding for the houses in this neighbourhood.

Box 3.5 (continued)

This software is called Visual Vulnerability and Risk (VISK) (see De Risi et al. 2013b) and allows the user to quickly identify and capture (on an orthophoto) the houses for which the risk assessment is going to be performed. Moreover, VISK characterises the uncertainties in various building features based on limited amount of field survey statistics. The results are reported both in terms of building vulnerability (e.g. fragility curves) and point-wise risk estimates (e.g. expected costs, expected number of people exposed, etc.). Once the necessary licenses are obtained and the product is registered, VISK and its user manual are going to be available on CLUVA web-site at the following URL: www.cluva.eu.

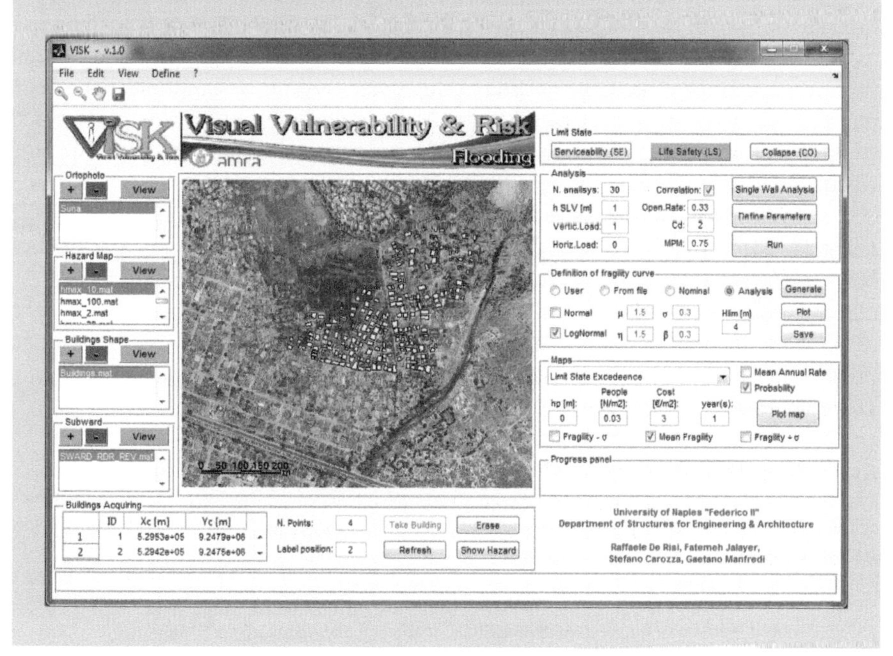

Conclusion

This chapter provides an overview of the activities carried out in the context of the CLUVA project for vulnerability assessment of the built environment in three case study cities. As with the rest of the project, the focus of the activities pursued in this working group has been on long-term assessment of vulnerability with the objective of providing supporting information to decision-makers in composing strategic measures for adapting the cities to climate change. Depending on the spatial resolution of interest, alternative methods with varying levels of accuracy have been adopted for vulnerability and risk assessment. These methods are referred to as *meso-scale* and *micro-scale* methods in order to further emphasise the importance of the spatial resolution in the choice of the optimal assessment method. Combining the meso-scale and micro-scale methods, a two-stage assessment strategy has been pursued. In particular, the meso-scale assessment methods have been used in the first stage as a screening tool for delineating the urban flooding hotspots for various urban morphologies identified as built environment, such as residential and major urban corridors. In the second stage and after the urban flooding hotspots are identified, a more refined micro-scale flood vulnerability and risk assessment is performed.

The application of the meso-scale methods in flooding hotspots delineation is illustrated for Ouagadougou, Dar es Salaam, and Addis Ababa. The micro-scale methods are adopted for flooding risk and vulnerability assessment for a portfolio of informal settlements in Suna in Dar es Salaam. Suna has been identified as a residential urban flooding risk hotspot by employing the meso-scale methodology. The flood risk assessment for Suna has been performed both by employing the historical rainfall time-series and the climate projections. Moreover, working assumptions have been made as much as it regards a change in the flooding run-off due to future urbanisation of the catchment area that drains to Suna. In particular, in the absence of specific calculations, a hypothetic increase in flood run-off (increasing curve number and decreasing permeability) is assumed. The impact of future climate trends on the resulting risk estimates is not significant, assuming the same land-cover characteristics between the present and the future (within around 50 years). However, pairing up the climate projections with a hypothetical urban expansion scenario for the city (i.e. a much more impervious terrain), leads to around 30 % increase in overall exposure to flooding in the neighbourhood studied (assuming that the buildings' characteristics are going to remain the same).

Acknowledgements and Data Sources This work was supported in part by the European Commission's Seventh Framework Program Climate Change and Urban Vulnerability in Africa (CLUVA), FP7-ENV-2010, Grant No. 265137. This support is gratefully acknowledged. The authors gratefully acknowledge the precious help of Prof. F. De Paola, Prof. M. Giugni and M. E. Topa for providing the rainfall curves, the inundation profiles and the TWI maps for the three case study cities considered. Moreover, the authors would like to gratefully acknowledge Dr. E. Bucchignani and the rest of the CMCC team for providing the Climate Projections for RCP 8.5 Scenario. The authors would like to acknowledge also the precious help of Prof. Kumelachew Yeshitela, Alemu Nebebe, Dr. Riziki Shemdoe, Dr. Deusdedit Kibassa, Dr. Sarah Lindley, Dr. Gina Cavan, Dr. Andreas Printz and Florian Renner for providing the Urban Morphology types for Addis Ababa and Dar es Salaam. Last but not least, the authors would like to acknowledge S. Carozza for his invaluable work in developing the software VISK.

References

Bellucci A, Bucchignani E, Gualdi S, Mercogliano P, Montesarchio M, Scoccimarro E (2012) Data for global climate simulations available for downscaling. Available: http://www.cluva.eu/deliverables/CLUVA_D1.1.pdf. Accessed 9 July 2014

Cavan G, Lindley S, Yeshitela K, Nebebe A, Woldegerima T, Shemdoe R, Kibassa D, Pauleit S, Renner R, Printz A, Buchta K, Coly A, Sall F, Ndour NM, Ouédraogo Y, Samari BS, Sankara BT, Feumba RA, Ngapgue JN, Ngoumo MT, Tsalefac M, Tonye E (2012) Green infrastructure maps for selected case studies and a report with an urban green infrastructure mapping methodology adapted to African cities. Available: http://www.cluva.eu/deliverables/CLUVA_D2.7.pdf. Accessed 9 July 2014

Connolly RD, Schirmer J, Dunn PK (1998) A daily rainfall disaggregation model. Agric For Meteorol 92:105–117

CRED, Centre for Research on the Epidemiology of Disasters (2012) Disaster data: a balanced perspective. Available: http://www.cred.be/publications. Accessed 9 July 2014

De Risi R (2013) A probabilistic bi-scale framework for urban flood risk assessment. PhD dissertation, Department of Structures for Engineering and Architecture, University of Naples Federico II, Naples

De Risi R, Jalayer F (2013) Identification of hotspots vulnerability of adobe houses, sewer systems and road networks. Available: http://www.cluva.eu/deliverables/CLUVA_D2.1.pdf. Accessed 9 July 2014

De Risi R, Jalayer F, De Paola F, Iervolino I, Giugni M, Topa ME, Mbuya E, Kyessi A, Manfredi G, Gasparini P (2013a) Flood risk assessment for informal settlements. Nat Hazards 69(1):1003–1032. doi:10.1007/s11069-013-0749-0

De Risi R, Jalayer F, Iervolino I, Manfredi G, Carozza S (2013b) VISK: a GIS-compatible platform for micro-scale assessment of flooding risk in urban areas. In: Papadrakis M, Papadopoulos V, Plevris V (eds) COMPDYN, 4th ECCOMAS thematic conference on computational methods in structural dynamics and earthquake engineering, Kos Island, Greece, 12–14 June 2013

FLO-2D Software, Inc (2004) FLO-2D® user's manual, Nutrioso, Arizona. www.flo-2.com. Accessed 9 July 2014

Gill SE, Handley JF, Ennos AR, Pauleit S, Theuray N, Lindley SJ (2008) Characterising the urban environment of UK cities and towns: a template for landscape planning. Landsc Urban Plan 87:210–222

Giugni M, Adamo P, Capuano P, De Paola F, Di Ruocco A, Giordano S, Iavazzo P, Sellerino M, Terracciano S, Topa ME (2012) Hazard scenarios for test cities using available data. Available: http://www.cluva.eu/deliverables/CLUVA_D1.2.pdf. Accessed 9 July 2014

Gualdi S, Somot L, Li L, Artale V, Adani M, Bellucci A, Braun A, Calmanti S, Carillo A, Dell'aquila A, Deque M, Dubois C, Elizade A, Harzallah A, Jacob D, L'hévéder D, May W, Oddo P, Ruti P, Sanna A, Sannino G, Scoccimarro E, Savault F, Navarra A (2013) The CIRCE simulations: regional climate change projections with realistic representation of the Mediterranean sea. Bull Am Meteorol Soc 94:65–81

Güntner A, Olsson J, Calver A, Gannon B (2001) Cascade-based disaggregation of continuous rainfall time series: the influence of climate. Hydrol Earth Syst Sci 5:145–164

IPCC (2007) Climate Change 2007: synthesis report. In: Core Writing Team, Pachauri RK, Reisinger A (eds) Contribution of Working Groups I, II and III to the fourth assessment report of the Intergovernmental Panel on Climate Change. IPCC, Geneva, Switzerland, 104 pp

Jalayer F, De Risi R, Elefante L, Manfredi G (2013a) Robust fragility assessment using Bayesian parameter estimation. In: Adam C, Heuer R, Lenhardt W, Schranz C (eds) Vienna congress on recent advances in earthquake engineering and structural dynamics 2013 (VEESD 2013), Paper No. 503, 28–30 August 2013, Vienna, Austria

Jalayer F, De Risi R, Manfredi G, De Paola F, Topa ME, Giugni M, Bucchignani E, Mbuya E, Kyessi A, Gasparini P (2013b) From climate predictions to flood risk assessment for a portfolio of structures. In: 11th international conference on structural safety & reliability, ICOSSAR 2013. Columbia University, New York, 16–20 June 2013

Jalayer F, De Risi R, De Paola F, Giugni M, Manfredi G, Gasparini P, Topa ME, Nebyou Y, Yeshitela K, Nebebe A, Cavan G, Lindley S, Printz A, Renner F (2014) Probabilistic GIS-based method for delineation of flood-prone areas and identification of urban hotspots. Nat Hazards 73:975–1001

Manfreda S, Sole A, Fiorentino M (2008) Can the basin morphology alone provide an insight into floodplain delineation? WIT Trans Ecol Environ 118:47–56

Manfreda S, Di Leo M, Sole A (2011) Detection of flood-prone areas using digital elevation models. J Hydrol Eng 16:781–790

MapAction (2011) Burkina Faso: flooding in Ouagadougou as of 03 Sept 09. http://www.mapaction.org/map-catalogue/mapdetail/1719.html. Accessed 9 July 2014

Mockus V (1972) National engineering handbook, section 4, hydrology. US Soil Conservation Service, Washington, DC

NBS, National Bureau of Statistics (2006) The 2002 population and housing census, analytical report, Dar es Salaam, Tanzania. Available: http://www.nbs.go.tz/projections/dsm_projections.pdf. Accessed 9 July 2014

O'Brien JS, Julien PY, Fullerton WT (1993) Two-dimensional water flood and mudflow simulation. J Hydraul Eng 119(2):244–261

Olsson J (1998) Evaluation of a scaling cascade model for temporal rain-fall disaggregation. Hydrol Earth Syst Sci Discuss 2:19–30

Pauleit S, Duhme F (2000) Assessing the environmental performance of land cover types for urban planning. Landsc Urban Plan 52:1–20

Qin CZ, Zhu AX, Pei T, Li BL, Scholten T, Behrens T, Zhou CH (2011) An approach to computing topographic wetness index based on maximum downslope gradient. Precis Agric 12:32–43

Rebora N, Ferraris L, Vonhardenberg J, Provenzale A (2006) RainFARM: rainfall downscaling by a filtered autoregressive model. J Hydrometeorol 7:724–738

Rockel B, Will A, Hense A (2008) The regional climate model COSMO-CLM (CCLM). Meteorol Z 17:347–348

UN-Habitat (2009) Slums: levels and trends, 1990–2005. Monitoring the millennium development goals slum target. Available: http://www.unhabitat.org/downloads/docs/9179_33168_Slum_of_the_World_levels_and_trends.pdf. Accessed 9 July 2014

UN-Habitat (2010) State of the world's cities 2010/2011 – cities for all: bridging the urban divide. Available: http://www.unhabitat.org/pmss/listItemDetails.aspx?publicationID=2917. Accessed 9 July 2014

Vega C (2011) Review of the IFRC-led shelter cluster Burkina Faso September 2009 floods response. Available: http://www.ifrc.org/docs/Evaluations/Evaluations2011/Africa/Burkina%20Faso%20Shelter%20Cluster%20Review%20(2).pdf. Accessed 9 July 2014

WCRP (2008) World climate research programme, CMIP5 Coupled Model Intercomparison Project [Online]. Available: http://cmip-pcmdi.llnl.gov/cmip5/. Accessed 9 July 2014

Chapter 4
Green Infrastructure for Climate Adaptation in African Cities

Sarah J. Lindley, Susannah E. Gill, Gina Cavan, Kumelachew Yeshitela, Alemu Nebebe, Tekle Woldegerima, Deusdedit Kibassa, Riziki Shemdoe, Florian Renner, Katja Buchta, Hany Abo-El-Wafa, Andreas Printz, Fatimatou Sall, Adrien Coly, Ndèye Marème Ndour, Rodrigue A. Feumba, Maurice O.M. Zogning, Emmanuel Tonyé, Youssoufou Ouédraogo, Saïdou Bani Samari, and Bakary T. Sankara

Abstract Green infrastructure is a core component of any city. The ecosystem services that it provides already make an important contribution to the health and wellbeing of urban dwellers and are considered to be vital for future urban sustainability. In the case of African cities, this argument is stronger still given that other forms of infrastructure are often lacking or seriously underperforming. This chapter

S.J. Lindley (✉) • S.E. Gill • G. Cavan
School of Environment, Education and Development, University of Manchester, Oxford Road, Manchester, UK
e-mail: sarah.lindley@manchester.ac.uk; susannah.gill@merseyforest.org.uk; G.Cavan@mmu.ac.uk

K. Yeshitela • A. Nebebe • T. Woldegerima
Ethiopian Institute of Architecture, Building Construction and City Development (EiABC), Addis Ababa University, Addis Ababa, Ethiopia
e-mail: kumeyesh@googlemail.com; mekonnin@yahoo.com; tekle_wgerima@yahoo.com

D. Kibassa • R. Shemdoe
Institute of Human Settlements Studies (IHSS), Ardhi University (ARU), P.O. Box 35124, Dar es Salaam, Tanzania
e-mail: dkibassa2000@yahoo.com; shemdoes@gmail.com

F. Renner • K. Buchta • H. Abo-El-Wafa • A. Printz
Chair for Strategic Landscape Planning and Management, Technical University of Munich (TUM), Emil-Ramann-Str. 6, 85354 Freising, Germany
e-mail: florian.renner@tum.de; buchta@wzw.tum.de; hanywafa@gmail.com; printz@wzw.tum.de

F. Sall • A. Coly • N.M. Ndour
Department of Geography, University of Gaston Berger, B.P. 234, Saint Louis, Senegal
e-mail: salltima@gmail.com; colyadrien@yahoo.fr; sodamareme2007@gmail.com

R.A. Feumba
Department of Geography in the Ecole Normale Supérieure, University of Yaoundé I, Yaoundé, Cameroon
e-mail: rfeumba@yahoo.fr

S. Pauleit et al. (eds.), *Urban Vulnerability and Climate Change in Africa*, Future City 4, DOI 10.1007/978-3-319-03982-4_4

discusses the potential role of urban ecosystem services for climate adaptation in African cities. It is based on an empirical assessment of the urban morphology, green structures and ecosystem services of five cities, with a particular emphasis on provisioning services from woody cover and temperature regulating services from evapotranspiring surfaces in two of them. An analysis of retrospective and prospective change helps to establish the extent of pressures to green structures – including in the context of climate change – and the prospects for using green infrastructure for achieving urban climate adaptation. The results show considerable losses in green structures and their associated ecosystem services; something set to continue under 'business as usual' development scenarios projected to 2025. Indeed, there is already a greater need for services than is currently satisfied, especially in the urban core. Our results suggest that, although climate change is an additional pressure to ecosystem services, it is development which poses the greatest immediate threat. It is therefore critical that green infrastructure planning is strengthened and brought into the core of urban development planning as part of climate adaptation and broader sustainability goals.

Keywords Urban ecosystem services • Urban heat island • Vegetation

Introduction

Urban ecosystems are increasingly recognised as a core component of the drive towards sustainable and healthy cities as a result of the services which they provide for human populations (Smit and Parnell 2012; Douglas 2012). The Millennium

---

M.O.M. Zogning
Department of Geography in the Faculté des Arts, Lettres et Sciences Humaines, University of Yaoundé I, Yaoundé, Cameroon
e-mail: zoliver18@yahoo.fr

E. Tonyé
National Advanced School of Engineering, University of Yaoundé I, Yaoundé, Cameroon
e-mail: tonyee@hotmail.com

Y. Ouédraogo
L'Unité de Formation et de Recherche en Sciences de la Vie et de la Terre (UFR SVT), Department BA/PA, University of Ouagadougou, 03 BP 7021 Ouaga 03, Burkina Faso
e-mail: issoufa@fasonet.bf

S. Bani Samari
Laboratoire de Système d'Information, de Gestion del'Environnement e de Développement Durable (LSI-GEDD), Université Aube Nouvelle (U-AUBEN), 06 BP 9283 Ouagadougou, Burkina Faso
e-mail: saidou.bani@yahoo.fr

B.T. Sankara
Géographe / aménagiste, University of Ouagadougou, 03 BP 7021 Ouaga 03, Burkina Faso
e-mail: sankbak82@yahoo.fr

Ecosystem Services Assessment (MEA) defines ecosystem services as "the benefits people obtain from ecosystems" (Millennium Ecosystem Assessment 2005: 10) where ecosystems cover a range of managed and unmanaged functional units, each of which is associated with both tangible and intangible goods and services (Costanza et al. 1997; Daily 1997). Urban ecosystem services are one of ten distinct categories of ecosystem services identified as part of the 2005 MEA. Considering that human society is now predominantly urban, the central premise of connecting goods and services to human wellbeing makes the assessment of urban ecosystem services particularly important.

There is a long history of academic interest in the connections between society and nature, and 'socio-ecological systems'. The concept of green infrastructure as "an interconnected network of green space that conserves natural ecosystem values and functions and provides associated benefits to human populations" (Benedict and McMahon 2001: 5) is closely aligned to the notion of ecosystem services. Its associated principles and frameworks make a connection to the planning process, something which is essential to achieve sustainable urban development. The term emphasises that green structures are as essential to cities as structures associated with energy, transport, water and waste, that these are interconnected and that they require active intervention in terms of protection, management and restoration (Benedict and McMahon 2002).

Globally, African cities have some of the greatest need for sustainable development and face some of the most demanding challenges, yet it is in these cities where the body of evidence to support decision-making is particularly scant (Smit and Parnell 2012). In the context of extreme inequalities in income and health, high rates of exposure and low adaptive capacity to hazards and pollution, and under-resourced governance structures, there is a danger that tackling issues associated with green structures may be seen as something of a luxury. However, the dangers of taking such a narrow view become clear once the wider implications of the term green infrastructure are considered and account is taken of the associated human health and wellbeing benefits. This chapter provides an assessment of urban ecosystem services as an inherent part of the wider notion of green infrastructure and constitutes part of the empirical foundation for an improved evidence-based approach to green infrastructure planning. The underpinning research draws on expertise in a wide range of disciplines, including geography, planning, environmental science, social science, ecology and landscape ecology.

Vulnerability and Adaptation Potential Associated with Urban Ecosystems: Research Objectives

The research associated with this chapter had three specific objectives, each of which are explained below.

1. **To analyse, quantify and map important ecosystem services of the urban green structure that increase the resilience of African cities to climate change** Objective 1 was tackled through the characterisation of each of the

five CLUVA case study cities. The spatial expression of each city needed to connect to recognised elements in green infrastructure planning (e.g. the patch-matrix-corridor model (Forman and Godron 1986)), and the environmental and geographical sciences (Pickett et al. 1997; Douglas 2012) as well as providing a basis for scientific enquiry. These requirements were satisfied through the use of Urban Morphology Types (UMTs). Essentially UMTs are the foundation of a classification scheme which brings together facets of urban form and function. Their application allows the delineation of geographical units which are functional in terms of their biophysical processes. Connections to urban functions (land uses) allows biophysical functions to be combined with a planning orientated perspective. Thus UMT units can be seen as "integrating spatial units linking human activities and natural processes" (Gill et al. 2008: 211). Such an approach is often necessary because biophysical units, such as discrete green spaces, may not be very well represented by existing administrative units. Similarly, existing land use frameworks do not normally consider aspects of urban form and structure together. For example, such frameworks do not differentiate areas within a residential land-use category that are predominantly associated with mud-wood construction compared to dwellings made of stone or concrete. The UMT approach has been increasingly adopted for urban ecological studies in Europe (Pauleit and Breuste 2011). However, its application in the context of African cities has considerable novelty. Examples of urban morphological assessments in African cities do exist, such as in North Africa, and whilst these conform to some of the same general principles, they do not have the scope of the current assessment (Moudon 1997).

Once created, UMT units can be used as a coherent spatial framework through which to derive a series of other indicators, such as associated with land cover (e.g. drainage or thermal properties), relative accessibility or population characteristics (see Chap. 3). UMT-based land cover profiles are particularly important because assessing city-wide ecosystem services requires a detailed understanding of the non-built structures within all land types – not just those which are wholly or mainly 'blue' (e.g. open water) or 'green' (e.g. parks). In other words, knowledge is needed about such structures within the wider urban matrix as well as those associated with patches and corridors. For example, residential areas include urban structures and impervious surfaces, such as roads, in addition to green structures in the form of street trees, household gardens, small community open spaces and strips of land associated with urban agriculture.

The UMTs and associated land cover profiles provide an understanding of the cities at the heart of this study and an input into methods for quantifying their urban ecosystem services. In order to do this effectively and also to provide a suitable basis for green infrastructure planning, the city was broadly conceived as the main urban zone and its peri-urban hinterland, i.e. the 'city-region'. The UMTs and associated mapped units provide a meso-scale perspective and allow some local refinement of neighbourhood characteristics, in this case through land cover assessment.

Table 4.1 List of ecosystem services considered

	Provisioning	Regulating	Cultural	Supporting
Primary set	Food	*Temperature control – shade & evaporative cooling*	*Recreation*	Species habitats
	Wood (non-fuel wood) and fibre *Fuel*		*Livelihoods*	Maintaining genetic diversity
		Temperature control – cool/fresh air corridors	Tourism	
			Spiritual/religious values	
	Water (irrigation)	*Carbon sequestration and storage*	Educational	
	Fresh water (drinking)	Flood – urban surface water regulation	Aesthetics and inspiration	
	Medicinal resources	Flood – river	Heritage/sense of place	
	Ornamental resources	Erosion regulation	Psychological/other aspects of health/wellbeing	
		Water purification/waste treatment		
Secondary set	Compost	Coastal hazard regulation	Other social meetings	Soil formation
	Minerals	Noise	Cultural diversity	Photosynthesis
	Genetic resources	Air quality	Knowledge systems (traditional/formal)	Primary production
		Pollination		Nutrient/water cycling
		Biological control of pests/diseases		

Italics denote services that were the subject of further research in the project

This study could not cover all urban ecosystem services identified in the literature. Instead, a set of possible services was constructed based on key sources (Millennium Ecosystem Assessment 2005; The Economics of Ecosystems and Biodiversity (TEEB) 2011). These lists were refined to a final set (Table 4.1). The set includes provisioning services, or products obtained from ecosystems, such as food and water; regulating services, such as regulation of floods, drought, land degradation, and disease; cultural services – non-material benefits obtained from ecosystems, such as those associated with recreational, spiritual and religious activities; and, supporting services necessary for the production of all other ecosystem services, such as soil formation and nutrient cycling. They informed a UMT-based assessment of multi-functionality to identify the current provision of services and the need for services across case study cities. Three case study cities were assessed in this way: Addis Ababa, Dar es Salaam and Saint Louis. Specific services were also selected for further research work, including representative services within the provisioning, regulating and cultural service groups for two case study cities, Addis Ababa and Dar es Salaam.

2. **To assess the impacts of climate change on urban green structure and its ecosystem services** Objective 2 was tackled through recognition that urban green structures and the functions that they provide are, in fact, threatened

by a range of processes, only some of which are explicitly climate-related. Consideration of climate in isolation from development would fail to capture the true nature of processes of change in African cities and result in an incomplete picture of issues to consider in the quest for a more sustainable future.

Addis Ababa and Dar es Salaam were selected as in-depth case studies. For these cities, UMT-based assessments were carried out for two time points to provide an assessment of retrospective change (2011 and 2006 for Addis Ababa and 2008 and 2002 for Dar es Salaam). The specific impacts of climate change were assessed through an expert-based analysis of the sensitivity of green structures, with a particular focus on provisioning services in Dar es Salaam. A range of climate drivers were considered which could be connected to a probabilistic assessment of climate-related risk associated with green structures (see Chap. 7 for the overall framework). The relative importance of climate and development related drivers were assessed for both in-depth case study cities for regulating ecosystem services, specifically surface temperatures, using climate projections developed as part of the wider CLUVA project (see Chap. 2).

Objective 2 was also investigated through the use of local case studies in two low income communities in Dar es Salaam: Suna in Magomeni Ward (situated within the Msimbazi river valley around 5 km north of the city centre); and Bonde la Mpunga in Msasani Ward (in a coastal location approximately 6 km north of the city centre). The case studies provided insights into climate-development-green structure relationships and supported an assessment of the potential for green structure related adaptation strategies. This aspect of the work allowed connections to the wider CLUVA research into vulnerability, governance and planning (see Chaps. 6, 9, and 10). Results from this part of the work are reported elsewhere (see Cavan et al. 2012; Lindley et al. 2013).

3. **To evaluate the prospects for urban green structure as a measure for adapting African cities to climate change** Objective 3 is based on the assessment of evidence from the delivery of Objectives 1 and 2. In addition, two spatial scenario models were developed and used to assess prospective change in Addis Ababa and Dar es Salaam (Abo El Wafa 2013a, b; Buchta 2013a, b; Pauleit et al. 2013). The models allowed the assessment of a range of future scenarios, including ones developed in collaboration with local stakeholders. This final element of the work incorporated evidence from the CLUVA research on governance and urban planning (see Chaps. 9 and 10).

Urban Morphology and Green Structures

African cities have a very heterogeneous character. Writing about Dar es Salaam, Lupala (2002: 60) notes "The mixture of Swahili and new high-rise house types has resulted into a mixed character of an urban type in terms of building heights and their relationships to the surrounding spaces". Much of this can be attributed to the fact that more than 70 % of the city is unplanned (URT 2004). Furthermore, development

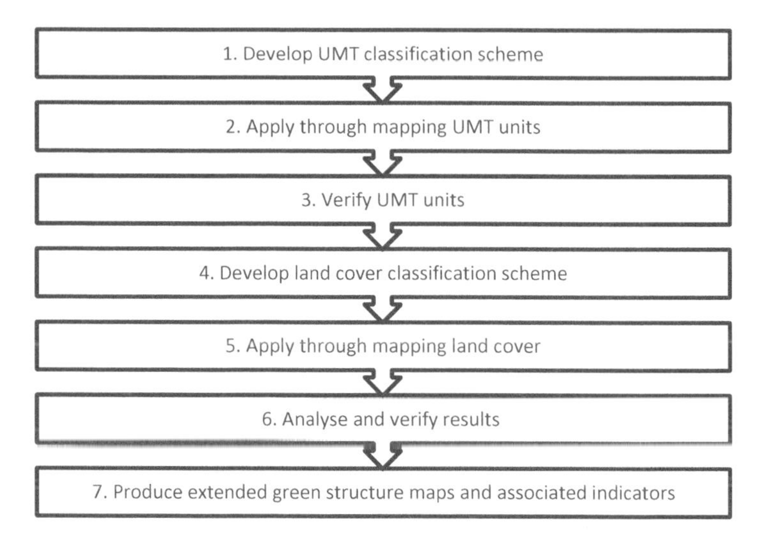

Fig. 4.1 Stages in the production of green structure maps for the case study cities

control compliance rates are reported to be as low as 35 %, leading to streetscapes which are "uncoordinated and chaotic" (Mosha and Mosha 2012: 236). These observations could be similarly attributed – to a greater or lesser extent – to the other CLUVA case study cities and many cities across the African continent. It raises the question of how far it is even possible to apply a concept such as the UMT to African cities given that it was developed largely in the context of European cities where urban form and function has evolved to be a good deal more uniform as a result of a stronger planning tradition, stricter planning regulations and the slower rates of change seen in more recent times. Accordingly, the development of an appropriate set of UMT categories was an important first step of the characterisation work (Fig. 4.1). This initial stage suggested that although categories tended to have more mixed character than in European cities it was still possible to produce meaningful thresholds which could be used to differentiate classes into particular geographical zones. It was also confirmed that the UMTs, in turn, enabled an appreciation of the type and extent of green structure cover in the CLUVA case study cities.

Urban Morphology

A first set of UMT classes was developed through collaboration between project researchers and stakeholders during a field visit to Addis Ababa. A series of thresholds were established in order to help differentiate the more problematic 'mixed' classes, such as those associated with residential zones. This template proved helpful for application across the other cities, although the final set of categories showed some local differences between cities (Table 4.2). Differences were a result of stakeholder and researcher perspectives (including language and

Table 4.2 Final urban morphology unit classification schemes, with % area associated with each UMT

UMTs	Addis Ababa UMTs	Dar es Salaam UMTs	Douala UMTs	Ouaga UMTs	Saint Louis UMTs	
Top	Sub	Sub	Sub	Sub	Top	Sub
1. Agriculture	1.1 Field crop (28.1%)	1.1 Field crops (4.3%)	1.1 Field crops and vegetable farm (4.2%)	1.1 Field crops (5.2%)	**1. Agriculture**	1.2 Orchards (1.2%)
	1.2 Vegetable farm (0.7%)	1.2 Horticulture (1.0%)		1.2 Horticulture (16.1%)		1.4 Irrigated agriculture (3.5%)
		1.3 Mixed farming (4.2%)		1.3 Mixed farming (26.8%)		1.5 Rain-fed agriculture (2.2%)
				1.4 Orchards (0.6%)		
2. Vegetation	2.1 Plantation (5.2%)	2.2 Mixed forest (0.4%)	2.1 Plantation (0.1%)	2.1 Plantation (0.4%)	**13. Natural wetlands**	13.1 Mangrove (5.7%)
	2.2 Mixed forest (3.1%)	2.3 Riverine (2.9%)	2.5 Grass/marsh/swamp (7.4%)	2.2 Mixed forest (5.3%)		13.2 Other aquatic vegetation (2.8%)
	2.3 Riverine vegetation (1.9%)	2.4 Bushland (1.8%)	2.6 Mangrove (10.2%)	2.3 Riverine (3.4%)		13.3 Former mangroves (1.8%)
	2.4 Grassland (1.6%)	2.6 Mangrove (0.8%)		2.4 Shrub savannah (5.6%)		13.4 Floodplain (9.7%)
		2.7 Marsh/swamp (0.8%)		2.5 Grass savannah (2.1%)		13.5 River banks (0.1%)
						13.6 Subtidal (0.2%)
						13.7 Intertidal (0.1%)
					16. Dry ecosystem	16.4 Wooded (1.8%)
						16.2 Bare soil (<0.1%)
						16.3 Beach (0.9%)
						16.4 'Zone de terroir' (30.3%)
						16.5 Vegetated dunes (0.3%)
						16.6 Plantation (<0.1%)

3. Mineral workings and quarries	3.1 Mineral workings and quarries (0.4%)	3.1 Mineral workings and quarries (1.0%)		3.1 Mineral workings and quarries (0.3%)		
4. Recreation	4.1 Park (0.1%)	4.1 Parks (<0.1%)	4.1 Parks (0.1%)	4.2 Stadium, sports and leisure (0.2%)	**4. Recreation and conservation**	4.3 Botanic garden (0.1%)
	4.2 Stadium and festival sites (0.2%)	4.2 Stadium and festival sites (<0.1%)	4.2 Stadium and festival sites (0.1%)			4.4 Other open space (<0.1%)
	4.3 Botanic garden (1.4%)	4.3 Sports grounds (0.2%)	4.4 Other open space (<0.1%)	4.3 Parks (1.2%)		4.5 Hotels (0.3%)
	4.4 Hotel (0.2%)	4.4 Other open space (0.2%)	4.5 Beach (<0.1%)			4.6 Sports ground (0.1%)
		4.5 Beach (0.1%)	4.6 Hotel (<0.1%)			
		4.6 Hotels (0.1%)				
5. Transport	5.1 Major road corridors (3.4%)	5.1 Major road corridors (0.5%)	5.1 Major road corridors (0.1%)	5.1 Major road corridors (<0.1%)	**5. Transport**	5.1 Major road corridors (0.3%)
	5.2 Bus terminals (<0.1%)	5.2 Bus terminals (<0.1%)	5.2 Bus terminals (0.1%)	5.2 Bus terminals (<0.1%)		5.4 Port (<0.1%)
	5.3 Train station (<0.1%)	5.3 Railway (0.2%)	5.3 Railway (<0.1%)	5.4 Airport (0.1%)		5.5 Bus terminal (<0.1%)
	5.4 Airports (1.2%)	5.5 Airports (0.3%)	5.4 Train station (0.5%)			5.6 Airport (0.6%)
		5.6 Port (0.1%)	5.5 Airport (1.9%)			5.7 Bridge (0.1%)
			5.6 Port (1.2%)			
			5.7 Bridge (over R. Wouri) (<0.1%)			
6. Utilities & infrastructure	6.1 Energy distribution (0.1%)	6.1 Energy production & distribution (0.1%)	6.1 Energy production and distribution (<0.1%)	6.2 Water treatment (0.2%)	**6. Utilities & infrastructure**	6.4 Cemeteries (0.1%)
	6.2 Water treatment (0.3%)	6.2 Water tanks & treatment (<0.1%)	6.3 Refuse disposal, inc. landfill (<0.1%)	6.3 Refuse disposal (0.1%)		6.5 Blocks (0.2%)
	6.3 Refuse disposal, inc. landfill (0.1%)	6.3 Refuse disposal, inc. landfill (<0.1%)		6.5 Canals (<0.1%)		6.6 Boat access areas (0.2%)
	6.4 Cemeteries (0.3%)	6.4 Cemeteries (<0.1%)	6.4 Cemeteries (0.1%)			

(continued)

Table 4.2 (continued)

UMTs	Addis Ababa UMTs	Dar es Salaam UMTs	Douala UMTs	Ouaga UMTs	Saint Louis UMTs	
Top	Sub	Sub	Sub	Sub	Top	Sub
7. Residential	7.1 Condominium & multi-storey (5.3%)	7.1 Condominium & multi-storey (0.2%)	7.1 Condominium & multi-storey (0.2%)	7.1 High class dwelling (1.7%)	**7. Residential**	7.1 Condominium & multi-storey (0.6%)
	7.2 Villa & single storey stone/concrete (13.2%)	7.2 Villa & single storey stone/concrete (7.8%)	7.2 Villa & single storey stone (7.6%)	7.2 Medium class dwellings (3.8%)		7.5 Scattered settlement (4.6%)
	7.3 Mud/wood construction (16.0%)	7.3 Mud/wood/sand brick construction (0.4%)	7.3 Mud/Wood construction (20.4%)	7.3 Low class dwellings (3.8%)		7.6 Area of regular dwelling (1.6%)
	7.4 Mixed residential (<0.1%)	7.5 Scattered settlement (24.9%)	7.9 Mixed construction (23.5%)	7.4 Very low class dwellings (1.3%)		7.7 Residential area subdivided (1.9%)
		7.9 Mixed residential (13.1%)		7.5 Scattered settlements (12.1%)		7.8 High density dwellings (0.6%)
						7.9 Precarious area of dwelling (1.7%)
						7.10 High standing residential (0.2%)
						7.11 Mixed urban area (0.5%)
8. Community service	8.1 Education (0.8%)	8.1 Education & culture (0.8%)	8.1 Education (0.7%)	8.1 Education (0.3%)	**8. Community services**	8.1 School (0.1%)
	8.2 Medical (0.2%)	8.2 Medical (<0.1%)	8.2 Medical (0.2%)	8.2 Medical (<0.1%)		8.5 University (1.5%)
	8.3 Religious (0.4%)	8.3 Religion (0.1%)	8.3 Religion (0.1%)	8.3 Religion (0.1%)		
		8.4 Institutional (0.1%)	8.4 Administrative centre (1.0%)	8.5 Military & fire brigades (0.3%)		
		8.5 Military (2.6%)	8.5 Military (0.5%)		**15. Military service**	15.1 Military (0.4%)

9. Retail	9.1 Formal shopping (0.3%) 9.2 Open market (<0.1%) 9.3 Mixed formal & open market (0.2%)	9.1 Formal shopping area (<0.1%) 9.2 Open markets (<0.1%) 9.3 Mixed formal & open (<0.1%) 9.4 Malls (<0.1%)	9.1 Formal shop area (<0.1%) 9.3 Mixed market: formal & open (0.1%)	9.2 Mixed formal & open markets (<0.1%)	**9. Retail**	9.4 Mixed formal & open market (0.1%) 9.5 Trade zone (<0.1%)
10. Industry & business	10.1 Industry & business (0.7%) 10.2 Office (2.0%) 10.3 Storage & distribution (0.3%) 10.4 Garage (1.8%) 10.5 Mixed industry & business (0.4%)	10.1 Manufacturing (0.9%) 10.2 Offices (0.3%) 10.3 Storage & distribution (0.2%) 10.4 Garages (<0.1%)	10.1 Manufacturing (2.5%) 10.3 Storage & distribution (0.1%) 10.6 Mixed (offices, storage, distribution) (1.1%)	10.1 Manufacturing (0.1%) 10.2 Commercial centres & offices (0.3%) 10.3 Storage & distribution (0.1%) 10.4 Mixed industry (0.1%)	**10. Industry & business**	10.1 Manufacturing (<0.1%) 10.5 Traditional processing of fish products (0.1%) 10.6 Services (0.1%)
11. Bare land	11.1 Bare land (8.7%)		11.1 Bare land (1.1%)	11.2 Dwellings under construction (0.8%)		
12. Water	12.1 River (1.6%)	12.1 River (0.1%)	12.1 River/scanty water (14.0%) 12.2 Lakes & reservoirs (0.2%)	12.1 Intermittent rivers (6.4%) 12.2 Natural lakes (0.3%) 12.3 Reservoirs (0.9%)	**12. Open water**	12.1 River (22.5%) 12.4 Lakes (0.3%)
Area (ha)	**51,966**	**150,193**	**24,213**	**305,074**		**12,737**

terminology) as well as genuine differences of urban type, geographical settings and the nature of green structures. For example, some teams favoured the identification of a discrete UMT to identify areas which were under construction whereas others did not. Similarly, some teams chose to use a more detailed identification of ecosystems associated with 'natural' UMTs than others. The outputs from this stage already underline the need to take a local approach and avoid over-generalising the characteristics of African cities, even before considering the respective environmental, socio-economic and demographic drivers of change in each location (Smit and Parnell 2012). Table 4.2 provides the final list of upper-level 'top' and lower-level 'sub' UMT categories following completion of the verification stage.[1]

Notwithstanding the fact that some variability is to be expected within towns and cities, even within the same countries and climate zones, the categories provide a useful starting point for conducting a similar exercise in other African cities. The UMT classification schemes were applied using reference aerial photography, local knowledge and additional geospatial data. The result was a geospatial data layer comprising adjacent units where the boundary of each unit delineates a zone largely conforming to a specific UMT sub-category. It largely disregards other sub-categories below a guideline threshold unit area of around 1–2 ha. Boundaries were usually taken from linear features, such as roads or rivers. The aim was not to reproduce a fine-scale land parcel dataset but rather to capture the broad urban morphological characteristics of the city in question at the meso-scale. As a result, features that are a part of the micro-scale mosaic of neighbourhoods, such as small open areas, minor access roads and drainage features, were not individually identified. Outputs instead show a city-scale overview which provides the starting point for additional analyses. UMT data were generated for:

- Addis Ababa (Fig. 4.2) – Twelve high-level UMTs and 36 detailed UMTs were mapped for 2011 and 2006, for an area covering a total of 520 km^2
- Dar es Salaam (Fig. 4.3) – Eleven high-level UMTs and 43 detailed UMTs were mapped for 2008 and 2002, for an area covering a total of 1,502 km^2
- Douala (2009) (Fig. 4.4) – Eleven high-level UMTs and 36 detailed UMTs were mapped for 2009, for an area covering a total of 242 km^2
- Ouagadougou (2009/10) (Fig. 4.5) – Twelve high-level UMTs and 36 detailed UMTs were mapped for 2009/2010, for an area covering a total of 3,051 km^2
- Saint Louis (2011) (Fig. 4.6) – Twelve high-level UMTs and 46 detailed UMTs were mapped for 2011, for an area covering a total of 127 km^2.

The numbers of UMT classes ranged from 36 to 46 with 11 or 12 high-level groups, and therefore achieved some consistency. However, these classes were applied to study areas which differed markedly in terms of their areal extent and

[1] The development of the final, verified UMT dataset was a lengthy process. It was sometimes necessary to use draft final versions to progress other parts of the work programme. As a result there is sometimes a slight mismatch of categories between different parts of the work – and the data cited in other Chapters in this book (e.g. Chap. 3) – but the impact on subsequent results is considered small.

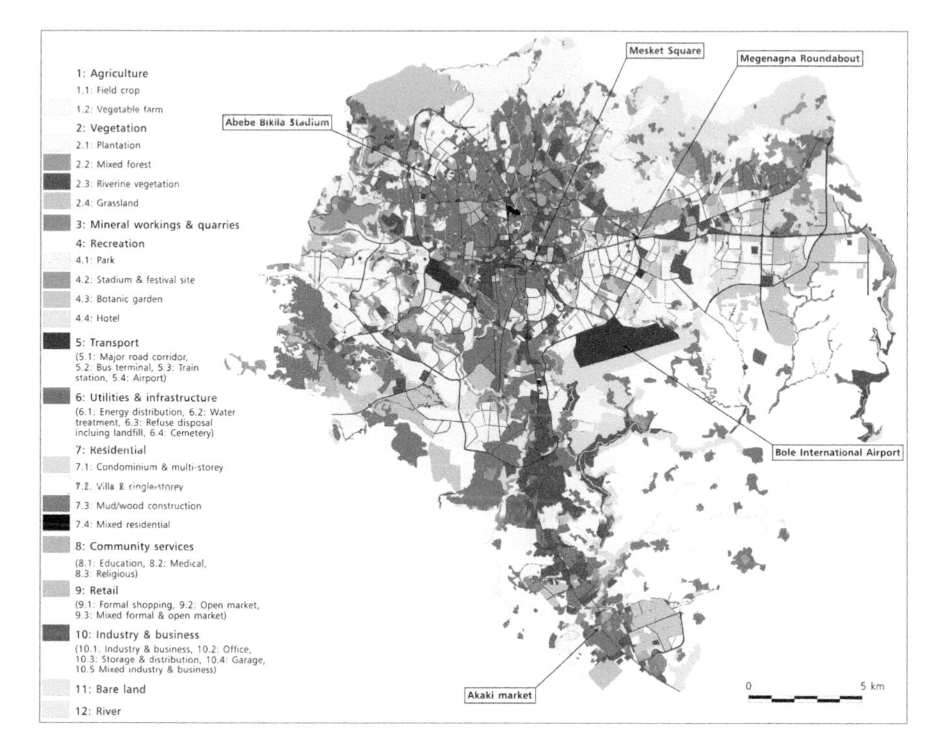

Fig. 4.2 High-level UMT categories and selected sub categories for Addis Ababa

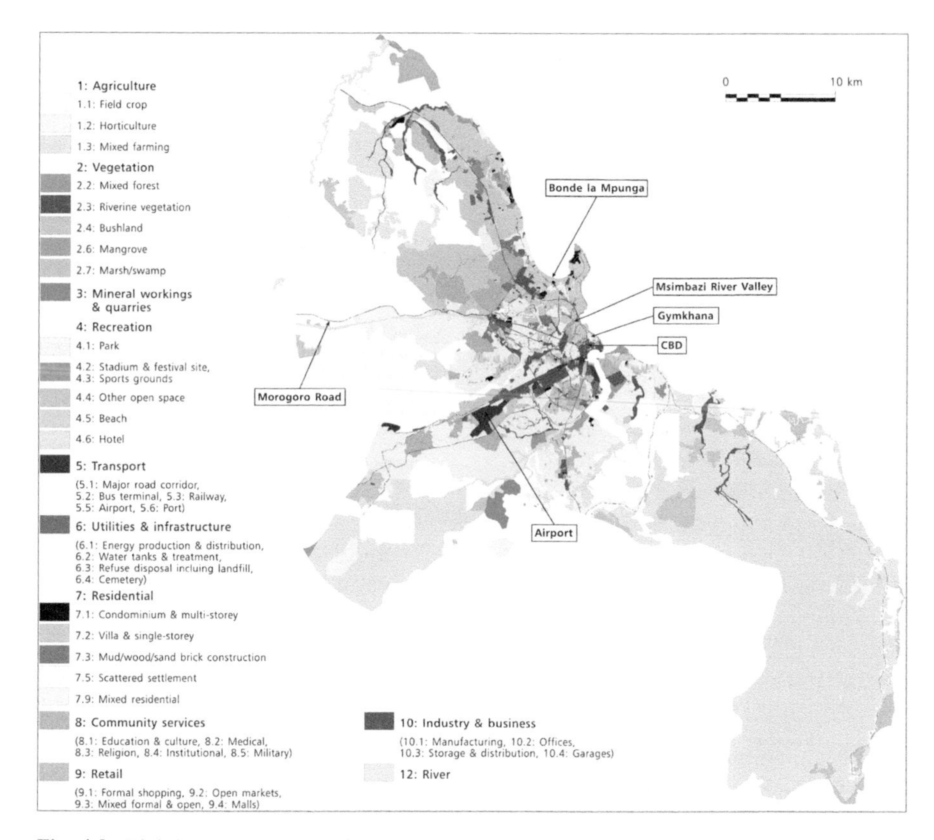

Fig. 4.3 High-level UMT categories and selected sub categories for Dar es Salaam

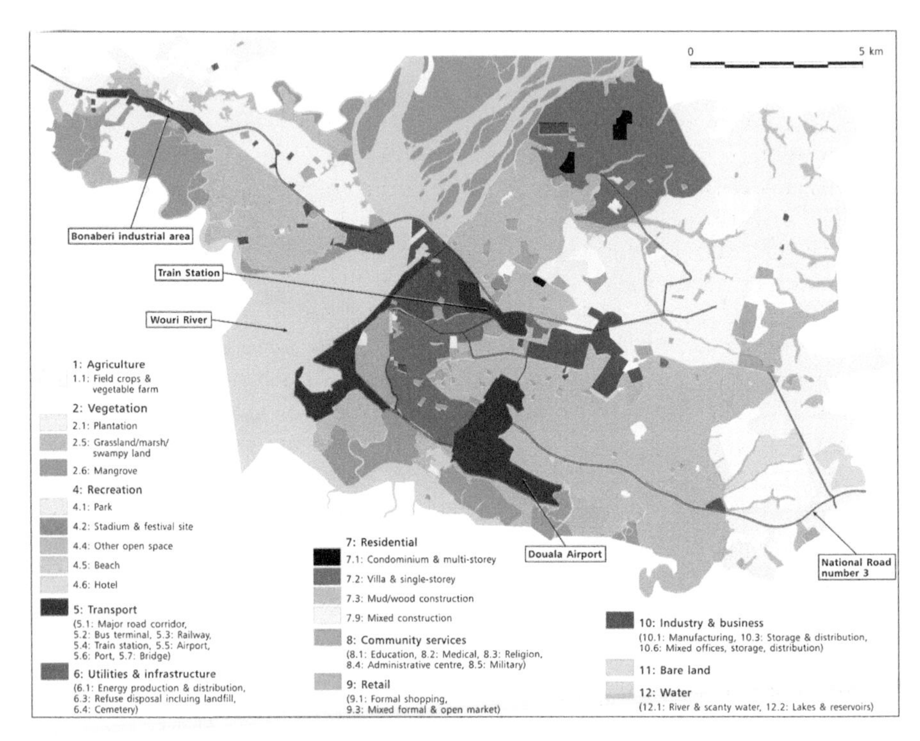

Fig. 4.4 High-level UMT categories and selected sub categories for Douala

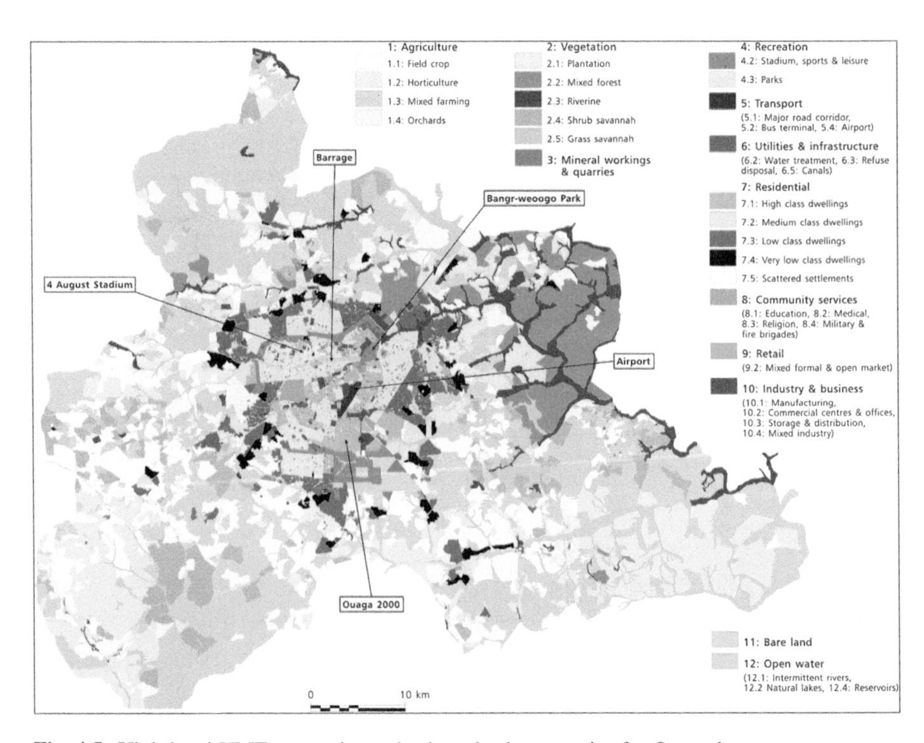

Fig. 4.5 High-level UMT categories and selected sub categories for Ouagadougou

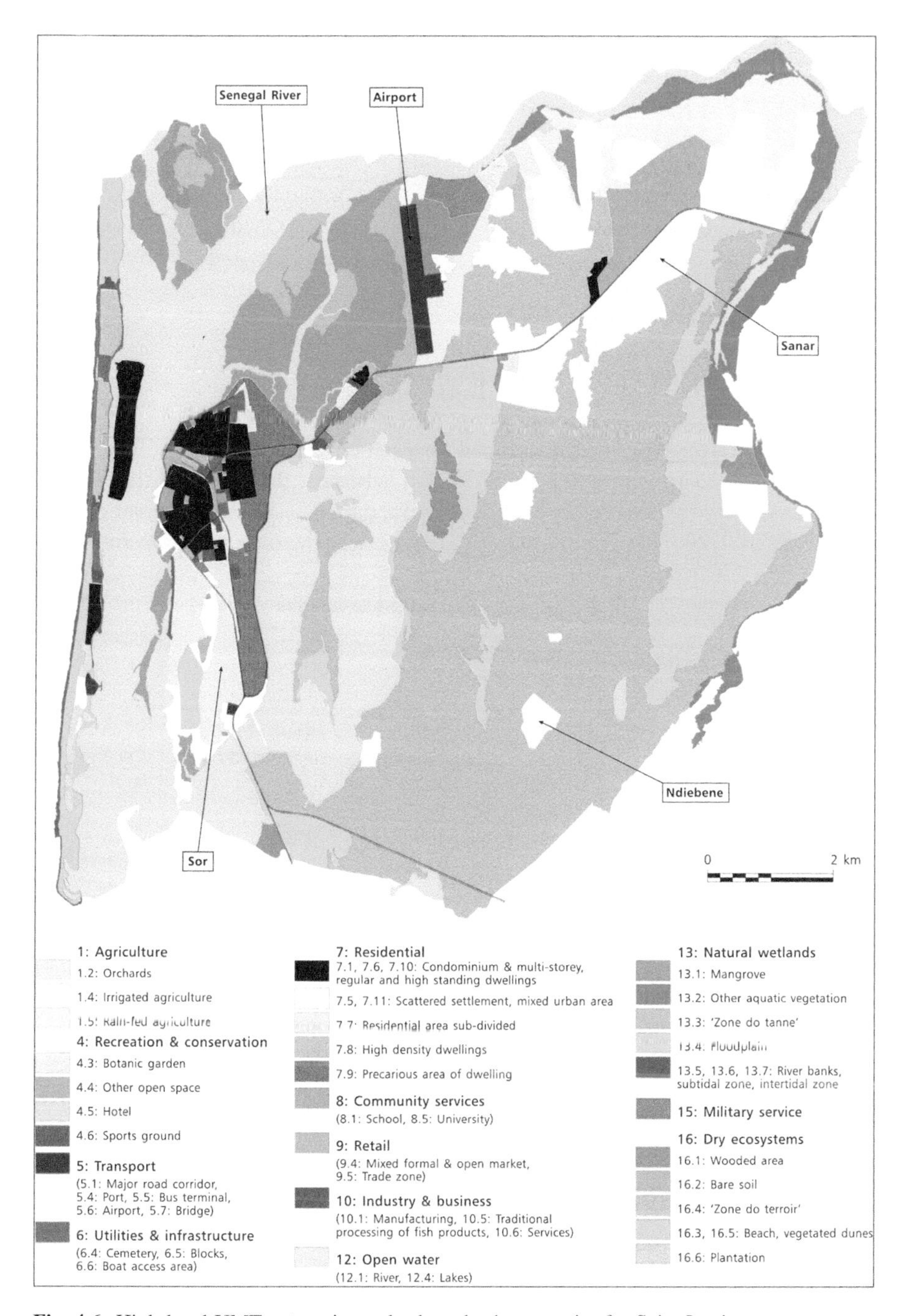

Fig. 4.6 High-level UMT categories and selected sub categories for Saint Louis

the degree to which they covered the city and its hinterland area. As a result, there were some differences in how the UMTs were mapped and the scale of the units generated. The overall mean size of a UMT unit was 69 ha with a standard deviation of 40 ha. Ouagadougou had the largest mean unit size of 117 ha and Addis Ababa the smallest at 13 ha. Despite this, unit sizes for classes which would be expected to be relatively similar between cities were somewhat uniform. For example, the mean size of units associated with medical services in Addis Ababa, Dar es Salaam, Douala and Ouagadougou were 4 ha, 6 ha, 6 ha and 4 ha respectively.

UMT-Based Land Cover Characteristics

A set of land cover types was developed for each city in a similarly collaborative way as described for the UMT classification schemes, this time starting with a template set of covers developed for Dar es Salaam. The final land cover classes are shown in Table 4.3. Again, there was some local variation. For example, in

Table 4.3 Land cover classification schemes used in the five case study cities completing the land cover assessment

Group	Addis Ababa	Dar es Salaam	Douala	Ouagadougou	Saint Louis
Buildings	Built structure I 'high rise & well constructed'	Urban structures I 'concrete/ stone'	Formal buildings 'large'	Built structure I (concrete)	Urban structure II 'hard construction'
	Built structure II 'informal and non-high rise'	Urban structures II 'mud/wood/ sand brick'	Informal buildings 'small'	Built structure II ('banco'- mud/adobe)	Urban structure III 'tile or slate'
					Urban structure I 'hut'
Other impervious	Other impervious	Other impervious	Other impervious	Other impervious	Other impervious
Bare ground	Bare ground – dark	Bare ground – dark	Bare ground dark	Bare soil	Bare ground dark
	Bare ground – light	Bare ground – light	Bare ground light	Wet soil	Bare ground light
Woody vegetation	Eucalyptus trees	Large trees	Large trees	Tree	Large trees
	Other trees	Small trees and shrubs	Small trees	Shrub	Small trees
	Shrubs	Palm trees	Palm trees		
Grasses	Grasses	Grasses	Grasses	Grass	Grasses
Crops	Field crop	Cultivated crops	Crops	Crop I	Cultivated crops
	Vegetable crop			Crop II (truck farming)	
Water	Water	Water	Water	Water	Water

Addis Ababa it was considered important to identify *Eucalyptus* trees as a species of particular economic significance (and one which could be relatively easily differentiated from other tree species types due to its crown characteristics). Similarly in Dar es Salaam, palm trees (Arecaceae) were differentiated, allowing them to be separated for the later merchantable wood and biomass estimation work. Although mangrove trees also merited differentiation, in practice this was too difficult given their visual similarities to other trees.

Land cover was then estimated via aerial photograph interpretation. Specifically, this involved a visual assessment of land covers associated with a random sample of point locations within each mapped UMT category (Gill et al. 2008). Once identification was complete an average profile for each category was derived. Although fairly labour-intensive, this method has the advantage of being relatively undemanding in terms of other resource and technical skill and it can produce a rich dataset in terms of the range of land covers which can be discerned, e.g. relative to other forms of remote sensing using readily available imagery (see Lindley et al. 2013). Disadvantages include the lack of ability to assess specific spatial variation within specific UMT categories.

A detailed reporting of land cover characteristics of the UMT classes across the CLUVA case study cities is beyond the scope of this chapter. However, a set of example UMT classes have been selected in order to illustrate the differences in average land cover profiles. These include the residential categories, which are important for interpreting some of the later work on ecosystem services. The extent of comparison is difficult given the generation of the average profiles based on areas which are considerably different in size and extent. Nevertheless, the analysis is still considered broadly reliable in a selected number of cases (e.g. for residential categories). Selected examples are given in Fig. 4.7 and maps of evapotranspiring cover are given for four cities in Fig. 4.8. As would be expected,

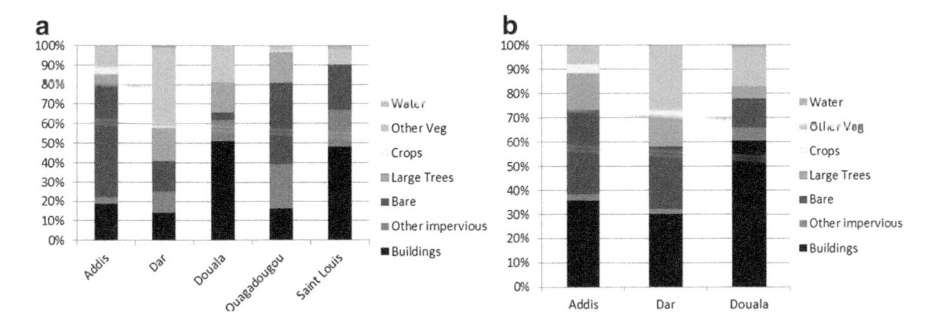

Fig. 4.7 Comparison of the land cover characteristics of selected UMT classes held in common. (**a**) Residential 7.1 (Condominium/multi-storey/'high class') (**b**) Residential 7.3 (mud/wood construction/'low class')

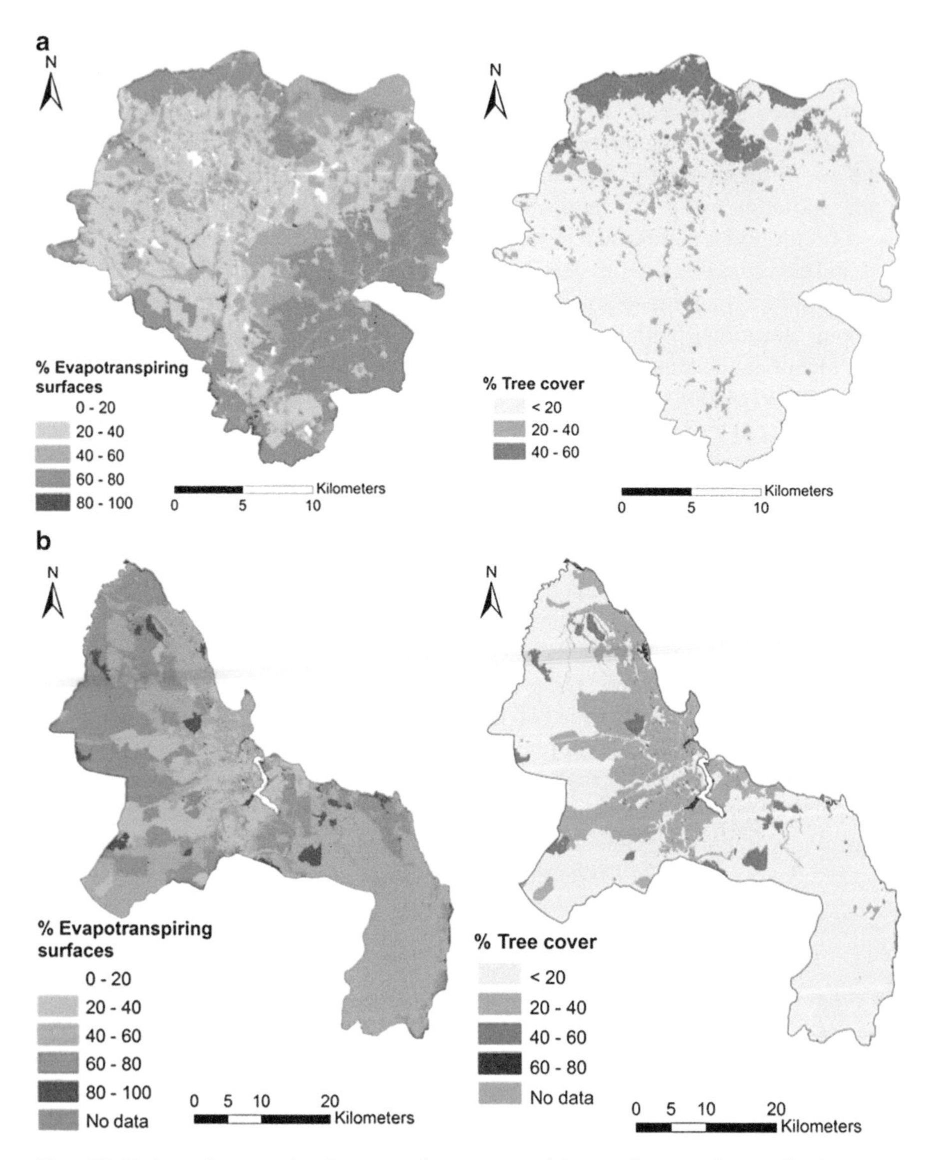

Fig. 4.8 Estimated proportional cover of evapotranspiring surfaces and trees in four case study cities (**a**) Addis Ababa, 2011 evapotranspiring surfaces (*left*) and trees (*right*) (**b**) Dar es Salaam, 2008 evapotranspiring surfaces (*left*) and trees (*right*) (**c**) Douala, 2009 evapotranspiring surfaces (*top*) and trees (*bottom*) (**d**) Saint Louis, 2011 evapotranspiring surfaces (*left*) and trees (*right*)

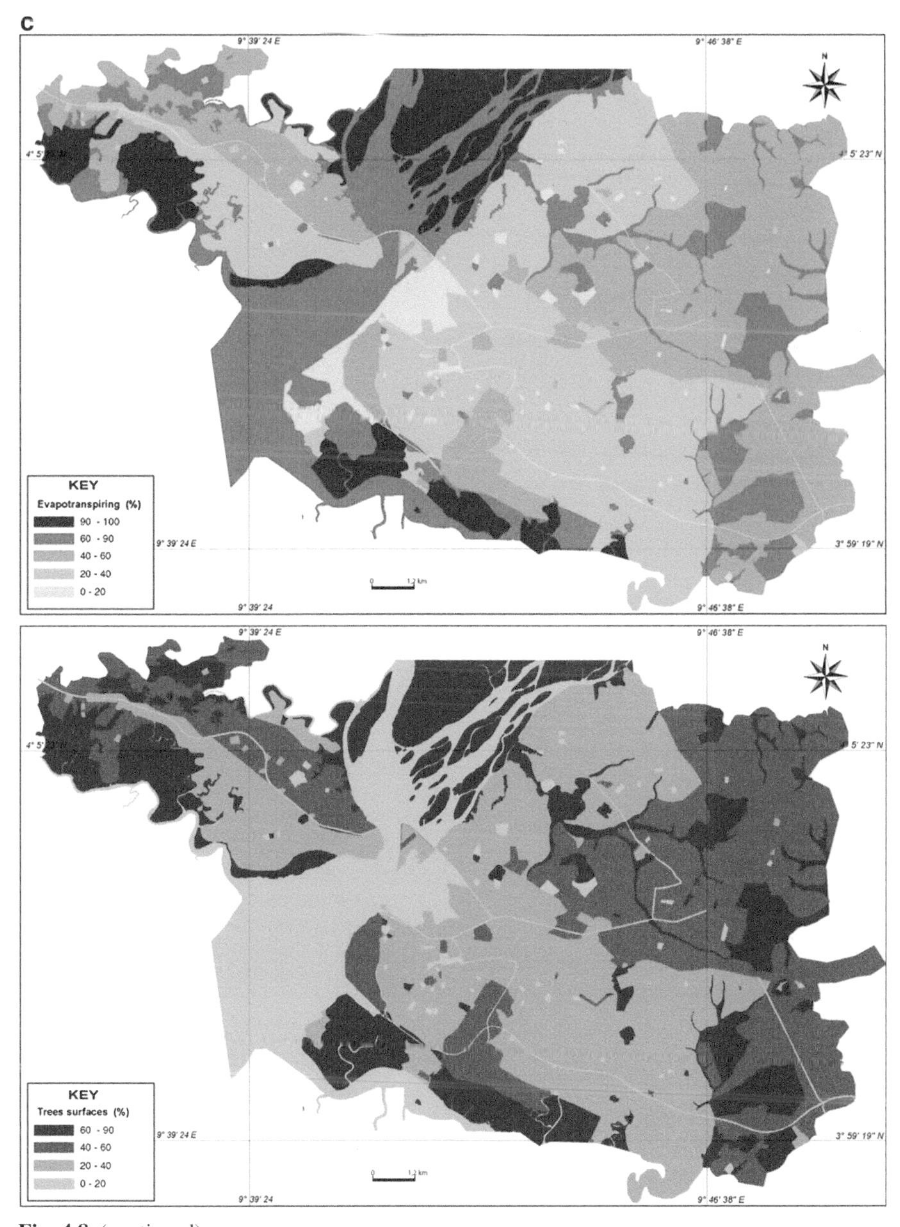

Fig. 4.8 (continued)

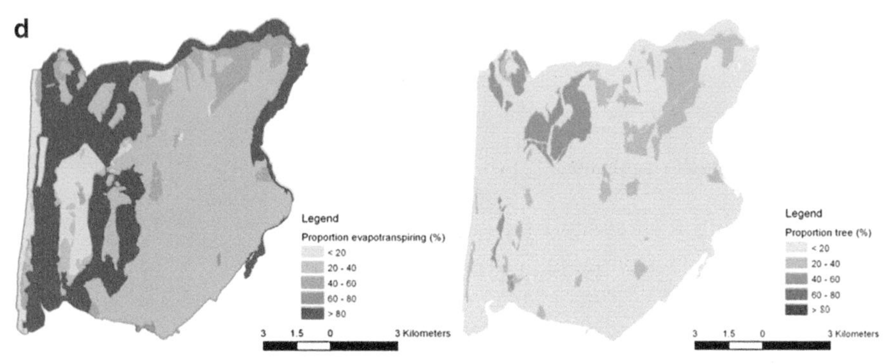

Fig. 4.8 (continued)

condominium and multi-storey zones are associated with a lower proportion of buildings compared to mud/wood[2] residential zones. It is interesting to note that in Addis Ababa, condominium areas are associated with a considerably lower mean percentage of vegetation cover and more bare ground compared to mud/wood construction areas. This may be due to many such constructions being partly or recently developed. This has a bearing on results presented later in the chapter.

Urban Ecosystem Multifunctional Assessment

The UMT framework was used as the thematic and spatial framework for an assessment of multiple ecosystem services (Table 4.4) in Addis Ababa, Dar es Salaam and Saint Louis. The local researcher teams decided whether to include a service based on the wider CLUVA objectives and local knowledge basis. Therefore, although services like air pollution abatement are important, they were not explicitly considered in this assessment.

The method is developed from green infrastructure planning approaches used by The Mersey Forest in the North West of England (The Mersey Forest 2010) and the ‘simple matrix’ for an initial assessment of multifunction potential (MacFarlane 2007). Each ecosystem service was considered in turn and a judgement made regarding the extent to which a particular UMT category (a) already supplies the service and (b) should ideally supply the service, i.e. there is an identifiable

[2] The mud/wood residential class is used to differentiate areas where the majority of dwellings are constructed using less formal construction methods and materials than is generally found with some other classes, e.g. condominiums. Dwellings in the mud/wood class may follow an irregular pattern and have few or no services. It should be noted that housing may not be constructed with mud/wood materials specifically – see Chaps. 3 and 6 – but materials are expected to be among the least resilient in each city. Finally, it should be noted that although this class is likely to be associated with informal settlements, not all areas associated with other residential classes can be assumed to be formally planned.

Table 4.4 Ecosystem services considered in Addis Ababa, Dar es Salaam and Saint Louis for the multi-functionality assessment

Service		Addis	Dar	Saint Louis
Provisioning	Food	X	X	X
	Wood (non-fuel wood) and fibre	X		
	Water (irrigation)	X	X	X
	Fuel	X	X	X
	Fresh water (drinking)	X		X
	Medicinal resources	X		X
	Ornamental resources			X
Regulating	Temperature control – shade	X	X	
	Temperature control – evaporative cooling	X	X	
	Temperature control – cool/fresh air corridors	X		
	Carbon sequestration and storage	X	X	X
	Flood – urban surface water regulation	X	X	X
	Flood – river	X		X
	Erosion regulation (non-coastal)		X	X
	Water purification/waste treatment			X
	Coastal erosion regulation			X
Cultural	Recreation	X	X	X
	Livelihoods		X	X
	Tourism		X	X
	Spiritual/religious values			X
	Educational			X
	Aesthetics and inspiration			X
	Heritage and sense of place			X
	Psychological/other aspects of health/wellbeing			X
Supporting	Species habitats	X	X	X
	Maintaining genetic diversity			X
TOTAL number of ecosystem services considered		14	12	22

need for it in the specific case of the UMT being considered. Scores were allocated according to the scheme shown in Table 4.5.

The judgements were made based on scientific and local evidence, local experience, and an understanding of the character of the UMT being considered, including its green structures. Where possible, conditional criteria were also specified as a means of differentiating the extent of supply of or need for the service within one or more UMTs, e.g. a higher need for erosion control in UMT units located on steep slopes compared to those located in flat areas. The results of the

Table 4.5 Scoring system used to assess the supply of, and need for, a service within the UMT classes

Score	Supply of service	Need for service
0	The service is never supplied by this UMT	The service is never needed in this UMT
1	The service is rarely supplied by this UMT, but may sometimes be supplied by individual UMT units or sub-sets of units	The service is rarely needed in this UMT, but may sometimes be needed in individual UMT units or sub-sets of units
2	The service is usually supplied by this UMT, but may not be supplied by all individual UMT units or sub-sets of units	The service is usually needed in this UMT, but may not be needed in all individual UMT units or sub-sets of units
3	The service is always supplied by this UMT	The service is always needed in this UMT

process can be mapped onto the UMT framework in order to give a broad view of the distributions of services. Finally, a three point scale qualitative confidence flag was also used in order to allow some estimation of the relative uncertainty in allocated scores. This identifies where further empirical evidence or other expert input may be required, but also where there was internal disagreement within the assessment team due to conflicting evidence or recognised conflicts of interest. A final quality check was undertaken by one researcher in order to identify possible inconsistencies and throughout the process a full documentation of decisions and their rationale was carried out. The results presented here are only a provisional 'first look' at using this assessment process generated by a rather small set of local experts. In each case, teams felt that the results would benefit from further revision, especially where perspectives on the supply of and need for services differed or where there was a lack of evidence about a particular service either generally or within particular UMTs. Nevertheless, the teams felt this was a useful exercise which could help inform green infrastructure planning activities in their cities.

On average, the assessment suggests that the supply of ecosystem services is lower than the need for the services in all cities. For Addis Ababa the mean score for supply of all services was 0.9 compared to a mean of 1.6 in terms of need. For Dar es Salaam the respective values were 0.8 and 1.2 and for Saint Louis 0.9 and 1.7. Although the services assessed had some local variation, it is possible to compare the mean scores for selected UMTs held in common. This is particularly the case for the mean scores for regulating services as these have the most similarity for the services considered. However, comparative scores for provisioning and total services are also included.

Figure 4.9 suggests that in both Addis Ababa and Dar es Salaam residential areas need to supply further services to their neighbourhood populations and that multi-functionality could be an important strand of a green infrastructure plan in these areas. This is particularly marked in relation to regulating ecosystem services in Addis Ababa. In contrast, many of the 'green' UMTs associated with patches of green structures within the urban matrix or green structures in the peri-urban areas (Fig. 4.10) already supply something near to requirements. An exception is riverine areas in Addis Ababa which were judged to be slightly underperforming at present.

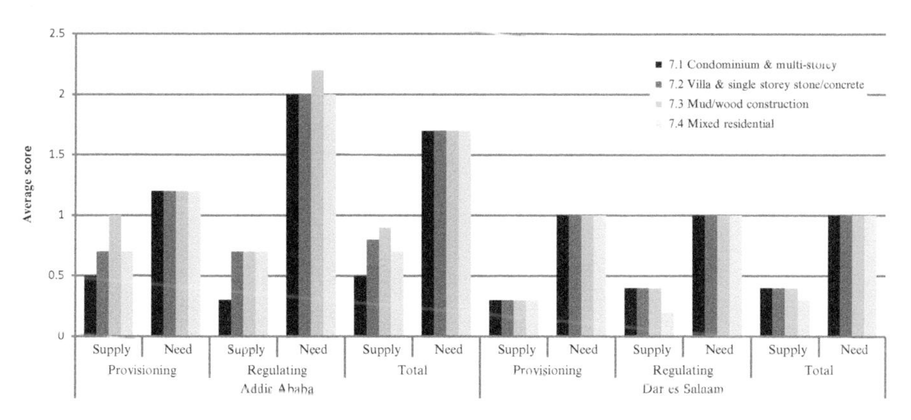

Fig. 4.9 Mean scores for ecosystem service supply and need for selected *Residential* UMT sub-categories in Addis Ababa and Dar es Salaam

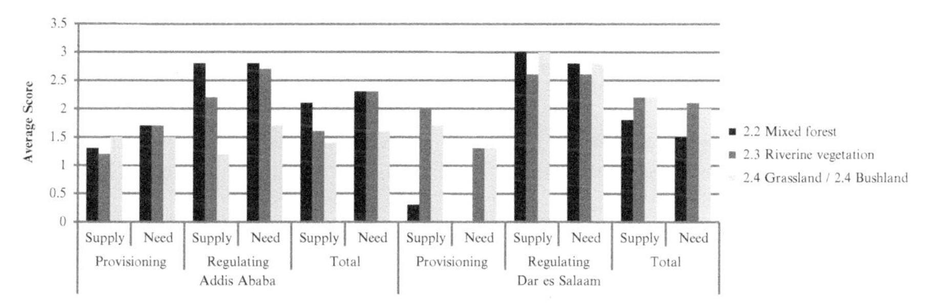

Fig. 4.10 Mean scores for ecosystem services supply and need for selected *Vegetation* UMT sub-categories in Addis Ababa and Dar es Salaam

It is also notable that while some provisioning services are required of mixed forest in Addis Ababa, the judgement of the team in Dar es Salaam was that their mixed woodland did not necessarily need to have a provisioning function, albeit in terms of the more limited set of services considered. Interestingly, in Dar es Salaam some 'green' UMTs were judged to be over-supplying services, which may be a reflection of the pressures on these areas and the desire to protect areas from deterioration.

Addis Ababa Based on the highest allocated scores, the most important functions of green infrastructure across all UMTs in the city were judged to be those associated with supporting (habitats) and regulating services. Regulating services and cultural services (recreation) show the largest gap between current supply of and need for services. The lower scores for provisioning services recognise that these functions are supplied by a smaller set of categories and that, while many categories need to supply these functions, this does not necessarily apply to all units within each of the UMT classes assessed. The *plantation*

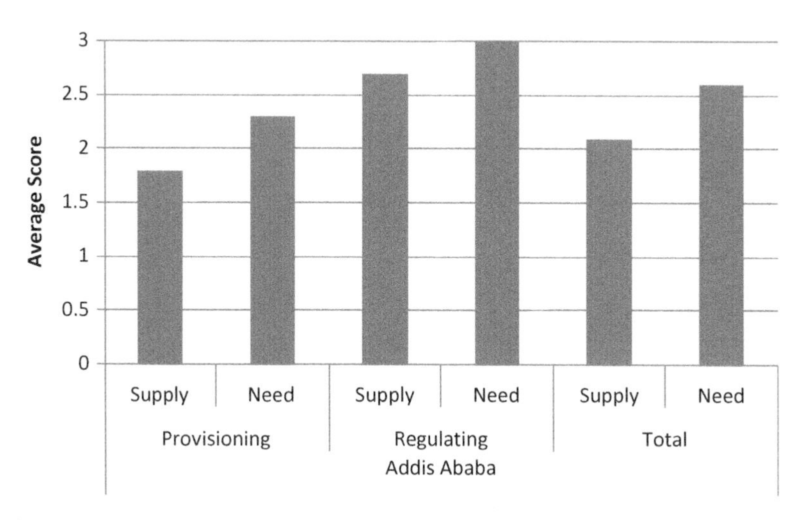

Fig. 4.11 Services supplied by and needed from the *Plantation* UMT in Addis Ababa

UMT is clearly shown as a vital part of any green infrastructure plan for Addis Ababa given the supply of multiple ecosystem services and the need to supply still more (Fig. 4.11).

In general, the North of the city and river corridors into the cities supply the most services, whilst the need for the services tends to extend more into the urban areas (Figs. 4.12 and 4.13). A high number of needs are met in the North of the city, and in *parks*, *botanic garden*, *mixed forest*, *plantation*, and *grassland*. This suggests that these UMTs, which are all green in nature, are generally good at performing what is needed of them. On the other hand, a high number of needs are not met especially in the urban areas, in the *Residential* UMTs, *mineral workings*, *cemeteries*, *manufacturing*, *garage* and *river* areas.

Dar es Salaam In general, the green structures in Dar es Salaam need to supply more services, or services to a greater level. The highest average scores are for regulating services, followed by supporting and cultural, with the lowest scores for provisioning services. The biggest gaps between the supply of and need for ecosystem services are seen for regulating services followed by the provisioning services. Overall, *riverine*, *bushland*, and *river* supply the most services and *riverine* has the greatest need. A high number of needs are met by *field crops*, *mixed farming*, *riverine*, *bushland*, *parks*, and *river*. As with Addis Ababa the 'green' UMTs generally supply what is needed of them, as can be seen in the case of *mangroves* (Fig. 4.14). Also, as with Addis Ababa, a clear unsatisfied need for services is found within the urban core (Fig. 4.15).

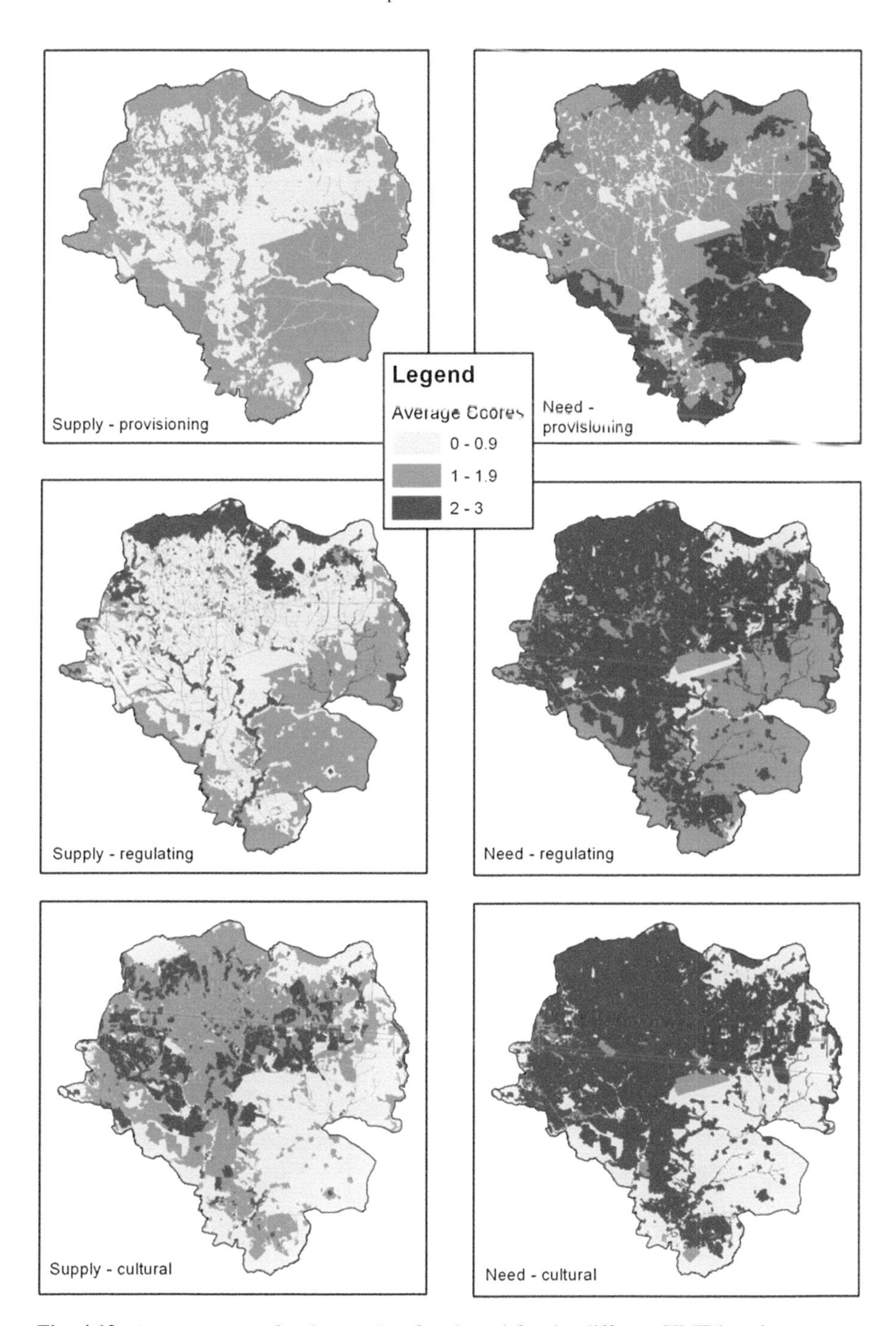

Fig. 4.12 Average scores for the supply of and need for the different UMT based ecosystem services (provisioning, regulating, cultural, supporting, as well as total services) in Addis Ababa

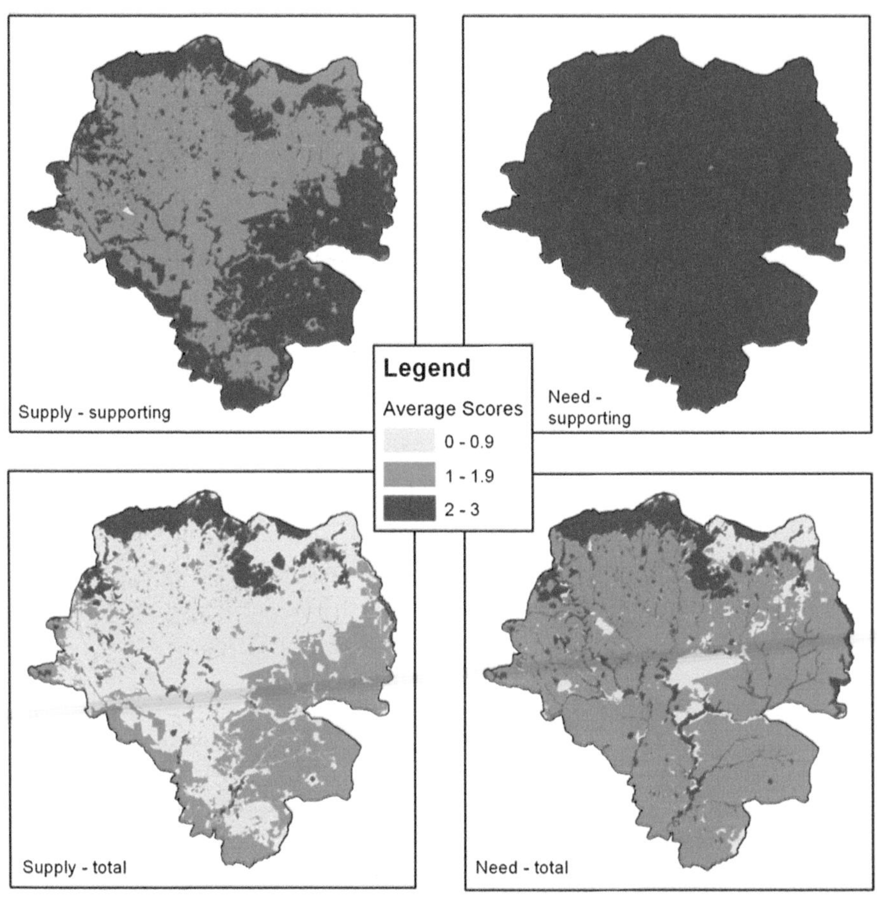

Fig. 4.12 (continued)

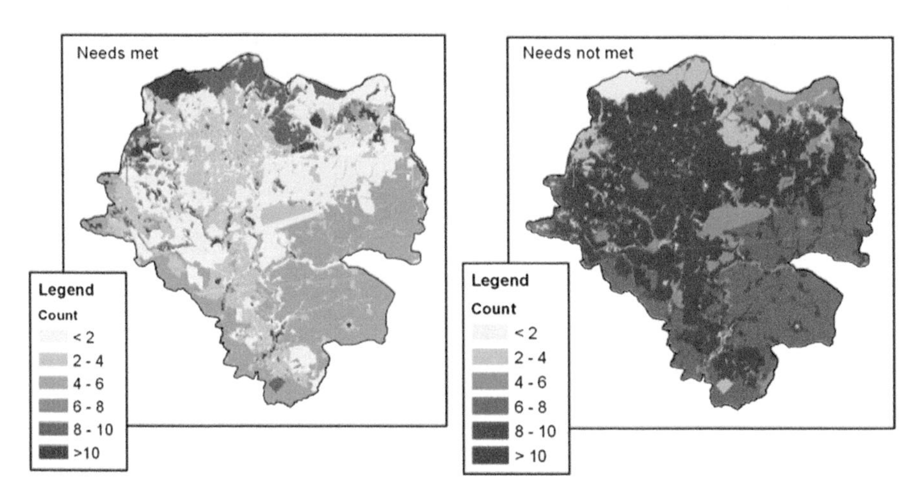

Fig. 4.13 Total number of ecosystem service needs that are met (i.e. there is a need for the service and it is supplied) or not met (i.e. there is a need for the service but it is not supplied) in Addis Ababa

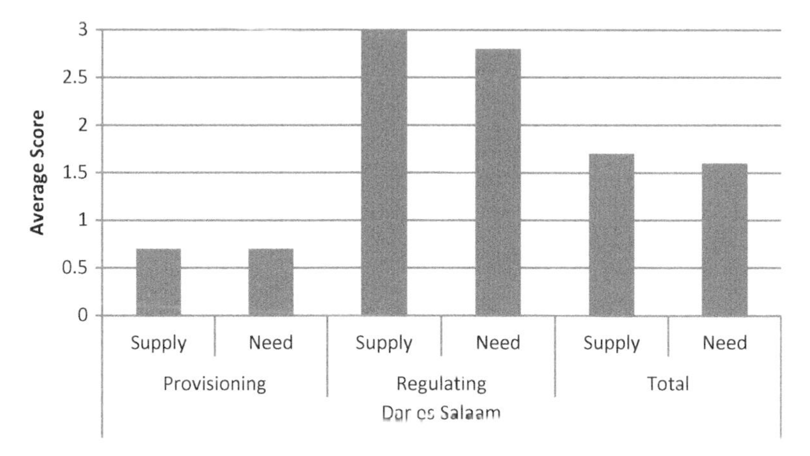

Fig. 4.14 Services supplied by and needed from the *mangroves* UMT in Dar es Salaam

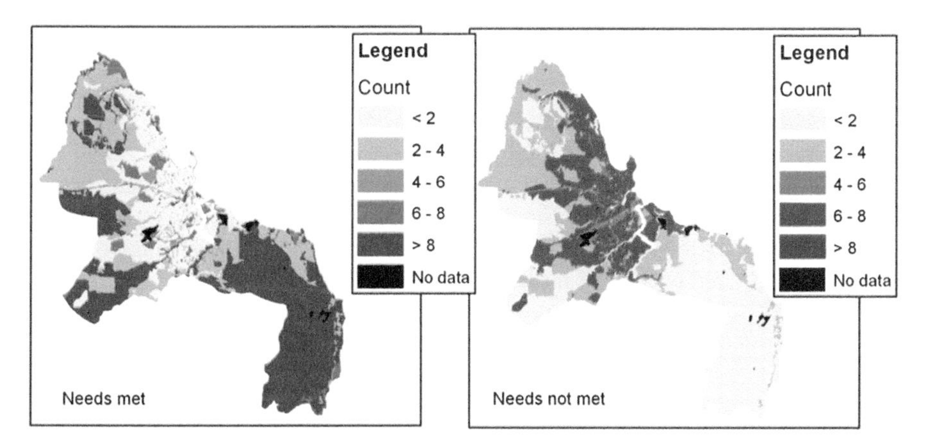

Fig. 4.15 Total number of ecosystem service needs that are met (i.e. there is a need for the service and it is supplied) or not met (i.e. there is a need for the service but it is not supplied) in Dar es Salaam

Building the Evidence Base

More detailed assessment work has been carried out for selected ecosystem services within the provisioning, regulating and cultural service groups.[3] The full details of this assessment work and the methodologies associated with each are given in Lindley et al. (2013) with examples of wood provisioning and regulating services

[3] The work in this section draws particularly from PhD work by Florian Renner (Technical University of Munich) (woody vegetation) and Deusdedit Kibassa (Ardhi University) (regulating services, specifically air temperatures). Work on cultural services is reported in Lindley et al. (2013).

(Dar es Salaam) and temperature regulation (Addis Ababa and Dar es Salaam) at the city-scale given here. The more detailed assessment work helps to evidence the judgements of the multifunctional ecosystem services assessment and provide a means through which it can be subsequently refined.

Ecosystem Services at the City Scale

Ecosystem Services Provided by Woody Vegetation – Dar es Salaam

Although technically within the vegetation zone of the Miombo woodlands, the vegetation of Dar es Salaam has long been described as 'Forest transition and other forests' (White 1983), and therefore may not necessarily replicate the same wide set of ecosystem services that are reported in the literature for Miombo woodlands (Syampungani et al. 2009). Accordingly, this element of the work involved further characterisation of the urban forest structure of Dar es Salaam, assessment of forest uses and estimates of the growing stock (total tree volume), building on the UMT-based land cover assessment work for Dar es Salaam. The growing stock estimation was carried out through adapting an approach developed by the Food and Agriculture Organisation (FAO 2011) (Fig. 4.16). The derived estimate was then converted into biomass by multiplying it by the mean wood density of ten of the most common non-palm tree species found in the city. Associated carbon storage was estimated using the Stock-Difference Method (Eggleston et al. 2006).

In 2008 Dar es Salaam had, on average, 15.6 m^3 ha^{-1} in merchantable wood stock related to tree cover (2.3 million m^3). Approximately 77 % of the calculated

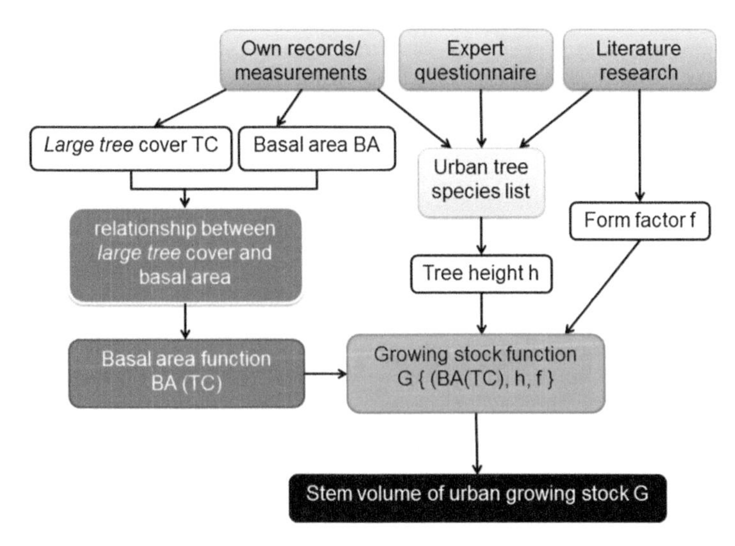

Fig. 4.16 Overview of the adapted FAO approach for urban growing stock estimation (FAO 2011)

growing stock was found within only four sub-category UMTs (*mixed farming*, *villa and single storey*, *scattered settlement* and *mixed residential*). Overall *Residential* UMTs were estimated to contain half of the growing stock, and *agriculture* around a third. The density of merchantable stocks was estimated to be higher in the study area than in the Miombo woodlands elsewhere in Tanzania, but with considerable variation between sub-category UMTs. The total urban tree biomass was estimated at 2.2 million tonnes of dry matter, corresponding to about one million tonnes of carbon. This represents an average of 15 tonnes ha^{-1} of dry matter (7 tonnes ha^{-1} of carbon) (lower than Miombo woodlands (Malimbwi et al. 1994; Chidumayo 1990, 1997)). With an annual sequestration rate of 34.5 thousand tonnes of carbon, the large trees in the assessment area alone are estimated to be able to compensate for more than a quarter of Dar es Salaam's total carbon emissions. This finding clearly links the climate adaptation and mitigation agendas, the need for which is discussed further in Chap. 11 of this volume.

In addition to the specific ecosystem services assessed in this exercise it is important to note that trees also provide a range of other ecosystem services. For example, the Neem tree (*Azadirachta indica*) has many beneficial qualities in the urban environment because it is drought tolerant, good for erosion control and soil conditioning and has medicinal properties (in Swahili it is known as 'the tree that cures forty [diseases]') (BioNET-EAFRINET 2011). It is also used as a shade and street tree and has properties which act as natural pesticides, making its wood pest resistant (Schmutterer 1990). The Neem tree is also an interesting case because this introduced species is also noted for its invasive character along Tanzania's coastal zone (BioNET-EAFRINET 2011). This serves as a useful reminder of the potential disservices associated with green infrastructure that must be effectively managed as part of any green infrastructure plan (Lyytimaki and Sipila 2009).

Temperature Control – A Regulating Ecosystem Service

The value of urban green spaces in providing local climate regulation services is widely recognised (Bolund and HunHammer 1999; Gill et al. 2007). However, relatively little previous research on local climate regulation services in African cities exists (Cavan et al. 2011). The issue of temperature regulation in urban areas is connected to the Urban Heat Island (UHI) phenomenon, whereby air and surface temperatures are elevated in relation to surrounding rural areas. This phenomenon has been recognised for many years (Oke 1982) and its causes – lack of evapotranspiring surfaces, differences in albedo and the thermal properties of urban surfaces compared to rural surfaces, anthropogenic heat emissions and the effect of urban structures like canyons – have been the subject of much research (Landsberg 1981; Taha 1997; Smith et al. 2009, 2011; Cheung 2011). While European cities have a long history of urban climatological studies (MacKillop 2012), relatively little is known about UHI in the Sub-Saharan cities of Africa.

The following sections explore the surface temperature and air temperature characteristics of Dar es Salaam and the surface temperature characteristics of Addis Ababa in baseline and future climates using the UMT map and UMT-based land cover data as key inputs. The importance of considering temperature regulation functions in the two cities is underlined when considering the context of climate change. The number of heat-wave events with a maximum length of 5 days could increase from 3 (during 1950–70) to 24–33 (during 2030–2050) in Dar es Salaam[4] and from 3 to 32–40 in Addis Ababa[5] depending on the IPCC scenario used (CSIR and CMCC 2012; Cavan et al. 2014). The expected persistence of long-lived heat-waves lasting approximately 1.5–2 weeks is also expected to increase in the future with respect to the climatological period 1961–1990. Communities in Dar es Salaam already recognise increased temperatures as a climate risk; a recent survey rated increased temperatures as second only to flooding and rated above drought and increased disease (START et al. 2011).

Surface Temperatures[6] Modelled land surface temperatures have previously been used as an indicator for calculating energy exchange in the urban environment (Whitford et al. 2001; Gill et al. 2007). Analysis of land surface temperature is advantageous because it enables a spatially explicit depiction of the thermal state over large areas and land surface temperatures are often a good indicator of human thermal comfort and heat stress outcomes (Laaidi et al. 2012). The results presented in this section are generated using a model which enables consideration of future conditions and urban scenarios. The model is based upon an original model developed by Tso et al. (1990, 1991), subsequently developed and customised by Whitford et al. (2001) and later translated into a freely available online tool ('STAR tools') (The Mersey Forest and The University of Manchester 2011). It expresses the energy balance of an area in terms of its surface temperature on a hot, cloud free day using a set of inputs associated with climate, building mass and evapotranspiring fraction.

Simulations of land surface temperatures were undertaken for 1981–2000 (baseline) and 2021–2050 (future) for Addis Ababa and Dar es Salaam. Climate projection data was obtained from CLUVA climate simulations, performed for the period 1961–2050 for the A2 IPCC emissions scenario (CSIR and CMCC 2012). Six projections were obtained from downscaling six different coupled models; all ensemble models were analysed, and the 50th percentile calculated. As the focus is on hot, cloud free

---

[4] Dar es Salaam has an equatorial savannah climate, generally hot and humid throughout the year with dry summers (Aw – Köppen-Geiger) (Kottek et al. 2006). Climate change projections for Dar es Salaam (for 2041–2050 relative to 1961–70) indicate no significant changes in the seasonality of rainfall, but potentially significant increases in rainfall in the March-May "long rains", and seasonal temperature increases around 1.5–2 °C (CSIR and CMCC 2012).

[5] Addis Ababa experiences a warm temperate climate with dry winters and warm summers (CwB – Köppen-Geiger) (Kottek et al. 2006). Climate change projections for Addis Ababa (for 2041–2050 relative to 1961–70) indicate no significant changes in the seasonality of rainfall, but slight changes in monthly rainfall and potentially significant increases in rainfall amounts during March to May (CSIR and CMCC 2012). Projected increases in seasonal temperatures are in the region of 1.5–2 °C (CSIR and CMCC 2012).

[6] The text in this section contains excerpts reprinted from Ecol Ind, 42, Cavan G, Lindley S, Jalayer F, Yeshitela K, Pauleit S, Renner F, Gill S, Capuano P, Nebebe A, Woldegerima T, Kibassa D, Shemdoe R, Urban morphological determinants of temperature regulating ecosystem services in two African cities, pp. 43–57, ©2014 with permission from Elsevier.

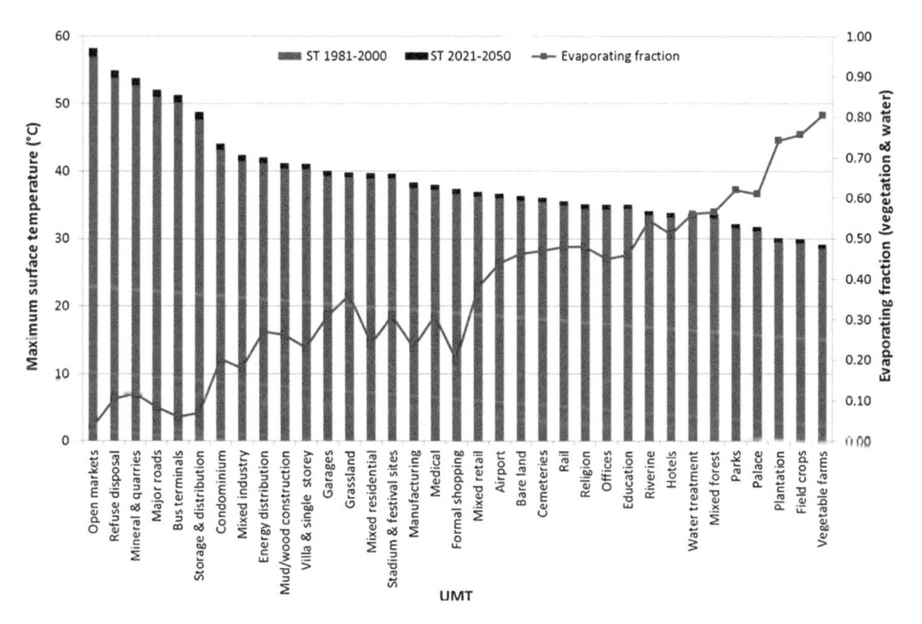

Fig. 4.17 Modelled maximum surface temperatures and evaporating fraction (green space and water) by UMT for Addis Ababa, for 1981–2000 and 2021–2050 (Reprinted from Cavan et al. 2014, ©2014 with permission from Elsevier)

days, the 98th percentile mean temperature in the short dry season (December-February) was used as the reference air temperature.[7] A weighted built mass was calculated for each UMT class. This used the proportion of roads, buildings (including those associated with formal and informal settlement areas), and impervious surfaces within each UMT category, as determined from the land surface cover assessment. Land cover data also provided the evapotranspiring fraction (Fig. 4.8) and it is assumed that vegetation is not affected by water stress. Broad differences in elevation were accounted for through the input climatological data for Addis.

The results for different UMTs are shown in Figs. 4.17 and 4.18. There are similarities across both cities, for example *open markets* and *major roads* are among the UMTs with the highest temperatures and *parks* among those with the lowest temperatures. In Addis Ababa, informal settlements (*mud/wood construction*) generally have higher proportions and better composition of green structures than other residential areas. As a result they are associated with the lowest modelled land surface temperatures and the most temperature regulating services. In Dar es Salaam, on average the *condominium* UMT has some of the largest proportions of green structures, and the best provision of temperature regulation services (Fig. 4.17). The quality of green structure is also important

[7] In Dar es Salaam this timeframe includes the end of the period of short rains termed in Swahili as *vuli* and the subsequent 'short dry season' when peak temperatures are expected. The main dry season is June-early October and the main rainy season March–May.

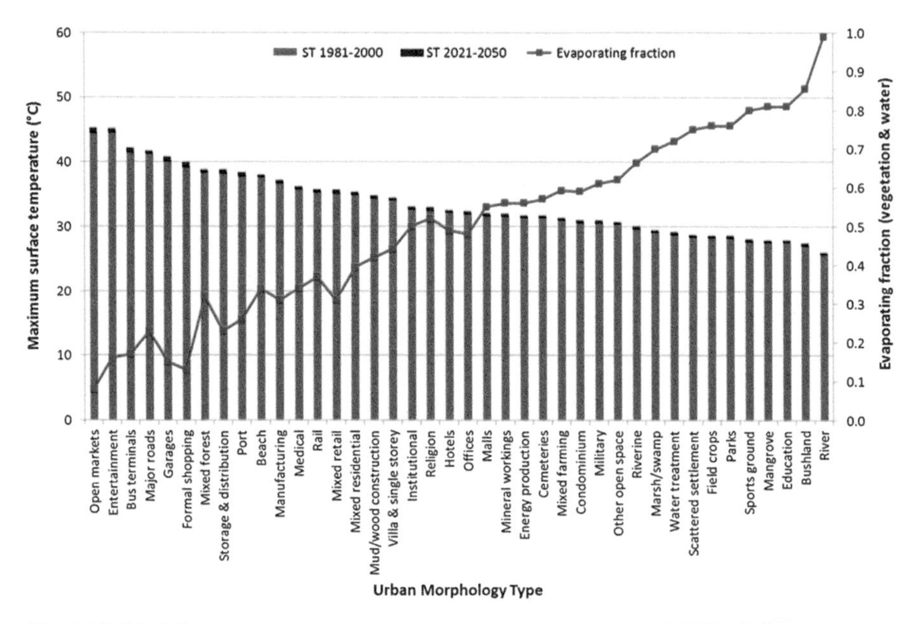

Fig. 4.18 Modelled maximum surface temperatures and evaporating fraction (green space and water) by UMT for Dar es Salaam, for 1981–2000 and 2021–2050 (Reprinted from Cavan et al. 2014, ©2014 with permission from Elsevier)

in determining the effectiveness of temperature regulation services provision. Particular combinations of green structures, such as trees over grass, are more effective as they provide cooling through evapotranspiration and also shade. The energy exchange model is not detailed enough to consider the composition of green structure types and their effect on surface temperatures, but the land surface cover assessment provides additional details about the relative proportions of different green structures within the UMTs. It is also recognised that there can be considerable variation of surface temperatures within UMT categories and that they are also affected by other local factors, which similarly affect the extent to which surface temperatures translate into air temperatures or physiologically equivalent temperatures. Nevertheless, the results suggest the existence of a surface temperature UHI in both cities.

Both climate change and urban development alter surface temperatures. However, land surface cover differences are associated with land surface temperature ranges of more than 25 °C, whereas the range for climate change projections is less than 1.5 °C (Figs. 4.19 and 4.20). Therefore, in terms of local temperature change, urban morphological change has the potential to have a much greater effect overall than temperature increase due to climate change. Increasing green structure cover in the 'urban matrix' is likely to considerably offset climate change induced increases in temperature. Green structures also provide other temperature regulating benefits for urban residents, including shade and the provision of cool air corridors, neither of which is considered in the model. Further details of the methods and results are given in Cavan et al. (2014).

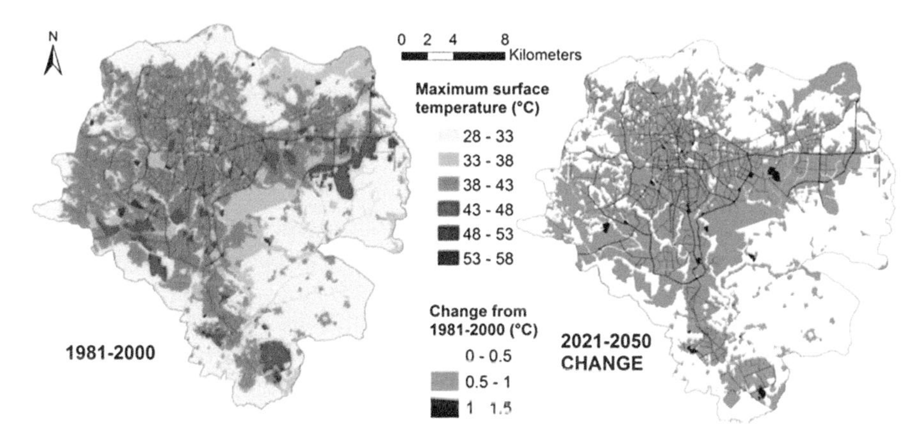

Fig. 4.19 Modelled maximum surface temperatures 1981–2000 and changes to 2021–2050, for Addis Ababa (Reprinted from Cavan et al. 2014, ©2014 with permission from Elsevier)

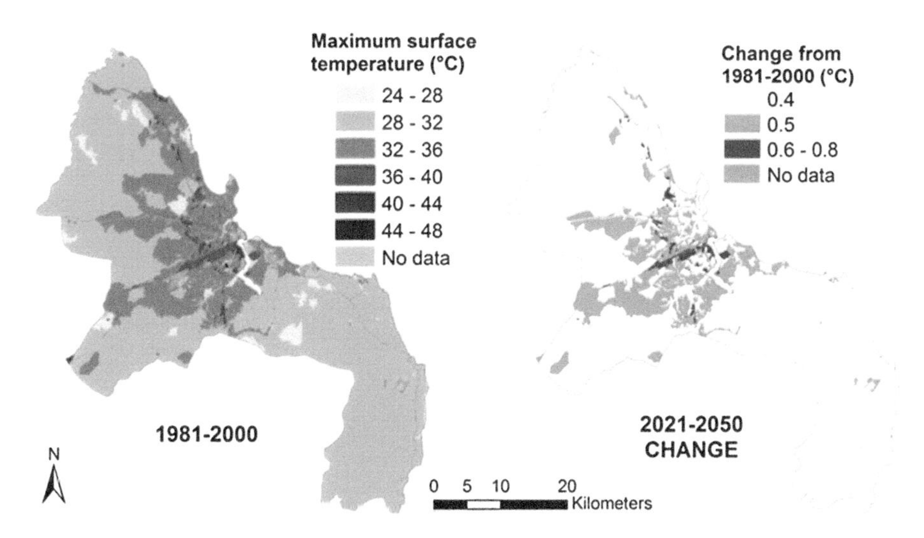

Fig. 4.20 Modelled maximum surface temperatures 1981–2000 and changes to 2021–2050, for Dar es Salaam (Reprinted from Cavan et al. 2014, ©2014 with permission from Elsevier)

Air Temperatures Given the existence of a surface temperature UHI, it is useful to explore the extent of any associated air temperature UHI, the variability of UMT-based air temperatures across the study area and how far these may be associated with green structures.

Some evidence has been published of an air temperature UHI in Dar es Salaam (Jonsson et al. 2004) and of the complex meteorology associated with the city's coastal location (Nieuwolt 1973). A study of the wet (taken as May) and dry seasons (taken as October) in 2001 using a 'garden' site, 1 km from the coast, as an urban reference and the international airport as a rural reference found a wet season UHI of 2–4 °C and a dry season UHI which was negative during the day (attributed to sea breeze) and around 4 °C during the night (Jonsson et al. 2004). In a study of three Dar es Salaam neighbourhoods, sky view factor[8] was found to be the most influential factor governing longwave incoming radiation with a green urban neighbourhood seeing slightly higher sky view factor and higher temperatures compared to a 'traditional' neighbourhood (Jonsson et al. 2006). However, to the authors' knowledge, there has been no wider monitoring of intra-urban temperatures using a network of in situ stations, nor a more systematic assessment of associations with surface cover.

Air temperature observations were made at 10 min intervals between 30th June 2012 and 31st January 2013 using 64 monitoring sites located in different UMTs across two transects of northern Dar es Salaam. Each site consisted of a calibrated low cost iButton sensor[9] in a wooden shield mounted on a pole at a height of 4 m above the ground following Cheung (2011). Sensors were sited at locations in the city centre and out to the peri-urban zone of the city. The international airport was used as the rural reference in previous studies (Jonsson et al. 2004), but is no longer outside of the city boundaries. In the absence of any other appropriate rural reference, the sensor located in Kibamba, a *scattered settlement* UMT unit located about 24 km from the city centre, was taken as the rural reference point. Data processing included dividing data into appropriate periods for analysis – namely, the main dry season (June to September/early October), and the period of short rains, termed in Swahili as *vuli*, from October to December/early January. Given that the air temperature UHI can be expected to be highest at night, particular attention has been paid to analysing nocturnal temperature data. Some results are presented using the Urban Heat Island Intensity (UHII) metric, where:

$$\text{UHI Intensity (UHII)} = \text{Temperature}_{\text{urban}} - \text{Temperature}_{\text{rural}}$$

Analysis of the collected data is ongoing and it is only possible to present some initial snapshots in this chapter. Figure 4.21 plots observed temperatures associated with different UMT units of similar average surface temperature characteristics. The units differ primarily due to their distance from the urban core along the Morogoro Road (A7).[10] The units are:

- Kariakoo, *mixed residential* 1.5 km (19 m a.s.l.) from the city centre;
- Magomeni usalama, *villa and single storey* 4 km (23 m a.s.l.);

[8] 'The ratio between radiation received by a planar surface and that from the entire hemispheric radiating environment' (Svensson 2004: 201), i.e. the portion of sky which is free from obstructions like buildings and therefore visible from a point on the ground (Watson and Johnson 1987).

[9] Sensors are 1-Wire (Maxim Integrated) battery operated loggers with a fixed and limited time span. Primarily designed for internal use, such as monitoring of container goods, they can be housed in appropriate environmental enclosures for external use. Their main benefit is that their very low unit cost allows a good coverage of sensors for a limited resource input.

[10] Local land cover characteristics may also vary and this is still to be investigated.

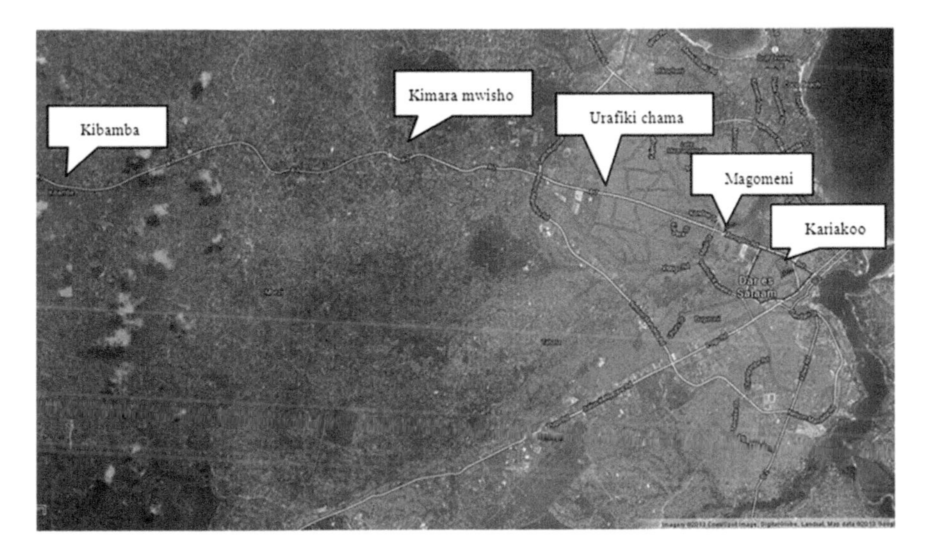

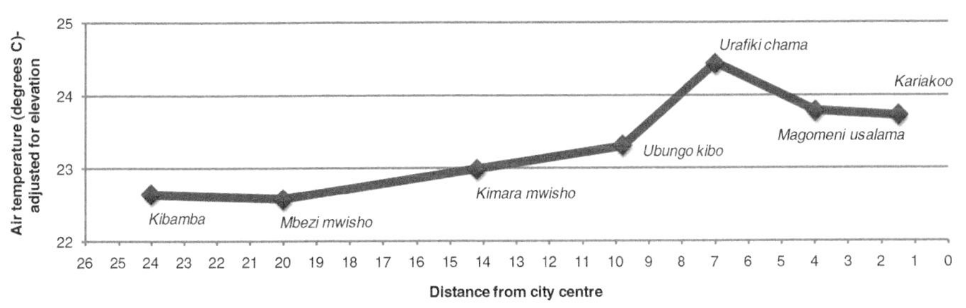

Fig. 4.21 Adjusted air temperatures in selected UMT units at different distances from the city centre at 1 am, 16/8/12 (*bottom*) and locations of selected sensors (*top*) (Data sources: Deusdedit Kibassa; Google Maps 2013)

- Urafiki chama, *mixed residential* 7 km (41 m a.s.l.);
- Ubungo kibo, *mixed residential* 10 km (67 m a.s.l.); and
- Kimara mwisho, *villa and single storey stone* 14 km (132 m a.s.l.).

These were all compared to the rural reference point in Kibamba, a *scattered settlement* UMT unit located at 24 km from the city centre (and at approximately 152 m a.s.l). The temperature data were recorded during the dry month of August on 16th August 2012 at 1 am. The data suggest a nocturnal air temperature UHII of at least 1 °C (1.1–1.8 °C) after adjusting for elevation. Adjustments were made using a standard atmospheric lapse rate of 6.4 °C km^{-1} under the assumption that this rate can adequately represent near-surface values and that it is representative of the atmospheric conditions in the study area. It is notable that the highest temperature is recorded at Urafiki chama, away from the coast but still within the main urban core of the city. A closer investigation of the land cover characteristics around the site was carried out using 50 randomly located points within a 50 m

Fig. 4.22 Location of the sensors and the surrounding UMTs Mikocheni nyumba za mawaziri (*left*) and Sinza madukani (*right*)

radius. This revealed that the area around the site is 30 % buildings with 20 % other impervious and 50 % dark bare ground cover. These characteristics would be consistent with expected elevated temperatures. However, further investigation of the characteristics of other sites, data taken for a longer time frame and supplementary data on the specific meteorological conditions are required to explain this further.

Some initial comparisons of UMTs with different land cover characteristics are also made. Here two different UMT units located at the same distance from the city centre in different directions are selected. They are at similar elevations and are estimated to have similar surface temperature profiles, on average, across Dar es Salaam as a whole. However, the two specific units themselves are in practice quite different (see Fig. 4.22). The UMT units are Sinza madukani which is a *mixed residential* unit located at 7.7 km to the western side of the city (Fig. 4.22, right) and Mikocheni nyumba za mawaziri which is a *villa and single storey stone* unit located at 7.7 km to the northern side of the city from the city centre (Fig. 4.22, left). Figure 4.23 presents the observed nocturnal temperatures for a day taken in the dry month of August 2012. Temperatures at Sinza madukani were up to 1 °C higher than in Mikocheni nyumba za mawaziri. It is also noteworthy that geographically Mikocheni nyumba za mawaziri is located nearer to the coast. It may also benefit from cooling from that source, except that during August night-time winds are likely to be very calm or affected by a weak nocturnal land breeze (Nieuwolt 1973). When active, sea breezes tend to be maximised in early afternoon, can persist through the night during the hottest parts of the year (January to March) and can reach well inland provided that sites are sufficiently exposed (Nieuwolt 1973). Beach zones (within 60 m of the coast) are publicly owned according to the National Environmental Management Act (NEMA) of 2004, and therefore, should be accessible to all and help in the process of ensuring ventilation. However, in reality, there are developments along the coast which act to both block the inland penetration of sea breeze and also restrict access.

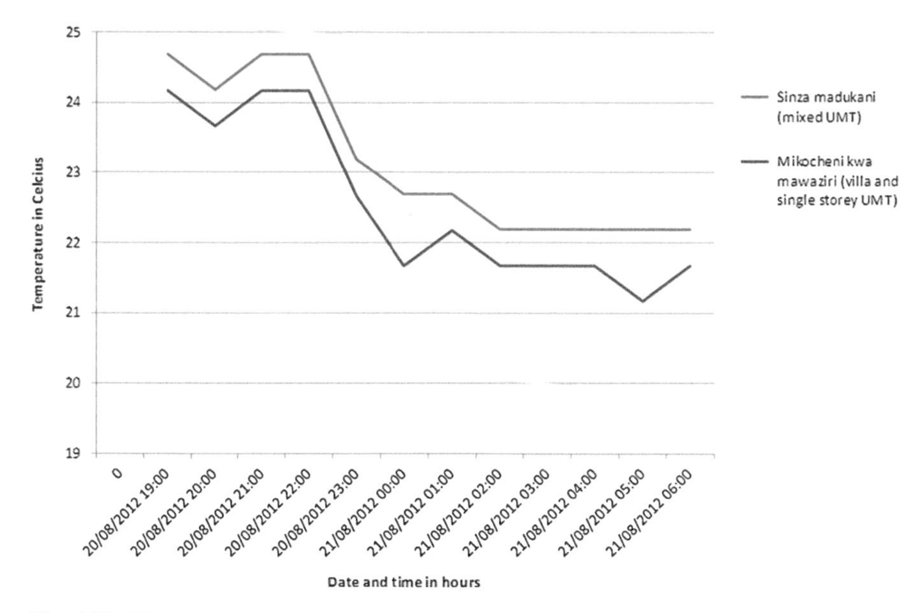

Fig. 4.23 Observed hourly nocturnal air temperatures in similar UMT units at equal distances from the city centre but different land cover characteristics

Threats to Ecosystem Services

Urban ecosystem services are made vulnerable through a range of human and physical factors and it is only through considering both that a full picture of the threats to ecosystem services can be assessed. This section discusses the implications of development and climate change pressures on green structures in Addis Ababa and Dar es Salaam.

Development Pressure

Past change and future prospects for development were assessed for Addis Ababa and Dar es Salaam. Past change was estimated by repeating the UMT and land cover assessments for two years: 2011 and 2006 for Addis Ababa; and 2008 and 2002 for Dar es Salaam. The retrospective assessment of change helped to inform the development of a grid-based spatial scenario model for each city aiming to estimate the impacts of several potential development and urban growth scenarios on green structures and their ecosystem services. Four scenarios were modelled in increments of five years based on two main factors that can be influenced by planning: the population density (low or high density) and settlement location in high flood risk areas (allowed or prohibited). The models follow the general principle of cellular automata, where the state of each grid cell depends on the state of the cells within a neighbourhood and changes are determined via a set of rules (White et al. 2000; de Smith et al. 2013). As well as being empirically informed, rules governing changes between each time step were also developed through a process of stakeholder engagement. This allowed the relative attractiveness of each cell for settlement to be established. Rules were based on: past land change dynamics, proximity to the city centre and sub-centres, proximity to the road network, proximity to existing settlements, location of known urban development projects and steepness of slope (Abo El Wafa 2013a, b; Buchta 2013a, b).

In terms of the investigation of retrospective change, there is clear evidence of continued depletion of green structures and a rapid pace of change in both cities. Although rates of change in some cities are not as large as have been reported in the past (e.g. Mung'ong'o 2004; Lupala 2002), this is set against a backdrop of higher populations and a lower baseline urban ecosystem resource, i.e. more need for ecosystem services and less opportunity to satisfy that need.

For Dar es Salaam, 'green areas' (consisting of the *mixed forest*, *riverine*, *bushland*, *mangrove*, *marsh/swamp*, *parks*, and *other open space* UMTs) saw losses of 3,476 ha, corresponding to 30.1 % of the 'green area' present in 2002. Within the next decade or two, discrete stands of *mixed forest* could disappear entirely given that only 176 ha are currently remaining and 74 ha (29.6 %) have been lost since 2002. Between 2002 and 2008 most of the 'green UMT area' which has been lost has been converted into *farmland* (3,060 ha), followed by *settlement areas* (738 ha). Both urban sprawl (increasing areal extent of low density development) and densification of the central zones have been identified. As a result, green structure has decreased in both the peri-urban zone (25–50 km from the city centre) and urban core (associated with the 'urban matrix') (<15 km from the city centre). The former has city-wide implications for provisioning services and future food security and the latter for regulating and cultural services, such as human livelihoods, health and wellbeing, including local capacity for provisioning services in some of the poorest neighbourhoods. Analysis of changes in land surface cover profiles within UMTs is important for understanding more subtle and gradual changes in vegetated cover, which can often go without detection. The results suggest that there is also a general loss of *woody cover* of between 9 and 12 %. This is independently corroborated by local people in Bonde and Suna who talk of the problems of high density living (e.g. people cutting down Neem trees which are thought to interfere with buildings). As a result of these changes there was one million m^3 less stem wood in 2008 compared to 2002, with the largest losses in *scattered settlement*, *mixed farming* and *mixed residential*. Around 64 % was lost from *riverine* UMTs. Around a third of growing stock was lost between 2002 and 2008 and the density of large stem wood declined by 8.5 $m^3\ ha^{-1}$. Large tree cover tended to be associated with patches and all tree cover was seen to be subject to a high rate of loss of 11.3 % per annum (2002–2008). These losses are clearly associated with a lower capacity for provisioning, regulating (such as local temperature regulation) and supporting ecosystem services. It is also likely that their removal will increase the intensity of hazard events, although this is not assessed here.

Future scenario modelling under the assumption of business as usual suggests that land associated with settlements could change from 47 to 53 % of the study area between 2008 and 2025, an increase of 9,769 ha. The expansion of settlements is estimated to be associated with a loss of 6,886 ha of farmland (which consists of *field crops*, *horticulture*, *mixed farming*) and 2,352 ha of green areas (which consists of *mixed forest*, *riverine*, *bushland*, *mangrove*, *marsh/swamp*, *parks* and *other open space*). The business as usual case suggests a high concentration of settlements in flood prone areas by 2025: 1,025 ha in areas classed as *riverine*. The largest total losses within a specific UMT sub-category are estimated to occur in *mixed farming* (6,485 ha), followed by *riverine* areas (1,025 ha) and *bushland* (898 ha). Under the assumptions of this scenario 238 ha of *mangroves* would be transformed into settlements. The business as usual case would therefore result in vast losses of green areas and farmland and the ecosystem services that they provide. In terms of relative changes, *riverine* areas would be expected to lose 60 % of their area, *mangroves* 51 %, and *other open spaces* 38 %. The loss of *other open spaces* has also been mentioned by local communities in Dar es Salaam. In reality losses can be expected to occur within different UMTs, such as those associated with the high-level *Residential* class, as well as within the *other open spaces* UMT sub-category itself.

In Addis Ababa there has also been a high rate of loss of green structures with a particularly dramatic reduction in agricultural land (loss of around one quarter of the agricultural land area from 2006 (19,639 ha) to 2011 (14,920 ha)). Much of this is concentrated into particular zones such as Bole to the east of the city. It should be noted that some of this is planned change.[11] The *vegetation* UMTs have also decreased markedly (by 1,440 ha), although within this category *mixed forest* has actually increased slightly. *Bare land* and *Residential* areas expanded, with evidence of both aggregation and sprawl. Within the city and in the uplands some losses and some gains in green structures were identified, including within some of the land which was bare in 2006. However, by 2011 bare land is estimated to have increased by more than 60 % of its 2006 area. This is likely to be caused by several drivers, development being one, but it is also possibly related to soil erosion.

A business as usual scenario modelled to 2025 suggests that a further 8,844 ha of the study area could be transformed to settlements by the year 2025; in other words, the spatial extent of the urban zone may increase by around half again from the situation in 2011. Between 2011 and 2025 this scenario would also see the loss of around a third of the city's agricultural land and almost a quarter of its other vegetated areas. *Field crops* could see the greatest reductions in land area (a loss of 4,738 ha, from a total agricultural land loss of 4,863 ha), followed by *grassland* (loss of 600 ha) and *riverine* areas (loss of 514 ha). The business as usual scenario

[11] This may also be the case elsewhere but a detailed assessment of the planning process has not been attempted.

also allows development in *riverine* corridors. All other things being equal, it would be expected that a further 31 % of the riverine corridor of 2011 would be lost, thus exacerbating problems of flooding, e.g. through the introduction of impervious cover and as a result of the increased exposure of the population within flood prone areas.

Green Structure Vulnerabilities to Climate Change

The results presented in the previous section suggest a considerable loss of provisioning services in the case study cities, especially related to food, wood, woodfuel, and the regulating service of carbon storage and sequestration. Furthermore, there are prospects for considerable additional losses under business as usual scenarios. In addition to development, climate change provides a further pressure. The literature suggests that, despite a complex picture of possible positive and negative climate change impacts on agriculture in Africa, "agriculture everywhere in Africa runs some risk to be negatively affected by climate change" (Müller et al. 2011: 4313). Over Sub-Saharan Africa as a whole, rain-fed agriculture is considered to be particularly susceptible and agricultural GDP could fall by 2–9 % under climate change scenarios (Dethier and Effenberger 2012).

An expert elicitation approach was used in order to understand some of the possible responses of green structures to climate drivers within Dar es Salaam, including as a result of heat-waves, drought, flood and wind-storm. Results suggest that in Dar es Salaam, *horticulture* is more sensitive than *field crops* to a range of hazard types, and therefore, would tend to have higher losses if exposed to a hazard event of the same magnitude. Some events have the potential to inflict more damage than others and some locations are more likely to be exposed. The potential for damage associated with some events, like heat-waves, can be adapted to through changes in management practice. However, even where this is possible a knock-on effect will follow with respect to demand for other resources, for example water and labour. Furthermore, if agricultural produce becomes scarcer, for whatever reason, then costs are likely to increase. The same principles may apply to other staples, too. As an indicator, Monela et al. (1993) reported that the cost of a bag of ex-kiln charcoal in the Dar es Salaam region in the rainy season was 38 % higher than the cost in the dry season. Costs within the city were six times higher than the ex-kiln price in the wider region. The experts consulted reported that weather-related drivers of differential costs still occur today, for example, during periods of flooding. Inevitably, improving the resilience of farming systems for current climate variability will have benefits, even where additional impacts and opportunities are not fully known (Cooper and Roe 2011).

It has not been possible within the scope of this study to assess impacts on other ecosystem services or generate a full database of the climate susceptibility of the green structures present in the CLUVA case study cities. However, it is known that regulating services provided by in situ and planned greening are also affected by climatic drivers; e.g. evapotranspiration is affected by water availability and vegetation health. Green roofs are a helpful example here as they are among the most susceptible to drying given their relatively shallow substrate. Studies in tropical and temperate climates show that cooling effects can be reversed if substrate is dry or vegetation damaged (Wong et al. 2007; Speak et al. 2013). Such possible restrictions on evapotranspiring functions are not taken into account in simple energy exchange models (Gill et al. 2007; Cavan et al. 2014) but this is likely to affect some vegetation structures more than others, e.g. those with shallower root systems like grasses in areas not artificially irrigated. It should also be acknowledged that basic data exist on climatic tolerances of individual species but further work is required to tie this data to the urban characterisation findings. Even basic data will help practitioners to determine which indigenous species may be best suited to greening strategies, e.g. for use within residential zones.

Prospects of Green Infrastructure for Climate Adaptation

Green infrastructure is an essential element of planning for a sustainable future for African cities. The CLUVA work has quantified some of the ecosystem services which are currently provided using a variety of techniques from modelling and empirical analysis through to qualitative techniques associated with local community and expert input. The work shows that an alarming rate of loss of green structures persists in Addis Ababa and Dar es Salaam and that this is leading to losses in a range of related services. The losses will also make the cities less attractive places to live, work and invest, contributing to the process of urban sprawl and thereby exacerbating the problem. Climate change is an additional pressure on green structures but it is not the greatest threat in the immediate term. Green infrastructure planning must seek to conserve existing patches and corridors in the urban matrix and peri-urban zone whilst enhancing multi-functionality elsewhere in order to satisfy the identified needs for services throughout the city. Conclusions of this nature are hardly new. However, this study provides a new set of evidence to demonstrate the benefits of retaining and enhancing green structures as part of a wider appreciation of green infrastructure and its ecosystem services. It demonstrates very clearly the role of green structures in creating sustainable African cities, which by extension will be better adapted to the threats of climate change. The study considered only a subset of ecosystem services and the notable exclusion is regulation of flood events which are given prominence elsewhere in this book and which are also connected to the ecosystem services provided by green structures.

This chapter has only covered part of the research carried out to tackle the objectives outlined at the beginning. It therefore only presents a sub-set of the full set of recommendations from the work, they are:

- *UMT datasets need to be recognised as a valuable spatial framework for characterising and mapping African cities*, especially in relation to identifying green patches and corridors. Related land cover assessments help to establish the essential properties of the different elements of the city-region and, importantly, the nature of the urban matrix. These in turn help to determine the ecosystem services associated with all elements of the urban fabric and not just UMTs which are wholly or predominantly green (i.e. the green patches and corridors). The associated frameworks and methods used in this study can be readily applied to other African cities. Indeed they may be helpful for decision-makers in smaller urban areas, too, as their roles within wider city-regions inevitably grow (see Chap. 5).

- *UMT frameworks should be considered for wider assessments of ecosystem services*. This chapter has demonstrated how they provide evidence to support the following conclusions, which in turn provides evidence for decision-makers for development, adaptation and mitigation agendas.
  - Development could be a much stronger driver of future urban temperatures than climate change.
  - Addis Ababa and Dar es Salaam have clear surface UHIs and initial analyses of air temperatures from a set of in situ sensors suggests that Dar es Salaam has a nocturnal UHII of up to 2 °C.
  - In 2008, Dar es Salaam had, on average, 15.6 m^3 ha^{-1} in merchantable wood stocks and the total urban tree biomass of the city held the equivalent of around one million tonnes of carbon. Large trees alone have been estimated to have an annual sequestration rate of 34.5 thousand tonnes of carbon and compensate for over a quarter of Dar es Salaam's total carbon emissions.
- *Green structure-related development control needs strengthening*. Current controls are having limited success; losses are occurring despite protection measures in both Addis Ababa and Dar es Salaam.
  - For example, 9–12 % of all woody coverage and around 30 % of woody biomass in Dar es Salaam has been lost between 2002 and 2008. At current rates of change, mixed forest stands could be lost entirely within the next few decades. This has implications for local temperature regulation, shading, timber, and carbon sequestration ecosystem services, amongst others. Although not part of the existing assessment, losses also have implications for the regulation of flood events.
  - Both cities have suffered considerable losses in provisioning services as a result of the reduction of land associated with agricultural uses. By 2025, 'business as usual' scenarios are estimated to be associated with losses of 6,886 ha of farmland (*field crops, horticulture, mixed farming*) in Dar es Salaam and 4,863 ha of farmland (*field crops, vegetable farms*) in Addis Ababa.
  - There is a consistent need to improve the supply of ecosystem services within the urban core, in particular. It is important for planners and decision-makers to be aware of the spatial dimensions of the need for

ecosystem services, and to manage potential trade-offs between the supply of different ecosystem services and possible ecosystem disservices.

- *The need for ecosystem services should be considered as well as their supply.* Development and climate change will impact on the supply of and need for ecosystem services within African cities and this needs to be taken into account in green infrastructure planning. The results in this chapter have suggested that climate change will not affect all green structures and their services equally – some are more sensitive than others. However, there is much more still to learn about particular sensitivities in urban contexts and how they should be managed.

Prospects for green infrastructure as an adaptation strategy for African cities may appear to be particularly bleak on the basis of the material in this chapter. However, the evidence of benefits provided in this study and engagement with local researchers, communities and practitioners must provide some opportunities to halt business as usual and support an alternative vision in which green structures are afforded the attention that they deserve. Indeed, the empirical data has already provided an improved evidence base for green infrastructure planning and some of the findings from the work are being incorporated into the Addis Ababa Master Plan process. With adequate enforcement the plan has the potential to ensure a more sustainable future for the city. It also provides a good practice example for decision-makers in other African cities seeking to similarly protect and enhance their own urban ecosystem services. Such activities are crucial for all cities across the continent, whether they are already large or whether they are just starting to emerge.

Acknowledgements The authors would like to acknowledge the input of the entire Task 2.2 team for their contributions to the research process, and the data providers, stakeholders and funders who have helped to make the research possible.

References

Abo El Wafa H (2013a) Urban settlement dynamics scenario modelling in Addis Ababa: background information. http://www.cluva.eu/CLUVA_publications/CLUVA-Papers/USSDM_Addis Ababa_Background Information_JULY2013a.pdf. Accessed 25 Sept 2013

Abo El Wafa H (2013b) Urban settlement dynamics scenario modelling in Addis Ababa: technical user guide. http://www.cluva.eu/CLUVA_publications/CLUVA-Papers/USSDM_Addis Ababa_Technical User Guide_JULY2013b.pdf. Accessed 25 Sept 2013

Benedict MA, McMahon ET (2001) Green infrastructure: smart conservation for the 21st century. The Conservation Fund. Sprawlwatch Clearinghouse monograph series. http://www.sprawlwatch.org/greeninfrastructure.pdf. Accessed 16 June 2014

Benedict MA, McMahon ET (2002) Green infrastructure: smart conservation for the 21st century. Renew Res J 20(3):12–17

BioNET-EAFRINET (2011) Invasive plants fact sheet – *Azadirachta indica* (Neem). http://keys.lucidcentral.org/keys/v3/eafrinet/weeds/key/weeds/Media/Html/Azadirachta_indica_(Neem).htm. Accessed 20 Sept 2013

Bolund P, HunHammer S (1999) Ecosystem services in urban areas. Ecol Econ 29:293–301
Buchta K (2013a) Spatial Scenario Design Modelling (SSDM) in Dar es Salaam – background information. http://www.cluva.eu/CLUVA_publications/CLUVA-Papers/USSDM_Dar es Salaam_Background Information_JULY2013.pdf. Accessed 25 Sept 2013
Buchta K (2013b) Spatial Scenario Design Modelling (SSDM) in Dar es Salaam – technical user guide. http://www.cluva.eu/CLUVA_publications/CLUVA-Papers/USSDM_Dar es Salaam_ Technical User Guide_JULY2013.pdf. Accessed 25 Sept 2013
Cavan G, Lindley S, Roy M, Woldegerima T, Tenkir E, Yeshitela K, Kibassa D, Shemdoe R, Pauleit S, Renner F, Printz A, Ouédraogo Y (2011) A database of international evidence of the ecosystem services of urban green structure in different climate zones. http://www.cluva.eu/deliverables/CLUVA_D2.6.pdf. Accessed 6 June 2014
Cavan G, Lindley S, Yeshitela K, Nebebe A, Woldegerima T, Shemdoe R, Kibassa D, Pauleit S, Renner F, Printz A, Buchta K, Coly A, Sall F, Ndour NM, Ouédraogo Y, Samari BS, Sankara BT, Feumba RA, Ngapgue JN, Ngoumo MT, Tsalefac M, Tonye E (2012) Green infrastructure maps for selected case studies and a report with an urban green infrastructure mapping methodology adapted to African cities. CLUVA deliverable D2.7. http://www.cluva.eu/deliverables/CLUVA_D2.7.pdf. Accessed 6 June 2014
Cavan G, Lindley S, Jalayer F, Yeshitela K, Pauleit S, Renner F, Gill S, Capuano P, Nebebe A, Woldegerima T, Kibassa D, Shemdoe R (2014) Urban morphological determinants of temperature regulating ecosystem services in two African cities. Ecol Indic 42:43–57. Accessed 20 Feb 2014
Cheung H (2011) An urban heat island study for building and urban design. PhD thesis, The University of Manchester, Manchester
Chidumayo E (1990) Above-ground woody biomass structure and productivity in Zambezian woodland. For Ecol Manag 36:33–46
Chidumayo E (1997) Miombo ecology and management: an introduction. Stockholm Research Institute, Stockholm
Cooper P, Roe R (2011) Assessing and addressing climate-induced risk in Sub-saharan rainfed agriculture. Foreword to a special issue of experimental agriculture. Exp Agric 47(2):179–184
Costanza R, D'Arge R, de Groot R, Farber S, Grasso M, Hannon B, Limburg K, Naeem S, O'Neill RV, Paruelo J, Raskin RG, Sutton P, van den Belt M (1997) The value of the world's ecosystem services and natural capital. Nature 387(6630):253–260
CSIR, CMCC (2012) CLUVA deliverable D1.5 regional climate change simulations available for the selected areas. http://www.cluva.eu/deliverables/CLUVA_D1.5.pdf. Accessed 18 July 2013
Daily GC (1997) Introduction: what are ecosystem services? In: Daily GC, Nature's services: societal dependence on natural ecosystems. Island Press, Washington, DC, pp 1–10
de Smith M, Longley P, Goodchild M (2013) Geospatial analysis: a comprehensive guide. http://www.spatialanalysisonline.com/HTML/?cellular_automata_ca.htm
Dethier J-J, Effenberger A (2012) Agriculture and development: a brief review of the literature. Econ Syst 36:175–205
Douglas I (2012) Urban ecology and urban ecosystems: understanding the links to human health and wellbeing. Curr Opin Environ Sustain 4:385–392
Eggleston S, Buendia L, Miwa K, Ngara T, Tanabe K (2006) 2006 IPCC guidelines for national greenhouse gas inventories, vol. 4 agriculture, forestry and other land use. Institute for Global Environmental Strategies, Hayama
FAO (2011) Global forest resources assessment 2010 – main report, FAO forestry paper no. 163. Food and Agriculture Organization of the United Nations, Rome. http://www.fao.org/docrep/013/i1757e/i1757e00.htm. Accessed 4 May 2013
Forman RT, Godron M (1986) Landscape ecology. Wiley, New York
Gill SE, Handley JF, Ennos AR, Pauleit S (2007) Adapting cities for climate change: the role of the green infrastructure. Built Environ 33:115–133
Gill SE, Handley JF, Ennos AR, Pauleit S, Theuray N, Lindley SJ (2008) Characterising the urban environment of UK cities and towns: a template for landscape planning in a changing climate. Landsc Urban Plan 87:210–222

Jonsson P, Bennet C, Eliasson I, Lindgren ES (2004) Suspended particulate matter and its relations to the urban climate in Dar es Salaam, Tanzania. Atmos Environ Int Afr Middle East 38:4175–4181

Jonsson P, Eliasson I, Holmer B, Grimmond CSB (2006) Longwave incoming radiation in the Tropics: results from field work in three African cities. Theor Appl Climatol 85:185–201

Kottek M, Grieser J, Beck C, Rudolf B, Rubel F (2006) World map of the Köppen-Geiger climate classification updated. Meteorol Z 15(3):259–263

Laaidi K, Zeghnoun A, Dousset B, Bretin P, Vandentorren S, Giraudet E, Beaudeau P (2012) The impact of heat islands on mortality in Paris during the August 2003 heat wave. Environ Health Perspect 120(2):254–259

Landsberg H (1981) The urban climate. Academic, New York

Lindley S, Gill SE, Cavan G, Yeshitela K, Nebebe A, Woldegerima T, Shemdoe R, Kibassa D, Pauleit S, Renner R, Printz A, Buchta K, Abo El Wafa H, Coly A, Sall F, Ndour NM, Ouédraogo Y, Samari BS, Sankara BT, Feumba RA, Ambara G, Kandé L, Zogning MOM, Tonye E, Pavlou A, Koome DK, Lyakurwa RJ, Garcia A (2013) A GIS based assessment of the urban green structure of selected case study areas and their ecosystem services CLUVA deliverable D2.8. http://www.cluva.eu/deliverables/CLUVA_D2.8.pdf. Accessed 6 June 2014

Lupala J (2002) Urban types in rapidly urbanizing cities: analysis of formal and informal settlements in Dar es Salaam, Tanzania. Unpublished PhD thesis, Department of Infrastructure and Planning, Division of Urban Studies, Royal Institute of Technology, Sweden

Lyytimaki J, Sipila M (2009) Hopping on one leg — the challenge of ecosystem disservices for urban green management. Urban Green 8:309–315

MacFarlane R (2007) Multi-functional landscapes: conceptual and planning issues for the countryside. In: Benson JF, Roe M (eds) Landscape and sustainability, 2nd edn. Routledge, London/New York

MacKillop F (2012) Climatic city: two centuries of urban planning and climate science in Manchester (UK) and its region. Cities 29:244–251

Malimbwi R, Solberg B, Luoga E (1994) Estimation of biomass and volume in Miombo woodland at Kitulangalo Forest Reserve, Tanzania. J Trop For Sci 7(2):230–242

Millennium Ecosystem Assessment (2005) Ecosystems and human well-being: a framework for assessment. Island Press, Washington, DC

Monela GC, O'Kting'ati A, Kiwele PM (1993) Socio-economic aspects of charcoal consumption and environmental consequences along the Dar es Salaam-Morogoro highway, Tanzania. For Ecol Manag 58:249–258

Mosha FM, Mosha LH (2012) Walking in transforming housing cityscape: a case of Kariakoo urban centre in Tanzania. Online J Soc Sci R 1(8):231–238

Moudon AV (1997) Urban morphology as an emerging interdisciplinary field. Urban Morphol 1:3–10

Müller C, Cramer W, Hare WL, Lotze-Campen H (2011) Climate change risks for African agriculture. Proc Natl Acad Sci USA 108(11):4313–4315

Mung'ong'o O (2004) A browning process, the case of Dar es Salaam city. s.l. Unpublished PhD thesis, Division of Urban Studies, Stockholm, Sweden

Nieuwolt S (1973) Breezes along the Tanzanian East Coast. Arch. Met. Geoph. Biokl., Ser. B, 21:189–206

Oke T (1982) The energetic basis of the urban heat-island. Q J R Meteorol Soc 108:1–24

Pauleit S, Breuste JH (2011) Land use and surface cover as urban ecological indicators. In: Niemelä J (ed) Handbook of urban ecology. Oxford University Press, Oxford, pp 19–30

Pauleit S, Buchta K, Abo El Wafa H, Renner F, Printz A, Kumelachew Y, Kibassa D, Shemdoe R, Kombe W (2013) Recommendations for green infrastructure planning in selected case study cities CLUVA deliverable D2.9. http://www.cluva.eu/deliverables/CLUVA_D2.9.pdf. Accessed 6 June 2014

Pickett STA, Burch WR, Dalton SE, Foresman TW, Grover JM, Rowntree R (1997) A conceptual framework for the study of human ecosystems in urban areas. Urban Ecosyst 1:185–199

Schmutterer H (1990) Properties and potential of natural pesticides from the Neem tree, Azadirachta Indica. Annu Rev Entomol 35:271–297
Smit W, Parnell S (2012) Urban sustainability and human health: an African perspective. Curr Opin Environ Sustain 4:443–450
Smith CL, Lindley SJ, Levermore G (2009) Estimating spatial and temporal patterns of urban anthropogenic heat fluxes for UK cities: the case of Manchester. Theor Appl Climatol 98:19–35
Smith CL, Webb A, Levermore GJ, Lindley SJ, Beswick K (2011) Fine-scale spatial temperature patterns across a UK conurbation. Clim Chang 109:269–286
Speak A, Rothwell J, Lindley S, Smith C (2013) Reduction of the urban cooling effects of an intensive green roof due to vegetation damage. Urban Clim 3:40–55
START, Pan-African START Secretariat, International START Secretariat, Tanzania Meteorological Agency, Ardhi University, Tanzania (2011) Urban poverty and climate change in Dar es Salaam, Tanzania: a case. http://start.org/download/2011/dar-case-study.pdf. Accessed 13 Sept 2013
Svensson MK (2004) Sky view factor analysis – implications for urban air temperature differences. Meteorol Appl 11:201–211
Syampungani S, Chirwa PW, Akinnifesi FK, Sileshi G, Ajayi OC (2009) The Miombo woodlands at the cross roads: potential threats, sustainable livelihoods, policy gaps and challenges. Nat Resour Forum 33:150–159
Taha H (1997) Urban climates and heat islands: albedo, evapotranspiration and anthropogenic heat. Energ Build 25:99–103
The Economics of Ecosystems and Biodiversity (TEEB) (2011) TEEB manual for cities: ecosystem services in urban management. www.teebweb.org. Accessed 5 Sept 2013
The Mersey Forest (2010) Liverpool green infrastructure strategy. Commissioned by Liverpool City Council Planning Business Unit and Liverpool Primary Care Trust. http://www.greeninfrastructurenw.co.uk/liverpool/. Accessed 31 July 2013
The Mersey Forest, The University of Manchester (2011) STAR tools: surface temperature and runoff tools for assessing the potential of green infrastructure in adapting urban areas to climate change. http://www.ppgis.manchester.ac.uk/grabs. Accessed 20 Sept 2013
Tso C, Chan B, Hashim M (1990) An improvement to the basic energy balance model for urban thermal environment analysis. Energ Build 14(2):143–152
Tso C, Chan B, Hashim M (1991) Analytical solutions to the near-neutral atmospheric surface energy balance with and without heat storage for urban climatological studies. J Appl Meteorol 30(4):413–424
URT (2004) The United Republic of Tanzania, Dar es Salaam City profile, document prepared by Dar es Salaam City Council and Cities and Health Programme. WHO Centre for Development Dar es Salaam, Tanzania
Watson I, Johnson G (1987) Graphical estimation of sky view-factors in urban environments. J Climatol 7:193–197
White F (1983) The vegetation of Africa, a descriptive memoir to accompany the UNESCO/AETFAT/UNSO vegetation map of Africa (3 Plates, Northwestern Africa, Northeastern Africa, and Southern Africa, 1:5,000,000). UNESCO, Paris
White R, Engelen G, Uljee I, Lavalle C, Ehrlich D (2000) Developing an urban land use simulator for European cities. In: Proceedings of the fifth EC GIS workshop: GIS of tomorrow. European Commission Joint Research Centre. Stresa, Italy pp 179–190
Whitford V, Ennos A, Handley J (2001) City form and natural process – indicators for the ecological performance of urban areas and their application to Merseyside, UK. Landsc Urban Plan 52(2):91–103
Wong NH, Puay Yok T, Yu C (2007) Study of thermal performance of extensive rooftop greenery systems in the tropical climate. Build Environ 42:25–54

Chapter 5
Small Cities and Towns in Africa: Insights into Adaptation Challenges and Potentials

Ben Wisner, Mark Pelling, Adolfo Mascarenhas, Ailsa Holloway, Babacar Ndong, Papa Faye, Jesse Ribot, and David Simon

Abstract This chapter is a counterpoint to those in the rest of this volume that treat Africa's large cities. As Simon (Int Dev Plann Rev 36(2):v–xi, 2014) has observed, most study of African urban climate change adaptation has focused on the challenges to large cities. So, by way of heuristic exercise, we attempt to approach a set of questions about small African cities and towns facing climate

B. Wisner (✉)
Aon-Benfield UCL Hazard Research Centre, University College London, London, UK

373 Edgemeer Place, Oberlin, OH 44074, USA
e-mail: bwisner@igc.org

M. Pelling
Department of Geography, King's College London, The Strand, London WC2R 2LS, UK
e-mail: mark.pelling@kcl.ac.uk

A. Mascarenhas
LInKS/REPOA, P.O. Box 35102, Dar es Salaam, Tanzania
e-mail: mascar@udsm.ac.tz

A. Holloway
RADAR (Research Alliance for Disaster and Risk Reduction), Stellenbosch University, Private Bag X1, Matieland 7602, South Africa
e-mail: ailsaholloway@sun.ac.za

B. Ndong
Fondation du Secteur Privé pour l'Education, 56, Avenue Lamine Gueye, 1er étage, Dakar, Sénégal
e-mail: babacarndong@hotmail.com

P. Faye
Institute of Anthropology, University of Bern, Lerchengweg 35, 3012 Bern 9, Switzerland
e-mail: papafay@gmail.com

J. Ribot
Department of Geography and Geographic Information Science, University of Illinois, 605 East Springfield Ave., Champaign, IL 61820, USA
e-mail: jesse.ribot@gmail.com

D. Simon
Department of Geography, Royal Holloway, University of London, Egham, Surrey TW20 0EX, UK
e-mail: D.Simon@rhul.ac.uk

S. Pauleit et al. (eds.), *Urban Vulnerability and Climate Change in Africa*, Future City 4,
DOI 10.1007/978-3-319-03982-4_5

change. What climate-related hazards are faced by small cities in Africa today and will be confronted in the future? What kind of enabling capacities should be strengthened so that staff in small cities can take the initiative to adapt to climate change? What obstacles do the governments and residents of small cities face in adapting to climate change? What potential is there for risk reduction and improved livelihood security even in the face of climate change? Reviewing literature and using case studies from Eastern, Southern and Western Africa, we find that small cities have potential not only to protect their infrastructure and residents from climate related hazards, but also to serve as catalysts of climate-smart development in their hinterlands. However, governance problems and a lack of finance severely limit the ability of small African cities to realise this potential. More research is urgently needed to inform feasible solutions to bridge these governance and funding gaps.

Keywords Tertiary urbanisation • Climate change adaptation • Urban-rural synergy • Governance • Climate-smart development • Water and sanitation • Flood

Introduction

'Africa has a distribution of the urban population by size of urban settlement resembling that of Europe, with 57 % of urban dwellers living in smaller cities (those with fewer than half a million inhabitants) and barely 9 % living in cities with over five million inhabitants' (UN 2012: 9). In terms of growth rate, the UNFPA (2008) concluded, '[w]hile mega-cities have captured much public attention, most of the new growth will occur in smaller towns and cities, which have fewer resources to respond to the magnitude of the change'. Climate change signals the need for public investment in Africa's smaller cities. However, public investment of the wrong sort can also be part of the problem. Some cities have already been affected irreversibly by the combination of climate stress and mega-projects. An example is Gorgoram in northern Nigeria. Dams were built on the Hadejia River, diverting 80 % of the water from the Hadejia-Nguru wetlands and lowering the water table, with profound consequences:

> . . . [O]nce a sizeable town in the heart of the wetland . . . [i]t had a famous fishing festival to which teams of youths came from hundreds of miles around to net the biggest catches in the last pools as the flood retreated. Dignitaries from across northern Nigeria came too, awarding prizes and lobbying for votes. But now Gorgoram is an emptying village. The festival is failing because the floods are poor and there are no fish. All around, the landscape is littered with fallen trees, victims of the falling water table. (Pearce 2006: 100)

Most of Africa's thousands of small cities are not so unfortunate. They still have a chance to adapt to climate change and also to stimulate support for 'climate-smart development' in their hinterlands (see Box 5.1).

Box 5.1: Climate-Smart Development

Mitchell (2010) suggests that stakeholders should work together in the following ways to create 'climate-smart disaster risk reduction (DRR)':

- Ensure good disaster risk assessments are conducted that factor in the best available climate, vulnerability and exposure information to work out future climate hazard risk. This will help to decide where best to site critical infrastructure or new settlements, for example, though such risk assessments will need to be updated as information is dynamic.
- Strengthen people's access to information (through education, media or dedicated early warning systems) about these risks and about the potential impacts of climate change.
- Create agencies and systems that are well connected across scales, can easily learn from each other, have space to innovate and experiment with approaches and conduct scenario planning exercises with regularity.
- Find ways to increase people's equitable access to markets and services, strengthen their ability to participate in decision-making and protect their rights.
- Initiate high standards of environmental protection in efforts to grow the economy and take advantage of international assistance to generate green jobs and, after disasters, to 'building back greener'. [Minor adaptations by authors to generalise Mitchell's points]

Note that these recommendations are also relevant to rural and urban development, livelihood enhancement (access to markets and services) and to investment strategies generally. Our question is whether small cities in Africa can catalyse such 'climate-smart development' which is simultaneously 'climate-smart DRR.'

For examples see: http://www.trust.org/spotlight/climate-smart-development/?tab=showcase.

Purpose of this Chapter

We present a counterpoint to the rest of the cases discussed in this book dedicated to Africa's large cities and climate change. Our purpose is to formulate questions about the future of small cities and towns facing climate change in Africa. We have fewer answers than questions, but we believe that clarifying and addressing such questions is an urgent task for Africa's citizens, policy makers, civil servants, researchers and civil society workers.

The specific questions are the following:

1. What climate-related hazards are faced by small cities in Africa today and will be confronted in the future?
2. What kind of enabling capacities should be strengthened so that staff in small cities can take the initiative to adapt to climate change?

Table 5.1 Example definitions for cities in Africa (Gough et al. 2013)

Country	Definition of "City"
Cameroon	Administrative headquarters with at least 2,000 inhabitants
Ghana	Settlements of at least 5,000 inhabitants
Rwanda	Those residents in the urban administrative units recognised as such by law
Tanzania	Regional and district headquarters; other areas where there is a concentration of houses and institutions [sic] police stations, post offices, health centres and streets

3. What obstacles do the governments and residents of small cities face in adapting to climate change and realising related opportunities?
4. What potential is there for risk reduction and improved livelihood security, even in the face of climate change?

By 'small city' we choose the range of population from 20,000 to 100,000 (Pelling and Wisner 2009). There is no universal definition of 'city' (see Table 5.1), so we have adopted what seems a reasonable one.[1] Towns are even smaller and sometimes are clustered with a small city in a municipal district, as in the South African case study that follows. There is considerable evidence that cities towards the bottom of the urban rank-size distribution may serve as organic parts of their hinterlands, providing valuable services for surrounding rural zones, if not acting as fully fledged 'growth poles' (Darwent 1969; Jacobs 1985; Parr 1999a, b; Ogunleye n.d.). Thus, an additional motivation for our team in contributing this chapter is to explore the potential developmental role of small African cities. We wish to understand the constraints on that developmental role due to climate change, as well as constraints due to the political and economic power of national and global systems. While managing their own growth and infrastructure in the face of climate change, can small cities in Africa also catalyse 'climate-smart' rural development?

Methodology

How does one try to generate relevant questions about climate change and small cities and towns in Africa with little or no data available? We estimate that there are at least 5,000 cities and towns in Africa that function as the centre of local government areas (LGA).[2] A major empirical research project would be needed

[1] The minimum population defining 'urban' varies worldwide from 200 in Sweden, 500 in South Africa, 5,000 in Nigeria to 30,000 in Japan (Hartshorn 1991). This makes it impossible to define 'city' and 'town' universally. Population alone isn't the only important criteria; others may include structure and function in the settled area.

[2] A crude estimate finds that cities and towns that are headquarters of local government areas in Tanzania, Uganda, Malawi, Ghana, Niger and Sudan number on average 120. Multiplying this by 46 Sub-Saharan African countries and adding the 1,000 local government areas' (LGA) centres in South Africa and Nigeria, results in a total of nearly 6,500. Thus 5,000 seems a reasonable, but possibly low, estimate. (Numbers of sub-regional centres of LGAs were obtained from various Wikipedia sites.)

even merely to visit, gather available data and to interview key officials in 10 % of these. Unable to mount such a study, we have combined a review of literature on decentralisation, governance and local economic development with intensive study of three small cities and towns. Three small settlements – one in each of three regions: Eastern, Southern and Western Africa – were chosen to provide some regionally diverse perspectives. The actual choices depended on personal contacts of the author team.

In Tanzania, Adolfo Mascarenhas, retired Tanzanian professor of geography, and Ben Wisner had been based intermittently in Mwanga from 2008 to 2014 while implementing a research project on climate change based at Sokoini University of Agriculture in Tanzania and Ohio University in the USA. They gathered market and business data, mapped and photographed environmental degradation and discussed urban development challenges with officials. In South Africa, Ailsa Holloway, director of the Research Alliance for Disaster and Risk Reduction (RADAR) at Stellenbosch University, had studied Oudtshoorn's flood problems while David Simon had also visited this city on numerous occasions since 1980 in the course of his own separate research. Simon is a member of the steering committee of an international research project, Urbanisation and Global Environmental Change (http://ugec.org/). For Senegal, we relied on the direct, firsthand knowledge of Babacar N'dong, supplemented with background knowledge on Senegalese governance by Papa Faye. N'dong is a staff member of a Senegalese civil society organisation, Fondation du Secteur Privé pour l'Education and Faye is based at the Institute of Anthropology at Bern University, Switzerland. Unreferenced evidence presented in these three case studies comes from firsthand observations by these individuals, key informant interviews or grey literature to which they had access.

The literature review was conducted by Mark Pelling and Ben Wisner, building on their earlier work (Pelling and Wisner 2009), along with Jesse Ribot and David Simon.

Background on Tertiary Urbanisation in Africa

The colonial administrative system left behind numerous primary outposts and garrisons that have become the district headquarter cities with administrative, service, market and transportation functions (Myers 2011). As third order cities – as opposed to national commercial/ political capitals and the regional headquarters – these central places link up primary settlements with the regional capitals and the national capital cities and national commercial-industrial hubs. These small cities usually feature secondary or even tertiary educational institutions and referral health facilities. They are strategic places in the networks that allow resources, people, information and investment to flow. Our argument is that any forces that undermine the vitality of these small cities will have systemic impacts on the urban and rural sectors. Yet they are also largely ignored.

We welcome the recent increase of empirical research on climate change affecting second order cities in Africa mentioned by Simon (2010, 2014). These include work in St. Louis, Senegal (Diagne and Ndiaye 2009; Silver et al. 2013) and Bobo-Dioulasso, Burkina Faso (Silver et al. 2013); however, these cities are larger – with populations of >300,000 and >500,000, respectively, and more important in regional governance than the third order cities that concern us.

Challenges to Small Cities Facing Climate Change

The defining characteristics and challenges of small cities are these: they are deeply embedded in the rural economy so are hybrid; they are very local in resource profile. This may have some advantages for self-organisation and utilisation of local knowledge, short decision-making chains, etc., but these advantages are probably overwhelmed by major disadvantages – spatial isolation and neglect when it comes to funding.

However, before considering the challenge of climate change for small cities and towns in Africa, one has to be clear about the word 'change'. Urbanites and rural dwellers alike deal with change in many forms. They experience change holistically as threats and opportunities in daily life (Wisner 1988, 1993; Smucker and Wisner 2008). Climate change is not the only change affecting small cities in Africa. Urban reality is punctuated by political, administrative, economic, social, technological, environmental, and demographic changes in addition to changes in the climate. At any given time, political leaders in small cities as well as civil servants, traders, bankers, artisans and all other residents will more easily recognise and emphasise one or another manifestation of these diverse, yet interconnected, kinds of change. Even farmers do not necessarily consider climate change to be the greatest threat they face (Mertz et al. 2009; Wangui et al. 2012). Disregarding these local perceptions and concerns, governments in Africa often echo former colonial/ metropolitan styles of top-down, technocratic planning (Scott 1998; Easterly 2014) and their current discourse of 'risk' and 'the climate imperative' (Pachauri 2009; Jasanoff 2012; Handmer and Dovers 2013; Smucker et al. 2015).

In addition to climate change, several other types of change preoccupy local people. World prices for exports and imports fluctuate, sometimes wildly – witness the grain price spike in 2008 and the price history of African export commodities, such as cocoa and coffee. Variable prices affect the trader's, transporter's and farmer's profit margin. National and municipal policies and collusive strategies of intermediaries and merchants may also be responsible for low producer prices that keep people in poverty – and vulnerable to, among other things, the impacts of climate change. Also of concern to local people is the arrival of large numbers of people displaced from the countryside by conflicts that are sometimes overlaid on drought or flooding (Haysom 2013). National government policies of all kinds may be proclaimed and dominate the labour, time and attention of civil servants who are unable to give adequate attention to routine matters. For example, a routine task of local government may be road maintenance. Intense rain – an increasing feature of the 'new normal' climate – more easily erodes poorly maintained roads away. Changes in the use of mobile phones (an example of technological change) may make it easier to sort out optimal marketing arrangements. However, annual changes in national government taxes on fuel and on mobile phone companies that take place during the 'budget season' may act as a further disruptive change. The complex, chaotically interacting array of changes that face residents, leaders and professionals in small African cities is suggested in Fig. 5.1, entitled 'Spaghetti of Doom'.

It is impossible to disentangle the various kinds of changes that have been taking place in Africa. Each of these changes may interact with others, strongly or weakly (see also Chap. 1).

Because such changes affect people's incomes and assets as well as African governments' ability to prepare and respond, they influence a city's or town's security, including its ability to address climate change. These changes increase or decrease the number and size of hazards, shocks and opportunities for different groups of people. The life and economy of small urban centres rely on bottom-up economic infusion from the rural sector in which they are embedded. Small cities

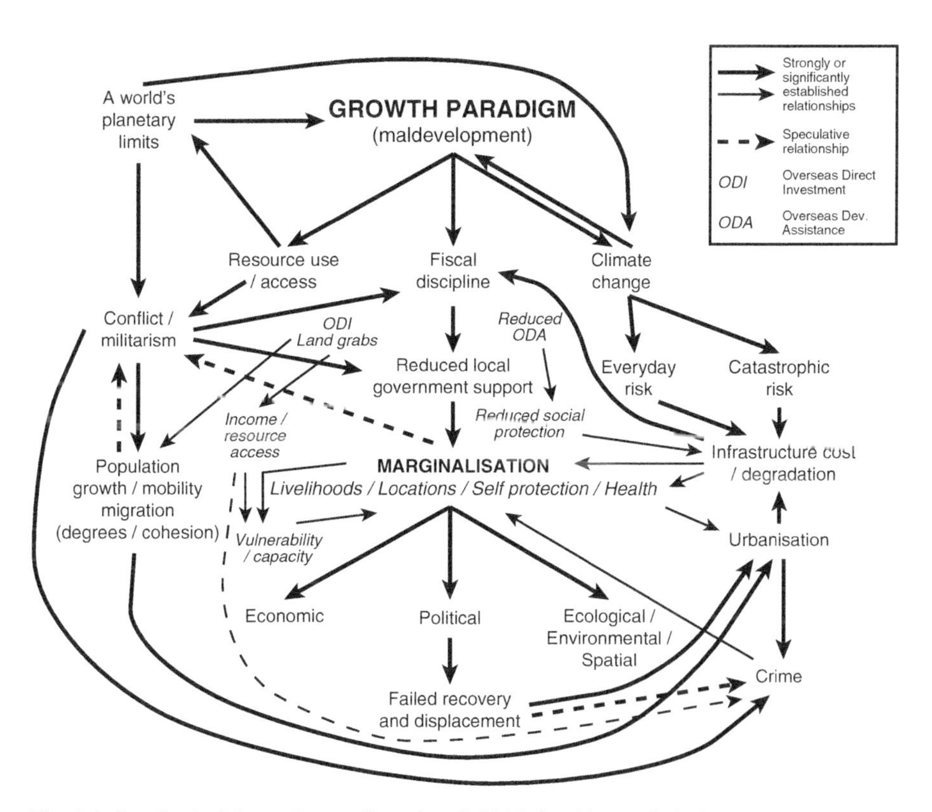

Fig. 5.1 Spaghetti of doom (Lopez-Carresi et al. 2014: 5, with permission)

are also dependent on top-down public sector investment over which they have little control. Managing and living with risk in such cities is a complicated hybrid of local action, often built on local social resources that extend into the surrounding countryside.

Administrative, Political and Organisational Challenges

Many small cities will not have local self-government but rather be governed by regional centres. Small cities are in a unique predicament with low visibility and often as a pawn in political games played out elsewhere. Where local government exists it will likely be basic, with limited capacity for strategy or powers of investment and decision-making. Even where local government is more capable and better resourced, its freedom to innovate and to seek solutions to the many challenges facing it, including climate change, is limited by the competence or incompetence at other levels. As Charles Sampford puts it (2013: 25):

> [M]any of the most intractable governance problems occur when inadequacies at one level of governance are reinforced and exacerbated by inadequacies at other levels… It is paradigmatically the case in the response to climate change…

The immediacy of local government, where it exists in small urban centres, provides an opportunity for direct accountability. Where there is will and collaboration among local government, businesses and civil society groups much can be done to reduce hazard exposure and enhance preparedness (GNDR 2009, 2011, 2013). As a hub for the surrounding rural economy, upon which much of the urban economy depends, the small city is positioned to help to protect upstream urban and downstream rural communities and economies from the impacts of climate change. The importance of this is the capacity to bring together information from rural producers and markets, as well as the national meteorology service. When weather and climate data are communicated in a context of trust and using appropriate metrics, its uptake and subsequent impact on agricultural yields can be startling. For example, the World Meteorology Organisation (WMO) is in the midst of a major campaign to help national governments push seasonal climate forecasts down to grassroots (Makoye 2013; WMO 2014a, b). In Africa, the small city will be a vital node in realising such penetration of climate science and its fine-tuning by discussion of local experts, including representatives of farmers and herders (Wisner 2010).

However, yields cannot come to market if rural roads are washed away or blocked by landslides. Small cities in Africa may not have jurisdiction over the building, maintenance and protection of transport infrastructure outside their boundaries. The ambiguous exception is where the small city is also a regional or district headquarters. As in nearly all matters, the small urban place must negotiate with higher order jurisdictions for attention and resources. Small cities in regions of a country that did not vote for the parliamentary majority in power may find themselves less visible and low on the list of priorities for infrastructural investment. They may also be unable to access grants from development agencies that work through national government channels. Such politically disfavoured small cities will usually not have the legal or financial management capacity to engage with sources of private finance.

Economic and Resource Access Challenges

The economy of small urban centres is typically based on two functions: (1) processing and marketing of agricultural produce generated from the surrounding countryside and sold to support local agriculture, and associated sectors such as transport; (2) public sector activity – schools, health centres, government agencies, and other public services. The balance of these two sectors will influence the vulnerability of the local economy to 'external' variables. Cities that are reliant on the agricultural economy will be exposed to market price fluctuations, government subsidies or guaranteed prices for agricultural produce and the impacts of climate variability on production. Cities that are overly dependent on public sector employment may experience some stability but are vulnerable to policy change, as many cities experienced from 1980 to 2010 in rounds of structural adjustment and public sector retrenchment. In these circumstances, there are few alternative employment opportunities in small urban centres, leading to out-migration and structural poverty.

Managing urban settlements depends on the ability of local government to derive rents and other income. This income can come from a variety of sources, including local property rates and taxes, central government subvention, private sector investment, public-private partnerships, and direct loans or grants from international financial institutions, non-governmental organisations or development donors. However, many of these potential sources of income are not available to smaller cities (IPCC 2014). One reason is the transaction cost of managing distributed projects with limited scale economies. Another reason is a legal barrier to such targeted external investments. Smaller cities rarely have the legal and management skills in house that are required to access significant funds or to manage any resulting projects. The result is that strategic funding is directed to higher order cities and regional centres, with only limited downstream investments, and even less opportunity to shape these investments. Similarly, the ability to enjoy international private sector investment requires strong human resource capacity and a population size large enough to attract these funds. Small cities rarely hold either. Opportunity may lie in networked approaches but there are only few examples of this (IPCC 2014). Simon (2014: ix) highlights successes in urban climate adaptation by the action of '[p]roactive local authorities – often driven by strategically placed climate/ environmental change "champions" in the political leadership ...' However, he is writing about cities one or two orders of magnitude larger than those considered here (for further discussion on importance of "local champions", see also Chaps. 9 and 10 of this volume).

The case studies in this research show the potential for a networked small urban development strategy. The South African case is an example of a District Municipality that embraces Oudtshoorn and six other local municipalities. In the Tanzanian case, we will show the possibility of developing a prosperous Greater Mwanga Urban Region, integrating the ribbon/ strip urbanisation between the two poles (sisal estate and ethanol factory) and seeing this urban growth as an opportunity and not as a problem. Most planners see strip urbanisation as 'sprawl'. However, any such spatial clustering would still exist at the bottom of the national budgetary food chain, and even networks of small cities and towns may not have the legal and managerial capacity to negotiate and supervise overseas investment. The danger is that they would be overwhelmed and unable to monitor and stop negative social or environmental impacts of such investments.

As regards national government grants and subventions, 'decentralisation' is a universal buzz word in Africa (Ribot 1999; Cohen and Peterson 1999; Bardhan and Mookherjee 2006); however, it is often a verbal exercise without corresponding devolution of financial, technological and professional resources or the transfer of authority over sectoral decisions, such as agricultural or natural resource management and use, and infrastructural investment. Decentralisation is often announced in laws and documents with no manifestation in practise (Ribot 2003; Ribot et al. 2008).

The widely legislated democratic decentralisations that have promised to increase the effectiveness and equity of local development policy, with few exceptions, have never materialised in practise. In their place have been many forms of privatisation and de-concentration of central government agents to local posts. When local governments are given powers, they are typically made to be accountable to higher levels of government, defeating the purpose of decentralisation. By contrast, when the local governments are accountable to the local population, they are systematically deprived of resources. In the end, few powers are transferred to local authorities that represent or are responsive to local people. The result is a form of local governance that is neither accountable nor responsive and a deep failure to improve local wellbeing (Ferguson 1994; Crook and Manor 1998; Harriss et al. 2004; Ribot et al. 2008; Poteete and Ribot 2011; Mascarenhas and Wisner 2012; Satterthwaite 2013; Smucker et al. 2015).

Potentials, Advantages and Opportunities of Small Cities Facing Climate Change

Political and Social Advantages

Political will to address climate change may be easier to establish in smaller cities than in the megacity context, where conflicting political and economic interests are more varied and intense. The professional and political elite residing in small cities and towns know each other and probably socialise. It is therefore easier for information to flow informally among key individuals, such as the water engineer, medical officer of health, agricultural officers, business owners, bankers, secondary school teachers, district councillors, religious leaders and senior staff of civil society organisations. Finally, small cities may be less anomic and anonymous than larger ones, the exceptions being those mentioned earlier where many traumatised people displaced by violence have settled or, in the case study from South Africa that follows, where strong racial fault lines or other sorts of discrimination persist. Residents may be easier to enlist in participatory planning than in the large informal settlements that characterise Africa's largest cities, and where less face-to-face interaction with leadership occurs.

Economic Advantages and Opportunities

Smaller urban centres have an organic relationship with their rural hinterlands (Nel and Binns 2002; Robinson 2005; Nel and Rogerson 2005). This is the basis for a solid and growing small manufacturing sector to serve the surrounding countryside. In principle it could provide opportunities to serve the rural hinterland with many products necessary for climate change adaptation such as windmills, irrigation pumps and pipes, water tanks (African Rainwater Harvesting Network 2014), specialist seeds and veterinary products, etc. However, increased climate variability and price fluctuations may reduce the purchasing power in the countryside to buy such products. This is a vicious circle since such inputs are key to climate-smart rural development.

Many small cities are located along transportation networks that nourish local markets. This market function can be important in sharing lowland and highland products, spreading the risk of climate change impacts. The spread of mobile communications and relatively cheap motorcycles and motor rickshaws, and increases in local bus services, are linking up rural producers in different agro-climatic zones as well as with cities of various sizes. Diverse foodstuffs available in local markets are a sign of how the market function can provide a buffer against climate change impacts (see Table 5.3 on p. 172).

Three Case Studies

Our three case studies certainly cannot claim to represent the great diversity of the thousands of small cities in Africa. However, in order to clarify and refine the questions we asked at the beginning, we look closely at these three small settlements, emphasising their physical site and exposure to hazards, governance, municipal finance and economic activity, as well as the challenges they face in planning for future climate impacts. There is little known about most small cities and towns in Africa: their histories, sites, infrastructure, functions, governance, economies, and relations with their hinterlands. All of this bears on the challenge of climate change for small cities and their capacity to adapt.

Case Study 1: Mwanga, Tanzania (Population 16,000 – Rounded)

Overview and History

Mwanga town (population 15,783)[3] is the administrative centre for Mwanga district (population 131,442) in Kilimanjaro Region, northern Tanzania (Fig. 5.2). When Mwanga district was created by dividing the very large Same district in 1982, Mwanga town was laid out according to an as-yet unfulfilled 1982 *Master Plan*. It was set back from the main road that connects Arusha, Moshi and the border with Kenya with the Tanzanian sea port of Tanga and the old German-built railway line (now abandoned). Formerly, considerable informal trading and some permanent restaurants and other shops were found at this junction, and despite attempts by the Mwanga district council to shift trading to the new interior market centre, the ribbon of roadside commerce both northwards and to the south has continued to elongate. Therefore, for planning purposes, one should consider a greater Mwanga urban area and its satellite settlements. Two such nodes along the main road are of particular interest. The first is the sisal estate owned by Mohamed Enterprise Tanzania Limited (METL). To the north is the newly created Kilimanjaro Bio-chemical industrial alcohol plant. Between them, they provide employment for nearly 1,000 men and women (more on these industries below).

The 2002 national population census showed that if one included Mwanga Urban and all the mixed 'urban/ rural' wards, they would account for about 50 % of the population of the district (see Table 5.2). The Agricultural Census completed the following year indicated that fewer than 40 % of the households listed agriculture, pastoralism or fishing as their core livelihood. Another study actually indicates that in rural Kirya Ward average household incomes are catching up with Mwanga Urban (Muzzini and Lindeboom 2008). Field work in Mwanga revealed that a large proportion of rural dwellers are in commercial activities, certainly not mere self-provisioning. This is more evidence that one should consider a greater Mwanga urban area as the operational unit for planning adaptation to climate change.

[3] Whilst Mwanga is smaller than our minimum 'small city' population, it is a district headquarters and has the administrative, service, and market functions of a city. It also has core central business district (CBD), manufacturing and processing zones, as well as residential suburbs and a rural-urban periphery (rurban edge). Nevertheless, for the sake of consistency, we will call Mwanga a 'town'.

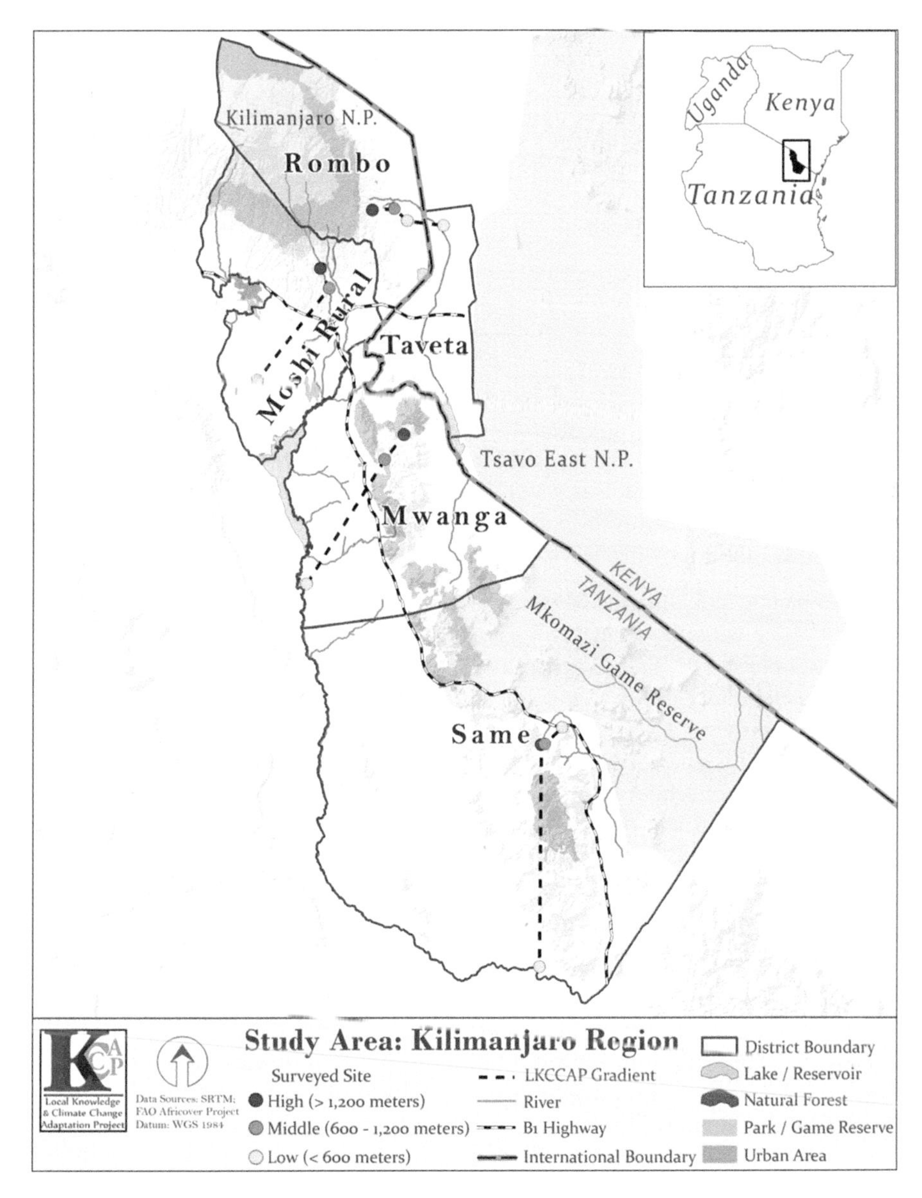

Fig. 5.2 Kilimanjaro Region, showing location of Mwanga District (Wangui et al. 2012, with permission)

Since its founding, Mwanga has grown steadily. A former Prime Minister, Cleopa Msuya, who lives in retirement in the North Pare Mountains above the town, was the moving force behind establishment of the new district and the new town. The early leadership of independent Tanzania agreed to ensure that growth

Table 5.2 Population of Mwanga District and the emergence of new centres

District ward	Total population	Male	Female	HH size
Mwanga District	131,442	63,199	68,243	4.35
Mwanga Urban Ward	15,783	7,414	8,943	4.07
Kileo (Mixed)	13,645	6,702	6,943	4.07
Mgagao (Miji = 'Town' surrounding a livestock market)	6.011	2,879	3,132	4.65
Toloha	3,055	1,484	1,579	4.45
Kigonigoni	2,610	1,364	1,246	4,47
Kivisini	1,694	880	814	5.01

was not only confined to Dar es Salaam, but to seven other designated urban areas and that basic services, such as water, health, and education, were widely available in the rural areas. Largely because Msuya worked hard to obtain basic services for the town, the following services are presently available:

- Two banks, including the Mwanga Community Bank;
- Post office with all services;
- Market place with shops, warehouses and bus stand with modern toilets;
- Mains electricity in Mwanga town and also in several of the mountain villages;
- Water supply;
- Several public and private primary schools;
- Five secondary schools, including three earmarked for girls;
- District hospital and the well-equipped Neema Health Centre operated by the Roman Catholic Church;
- Police post.

Mwanga's considerable commercial activity includes trading, service provision and small scale manufacturing. The activities comprise a large number of wholesalers and retailers, pharmacies, restaurants, photocopy and computer service enterprises, as well as small scale enterprises specialising in carpentry, cement block production and welding. Two tree seedling nurseries exist. There are also two petrol stations and a number of automobile workshops.

In addition to being the administrative headquarters for district political governance and professional services, Mwanga also hosts a number of non-governmental organisations and training centres. Religious society is represented by an active mosque, as well as Lutheran, Catholic and other churches. The Lutheran diocese runs a school for the deaf while the Catholic dispensary (the Neema Health Centre) complements the services provided by the district hospital.

Four hotels and several guest houses offer accommodation for visitors and a Catholic convent that doubles as a rest house serves some visitors. At least one of the hotels, in collaboration with the Tanzania Tourist Board, is running a hands-on training for hotel staff.

Administration

Founding president Julius Nyerere's administration transformed the colonial administration and also set up a parallel hierarchy of one-party influence. The old chiefdomship system was abolished, and by 1969 the party structure was found throughout the country. The *balozi ya nyumba kumi*, leader for every ten households, was found from the city of Dar es Salaam to the most remote villages in the country. The conspicuous green TANU flag near the house of the leader was a symbol of authority and unity. The *balozi* were never formally replaced. In urban areas this function has been taken over by ward government (*serikari ya mtaa*) and in rural areas by elected chairs of sub-villages (*vitongoji*). These are the most local administrative officials (Box 5.2).

Box 5.2: Tanzania's Administrative Structure
The administrative structure of Tanzania has many levels: nation (*taifa*), region (*mkoa*), district (*wilaya*), ward (*kata*), village (*kijiji*), sub-village (*kitongoji*) and clusters of rural homesteads known as *khaya*. At the ward and village level an executive officer who represents the government is appointed by the district administration. Elected representatives constitute a parallel set of institutions – elected councils, a committee structure, and chairs. At the sub-village scale representatives are elected to sit on the village council. Such parallel structure is replicated at each of the levels of governance. There are also gazetted cities and municipalities (see also Chaps. 9 and 10).

Physical Site and Climate-Related Hazards

Mwanga is situated where semi-arid lowland meets the base of the North Pare Mountains (Fig. 5.3). It is set back approximately one km from the main road on a site that abuts the North Pare Mountains and includes a central town area between two water courses that run West in the direction of the main road. The town also rises in three directions from this central zone onto the lower slopes of the mountains and up onto two ridges.

Over the 40 years since its founding, Mwanga has witnessed extreme deforestation of the western escarpment of the North Pare Mountains. This and the inexorable expansion of the town has meant that the water courses from the mountains no longer flow permanently, and massive gullying endangers bridges, pipelines and also peri-urban gardens and potentially house plots (Fig. 5.4). Flooding is becoming more and more of a challenge and this is made worse by an increased intensity of rainfall, likely due to climate change.

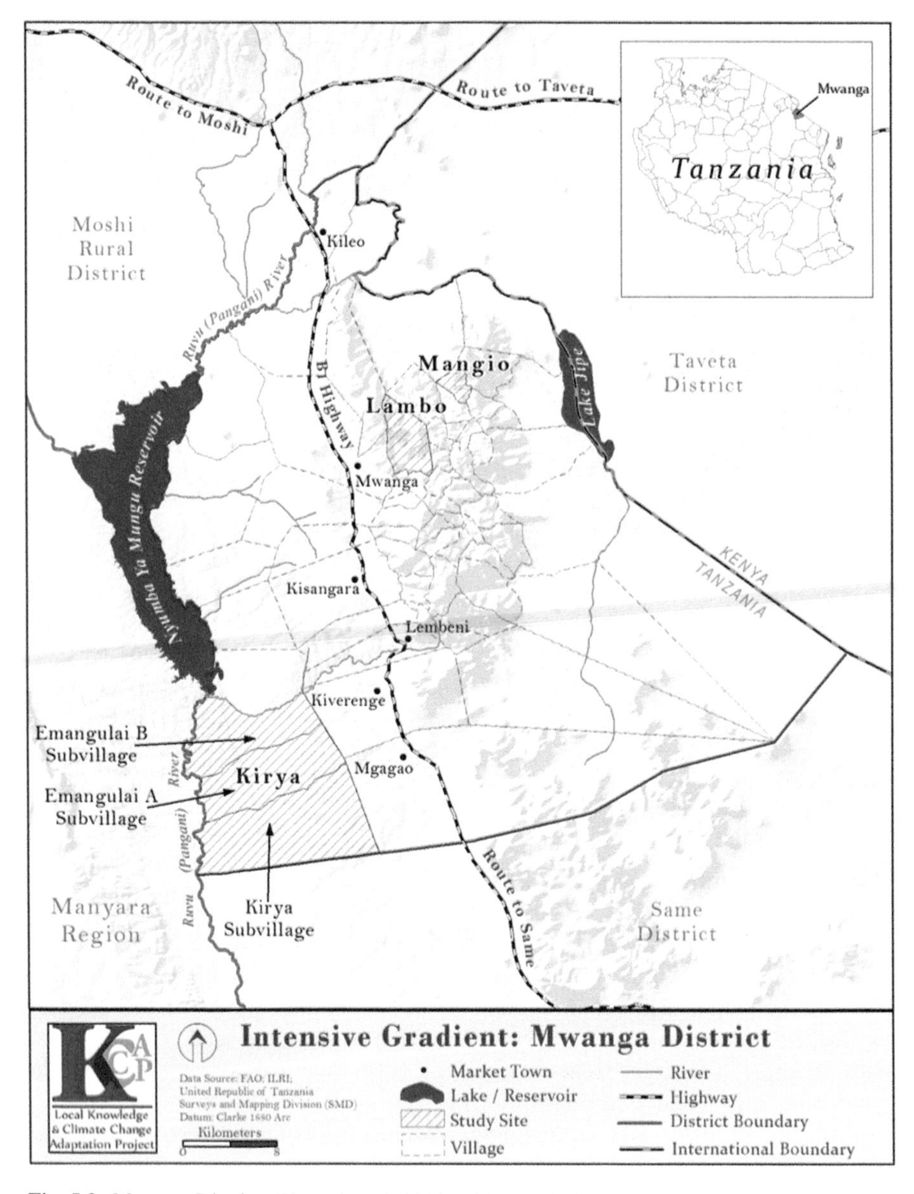

Fig. 5.3 Mwanga District (Wangui et al. 2012, with permission)

Hazards to Mwanga's Water Supply

The two wells that provide the town with water are limited by an insufficient pumping capacity. People also pay water sellers by the 20 litre container; while some wealthier people and institutions, such as guest houses, pay drivers to bring

Fig. 5.4 Gullying in Mwanga (Picture: Ben Wisner 2012, University College London)

5,000 litre containers that they then transfer to their own storage tanks. This highlights the precarious water supply situation that can only become more severe as deforestation and gullying reduce infiltration of seasonal rain into aquifers, as rainfall becomes more erratic, as the population of water users increases and, assuming a degree of economic growth, the water consumption per capita increases. An additional threat to the water supply is contamination. To date, no municipal planning has been undertaken for the disposal of solid and liquid waste. As population densities increase, the risk of contaminated sewerage water polluting domestic water will also increase.

Flooding, Destructive Winds and Hail

Sited at the base of the North Pare Mountains, Mwanga is normally windy at dawn and at night due to differential heating of the plains below and mountains above. Convectional storms bring even windier conditions that can damage roofs and crops in peri-urban gardens and the surrounding countryside, as can hail storms. Seasonal floods are normally expected with damage to rural roads in the mountains leading to Mwanga town and even landslides. Changing climate is likely to exaggerate the extremes of drought and flooding and bring more severe thunderstorms, micro-bursts (known locally as dust devils) and hail (IPCC 2012).

Piecemeal Planning in Mwanga and Its Hinterlands

To date, urban planning in Tanzania has been passive and uncoordinated. The 1982 *Master Plan* was never fully implemented. While Mwanga town is laid out with a grid of roads, these are unsealed and erode easily. Street lighting and solid waste collection are non-existent. Waste is burned alongside roads, contributing to poor air quality. Although abundant open space can be found in the central zone, unless creative efforts are made to raise financial resources, it will be hard to develop it into parks or recreational areas. Currently, livestock graze and thorny vegetation has colonised much of the open space. So far, the town's approach to climate change is 'piecemeal'. Even for critical issues, such as the town's water supply, a more comprehensive understanding is needed for effective planning and intervention.

People need public open spaces, and this means planning for playgrounds, parks and related amenities. Residents in Mwanga already suffer the health effects of smoke from burning of uncollected refuse (Nordhaus 2013: 91–99, citing McMichael et al. 2003) and dust from fallow open spaces trampled by livestock. Added to this will be the risk of heat waves exacerbated by the Urban Heat Island Effect. All these hazards can be reduced by the cooling and air filtering function of trees in open spaces and parks (Chap. 4).

Capacity of Mwanga to Adapt to Climate Change

Despite legislation that, in principle, decentralises many governmental functions from the national level to regions and districts, funding does not necessarily follow. Mwanga town has very few sources of revenue – principally, a portion of the commercial licensing fees paid by business and a fee collected from people using the formal market on Thursdays. Even this small amount of revenue suffers from misappropriation.

The district administration is mandated to manage the whole territory of the district, not just the town. Thus the available work time of district professional staff, such as the water engineer, forestry officer, legal officer, etc., is not sufficient to focus as a priority on the needs and problems of Mwanga town. These professionals are also transferred frequently and, as a result, vital local knowledge, contacts and institutional memory are lost. In addition, the office of the town manager is small and poorly funded and, in fact, relies heavily on donations via German and Swedish sister city relationships.

The Untapped Potential of Mwanga

Comparing Mwanga district to most of the other districts of Tanzania, the 2007/2008 *Agricultural Sample Survey* showed that fewer than 40 % of the households were engaged in subsistence farming, herding or fishing. So what does the 60 % majority do for a living? The Pare people, who are the majority ethnicity in Mwanga district, respect the value of education. Because of higher educational attainment, both young men and women from Mwanga district have migrated to jobs in many parts of Tanzania. Links to their homes have been maintained, and this explains why there is improved housing in many villages in the district. Remittance of income to homes in the countryside surrounding Mwanga town is a source of demand for products that are produced in Mwanga town, or could be produced with the existing manufacturing base, including the hardware necessary for climate-smart development (e.g. windmills, fuel-efficient cooking stoves, tree seedlings, etc.).

Purchasing power has also increased because of the entry of rural women into the formal labour market. An example is the transformation that has taken place at the Mohamed Enterprises sisal factory a few miles from Mwanga town. Until the 1970s the overwhelming majority of workers in a sisal estate were men. Currently, at the Mohamed Enterprises estate more than three-quarters of the workers, both in the sisal processing factory and in the field are women. Flexibility of working conditions and incentives, such as permission to grow beans or maize between sisal rows, means that women can have an income, be closer to their children and pay for education.

Mwanga town's economic vitality has stimulated economic activity up and down the gradient that leads up into the villages in the North Pare Mountains and down into the semi-arid plains that end eventually at the Ruvu River and the Nyumba ya Mungu hydro-electric dam and reservoir. The Thursday market is very large (Fig. 5.5), and market surveys have found a great variety of products from mountains and plains, including products made with the grasses and reeds that grow in the wetlands along the Ruvu River at nearby Kiboko. The diversity of foodstuffs available mirror the way the market integrates highlands and lowlands and enhances resilience to climate change (Table 5.3). A fleet of Chinese-built motorcycles serve as taxis on market days, and buses and lorries piled high with produce come and go. In nearly every case, speaking to those involved, one detects a spirit of self-reliance. These new activities have added value to local produce, increased incomes, and attention has been paid to changes to make livelihoods more sustainable.

Fig. 5.5 Mwanga's Thursday market (Picture: Ben Wisner 2012, University College London)

Table 5.3 Food basket available in Mwanga market place

Staples: maize, cooking bananas, rice, wheat, millet, sweet potatoes, yams, cassava
Leafy Vegetables: cabbage, several varieties of traditional and modern spinach, chard, cassava leaves, pumpkins, squash, several varieties of wild leafy vegetables
Beans, pulses, lentils: several varieties of beans including pigeon peas, gram, groundnuts, bambara nuts
Fruits: oranges, mangoes, fruit bananas, avocado, baobab pods, jackfruit, oysternuts, pineapples, pawpaw, passionfruit, fruit bananas, various berries
Meat/milk: beef, lamb, goat, chicken, duck & milk and milk products
Fish: including catfish, tilapia, various types of fresh water sardines
Spices /medicinal plants: ginger, garlic, onions, pepper, chillies, cardamom, tamarind
Beverages: tea, coffee, cane juices

Mwanga's manufacturing and wholesaling provides its hinterland with materials for irrigation, construction, grain milling and other rural industries. Carpenters use local hardwoods; restaurants consume produce from mountains and plains.

If planned well, a Greater Mwanga Urban Area of up to 40–60,000 people could dramatically tap into and expand on the potential to catalyse climate-smart growth in the district. In this scenario the incipient satellite settlements, such as those near the sisal estate and a large scale alcohol distillery along the main road in the opposite direction, would grow in a coordinated and mutually beneficial manner. A new private vocational centre, which opened in 2012 at a site along the main road between Mwanga and the sisal estate, is an example of the tertiary activities that could support such economic growth and adaptation to climate change.

Case Study 2: Oudtshoorn, South Africa (Population 29,000)

Overview and History

Oudtshoorn is an inland municipality (Greater Oudtshoorn Municipality), located approximately 55 km from South Africa's southern coast, covering an area of more than 3,500 km^2, and with a population of 29,000 (Fig. 5.6). Since 2000, it has comprised the larger towns/small cities of Oudtshoorn, Dysselsdorp and De Rust, along with smaller settlements (Oudtshoorn Greater Municipality 2013). The city serves as the primary administrative and commercial centre for the Klein ('Little') Karoo, a semi-arid area bordered by the Langeberg, Outeniqua, Kammanassie and Swartberg Mountains.

Evidence of human settlement in Oudtshoorn by ancestral Hottentot herders can be traced as early as the second century A.D. The arrival of European settlers in the

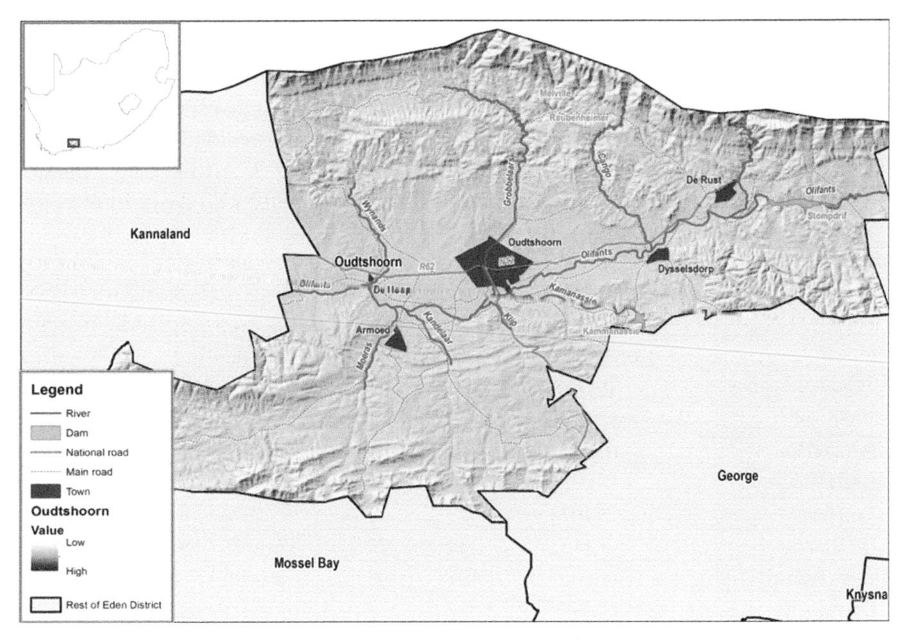

Fig. 5.6 Oudtshoorn Local Municipality, showing the location of cities, roads and major rivers (Design by Gillian Fortune 2013, RADAR, University of Stellenbosch, with permission)

early nineteenth century resulted in marked changes to vegetation and water courses. These were further accelerated from the 1870s onward, due to the expansion of the ostrich farming industry. In 2011, Oudtshoorn was reported to have the highest concentration of ostriches globally (with 450 registered ostrich farms within the greater municipality). The area is also a major eco-tourism destination, due to the area's proximity to the historic Boomplaas and Cango Caves and the Gamkaberg Nature Reserve. Reportedly, 184,290 tourists visited the Cango Caves in 2010 alone (Urban-Econ 2011: 158). Since 1994, Oudtshoorn has hosted the Klein Karoo Arts Festival annually in March/April – an event estimated to generate USD 11 million and 600 temporary jobs (Oudtshoorn Greater Municipality 2013). In addition, the municipality hosts four ostrich 'show farms,' dating back as far as the 1920s.

The city is the administrative centre for the Greater Oudtshoorn Municipality, containing a wide range of government, retail, tourism and commercial services. Numerous government services are provided, including five community clinics, a public hospital and a local office of the South African Social Security Agency (SASSA). A private medical clinic also operates within the city.

The 2011 national census recorded approximately 96,000 residents living in the Greater Oudtshoorn Municipality, of whom 29,000 lived within the actual city. Census findings also showed that 21,910 households resided in the greater municipality, with 34.3 % of all households headed by women and an average household size of 3.1 members (Oudtshoorn Greater Municipality 2013). In 2011, the municipality's demographic profile included 8,739 residents of Black African origin, 74,202 who were Coloured, 11,983 who were White and 317 of Indian/Asian descent (Statistics South Africa 2011) (Table 5.4).

A closer examination of Table 5.4 provides insight on the persisting effects of historic, racially-based relocations enforced by the previous government. It shows, for instance, that in 2011, 80 % of the white population resided in the original Oudtshoorn city (Fig. 5.7), constituting nearly a third (32 %) of the city's inhabitants. This contrasted with only 21 % of coloured residents who lived in the city,

Table 5.4 Demographic composition, Oudtshoorn Greater Municipality as of 2011 (Statistics South Africa, Census 2011)

	2011					
Town/city name	Coloured	Black	Other	Asian	White	Total pop
Volmoed	424	31			17	472
Bongolethu	10,931	3,640	77	43	33	14,724
Bridgton	17,073	320	177	38	32	17,640
De Hoop	107		1		43	151
De Rust	3,115	88	29	17	317	3,566
Dysselsdorp	11,906	494	48	50	46	12,544
Oudtshoorn	15,609	3,699	338	161	9,336	29,143
Oudtshoorn (non-urban)	15,037	467	23	8	2,159	17,694
Total	74,202	8,739	693	317	11,983	95,934

Fig. 5.7 High income inner suburb of Oudtshoorn (Picture: David Simon 2008, Royal Holloway, University of London)

with the majority dispersed across the settlements of Bongolethu, Bridgton, De Rust and Dysselsdorp. White settlement patterns also diverge from those of the municipality's black residents, of whom 82 % were living in either the original Oudtshoorn city or in Bongolethu, an under-served township established in 1966.

Such social-spatial divisions reflect the consequences of apartheid laws harshly applied in Oudtshoorn from the 1960s–1980s and that involved the removal of thousands of Africans and Coloureds from town (TRC 2001). Beginning in 1961, Oudtshoorn's coloured residents saw their homes demolished and their systematic removal to Bridgton, Dysselsdorp and De Rust. Similarly, the city's African inhabitants were forcibly resettled in Bongolethu.

The historical period that entrenched racial segregation in both Oudtshoorn city and its hinterlands was marked by government-sanctioned detentions, demolitions and deaths, as well as violent local resistance and protest action. This was especially evident in Bongolethu during the 1980s, reflected in 1985 alone in the arrests of 155 Fisikele high school students and shooting deaths of three teenagers by police (TRC, *ibid).* The widespread devastation of the black and coloured areas during this period bequeathed a legacy of difficult socio-economic and governance challenges that persist today. Oudtshoorn's 2013/2014 draft *Integrated Development Plan* acknowledged the persistence of both high unemployment and school drop-out rates, while the SASSA reported that 20,460 social grant recipients were registered in the municipality. Protracted poverty was further signalled by a staggering tuberculosis prevalence rate of 1,085/100,000 (Oudtshoorn Greater Municipality 2013). This well exceeds South Africa's national rate of 857/100,000 and is more than six times higher than the global rate of 169/100,000 (World Health Organisation 2013). Rising concerns about drug and alcohol abuse have resulted in initiatives, such as the establishment in Dysselsdorp of a crèche for children with special needs and foetal alcohol syndrome (Oudtshoorn Municipality 2011).

These social challenges have prevailed in a climate of disabling local governance for much of the past decade, characterised by local council acrimony and contestation. Significantly, for six months in 2007, the municipality was placed 'under administration' by the provincial government (IOL News 2007), and in 2013, the Special Investigations Unit submitted a detailed report to the Office of the President on alleged malpractice and corruption within the municipality (Mtyala 2013). In July 2013, the Provincial Cabinet was required to intervene and directly authorise a temporary budget for Oudtshoorn, due to the local council's failure by 1 July to approve its 2013/2014 budget (IOL News 2007).

Administration

Oudtshoorn constitutes one of seven local municipalities that comprise the Eden District Municipality, a district that sweeps across the southern band of South Africa's Western Cape Province. Its administrative arrangements illustrate the governance systems established to ensure alignment of services across multiple spheres: municipal, provincial and national. The municipal sphere specifically is differentiated into metropolitan, district and local municipalities with their associated councils constituted by elected representatives (Leck and Simon 2013). Municipalities are further subdivided into wards, South Africa's smallest geopolitical unit, with each ward represented on the local council by an elected councillor. Governance of local municipalities is further enabled by the election of an executive mayor and appointment of a municipal manager, as is the case in Oudtshoorn.

District Municipalities are an intermediate spatial and administrative unit between the Local Municipalities and South Africa's nine Provinces. These Provinces are sizeable and are largely self-governing. Of course, there are a host of other administrative entities that cut across these governance arrangements, such as river catchment management agencies, fire protection agencies, etc.

Physical Site and Climate Related Hazards

Oudtshoorn is located in the Klein ('Little') Karoo valley, ringed by the Langeberg, Outeniqua, Kammanassie and Swartberg Mountains. Although mountain rainfall can range as high as 900–1,600 mm annually, the Klein Karoo is situated in a rain shadow, so that Oudtshoorn's annual average rainfall is between 600 and 800 mm. It is supplied by 20 perennial rivers. The river names 'Olifants River' ('elephants') and 'Moeras' River ('marsh') indicate that these once supported fertile wetlands, inhabited by buffalo, elephant and hippopotamus. The Greater Oudtshoorn Municipality is also served by three storage dams, with two of these dedicated to irrigation agriculture. The smallest (Koos Raubenheimer Dam) supplies water primarily for city consumption (Oudtshoorn Greater Municipality 2013).

The Klein Karoo faces recurrent hydro-meteorological hazards, due to climate, topography and land use patterns. Within the past decade, Oudtshoorn has borne the brunt of severe floods in 2006, 2007 and 2012. In addition, it is recurrently drought exposed (most recently in 2009–2010), and it faces wildfire hazards. The local economy is also exposed to the recurrent threat of livestock disease, particularly avian influenza. In the most recent outbreak (2011–2012), more than 38,000 ostriches were culled (Urban-Econ 2012), with export bans on raw meat to crucial markets (such as the European Union) imposed and sustained through 2013 (Erasmus 2013). The municipality's recent flood and drought experience also illustrates the 'see-saw' and highly variable character of the weather conditions that face many small and medium-sized cities in the Western Cape – and that pose serious challenges to local development (Fig. 5.8).

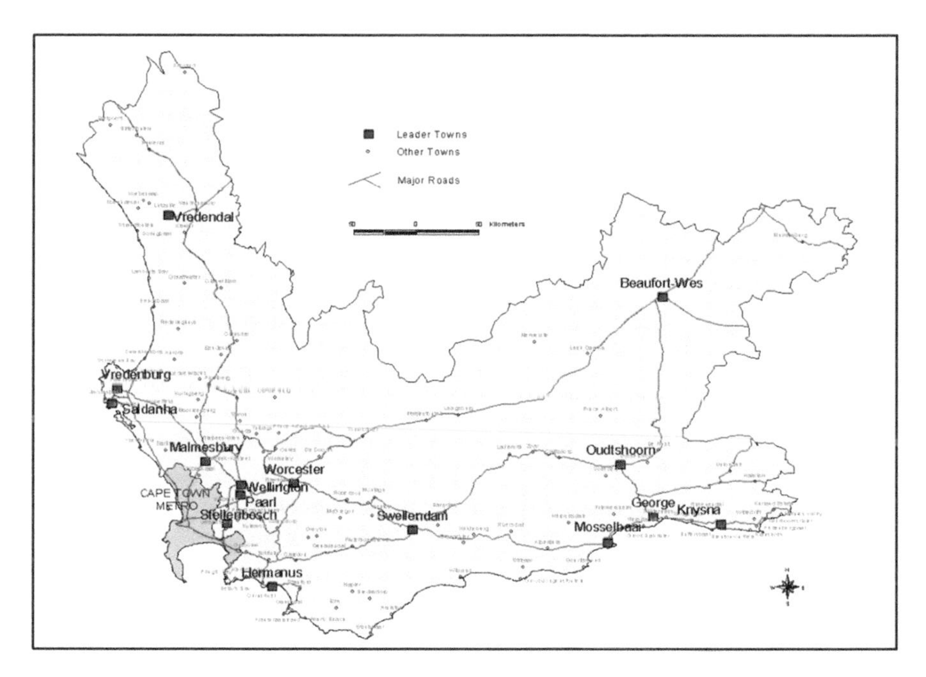

Fig. 5.8 Western Cape Province, showing 'leader towns' including Oudtshoorn (Centre for Geographical Analysis 2004: 138; Oudtshoorn Greater Municipality 2013: 35; Courtesy of Adriaan van Niekerk, Centre for Geographical Analysis, University of Stellenbosch, with permission)

Table 5.5 Oudtshoorn Municipality: budget for civil and technical services 2006/07 and flood-related losses for August 2006 (in USD) (DiMP 2007)

Planned expenditure budget 2006/07 USD	Maintenance & repair budget 2006/07 USD	Maintenance & repair as % of planned expenditure 2006/07	Total flood damage costs August 2006 USD	Flood damage as % planned expenditure	Flood damage as % of maintenance & repair budget
1,330,532	56,022	4.21 %	976,751	73.4 %	1,743.5 %

In 2006 and then again in 2007, Oudtshoorn sustained direct losses of around USD 1.2 million due to flash flooding associated with three intense weather systems. The economic impact of the 2006 rainfall events alone is illustrated in Table 5.5. It shows that Oudtshoorn's planned budget for civil and technical services (municipal engineering) for 2006/2007 was projected at USD 1.33 million. Of this, a mere USD 56,022 (4.21 %) was earmarked for maintenance and repairs (DiMP 2007). Yet, in August that year, USD 976,751 in damage was sustained by municipal bridges, roads and storm-water services due to two 'back-to-back' storm systems. These flood-related repair costs represented 73.4 % of the municipality's entire annual planned budget for civil and technical services (DiMP *ibid)*. Just over a year later, in November 2007, flooding attributed to yet another intense storm led to around USD 261,000 in direct damage to municipal infrastructure (Holloway et al. 2010). For smaller cities like Oudtshoorn, the fiscal shock of one million USD damage to infrastructure means a major development set-back, especially when it is followed by similar shocks in following years.

These losses, while significant for a small inland municipality, still under-represent the directly recorded flood impacts sustained by non-municipal entities within the greater Oudtshoorn area, including those that offer essential services. A detailed study of the 2006 storms indicated that as much as USD 14.9 million in losses were borne by farmers as well as public entities operating within the local municipality's jurisdiction (Holloway et al. 2010). Provincial roads alone sustained damage approximating USD 11 million; whilst local farmers reported impacts greater than USD 3.5 million (ibid). In 2011, and then again in 2012, storm-related flood damage was also reported. In 2012, this exceeded USD 44.6 million in agricultural losses alone.

In rural administrative nodes such as Oudtshoorn, complementary governmental services beyond the local authority, such as Provincial Roads and the National Department of Water Affairs, play crucial protective roles in ensuring business continuity in times of duress. When these essential non-municipal services are disrupted or operate inefficiently, this has serious knock-on consequences for residents and businesses living in the cities they serve.

Other Hazards and Environmental Challenges

From 2009 to 2010, Oudtshoorn successfully withstood a major drought, only to be confronted by a far more complex emergency due to an outbreak of avian influenza in its ostrich population. This has had profound consequences for the livelihoods of residents within the city and beyond, largely due to the forced discontinuation of raw ostrich exports to key markets, including the European Union. Since flooding has been implicated in the spread of the H7N7 bird flu virus from farm to farm, this animal health and economic disaster is also climate-related.

The outbreak was first reported in April 2011, resulting in the culling of more than 38,000 ostriches by December (Urban-Econ 2012) and the loss of a further 12,000 birds during 2012 (Awetu Traders 2012). This had direct consequences for farmers and their workers, and despite government-supported efforts to retain farm labourers, thousands were laid off. The imposition of import bans, particularly by the European Union, also exacted industry-wide implications, including effects on abattoirs and tanneries. For instance, the Klein Karoo International abattoir group anticipated retrenchment of up to 500 of its 950 general labourers from the poorer settlements of Bridgton and Bongolethu by December 2012, if the EU import ban remained in effect (Urban-Econ *ibid*).

The scale of the economic impact also extended well beyond Oudtshoorn's boundaries. For example, one social accounting estimation of the avian influenza's direct and indirect impacts projected that from 2011 to 2014, the ostrich industry's annual contribution to the broader Western Cape economy would drop from USD 115.6 million to USD 30.5 million (Urban-Econ *ibid).* As late as December 2013, the European Union ostrich meat import bans were still in effect, with a reported case detected as recently as October of that year (The Poultry Site 2013).

Capacity of Oudtshoorn to Adapt to Climate Change

The avian influenza outbreak in Oudtshoorn and the resultant 2011–2013 export ban coincided with an intense drought and two severe weather events. These shocks took place under conditions of protracted local government turmoil. This 'perfect storm' illustrates the interaction between different hazards, and the complexity of managing multiple risks under constrained resources.

Widespread transmission of avian influenza across Oudtshoorn's ostrich population was attributed to flooding associated with the severe weather events of 2006 and 2007 which destroyed agricultural water impoundment structures (as well as farm canals). As financial resources were not released to repair the damaged structures, flood-waters during the 2011 and 2012 storms flowed into natural water courses. This allowed contaminated excreta and other secretions from infected ostriches and wild fowl to move unimpeded across catchments and farms (Awetu Traders 2012) and to increase the scale of ostrich exposure to the avian influenza virus. Failure to release government funds to assist farmers' repair of impoundment structures and canals exacerbated the situation.

The complexities illustrated by these recent events in Oudtshoorn illustrate the real challenges this specific local municipality faces in adapting to future climate risks. They show how climate variability is only one of many concurrent hazards the city must address. Encouragingly, the draft *Integrated Development Plan* (IDP) gives explicit priority to addressing poverty, underlining that 'our future depends on how we deal with the poor and disadvantaged citizens in our areas and that our interventions must be pro-poor' (Oudtshoorn Greater Municipality 2013: 36). The IDP describes a wide range of interventions that would address social development needs as well as longer-term climate adaptation requirements. An example of a local adaptation measure is planned upgrading of local infrastructure to improve storm water drainage capacity as well as reduce leaks and urban water losses. Introduction of drip irrigation onto farms is also planned as an initiative to relieve pressure on constrained urban water supplies. Awareness is growing of the importance of diversifying beyond ostrich meat exports, and there is an explicit emphasis on rural economic development, especially in the outlying and historically disadvantaged settlements.

Despite such measures, the sheer complexity and interactions among climate-related, veterinary and economic shocks (and their social consequences) in recent years underline the urgent need for coherent and stable local governance. This implies the engagement of a committed, functional local government, whose attention is not diverted by continuing internal dissent (Meyer 2013; Botha 2013).

The Untapped Potential of Oudtshoorn

The Greater Oudtshoorn Municipality has well established facilities for adding value and exporting livestock and livestock products. It also is the gateway to considerable eco-tourism resources at a time when that is a rapidly growing type of tourism. Furthermore, even a small city in the Republic of South Africa has a great deal of technical expertise available. If political dissent subsides and Oudtshoorn's *Integrated Development Plan* is implemented, some of the tensions underlying political conflict could also ease – a benign circle. Technical engineering capacity and other technical skills exist for diversifying the economy of Oudtshoorn's hinterland through a variety of climate-smart initiatives, including the recently approved Oudtshoorn Water Project that will provide piped groundwater from the Blossoms well-field (Oudtshoorn Municipality 2014); however, local retention of skilled technical personnel may be more difficult under conditions of sustained political turmoil and disruptive leadership.

Case Study 3: Tambacounda, Senegal (Population 88,000)

Overview and History

The city was founded by a Bamana hunter named Tamba Waly. He first established the village of Tamba Soce around 1900, a village located five km from the current site of Tambacounda. During the French colonial period, Tambacounda rapidly became a trade centre because of the Dakar-Bamako railway connecting Tambacounda to the coast and to neighbouring Mali in 1923. The railway infrastructure was the first modern infrastructure, and officials who worked for the state-owned railways were the first professionals to live there. With the railway came more intensive cultivation of grains and cotton by Wolof and Sereer peoples seeking arable land.

Tambacounda became capital of the *cercle* of Niani-Ouly in 1920 and later a mixed urban/ rural jurisdiction (*commune mixte*) in 1952. After independence in 1960, Tambacounda became the capital of the region of Eastern Senegal (*Senegal Oriental*), and in 1984, when regions began officially to take their names from their capital cities, Eastern Senegal became known as the region of Tambacounda. Currently, Tambacounda refers to three levels of administrative entities: the capital of the administrative region of Tambacounda, the capital of one of the three departments of that region, and the municipality (*commune*) of Tambacounda (Fig. 5.9). The city has grown from 41,885 inhabitants in 1988, 67,543 in 2002 to an estimated at 78,800 in 2007. Currently (2013), the municipal population is estimated at 87,938 inhabitants, made up of Pular (46 %), Sereer (19 %), Manding (17 %), Wolof (14 %), Bambana (13 %), Soninke (9 %) as well as smaller numbers of Diola, Bassari, Diakhanké and Moors (NRC 1995).

Fig. 5.9 Location of Tambacounda Region in Senegal (http://fr.wikipedia.org/wiki/Tambacounda_(region), accessed 28 Jul 2014)

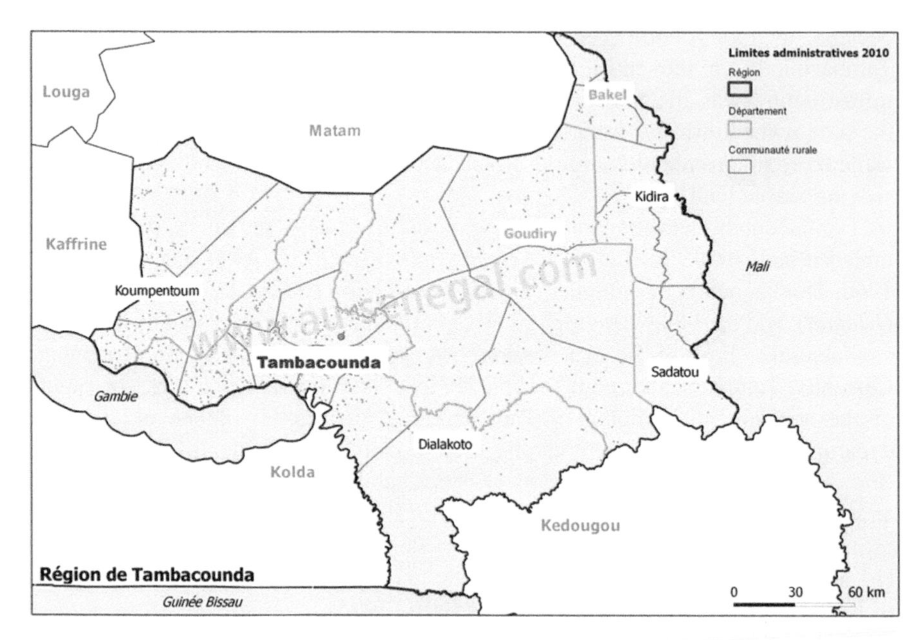

Fig. 5.10 Map of Tambacounda Region in 2010, also showing location of Tambacounda City (http://www.au-senegal.com/carte-administrative-de-la-region-de-tambacounda,039, accessed 28 Jul 2014)

Tambacounda city is an entrepot as well as administrative and service centre (Fig. 5.10). The region administered from the city borders Mali, Gambia and Guinea Bissau. The location of the city on the rail line and secondary roads leading to these other countries shapes its role as a crossroads. This position provides commercial opportunities, and the city has a daily central market.

Agriculture and agro-industrial activity are the most important sectors of the regional economy. The availability of arable lands has made it a destination for in-migration of farmers who cultivate cotton, groundnuts (peanuts), maize and sorghum. Together with Kolda region to the Southwest of Tambacounda region, Tambacounda city is host to units of the major textile company, SODEFITEX (*Société nationale de développment des fibres textiles*). SODEFITEX encourages cotton farmers with subsidies and offers temporary employment to people in the city and surrounding villages during the dry season. Livestock herding remains mostly traditional, with only a few young people interested in improved breeds. Three small dairies are operating in Tambacounda, but milk yields are low – no more than two litres per day. Some feed cotton seed cake from the textile mill to their animals. Despite tensions between farmers and herders (subject of a new pastoral law under consideration), negotiation allows herders to graze crop residue after the harvest. Despite abundant rivers, the local population does not have a taste for fresh water fish, and the fishery resource remains largely untapped. The wealth of diverse tree species in nearby forest on all sides of the city and a longstanding craft tradition have made Tambacounda a magnet for carpenters and cabinetmakers from all over Senegal. Thirty timber mills have been established in and around Tambacounda.

Today the city of Tambacounda has several facilities in the areas of health, education, worship and business and commerce (see Box 5.3).

Box 5.3: Facilities Found in Tambacounda Town

Health	Education	Business and Commerce
1 regional hospital	3 preschools	5 banks
1 health centre	21 primary schools	4 savings and loan facilities
2 clinics	4 high schools	2 post offices
5 health posts	8 colleges	
5 infirmaries		1 central market
3 private clinics		6 neighborhood markets
8 pharmacies		2 bus stations
		1 railway station
		1 airport (weekly to Dakar)
Worship		1 craft village
		1 slaughter house
>20 mosques		1 fairground
4 Christian churches		8 petrol stations
2 Muslim cemeteries		5 hotels
2 Christian cemeteries		3 hostels
		2 radio stations
		5 national press branches

Administration

Senegal has two overlapping governance structures – the administrative/territorial and the political. The administrative/territorial is composed of *regions*, *departments* and *arrondissements*. Administrators are appointed civil servants who represent the ministries and the President of the Republic. A hierarchy of democratic and decentralised authority is organised along political lines. Organisational units are the *region*, the *communes* (municipalities), and rural communities (smaller local governments) made up of groups of villages sharing common geographical and socio-cultural patterns. At each level is a council whose members are politically appointed as candidates by political parties but elected democratically. There are no independent candidates although this has become a strong social demand. In addition a central government administrator is appointed at the level of the region (the *Préfect*) and of the *arrondissement* (between region and rural community/ municipality), who is the *Sous-Préfect*. They play the role of 'legal control,' ensuring that national laws are being followed. Thus the region has a double status:

it is an administrative and territorial entity led by a Governor (*Gouverneur de région*) and political jurisdiction led by a President of the Regional Council and its council members, composed of the representatives of the communes and rural communities within the territory of the region.

The complexity and overlap of this administrative and governance system leads to delays, inefficiency and also invites informal shortcuts in access to state resources and permits (see Chaps. 9 and 10).

Physical Site and Climate-Related Hazards

Tambacounda lies on a low altitude plateau, between 25 and 60 m above sea level. The city covers an area of 7,755 ha, of which, 2,670 ha is occupied. Its density is nine inhabitants per km^2. The city has 28 neighbourhoods, of which ten are older, historic districts, together with 18 newer, outlying districts. The housing and settlement patterns in Tambacounda are a mixture of informal settlements (62 %), modern villas (21 %), and the village-like peri-urban settlements (9 %).

Agriculture provides almost 70 % of the employment in the surrounding region and contributes about 40 % of regional GDP. The city is surrounded by fields and peri-urban market gardens. While there is some small scale irrigation, principally for vegetables and bananas, most farmers depend on rainfall. Cycles of drought and flooding constitute a persistent threat to livelihoods and the regional economy. The city itself is crossed by a branch of the Mamacounda River, and some settlement has taken place in and near the flood plain of this river, creating conditions exposed to flooding. Faced with problems related to the disposal of solid waste (see below), people dump garbage in the Mamacounda River and along the main road (entrance and exit of the city). Solid waste from the saw mills is also dumped into the river. Flooding is therefore exacerbated by this added solid waste that comes on top of soil erosion, depositing clay and sand in the river. Very intense convectional rainstorms (thunderstorms) and lack of a plan for dredging the Mamacounda add to the flooding hazard. Thus, each year the city faces flooding due to narrowing of the riverbed.

Other Hazards and Environmental Challenges

Tambacounda faces a serious sanitation problem. The city produces about 1,650 tonnes of solid waste per month and does not have an efficient system of garbage collection. Odd as it may sound, the proliferation of plastic bags constitutes a serious veterinary problem because animals that graze in peri-urban and other parts of the city eat them and suffer intestinal blockage that may lead to death (Simpkin 2005: 167; Bashir 2013: S9).

In Tambacounda city only 45 % of households have access to drinking water provided by the National Water Company. Many people use surface waters (the river, creeks and ponds) for drinking, bathing and laundry, thus exposing

Fig. 5.11 Woman bathing a child in Tambacounda (Picture: Water Charity, with permission)

themselves to serious diseases (Fig. 5.11). Added to this is the lack of sanitation facilities, with only 52 % of households having pit latrines. The city also faces the challenge of solid waste management. Medical waste from the regional hospital constitutes a special hazard because this hospital is not only the referral facility for its region, but also receives patients from neighboring countries. The hospital does not have an incinerator and has only a pit for disposal of all of its medical waste. As the patient load increases, this problem is becoming urgent.

The forests surrounding Tambacounda are robust, despite half a century of discourses about deforestation. While they are cut for charcoal and firewood, as well as some timber, these forests regenerate. It is difficult to call the woodcutting in this region 'deforestation' (Ribot 1999). Wurster (2010) found no biodiversity declines in cut areas, and species are not yet disappearing. Woodcutting is nevertheless a contentious social issue. Forest villagers do not like the woodcutting, which is organised by the national Forest Service against their will. They are unhappy with this system which denies them any share in profits, forces them to go further for woodfuel and leads to women being chased and abused by migrant woodcutters who work in their forests. Wood is harvested for the production of charcoal for use as fuel in the capital city Dakar, and more recently in Tambacounda. Ribot's 1986–1987, 1989, and 1994 studies show that charcoal consumption in Dakar is double the amount the Forest Service claims is harvested each year. Nevertheless, regeneration is sufficient for continued harvest. Ribot (1998) describes a commodity chain in which charcoal merchants pay migrant woodcutters low wages to cut forests. The merchants gain access through illegal arrangements with the Forest Service. The merchants draw large profits through price-fixing made possible by government supported licenses and permits. Until recently, rural dwellers in the region did not benefit from employment as woodcutters.

Capacity of Tambacounda to Adapt to Climate Change

The region surrounding the city has a large untapped agricultural and pastoral potential, as well as potential as a centre of carpentry and craft work. However, as in many other countries, investors from the national elite and overseas are buying land for speculation, money laundering and agri-business (World Bank 2010). Tambacounda's hinterland is not immune to land grabbing (Pearce 2012). In order to tap this potential and catalyse economic activity, the city and region of Tambacounda need a comprehensive land use and natural resource management plan that divides benefits among all stakeholders equitably and sustainably while also addressing climate change. It cannot simply be an elegant paper plan but something that will actually be implemented and whose regulations over various kinds of forest use and outside investment in land are enforceable and enforced.

As shown earlier, the city and region find themselves in a complex centralist system of national administration in which political and economic power (e.g. the influence of the charcoal merchants) may trump the planning process and laws. An example of the problems faced by administrators and professionals at city and regional scale who want to plan and take action to reduce the impacts of climate change is the taxation system. Located in the middle of an area rich in natural forest resources, the city does not benefit enough because the tax system (including taxes on marketing of charcoal) excludes cities, in favour of the national state and rural communities (communes – clusters of villages that are located closer to the forests). Some reassessment and redistribution/sharing of tax revenues are necessary to allow the city and region of Tambacounda to plan for and adapt to climate change and other challenges. In the end, rural dwellers would also benefit.

Without comprehensive planning and decisive action, the well-established flow of poor rural migrants to Dakar and abroad to find work will accelerate, driving Tambacounda region further into a spiral of economic and environmental decline. However, one hint of optimism is the recent interest in regional development expressed by a new government of Senegal at the national level. Could a major change in policy overcome decades of centralism, client-patron relations and corruption in order to reverse the declining trend? A hint that this might occur is the plan to build a University of Natural Resources in the city of Tambacounda. Such was a political promise of the former President in 2004. The new president has also promised that the university will be functioning in January 2015, and the Ministry of Higher Education and Research has confirmed that the university will open on schedule.

Untapped Potential of Tambacounda

After agriculture and livestock husbandry, crafts work is the second largest employer in the city. Tambacounda has access to natural capital in the form of biodiverse forests that contain some of the trees most desired for craft work such as *venn* (pterocarpus erinaceus), *dimb* (cordyla pinnata), *nere* (parkia biglobosa), *caïlcédrat* (khaya senegalensis), *kapokier* (ceiba pentandra) and *rônier* (borassus aethipum). However, deforestation as part of the highly organised and well-established charcoal trade may threaten the craft industry, as may smuggling of wood to neighbouring countries. Currently, much rosewood is being exported via Gambia to China. Nevertheless, the Chamber of Crafts attempts to organise the craft sector and help professionals to overcome a number of other constraints. These include poor access to credit, lack of a well organised market, limited upgrading of craft skills and little organisation among the artisans themselves.

Regarding the charcoal trade, an effort is being made to engage villagers themselves in sustainable, non-destructive charcoal production and marketing as a way of countering the non-sustainable extraction of biomass by powerful urban charcoal merchants whose activities threaten the interests in farmers and herders as well as craftspeople. The USAID-Wula Nafaa project has generated forest management plans for communities surrounding Tambacounda city. The project aims to produce alternative energy, alleviate poverty and maintain forest cover allowing regrowth of trees after cutting, returning to that section only when the forest has grown. However, the project's access to the urban market, the most lucrative, is still insignificant when compared to that of private urban charcoal merchants.

Discussion

Visions of Africa's urban future take extreme forms. On the dark, even apocalyptic side, there is *Planet of Slums* (Davis 2006). On the optimistic, UN-Habitat (2013: x) invites us to 're-think urban prosperity'.

> The fostering of prosperity has been one of the main reasons that explain the existence of cities. They are the places where humankind fulfils ambitions, aspirations and dreams, fulfils yearning needs . . . .

A more likely future lies between pessimistic and such optimistic extremes. Returning to Mitchell's exposition of the principles of 'climate-smart development' (Box 5.1), the challenge for Africa's small cities and towns could be large indeed. Mitchell emphasises five pillars.

1. *Ensure good disaster risk assessments are conducted that factor in the best available climate, vulnerability and exposure information to work out future climate hazard risk.* All three case study settlements had public works

departments and other civil servants aware of the flood chronology over years and other specific climate-related hazards such as land slide and potential water shortage. However, this information and knowledge was not systematised and not updated.

2. *Strengthen people's access to information (through education, media or dedicated early warning systems) about these risks and about the potential impacts of climate change.* Little public outreach about climate-related hazards was evident in Mwanga, Oudtshoorn and Tambacounda. Shortage of municipal finance and human resources seem to be the reason.
3. *Create agencies and systems that are well connected across scales, can easily learn from each other, have space to innovate and experiment with approaches and conduct scenario planning exercises with regularity.* Such a 'networked' approach is incipient in the South African case, at least to the extent that so-called 'district municipalities' are structurally linked to higher order and lower order administrative units. Legally municipalities are required to conduct disaster risk assessment and reduction. Implementation is the challenge.
4. *Find ways to increase people's equitable access to markets and services, strengthen their ability to participate in decision-making and protect their rights.* At best, one could observe some modest efforts by local government to provide access to expanded livelihood options and municipal services. The problem one sees in all three cases, however, is that national and regional political economy impinges on and limits these local efforts.
5. *Initiate high standards of environmental protection in efforts to grow the economy and take advantage of international assistance to generate green jobs and, after disasters, to 'building back greener'.* The three local government centres that served as our case examples are nowhere near even beginning to consider 'high standards of environmental protection'. Water pollution, lack of solid waste management, erosion, and in the case of Mwanga, serious deforestation, are all evident.

While Mitchell's principles are based on much experience and research, they are nonetheless normative, even exhortatory in the case of small African cities. The situation in our three case study settlements, as just noted, are far from the ideal Mitchell advocates. In Tanzania, Mwanga currently appears to be stimulating rural livelihoods and development. Yet shifting national policies, cross-cutting use of power by elites and the threat of land grabs do not provide the incubating environment in which this town can solve its own climate change challenges and take its positive influence over the hinterland a step further. Senegal's Tambacounda seems to show the other face of the city. Rural producers are exploited until nearly every drop of their surplus income is taken. Local producers cannot get decent prices for their goods because there are fixed prices for grains. In Senegal, as elsewhere, urban-rural systems include legal means of extraction, such as permits and licences for trade that go to merchants, a practise that keeps small producers out of all of the significant markets. In addition, the police force, forest service and other groups extort and tax all local enterprises, leaving them at a subsistence level. Oudtshoorn in South Africa's Western Cape Province demonstrates the importance of local governance. Whatever the role as economic catalyst for its greater region the city might have, the dysfunction of the local council in Oudtshoorn and persistent social impact of the former apartheid regime thoroughly block it.

Our case studies hint at a variety of urban futures. The three case studies can help guide the formulation of the questions that need to be asked about small African cities facing climate change, among other changes. They cannot answer such questions.

What climate-related hazards are faced by small cities in Africa today and confronted in the future?

Flooding emerges in all three case studies as a major climate-related hazard. However, the knock-on effects are complex and different. In Mwanga, there have been flood related landslides in the surrounding mountains and severe gullying in the city itself. In Oudtshoorn, severe flooding has had a heavy financial impact by destroying infrastructure and an even heavier cost to the local ostrich industry due to the role flood played in transmitting avian influenza. Tambacounda also experiences flooding and this combines with poor water and sanitation infrastructure to undermine human health.

What kind of enabling capacities should be strengthened so that staff in small cities can take the initiative to adapt to climate change?

What obstacles do the governments and residents of small cities face in adapting to climate change and realising related opportunities?

All three case studies show that local officials and professional staff require two things: clear coordination of their efforts and decentralisation of finance. The former is required because of the complex administrative structures documented in the three cases. Governance takes dual forms. National and sub-national (regional) actors interfere with local decision making. There is confusion and even paralysis to be seen in the case studies. In the absence of the power to raise local revenue or to share all but a meagre portion of national budgetary resources, small cities simply do not have the funds required to partner with innovative residents in the city and its hinterland or with entrepreneurs for starting on climate-smart development.

What potential is there for risk reduction and improved livelihood security even in the face of climate change?

The case studies show the potential for these three small cities to engage further with their hinterlands as catalysts of smart development. Each of these urban-rural zones has agro-industrial and other economic potential. Two of the cities, Mwanga and Tambacounda, are entrepots, and the market/ transportation functions produce employment and help to circulate ideas and aspirations. The wood working industry in Tambacounda could grow, as well as irrigated farming and livestock production. Mwanga, too, has a wide spectrum of under-utilised potentials to service its hinterland with inputs and innovations in agriculture, herding, fishing, planting orchards and horticulture. Oudtshoorn could further develop its tourist industry and high prized ostrich meat export. However, we have also seen that these economic activities are the very ones that are most threatened by climate change. The cities at the centres of these economic zones need to help rural people protect their livelihoods from climate change so that there is surplus production for the cities to tax. With powers of taxation and a robust, climate-smart development taking place nearby, these cities would have revenue to work on their own climate adaptation programmes, especially urban water supply, drainage and sanitation.

Conclusion

Small cities in Africa have a chance to respond to the challenge of climate change because they are capable of stimulating climate-smart resource management, land use, production, processing and consumption in their hinterlands. The small urban place has the administrative, service and market functions necessary. The small city may also not yet suffer from grossly over-burdened lifeline infrastructure, and the complex of climate-related hazards they face may still be within the range of magnitude that thoughtful planning and intervention can address. In addition, small size means that members of the local intelligentsia know one another and socialise together – district water engineer, agricultural officer, medical officer, business owners, district councillors, etc. Therefore, comprehensive solutions have a base in social networks and informal sharing of ideas. Above all, it means that bureaucrats and the private sector are responsible and accountable for decision making and use of funds and resource utilisation.

However, some large obstacles have been identified that prevent small African cities from achieving this potential. The greatest obstacle is failure of decentralisation of decision making and resources to small cities. This is not merely a matter of party political manipulation and regionalism, but built into an economic system that extracts surplus from the countryside via hierarchies of cities without leaving much at all for development (climate-smart or otherwise) in either countryside or small city.

Another uncomfortable issue that most researchers and policy advisors do not mention in polite company is whether the development vision of ruling elites in Africa is compatible at all with a bottom-up strategy that would empower local authorities properly to serve urban residents and to support rural livelihoods. The framing of 'the urban question' in the context of climate change in Africa needs to be wider in scope than 'urban', 'Africa' and 'climate change' (Wisner et al. 2004; Shipper and Burton 2009; Bicknell et al. 2010; Baker 2012; Wisner et al. 2012a; Sygna et al. 2013; O'Brien and O'Keefe 2014; Smucker et al. 2015). Firstly, it is clear that many interlinked changes are occurring, not just climate change. Secondly, many of the drivers of change are global – climate obviously, but also geo-political and economic – and they are national as well. In addition, the small

African city cannot be considered without reference to its hinterland, the countryside, or, one might say, the greater municipal region of these small urban centres. The process of extraction of surplus from periphery to centre established in the pre-colonial period, further developed in colonial Africa and today perfected by national elites seems to be the foremost obstacle to any rural or small urban development, climate-smart or otherwise. There is a real danger that the call for climate-smart development will orient climate change adaptation to ring fencing existing development visions and trajectories even when these are manifestly part of the root causes of risk (UNISDR 2015). This would be a retrograde step given existing risk management's growing focus on leveraging risk and adaptation as opportunities for critical reflection on development and consideration of transformative change (Paavola 2006; Pelling 2011; Mascarenhas and Wisner 2012; IPCC 2012, 2014; Smucker et al. 2015).

From below, opposition politics is at play as well, as demonstrated by rising discontent with the limited access to resources the majority of African citizens have in the face of monopoly by a small elite. Quoting *The State of the World's Cities* (UN-Habitat 2013: x):

> [W]hen prosperity is absent or confined to some groups, when it is only enjoyed in some parts of the city, when it is used to pursue vested interests, or when it is a justification for financial gains for the few in detriment of the majority, the city becomes the arena where the right for a shared prosperity is fought for.

An open question is whether this discontent will converge on transformative changes in government and the allocation of resources. Without such transformation the political, economic and administrative preconditions for climate change adaptation will not be met. At the moment, African national elites are adept at using climate discourse to excuse development failures, blaming nature or the 'feckless, ignorant peasant' who wantonly cuts down trees, etc. (Wisner 2010; Smucker et al. 2015). The elite also use climate as the reason why they must exercise eminent domain to allocate land to overseas direct investors who will 'modernise' farming and 'save' the nation from famine created by climate change impact on crop yields grown on small farmers' fields. The reallocation of arable land and displacement of small farmers can also be viewed as land grabs (Pearce 2012; Wisner et al. 2012a, b).

Mitchell's framework, with which we began this exploration and to which we returned in the discussion section, highlights the considerable structural and resource barriers to climate-smart development, and indeed to wider sustainable and equitable development aspirations in small cities. Despite the potential advantages offered by the proximity of decision-making and breadth of livelihood which reaches into rural hinterland and wider urban systems, this study provides a reality check. Not only are small cities at the forefront of combined social and environmental change but their capacity to cope with resultant risks, let alone to achieve climate-smart development in the context of national development trajectories that are unsustainable and inequitable, is very constrained. Climate change, when seen in this context offers scope to highlight and raise the profile of small cities and the inadequacies of piecemeal, contradictory and centralised governance regimes as revealed in Mwanga, Oudtshoorn and Tambacounda. A considerable research and policy agenda lies ahead to assist local residents to navigate and to challenge maldevelopment and to realise the opportunity transformative adaptation offers to confront structural and resource barriers.

Acknowledgements The authors would like to acknowledge the assistance of Gillian Fortune, Research Alliance for Disaster and Risk Reduction (RADAR), Stellenbosch University and Adriaan van Niekerk, Centre for Geographical Analysis (CGA), Stellenbosch University, and to thank Girma Kebbede of Mount Holyoke College (USA) for valuable comments. Ben Wisner and Adolfo Mascarenhas are grateful for the years spent with their colleagues on LKCCAP (the Local Knowledge and Climate Change Adaptation Project) that laid the foundation for the Mwanga case study (work supported by the U.S. National Science Foundation under Grant No. 0921952).

References

African Rainwater Harvesting Network (2014) Available via http://whater.eu/mod/book/view.php?id=49&chapterid=20). Accessed 4 Apr 2014

Awetu Traders (2012) Impact of animal diseases in the agricultural sector, with a focus on avian influenza. Draft, unpublished report for the Western Cape Disaster Management Centre, Cape Town

Baker J (ed) (2012) Climate change, disaster risk, and the urban poor: cities building resilience for a changing world. The World Bank, Washington, DC

Bardhan P, Mookherjee D (2006) Decentralization and local governance in developing countries: a comparative perspective. MIT Press, Cambridge

Bashir N (2013) The plastic problem in Africa. Jpn J Vet Res 61(Supplement):S1–S11. Available via http://eprints.lib.hokudai.ac.jp/dspace/bitstream/2115/52347/1/JJVR61-S_REVIEW_01.pdf. Accessed 6 Apr 2014

Bicknell J, Dodman D, Satterthwaite D (eds) (2010) Adapting cities to climate change. Earthscan, London

Botha T (2013) ANC won't be able to cling to power for too much longer. The Gremlin, 14 November. Available via http://thegremlin.co.za/oudtshoorn-news/wordpress/2013/11/14/anc-wont-be-able-to-cling-to-power-in-oudtshoorn-for-too-much-longer/. Accessed 29 Apr 2014

Centre for Geographical Analysis (2004) Growth potential of towns in the Western Cape. A research study undertaken for the Department of Environmental Affairs and Development Planning of the Western Cape Provincial Government. Available via http://www.westerncape.gov.za/text/2005/12/growth_potential_findec05.pdf. Accessed 14 Apr 2014

Cohen J, Peterson S (1999) Administrative decentralization: strategies for developing countries. United Nations, New York

Crook R, Manor J (1998) Democracy and decentralization in Southeast Asia and West Africa: participation, accountability, and performance. Cambridge University Press, Cambridge

Darwent D (1969) Growth poles and growth centers in regional planning – a review. Environ Plann 1:5–32

Davis M (2006) Planet of slums. Verso, London

Diagne K, Ndiaye A (2009) History, governance and the millennium development goals: flood risk reduction in Saint-Louis, Senegal. In: Pelling M, Wisner B (eds) Disaster risk reduction: cases from urban Africa. Earthscan, London, pp 147–168

DiMP (2007) Severe weather compound disaster: August 2006 cut-off lows and their consequences in the Southern Cape, South Africa. Unpublished report for the Western Cape Disaster Management Centre and Provincial Department of Public Works, Cape Town. Available via http://riskreductionafrica.org/en/rra-docs-docs. Accessed 16 Apr 2014

Easterly W (2014) The tyranny of experts: the forgotten rights of the poor. Basic Books, New York
Erasmus D (2013) Ostrich industry produces new products, new hope. Farmers Weekly, 30 August. Available via http://www.farmersweekly.co.za/article.aspx?id=45142&h=Ostrich--industry--new-products,-new-hope. Accessed 12 Apr 2014
Ferguson J (1994) The anti-politics machine: development, depoliticization, and bureaucratic power in Lesotho. University of Minnesota Press, Minneapolis
GNDR (Global Network of Civil Society Organisations for Disaster Reduction) (2009) Clouds but little rain ... —views from the frontline: a local perspective of progress towards implementation of the Hyogo Framework for Action. Global Network of Civil Society Organisations for Disaster Reduction, Teddington. Online www.globalnetwork-dr.org/reports/VFLfullreport0609.pdf. Accessed 4 Sept 2014
GNDR (Global Network of Civil Society Organisations for Disaster Reduction) (2011) 'If we do not join hands...'—views from the frontline: a local perspective of progress towards implementation of the Hyogo Framework for Action. Global Network of Civil Society Organisations for Disaster Reduction, Teddington. Online http://www.globalnetwork-dr.org/images/documents/vfl2011_report/VFL2011_Core_Report_en.pdf. Accessed 4 Sept 2014
GNDR (Global Network of Civil Society Organisations for Disaster Reduction) (2013) Views from the frontline— beyond 2015: a local perspective of progress towards implementation of the Hyogo Framework for Action. Global Network of Civil Society Organisations for Disaster Reduction, Teddington. Online http://www.globalnetwork-dr.org/images/documents/VFL2013/vfl2013%20reports/GNFULL%2013%20ENGLISH%20FINAL.pdf. Accessed 4 Sept 2014
Gough K, Esson V, Andreasen J, Singirankabo A, Yankson P, Yemmafouo A with Mainet H, Ninot O (2013) City dynamics. RurbanAfrica briefing no. 3. University of Copenhagen, Copenhagen. Available via http://rurbanafrica.ku.dk/outreach/briefings/2013/RurbanAfrica-Briefing-Paper-3.pdf/. Accessed 7 Apr 2014
Handmer J, Dovers S (2013) Handbook of disaster policies and institutions: improving emergency management and climate change adaptation, 2nd edn. Earthscan, London
Harriss J, Stokke K, Törnquist O (eds) (2004) Politicising democracy: the new local politics of democratisation. Palgrave Macmillan, New York
Hartshorn T (1991) Interpreting the city: an urban geography, 2nd edn. Wiley, New York
Haysom S (2013) Santuary in the city? Overseas Development Institute (ODI), London. Available via www.odi.org.uk/resources/docs/8444.pdf. Accessed 2 Apr 2014
Holloway A, Fortune G with Chasi V (2010) Risk and development annual review. Research Alliance for Disaster and Risk Reduction (RADAR). Western Cape and Periperi Publications, Cape Town. Available via http://riskreductionafrica.org/en/rra-pubs-pubs. Accessed 14 Apr 2014
IOL News (2007) Outdtshoorn municipality to function again. IOL News, 18 September. Available via http://www.iol.co.za/news/politics/oudtshoorn-municipality-to-function-again-1.371220#.U1qP0KKoGSq. Accessed 2 May 2014
IPCC (Intergovernmental Panel on Climate Change) (2012) Managing the risks of extreme events and disasters to advance climate change adaptation. A special report of working groups I and II of the IPCC. Cambridge University Press, Cambridge
IPCC (Intergovernmental Panel on Climate Change) (2014) Urban issues. In: IPCC working group II, impacts adaptation and vulnerability. Fifth assessment report. Cambridge University Press, Cambridge
Jacobs J (1985) Cities and the wealth of nations. Vintage, New York
Jasanoff S (2012) The songlines of risk. In: Jasanoff S (ed) Science and public reason. Earthscan, London, pp 133–149
Leck H, Simon D (2013) Fostering multiscalar collaboration and co-operation for effective governance of climate change adaptation. Urban Stud 50(6):1221–1238
Lopez-Carresi A, Fordham M, Wisner B, Kelman I, Gaillard JC (2014) Introduction: who, what and why. In: Lopez-Carresi A, Fordham M, Wisner B, Kelman I, Gaillard JC (eds) Disaster management: international lessons in risk reduction, response and recovery. Earthscan, London, pp 1–9

Makoye K (2013) Weather info project helps farmers to adapt, Thompson-Reuters AlertNet. Available via http://www.trust.org/item/20131213115309-u2rxt/. Accessed 2 May 2014
Mascarenhas A, Wisner B (2012) Politics: power and disaster. In: Wisner B, Gaillard JC, Kelman I (eds) The Routledge handbook of hazards and disaster risk reduction. Routledge, London, pp 48–60
McMichael A, Campbell-Lendrum H, Corvalán C, Ebi K, Githeko A, Scheraga J, Woodward A (2003) Climate change and human health. World Health Organisation, Geneva
Mertz O, Mbow C, Reenberg A, Diouf A (2009) Farmers' perceptions of climate change and agricultural adaptation strategies in rural Sahel. Environ Manag 43:804–816
Meyer W (2013) Infighting behind Oudtshoorn violence. IOL News, 4 May. Available via http://www.iol.co.za/news/politics/infighting-behind-oudtshoorn-violence-1.1510392#.UpDyttJmisE. Accessed 12 Apr 2014
Mitchell T (2010) Climate-smart disaster risk reduction. Overseas Development Institute Opinion, 12 October. Available via http://www.odi.org.uk/opinion/5076-climate-smart-disaster-risk-reduction. Accessed 13 Mar 2014
Mtyala Q (2013) Cape town's face being taken over. Times Live, 22 February. Available via http://m.timeslive.co.za/?articleId=8191187. Accessed 2 Apr 2014
Muzzini E, Lindeboom W (2008) The urban transition in Tanzania: building the empirical base for policy dialogue. World Bank, Washington, DC. Available via http://siteresources.worldbank.org/CMUDLP/Resources/tanzania_wp.pdf. Accessed 11 Apr 2014
Myers G (2011) African cities: alternative visions of urban theory and practice. Zed, London
Nel E, Binns T (2002) Putting developmental local government into practice: the experience of South Africa's towns and cities. Urban Forum 14(2.3):165–184
Nel E, Rogerson C (2005) Local economic development in the developing world: the experience of southern Africa. Transaction, New Brunswick
Nordhaus W (2013) The climate casino: risk, uncertainty and economics for a warming world. Yale University Press, New Haven
NRC (National Research Council) (1995) Population dynamics of Senegal. NRC, Washington, DC
O'Brien G, O'Keefe P (2014) Managing adaptation to climate risk: beyond fragmented responses. Routledge, London
Ogunleye E (nd) Transformation through development of growth poles in SSA. UN Economic Commission for Africa, Addis Ababa. Available via http://www.uneca.org/sites/default/files/page_attachments/ogunleye-ssa_economic_transformation_through_growth_poles_1.pdf. Accessed 2 Apr 2014
Oudtshoorn Greater Municipality (2013) Draft integrated development plan 2013. Oudtshoorn Greater Municipality, Oudtshoorn. Available via http://www.westerncape.gov.za/assets/departments/local-government/oudtshoorn-draft-idp-2013_-_2014.pdf. Accessed 11 Mar 2014
Oudtshoorn Municipality (2011) Construction of a creche for children with special needs commences in Dysselsdorp. Oudtshoorn Municipality, Oudtshoorn. Available via www.oudtmun.gov.za/. Accessed 11 Mar 2014
Oudtshoorn Municipality (2014) Oudtshoorn water project wins design accolade. Oudtshoorn Municipality, Oudtshoorn. Available via http://www.oudtmun.gov.za/index.php?option=com_content&view=article&id=409:oudtshoorn-water-project-wins-design. Accessed 15 Apr 2014
Paavola J (2006) Justice in adaptation to climate change in Tanzania. In: Adger N, Paavola J, Huq S, Mace M (eds) Fairness in adaptation to climate change. MIT Press, Cambridge, pp 201–222
Pachauri R (2009) The climate imperative. Project Syndicate, 28 April. Accessible via http://www.project-syndicate.org/commentary/the-climate-imperative. Accessed 13 Mar 2014
Parr J (1999a) Growth-pole strategies in regional economic planning: a retrospective view (Part 1: Origins and advocacy). Urban Stud 36:1247–1268
Parr J (1999b) Growth-pole strategies in regional economic planning: a retrospective view (Part 2: Implementation and outcome). Urban Stud 36:1195–1215

Pearce F (2006) When the rivers run dry: what happens when our water runs out? Transworld/Eden Project Books, London
Pearce F (2012) The landgrabbers: the new fight over who owns the Earth. Transworld/Eden Project Books, London
Pelling M (2011) Adaptation to climate change: from resilience to transformation. Earthscan, London
Pelling M, Wisner B (eds) (2009) Disaster risk reduction: cases from urban Africa. Earthscan, London
Poteete A, Ribot J (2011) Repertoires of domination: decentralization as process in Botswana and Senegal. World Dev 39(3):439–449
Ribot J (1998) Theorizing access: forest profits along Senegal's charcoal commodity chain. Dev Chang 29(2):307–338
Ribot J (1999) Decentralization and participation in Sahelian forestry: legal instruments of central political-administrative control. Africa 69(1):23–43
Ribot J (2003) Democratic decentralization of natural resources: institutional choice and discretionary power transfers in Sub-Saharan Africa. Public Adm Dev 23(1):53–65
Ribot J, Chhatre A, Lankina T (eds) (2008) Special issue introduction: the politics of choice and recognition in democratic decentralization. Conserv Soc 6(1):1–11
Robinson P (2005) From rural service centres to systems of rural service delivery. Int Dev Plann Rev 27(3):359–384
Sampford C (2013) Sovereignity: changing conceptions and challenges. In: Cooper A, Kirton J, Lisk F, Besada H (eds) Africa's health challenges. Ashgate, Surrey, pp 9–40
Satterthwaite D (2013) The political underpinnings of cities' accumulated resilience to climate change. Environ Urban 25(2):381–393
Scott J (1998) Seeing like a state: how certain schemes to improve the human condition have failed. Yale University Press, New Haven
Shipper L, Burton I (eds) (2009) Adaptation to climate change. Earthscan, London
Silver J, McEwan C, Petrella L, Baguian H (2013) Climate change, urban vulnerability and development in Saint-Louis and Bobo-Dioulasso: learning from across two West African cities. Local Environ 18(6):663–677
Simon D (2010) The challenges of global environmental change for urban Africa. Urban Forum 21(3):235–248
Simon D (2014) New thinking on urban environmental change challenges. Int Dev Plann Rev 36(2):v–xi
Simpkin S (2005) ICRC livestock in the Greater Horn of Africa. International Committee of the Red Cross, Nairobi. Available via http://www.icrc.org/eng/assets/files/other/regional-livestock-study-book.pdf. Accessed 6 Apr 2014
Smucker T, Wisner B (2008) Changing household responses to drought in Tharaka, Kenya: persistence, change, and challenge. Disasters 32(2):190–215
Smucker T, Wisner B, Mascarenhas A, Munishi P, Wangui E, Sinha G, Weiner D, Bwenge C, Lovell E (2015) Differentiated livelihoods, local institutions, and the adaptation imperative: assessing climate change adaptation policy in Tanzania. Geoforum 59:39–50
Statistics South Africa (2011) Population census data from South Africa National Census 2011. Available via http://beta2.statssa.gov.za/. Accessed 11 Mar 2014
Sygna L, O'Brien K, Wolf J (eds) (2013) A changing environment for human security: transformative approaches to research, policy and action. Routledge, London
The Poultry Site (2013) H7N7 bird flu virus found on South African farm. The Poultry Site, 23 December. Available via http://www.thepoultrysite.com/poultrynews/31010/h7n7-bird-flu-virus-found-on-south-african-ostrich-farm. Accessed 11 Mar 2014
TRC (Truth and Reconciliation Commission) (2001) Oudtshoorn post hearing community programme. Available via http://www.justice.gov.za/trc/reparations/oudtshoo.htm. Accessed 5 Mar 2014
UN (United Nations Department of Economic and Social Affairs/Population Division) (2012) World urbanization prospects: the 2011 revision. United Nations, New York. Available via http://esa.un.org/unup/pdf/WUP2011_Highlights.pdf. Accessed 11 Apr 2014

UNFPA (United Nations Fund for Population Activities) (2008) Linking population, poverty and development. UNFPA Population Issues. Available via https://www.unfpa.org/pds/urbanization.htm. Accessed 13 Mar 2014
UN-Habitat (2013) State of the world's cities 2014. UN-Habitat, Nairobi
UNISRD (United Nations International Strategy for Disaster Reduction) (2015) Global assessment of disaster risk reduction 2015. UNISDR, Geneva (in press)
Urban-Econ (2011) Eden district municipality: regional economic development strategy (draft). Urban-Econ, Cape Town. Available via http://www.edendm.co.za/index.php?option=com_remository&func=select&id=29. Accessed 13 Mar 2014
Urban-Econ (2012) Economic impact assessment of the impact of the avian influenza virus on the ostrich industry in the Western Cape. Final draft report for the Department of Economic Development and Tourism. Urban-Econ, Cape Town
Wangui E, Smucker T, Wisner B, Lovell E, Mascarenhas A, Solomon M, Weiner D, Munna A, Sinha G, Bwenge C, Meena H, Munishi P (2012) Integrated development, risk management and community-based climate change adaptation in a mountain-plains system in Northern Tanzania. J Alp Res 100(1). Available via http://rga.revues.org/1701. Accessed 1 Mar 2014
Wisner B (1988) Power and need in Africa. Earthscan, London
Wisner B (1993) Disaster vulnerability: scale, power, and daily life. GeoJournal 30(2):127–140
Wisner B (2010) Climate change and cultural diversity. Int Soc Sci J 61:131–140
Wisner B, Blaikie P, Cannon T, Davis I (2004) At risk: natural hazards, people's vulnerability and disasters, 2nd edn. Routledge, London
Wisner B, Gaillard JC, Kelman I (2012a) Framing disaster: theories and stories seeking to understand hazards, vulnerability and risk. In: Wisner B, Gaillard JC, Kelman I (eds) The Routledge handbook of hazards and disaster risk reduction. Routledge, London, pp 18–34
Wisner B, Mascarenhas A, Bwenge C, Smucker T, Wangui E, Weiner D, Munishi P (2012b) Let them eat (maize) cake: climate change discourse, misinformation and land grabbing in Tanzania. Contested global landscapes. A multidisciplinary initiative of the Cornell Institute for the Social Sciences. Cornell University, Ithaca. Available via http://www.cornell-landproject.org/download/landgrab2012papers/wisner.pdf. Accessed 14 Mar 2014
WHO (World Health Organisation) (2013) Global tuberculosis report. WHO, Geneva. Available via http://apps.who.int/iris/bitstream/10665/91355/1/9789241564656_eng.pdf. Accessed 10 Mar 2014
WMO (World Meteorology Organisation) (2014a) Global framework for climate services. WMO, Geneva. Available via http://www.gfcs-climate.org/. Accessed 2 May 2014
WMO (World Meteorology Organisation) (2014b) Roving seminars – West Africa (METAGRI Project). WMO, Geneva. Available via http://www.wmo.int/pages/prog/wcp/agm/roving_seminars/west_africa_en.php#background. Accessed 2 May 2014
World Bank (2010) Rising global interest in farmland: can it yield sustainable and equitable benefits? World Bank, Washington, DC. Available via http://siteresources.worldbank.org/INTARD/Resources/ESW_Sept7_final_final.pdf. Accessed 9 Apr 2014
Wurster K (2010) Management matter? Effects of charcoal production management on woodland regeneration in Senegal. Dissertation. University of Maryland. Available via http://drum.lib.umd.edu/bitstream/1903/10307/1/Wurster_umd_0117E_11139.pdf. Accessed 13 Apr 2014

Chapter 6
Assessing Social Vulnerability of Households and Communities in Flood Prone Urban Areas

Sigrun Kabisch, Nathalie Jean-Baptiste, Regina John, and Wilbard J. Kombe

Abstract This chapter deals with the assessment of social vulnerability in flood prone areas in Dar es Salaam, Tanzania, which recently experienced intensified flood events. While recognising that no single concept of vulnerability is widely accepted, the CLUVA project has developed an overall understanding that social vulnerability refers to the *ability of an actor/group or system to anticipate, cope with and recover from the impact of a hazard*. Our research focuses on households and communities and their adaptive capacities towards flooding events. The levels of responses have a direct effect on their livelihood. Our main concern is to investigate the conditions of households and communities when facing a particular climatic event. For our empirical analysis we developed a conceptual frame consisting of a vulnerability ladder including four main dimensions: asset, institutional, attitudinal and physical. These distinct dimensions offer a base for identifying and developing an indicator set through a participatory approach with different stakeholders. This chapter presents the example of Bonde La Mpunga in Dar es Salaam and demonstrates the different kinds of coping and adaptation measures being adopted with respect to local flooding issues.

Keywords Social vulnerability framework • Indicators • Household • Community • Flooding • Adaptation

---

S. Kabisch (✉) • N. Jean-Baptiste
Department of Urban and Environmental Sociology, Helmholtz Centre for Environmental Research-UFZ, Permoserstraße 15, 04318 Leipzig, Germany
e-mail: sigrun.kabisch@ufz.de; nathalie.jean-baptiste@ufz.de

R. John
School of Urban and Regional Planning, Ardhi University, P.O. Box 35176, Dar es Salaam, Tanzania
e-mail: lyakurwa_gina@yahoo.co.uk

W.J. Kombe
Institute of Human Settlements Studies, Ardhi University, P.O. Box 35176, Dar es Salaam, Tanzania
e-mail: kombewilbard@yahoo.com

S. Pauleit et al. (eds.), *Urban Vulnerability and Climate Change in Africa*, Future City 4,
DOI 10.1007/978-3-319-03982-4_6

Introduction

Millions of people worldwide are affected by climate change related extreme events such as flooding, drought, heat stress, storms or landslides. While attention from decision-makers as well as scientists was primarily given to rural areas in the past, urban regions have gained in importance in recent years. This is due to the increase of urban population, the growth of cities and their susceptibility towards climate change impacts (Brauch et al. 2011; UN-Habitat 2011; Jha et al. 2012; Heinrichs et al. 2012; Pelling and Blackburn 2013; Wamsler 2014). The consensus is that Africa has entered into the twenty-first century facing challenges of unprecedented urbanisation, which is highly related to poverty. This includes not only poor social and technical infrastructure provision; in many cases the absence of basic services such as potable water supply and storm water drainage in most informal settlements intensifies aspects of social vulnerability. Haphazard growth of housing areas, especially unregulated infill, extensions and rapid densification in informal settlements, which often accommodate the very poor population, compound problems of urban management and governance. In addition, climate change induced extreme events exacerbate the problems urban dwellers already face. African cities need to simultaneously deal with development issues of today while facing future climatic uncertainty.

In situations of relative resource scarcity, severe governance deficits, unregulated informal urbanisation and escalating threats emanating from climate change, the increasing social vulnerability especially among marginalised settlers in informal settlements emerges as a key priority. How do households and communities in CLUVA case study cities deal with these challenges? What mechanisms do they use to cope with and adapt to climate related events? In order to answer these questions we developed a specific social vulnerability framework following a tradition of analysis based on a "multidimensional view of climate – society interactions" (O'Brien et al. 2007: 76) as well as asset adaptation frameworks (Moser 1998; Moser and Stein 2011). Our framework substantiates the social science approach concerning actor involvement as a key component in climate adaptation strategies. We applied it in Ouagadougou, Addis Ababa and Dar es Salaam to different degrees because of restricted available resources. This chapter highlights the findings from the community Bonde la Mpunga in Dar es Salaam.

The results strengthen the perspective on social aspects in addition to the high-resolution climate models consistent with global change scenarios, presented in Chap. 2 of this volume. In doing so, we complete the multi-level concept of CLUVA. By focussing on the community and household level we offer particular insights on how people deal with the vulnerability of their livelihoods. We illustrate low-level institutional processes in contrast to the multi-level governance analytical framework discussed in Chap. 9 of this volume. Thus, the assessment presented below contributes to a holistic vulnerability appraisal, which was intended within the CLUVA project on "CLimate Change and Urban Vulnerability in Africa" (for a general introduction to CLUVA, see also Chap. 1).

This chapter focuses on the question of how individuals, households and communities anticipate, cope with, resist, as well as recover from frequent seasonal floods, taking into account the recent experiences of heavy rainfall and extreme flooding. The first part of the chapter offers a theoretical overview of the CLUVA social vulnerability framework and specifications on a set of relevant context-based indicators. The second part introduces the case of Bonde la Mpunga and offers empirical findings gained by using qualitative and quantitative methods.

Social Vulnerability

Vulnerability is a multifaceted and contested construct. It is related to terms such as risk, natural hazards, coping and adaptive capacity, sensitivity, resilience, poverty and even food security. Over the years, vulnerability definitions have emerged among the disaster risk community (Hewitt 1998), in the development context (Chambers 1989; Watts and Bohle 1993), in sustainability (Turner et al. 2003), and climate change research (Füssel and Klein 2006), as well as in people-environment studies (Kabisch et al. 2012). The proliferation of existing definitions is due to the range of different approaches and perspectives of what vulnerability represents (Birkmann 2006). However, it is most often conceptualised as being constituted by elements of exposure, susceptibility,[1] and coping and adaptive capacity. Thereby, exposure describes the physical precondition being affected; susceptibility involves preconditions of fragility; and finally, coping and adaptive capacities refer to mechanisms and resources to prepare, to resist and to recover (Fuchs et al. 2011). The review of the existing literature shows that the term vulnerability stretches from being considered as an internal risk factor to being viewed as a multiple structure concept, which integrates different spheres of knowledge (Vogel and O'Brien 2004; O'Brien et al. 2007). Such spheres include physical, environmental, institutional and social factors that investigate the sensitivity level of a certain group or population.

Regarding the concept of vulnerability, it is related to the exposure and susceptibility of a household towards an extreme event and also to people's coping and adaptive capacities. This capacity has an important implication. It considers affected people to be capable of acting. It implies a dynamic relation of the level of awareness of a group, their knowledge about natural hazards, their motivation and their attitude to act. Furthermore, it includes how groups or individuals take responsibility for their safety and describes their ability to access different types of

[1] The term *susceptibility* used by the disaster risk community holds similar meaning as the concept of *sensitivity* which is predominantly used among climate change scholars. It refers essentially to the predisposition of a system (people, infrastructure and environment) to be affected or to suffer harm from a disastrous event (Cardona et al. 2012).

resources (i.e. financial aid information). We therefore propose to understand vulnerability in accordance to Blaikie et al. (1994: 9) as "the characteristics of a person or group in terms of their capacity to anticipate, cope with, resist, and recover from the impact of a natural hazard".

With regards to the approach of vulnerability and also echoing the basic principles of the 'Sustainable Livelihoods Frameworks', Wisner provides basic considerations to our understanding of social vulnerability (Wisner and Luce 1993; Wisner 2004). He emphasises how climatic stressors appear as 'normal' events that need to be repeatedly dealt with in 'everyday life' of individuals and communities. This 'normality' is found to be a trigger for different types of self-protection and adaptation measures undertaken periodically. Wisner attributes *social vulnerability* to a "combination of factors that determine the degree to which someone's life, livelihood, property and other assets are put at risk" (2004: 11). This understanding is focused on social dimensions of a hazard or disaster, respectively.

There are two basic assumptions at the core of social vulnerability. First, as stated above, it is related to exposure, susceptibility and coping capacities. Second, it is a relational construct as it relates "something or someone who is vulnerable to something else as a source of potential harm because of some property of the subject or the object" (Green 2004: 323). It hence describes the complex relation between *place* and *extreme event* (Bohle and Glade 2008). Therefore, individuals, households, communities, organisations, regions or entire states can be vulnerable to natural hazards.

The term *social vulnerability* describes the capability of people to deal with a hazard. For understanding and explaining their behaviour, the hazard itself (e.g., flood, earthquake, fire) is of subordinate interest. The main focus of social vulnerability research is not the height of a flood or the intensity of an earthquake that defines its social, psychological, health and economic consequences; it is rather the *societal context* that explains the severity of the consequences (Steinführer and Kuhlicke 2007; Kuhlicke et al. 2012). Within social vulnerability research it is argued that it makes a difference whether a flood hits a wealthy or a poor community, an elderly or a young person. It is a matter of resources and possessions that intensify or reduce the level of vulnerability. Kithiia stresses "that the greatest burden of the impacts of climate change in major urban areas in East Africa is likely to fall disproportionately on the urban poor who lack the most basic services … Climate change impacts are bound to undermine the livelihoods of these urban poor communities that are already on the edge of coping capacity" (2011: 176). Individuals, households and communities who fall into more than one category of vulnerability (e.g. poverty, marginalised minority) are confronted with huge obstacles to prepare for and respond to a hazard. Thereby, women, young children and older persons share vulnerabilities in terms of being excluded from decision making. Children and the elderly are particularly dependent on support by relatives and community based social networks (UN-Habitat 2011: 79–84). Reflecting the several factors, the level of social vulnerability is related to specific social inequalities, such as financial insecurity of people facing a natural hazard.

Assessing Social Vulnerability in the African Context

Assessing social vulnerability within the African urban context implies considering existing challenges that make the region particularly sensitive to climate related changes (Pelling and Wisner 2009). Recent intensified flooding events, such as the following: 2009 in Ouagadougou; 2013 in Saint Louis; 2011 and again quite recently in April 2014 in Dar es Salaam, have added a new layer of problems in the Sub-Saharan region. It is an area already known to host the largest number of water-stressed countries of any other place on the globe as 300 million out of an estimated 800 million people live in a water stressed environment (NEPAD 2006; United Nations World Water Assessment Programme 2012). Communities frequently exposed are those consolidated in coastal settlements (Pelling and Blackburn 2013) and/or located in proximity to a river or water channel and with their placement dependent on fresh water sources. These are affected by recurrent flooding, characterised by changed intensity and frequency, which causes substantial physical damages, such as the destruction of houses and their sanitation facilities, as well as damage to roads, and infrastructure, leading to the disruption of daily social lives. Evidences from these catastrophes highlight that the anticipation of and preparation for hazardous events have not been appropriate or sufficient. As a consequence, the resistance and the coping capacities of local residents were restricted and people's recovery and reconstruction took both a long time and required a high level of support (John et al. 2014). Thus, to reduce the social vulnerability to natural hazards, the variation of risk events has to be taken into account. Changed intensity and frequency of natural hazards require appropriate awareness of residents and decision-makers to develop and implement suitable adaptation strategies (Dodman et al. 2011; Dodman 2013). The attitude and perception of risks in hazard prone settlements influence coping capacities in case of extreme events (Jean-Baptiste et al. 2013).

Furthermore, the complexity of land management as well as the interactions of formal and informal institutions, shape local government responses and long-term adaptive strategies. Local governmental authorities often lack the mandate and financial means to take long-term flood reduction actions. In that regard, unregulated urban consolidation and crowded informal settlements on marginal land, including wetlands, river valleys and other flood prone areas, provide ample evidence that governmental institutions are ill-equipped to fulfil their critical urban management functions, let alone to prepare for and effectively respond to climate change induced disasters such as floods. Finally, the conditions of the physical environment, particularly man-made structures, influence the degree of severity of the physical damages reported in recent years (De Risi et al. 2013). In poor communities, the exposure to flood not only depends on their location but also on the quality of the settlements and is closely connected with adequate technical infrastructure such as water access, sanitation and drainage. The provision of such infrastructure remains challenging in informal or gradually consolidated settlements, which face cumulative losses and increasing needs.

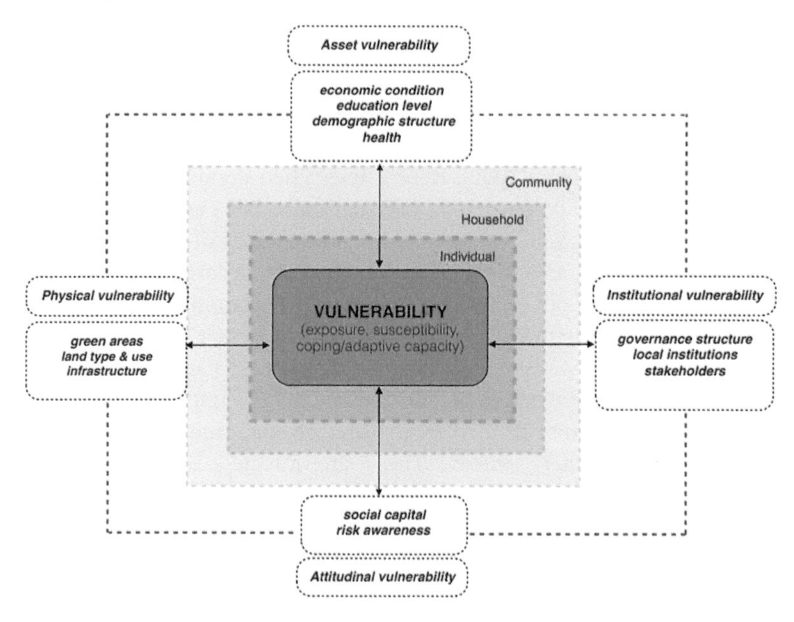

Fig. 6.1 CLUVA framework for assessing social vulnerability (Jean-Baptiste et al. 2011, 2013)

In light of these contextual particularities, we argue that vulnerability assessment in African cities requires highlighting and interlinking four main dimensions: Asset, institutional, attitudinal and physical (Jean-Baptiste et al. 2011, 2013; see Fig. 6.1):

1. Asset vulnerability encompasses the human livelihood and material resources of individuals and groups.
2. Institutional vulnerability refers to the state of local authorities and civil action groups that operate to prevent, adapt or mitigate the effect of extreme weather events.
3. Attitudinal vulnerability conveys the perception and risk management attitude of individuals and groups.
4. Physical vulnerability is determined by the characteristics of the land cover, whether natural or man-made.

We subscribe to the idea that assessment of vulnerability cannot be studied in isolation from contextual circumstances (Cannon 1994; Wisner et al. 2004; Pelling 2006). It is indeed the context that frames and influences the exposure to climatic changes as well as potential responses. It is generally agreed upon that the main reasons for a more acute impact of floods are related to very dynamic socio-economic, demographic and technical conditions. Unplanned and rapid urbanisation in high risk areas as well as environmental degradation is at the centre of causes of climate related disruptions.

The asset, institutional, attitudinal, and physical dimensions proposed in Fig. 6.1 offer a framework for assessing the vulnerability of the CLUVA case study cities. The described approach characterises the four vulnerability dimensions at individual, household and community level, and stresses the interlinkages and dependencies between them. It contributes to a context-sensitive assessment of social vulnerability, differentiating between exposure and susceptibility as well as coping and adaptive capacity. It sheds light on the available resources, detects deficiencies, obstacles and constraints, and supports prioritisation of decision-making and action.

Defining the Asset, Institutional, Attitudinal and Physical Dimensions of Social Vulnerability

Asset is understood as the human livelihood including material resources of individuals and groups. As Fig. 6.1 illustrates, the economic condition of individuals or groups, including housing conditions along with their education level, demographic structure, and health can be described as asset-based conditions. Asset is seen as the margin of well-being. It allows those at risk to resist potential hazards. Asset can be understood as a stimulus of vulnerability or capacity to act. Asset-based assessments with regard to vulnerability are conducted to identify the different resources individuals are equipped with (Chambers 1995; Moser and Stein 2011). It is assumed that the availability and diversity of these assets can reduce vulnerability. Since the 1990s a number of frameworks and approaches emerged adding to an already extensive literature on asset and rights and entitlements. For example, the "Asset Vulnerability Framework" (Moser 1998) regroups an extensive household asset portfolio, distinguishing asset types such as labour, human capital, productive assets, household relations and social capital. The dependency on informal sources of income plays a vital role in this context. These conditions can be regrouped in tangible (i.e. material possessions) and intangible (i.e. well-being and knowledge) resources. These influence the degree of resistance of a household or community when dealing with seasonal and/or extreme natural events.

We understand *institutions* as societal manifestations representing "formal constraints (rule, laws and constitutions) or informal ones (norms of behaviour, conventions) that mould interaction in a society." (Badjeck et al. 2009: 211). They influence and shape formal and informal procedures, routines, norms and conventions of people embedded in the daily interaction of groups, organisations and other collective bodies. Institutional vulnerability is thus the local governance mechanisms as well as formal or informal modes of interaction that put people at risk or help them to adapt to and cope with risks. It can include the state of local authorities and civil groups that operate to resist, adapt or mitigate the effect of extreme weather events. In the absence of institutional capacity there is limited knowledge of an event, poor dissemination opportunities, low emergency preparedness and ability to absorb impacts. As a result, any biophysical event has the potential to turn into a serious threat to communities. In African cities, the importance of identifying and recognising the role of formal and informal institutions and actors lies in the fact that these two are intertwined. In CLUVA case study cities we observed that formal hierarchies are linked to legitimate channels of local government, whereas informal hierarchies are enforced outside legally binding mechanisms (for detailed examination of multi-level governance structures in the CLUVA case study cities see Chap. 9). However, the two are often functionally inter-linked and complementary.

The *attitude* towards risk includes both social capital of people as well as their risk awareness. Social capital is understood as an individual resource. It is related to the various social networks a person creates and belongs to; and the economic, social and cultural resources these networks provide. Social capital also describes a collective asset in terms of community resources for which trust and shared norms are basic requirements. In addition, we consider the availability of network capital such as family support, access to resources through information, access to help through social ties to positively affect susceptibility. In contrast to social capital, risk awareness denotes the process of collecting, selecting and interpreting signals about uncertain impacts of events, activities or technologies. These signals can refer to direct experience (e.g. witnessing a flood) or indirect experience (e.g. information from others, such as reading about a natural disaster or having heard a story) (Wachinger et al. 2013). In the context of CLUVA, we view risk awareness as the everyday processes by which settlers in flood prone areas perceive risk. Our understanding relates to a certain extent to intuitive awareness of risks based on attitudinal attributes and residents' own evaluation of the consequences of climatic events.

Physical components such as the urban ecosystem, existing green areas, use of land as well as the condition of the technical infrastructure (including buildings and sanitation) are important to consider as they are particularly vulnerable to climatic related events. The green structure is relevant to increase the coping and adaptive capacities of societies as it can protect urban neighbourhoods through flood and storm water retention, soil protection and mitigation of heat. However, ad-hoc and very quick urban zoning initiatives from local governments and weak urban planning law enforcement cannot prevent the expansion of settlement into green and/or hazardous areas. In Dar es Salaam, it is estimated that 80 % of the city consists of informal, unplanned settlements (Kombe 2005).

Putting Theoretical Reflections into Practice – Identifying Relevant Indicators for Assessing Social Vulnerability

The CLUVA social vulnerability framework is operationalised through the identification and development of a set of appropriate indicators. Following a literature review, we conducted two working sessions with local stakeholders to exchange views on the contextual relevance of pre-identified indicators. Discussions were held in Ouagadougou, Addis Ababa, and Dar es Salaam. These exchanges offered a strong local reference to initial desk-based research. For instance, the discussions surrounding indicators to evaluate the economic condition in flood prone areas led to insightful comments on the relevance of an indicator such as 'source of income' against another such as 'level of income'. The former would take into account the variety of earning opportunities that play a key role in ability to recover in the aftermath of a flood. From these initial exchanges, a total of 39 indicators for all four vulnerability dimensions – asset, attitudinal, institutional and physical – were identified during stakeholder meetings (for detailed information about the selection procedure of indicators and the whole set of indicators see Jean-Baptiste et al. 2011, also available at: http://www.cluva.eu/deliverables/CLUVA_D2.11.pdf).

Indicators for Assessing Asset Vulnerability

Several indicators were regrouped to evaluate asset-based conditions. They refer mostly to socio-demographic factors that have been long known to have an important contribution to the level of vulnerability of affected populations. The variables most commonly utilised are those addressing the level of education, age structure and source of income (Adger et al. 2004; Vincent 2004). The *level of education* is linked to access information and resources as well as risk acknowledgment, which ultimately reduces vulnerability. The *age structure* offers some indication on the percentage of young and old members of households as they represent the most dependent groups unable to respond to a critical event on their own. Finally, the *source of income* (formal or informal) indicates the level of employment or type of economic asset in the household. An important consideration is evaluating whether incomes reported are 'stable,' regardless of local hazards. This implies a certain level of preparedness or capacity to take measures against a potential hazard.

Indicators for Assessing Institutional Vulnerability

The indicators regrouped to evaluate the institutional landscape refer mostly to the local governance mechanism, including formal or informal modes of interaction, that influences positively or negatively people's ability to adapt to and cope with risks. The *type of local government* includes the state of local authorities and civil groups that operate to resist, adapt or mitigate the effect of extreme weather events. It also indicates the existing structure through which residents' voices are heard and the degree to which local residents can express their needs. The *existence of community based organisations (CBOs), non-governmental organisations (NGOs) and other local institutions* indicates the degree to which households have access or contact with local institutions and whether they have benefited from them. At a community level, this indicator refers to numbers of active local organisations in the community and their level of contact with local residents and/or local leaders. This can be commonly achieved through stakeholder mapping exercises, allowing an overview of active organisations that can influence responses on the ground.

Indicators for Assessing Attitudinal Vulnerability

The indicators measuring household and community attitudes were regrouped to evaluate the individual and collective network resources that allow those exposed to hazards to better cope with a potential event. The *number of networks* and *trust* are linked with the existing social capital described above. The *hazard experience* relates to any previous experiences (e.g. death, damages, injuries, material and non-material losses, evacuation).

Indicators for Assessing Physical Vulnerability

The indicators measuring the physical vulnerability are closely linked to anthropogenic causes of flooding. They refer largely to the condition of existing *buildings and infrastructure,* but also to urban ecosystems and related ecosystem services (Chap. 4), existing *green areas* and the use of land and its degree of degradation. Particularly, the existence and condition of *storm water drainage* provide information on the pressure on the urban infrastructure, which has a direct impact on community risk.

The Case Study Approach and Empirical Findings[2]

Selection Criteria for the Case Study Area and Research Questions

The indicators mentioned above were used to measure the different vulnerability dimensions in the selected study location in Dar es Salaam. Five criteria are crucial for selection: The first criterion is *exposure*; the settlement must have been located in a flood prone area such as a low-lying basin with little to no elevation, or adjacent to a river with the risk of water sprawl. Second, the settlement must have been experiencing *recurrent flooding* and more intense flood impact in recent years. Thirdly, some *adaptation measures* had to exist to deal with floods, enabling the assessment of the fragility of different adaptation strategies. Fourth, *local institutions,* such as CBOs or NGOs must have been active in the settlement. Fifth, the settlement must be either at a *booming and/or saturation stage of development* so as to have the potential for sufficient information to exist regarding household experiences of flood hazards. In an initial scoping exercise 20 settlements particularly

[2] The case study was carried out by PhD candidate Regina John from Ardhi University. Selected findings presented in this chapter are part of her current dissertation which compares the social vulnerability of households of two flood prone settlements in Dar es Salaam: Magomani Suna and Bonde La Mpunga.

exposed to floods were identified. Among them was Bonde la Mpunga, a low-lying settlement facing the Indian Ocean. It was selected for investigation because it met all listed criteria.

Four major considerations framed the empirical study: To begin with, the overall socio-economic status of households was investigated to determine the asset vulnerability. Following, the existing local institutions and the actors involved were identified. Afterwards we analysed how the residents perceived and evaluated their activities, thus how they become aware of the hazard. And finally, we looked at the physical components and put particular focus on the drainage system and how people take care of this infrastructure facility. Taking this knowledge into consideration, we draw attention to existing coping and adaptation strategies to reveal current actions undertaken to reduce negative consequences of floods in the community. Five questions helped to guide the empirical work; they address the context of social vulnerability in Bonde la Mpunga. The questions are:

1. What are the asset features that influence the vulnerability of households in Bonde la Mpunga?
2. Which local community institutions exist and how do residents perceive their activities in case of flooding events?
3. Which attitudinal characteristics are essential in coping with flooding?
4. What are the major characteristics of physical vulnerability?
5. What are the most recurrent coping and adaptation measures to flooding in the community?

The Methodological Design of the Case Study Approach

Characteristics of the Case Study Area

Bonde la Mpunga is one of Dar es Salaam's coastal Sub Wards (also known as Mtaa) facing the Indian Ocean in the North-East. Bonde la Mpunga belongs jurisdictionally to the Msasani Ward, which is part of the Kinondoni Municipality (Fig. 6.2). The Sub Ward covers an area of about 120 ha with a population of about 2,820 households. The total population was 16,430 according to the 2002 National population census (URT 2005).

Bonde la Mpunga is a low-lying flat area, which lies between 0 and 3 m above sea level. Its central part is a wetland used for paddy production back in the 1970s (Kiunsi et al. 2009a). The area was part of the drainage basin of the city and an outlet of the Kijitonyama River towards the Indian Ocean (ibid.). For that reason, the 1979 Dar es Salaam Master Plan earmarked it as hazardous land with certain planning restrictions. However, despite these restrictions, the settlement has grown and gained significant institutional and economic value in the last 40 years. It has both planned and unplanned land use characteristics, including governmental and institutional buildings and a number of new residential development areas, as well as a private health and shopping centre. Bonde la Mpunga is already a saturated

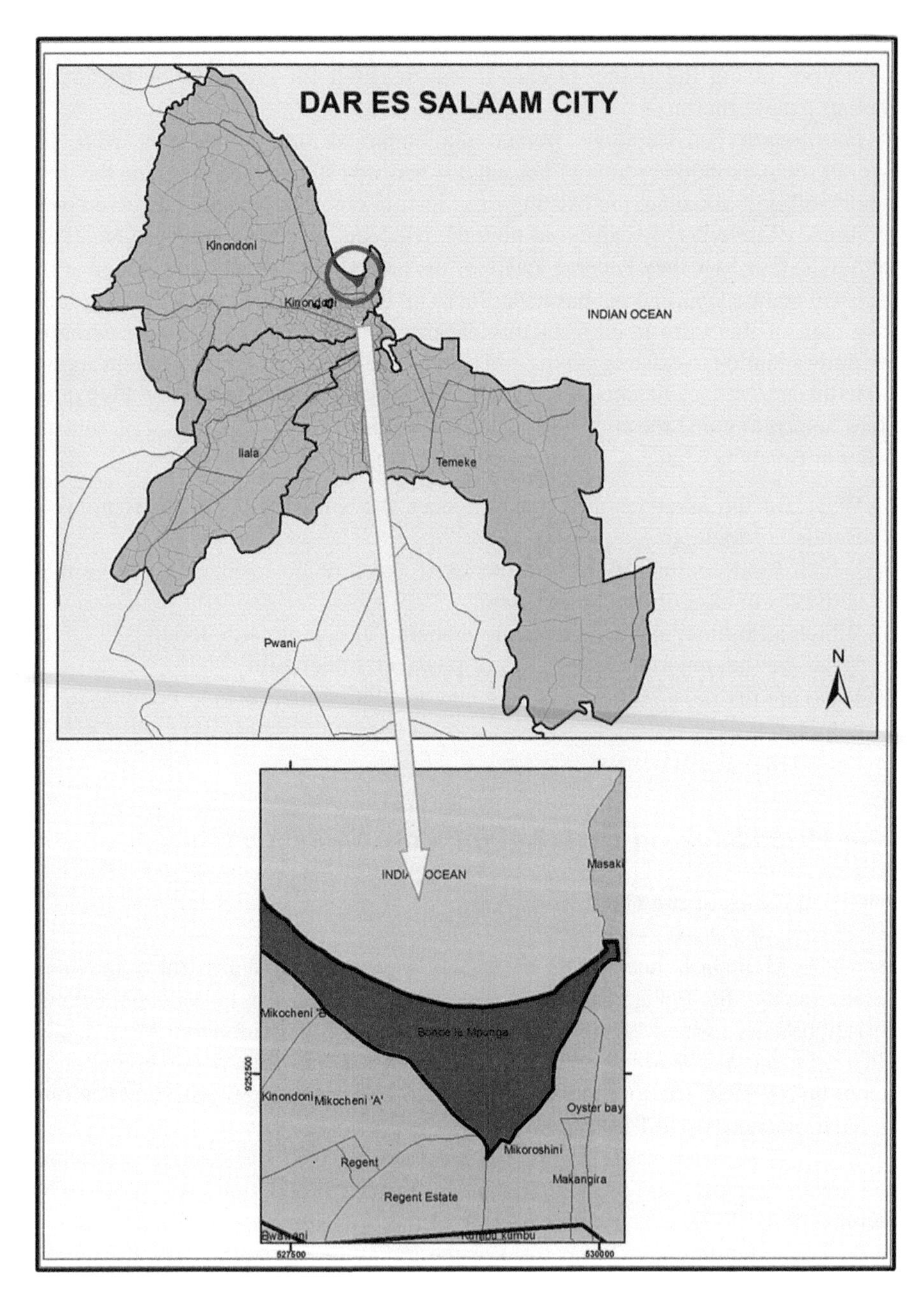

Fig. 6.2 Location of Bonde la Mpunga in Dar es Salaam, modified from National Bureau of Statistics (NBS) Census Data 2012 (Design by Regina John 2013)

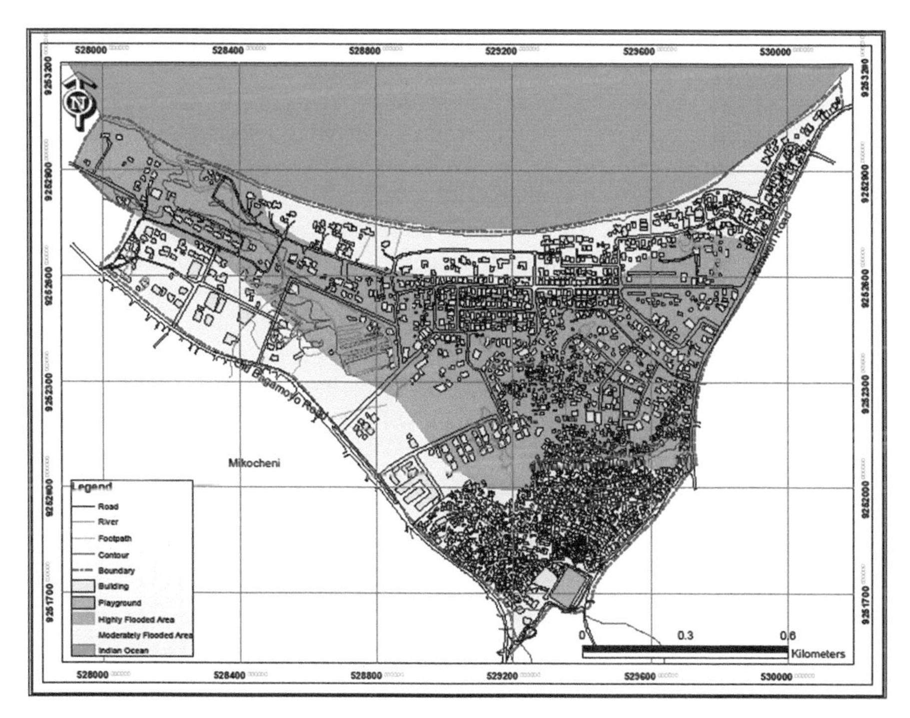

Fig. 6.3 Map of Bonde la Mpunga settlement, based on satellite images 2008 from the National Bureau of Statistics (NBS) (Design by Regina John 2013)

settlement; however, it faces further housing and commercial development at its fringes. Thus, economic pressure on the settlement due to its favoured location pushes the poor residents to sell their estates and move out. The poorer households, which are located in the lower and flood prone basin (Fig. 6.3), have found themselves enclosed in an area that brings economic revenues as well as recurrent flood hazard related challenges.

Figure 6.3 shows a map of moderate and high flooded areas as well as the existing exposed buildings in Bonde la Mpunga. The settlement is generally exposed to floods. All areas are affected, except the southern portion of the settlement. Most residential buildings are made of sand cement blocks roofed with corrugated iron sheets. We recorded a total of 2,085 constructions, of which only very few were built of adobe and roofed with thatches or iron roofs. This implies that the settlement has been consolidated over several years, transitioning from low quality materials to more resistant ones despite recurrent seasonal flooding. The majority of households experiences poor access to basic social and infrastructure services, including sanitation, energy, communication, solid waste management, water, education, health, and green structures. An estimated two-thirds of the households use pit latrines, while one-third has water closets.

Bonde la Mpunga is one of the typical flood prone settlements in the city of Dar es Salaam that experience frequent floods. In most flooding events, water reaches the knee level, which causes risk to human life, as well as material assets. Despite previous research conducted in the area (Kiunsi et al. 2009b; START 2011), as well as various measures taken by the residents to alleviate the problem, the settlement has been experiencing recurrent flooding and more intense flood impacts in recent years. In particular, the flood in December 2011 caused serious damages, including fatal casualties. These experiences are quite fresh in the memories of the residents. Therefore, the presence of various institutions within the settlement that deal with community issues towards coping with flooding events and preventing damages signifies some adaptive capacity worth exploring. These institutions, in particular, focus on management activities to keep storm water drains and solid waste collection functioning.

Furthermore, the settlement is characterised by high housing density. Vacant land is no longer available. The disappearance of retention areas and the increase of population and buildings have led to more challenges in terms of managing the flooding problem. It is therefore of interest to understand how vulnerability and adaptive capacities change in such a situation.

Research Methods

We applied several quantitative and qualitative approaches using a variety of data collection instruments. We combined tools to gather relevant information, as statistical data is lacking at household level. Qualitative data collection tools, such as structured household interviews and focus group discussions helped us to manage and balance available information retrieved from different field activities.

Structured household interviews were carried out based on a questionnaire in 2012 between June and December. The research team conducted 309 interviews to investigate residents' livelihood assets, their social and demographic characteristics, the impact of flooding as well as their experiences and capacities to cope with flooding. The number of houses interviewed depended largely on the extent of densification and severity of flooding. Consequently, more households were selected from the highly flooded zones. Respondents were household heads or elder household members with sufficient experience, preferably one who has stayed with the household since its establishment in the settlement.

We used questionnaire survey data, official statistics, quantitative information, and focus group discussion to get an overview on the geographical situation, socio-economic characteristics and behavioural routines towards flooding of the local population. The *focus group discussions* were conducted between June and September 2012 with different groups. Three focus group sessions were organised consisting of a women's group, a men's group and a fishermen's group (Fig. 6.4). The discussions centred around the role of local leaders in advocating and driving social mobilisation and change, the socio-cultural issues that influence adaptation to flood, and the risk management at community level. During the discussions, the impact of the floods of December 2011 were once again described with emphasis on household experiences of flood hazards and the various adaptation strategies employed.

Fig. 6.4 Focus group discussion with fishermen (Picture: Sigrun Kabisch 2012)

Insights into the livelihood of the community and their context sensitivity were gained through a qualitative approach. This was highly effective to have a more in depth knowledge of the constraints of households and their local perception of vulnerability. Time investment and the establishment of trust were crucial for obtaining the data on the ground. We prioritised depth of knowledge over breadth of experiences and therefore concentrated on the results obtained from one flood prone community.

By transparent description of the selection criteria we underline the transferability of key findings to other communities with similar characteristics. We would like to mention, however, that restricted resources (finances, time, personnel) and obstacles in field work, such as language problems (translation between English and Swahili) did not fully allow reaching our scientific goals. These obstacles relate to two aspects: First, to explore more than one settlement in depth for a comparative study in Dar es Salaam, and second, for detailed empirical investigation of connections between the dimensions of the theoretical framework.

Assessing Asset Vulnerability

To assess the asset characteristics of vulnerable households in Bonde la Mpunga the following important socio-economic indicators have been analysed by interviews in the questionnaire survey: household size, age structure, education, source of income, income level and tenure arrangement. These indicators are related to the resources and competences of the different household compositions to cope with flood hazards.

In regard to the household size, Bonde la Mpunga consists of a large proportion of households with more than five persons. The smallest household size consists of one person, while the largest has 11. Among the 309 households involved in the survey, those with five members and more represent the highest proportion (41 %), followed by households with four members (24 %), three members (19 %), two members (13 %) and one member (3 %). The mean household size is 4.2. Although the household size in Bonde la Mpunga is below the national average of 4.8, it is rather above the Kinondoni Municipal household size which is 4.0 and the Msasani Ward, which is 3.9 (URT 2013). A large household size is associated with a low social status and an increased number of dependent children or elderly (Cutter et al. 2003). Thus, the studies' findings suggest that the majority of the households in Bonde la Mpunga are characterised by large household sizes that are particularly vulnerable to floods. The reasons are the relatively large number of dependent household members and the restricted economic resources to prepare for and resist flooding events.

The household sizes in Bonde la Mpunga are closely linked with a diverse age structure. The age of the household members ranges from below 18 to over 65 years. However, the households of particular interest are those with members aged below 18 and above 65 years. The findings show that 50 % (146 out of 309) of the interviewed households have two to three persons below the age of 18. In contrast, the oldest age group constitutes the smallest share. Only 44 out of 309 interviewed households have one or two members above 65 years. But households with persons in these age groups are more vulnerable than households in which every person can rely on her-/himself in case of a hazard event. People in extreme ages are reliant on other people in case of evacuation during flooding. In the interviews, it has been reported that old people drowned because rescuing them was not possible.

A low education level characterises the respondents from Bonde la Mpunga because more than 50 % of the households have only attained primary level education. A very small proportion (8 %) has attained a university degree. Low education level is associated with poor remuneration of employees and restricted chances of formal employment. In consequence, these households have fewer economic resources to anticipate, cope with and recover from the effects of floods (Adger et al. 2004; Cutter et al. 2009).

The *source of income* is connected with the level of education. Of the 309 households involved, 73 % derive their income from informal economic activities, largely small businesses, which substantiate the importance of the informal sector. Informal activities include food vending, such as selling vegetables and fruits, or retailing. These activities generally take place in the vicinity of the house. Therefore, they are susceptible to damages in case of flooding in their settlement in a twofold way: first, with regard to housing; and second, with regard to employment. The status of employment and the stability of income influence the capacity to prepare and to adapt to flood events. The findings of the study reveal a low adaptive capacity of the majority of the households in Bonde la Mpunga. Poor and low-income households have fewer monetary resources to spend on prevention, emergency supplies, and recovery (Cutter et al. 2009). Our survey findings, furthermore, reveal significant variations of income among the households. A significant percentage (39 %) of the population has a monthly income level of between Tanzanian Shillings (TShs) 100,000–300,000 (~USD 62.5–USD 181). Another 24 % earn between 300,001–500,000 TShs (~USD 181–USD 300). Only 9 % have an income level of TShs 1,000.000 (~USD 625) and above. These findings confirm the assumptions of a diverse social-economic status.

It is also worth noting that the interviewed households consist of both home-owners (70 %) as well as tenants (30 %). The two tenure arrangements inherit different opportunities and challenges in terms of dealing with the flood hazards. Our study findings show that tenants are less able to take measures to reduce flood impacts than the house owners. Due to their lack of decision-making authority concerning their residency and the build structures, tenants are more vulnerable.

Furthermore, 36 % have title deeds, while 64 % of the households have no title deeds. This stresses the fact that most of the households lack formal tenure. On the other hand, the appearance of formal tenure in a settlement marked as informal and flood-prone adds another layer of complexity to the study of social vulnerability.

The analysis of the socio-economic features confirmed the importance of indicators such as household size, age structure, education level and source of income in relation to the employment status and stability, as well as tenure arrangement to evaluate asset vulnerability of residents in Bonde la Mpunga. These household characteristics overlap and are interlinked; it is therefore crucial to understand the interdependence between them in the context of asset vulnerability towards flooding.

Assessing Institutional Vulnerability

Based on focus group discussions in the settlement under investigation, we found various institutions engaged in flood reduction activities. While formal institutions are registered and recognised by law, informal ones operate outside legally binding frameworks. Formal institutions include local government authorities, such as the Sub Ward government and NGOs, while informal institutions are composed largely of CBOs and other informal groups. The local government authorities, in particular the Sub Ward government and Mtaa-leaders as well as Ten cell units[3] with their representatives as the lowest level of government, appear as the most important institutions. In the focus group discussions these institutions were praised for their efforts to mobilise residents to construct and clean storm water drains, albeit at a small scale. In addition to environmental cleanliness campaigns and cleaning of

[3] Ten cell units consist of ten adjacent plots in the settlement.

Table 6.1 Identified institutions and their role in Bonde la Mpunga based on focus group discussions

Type of institution	Name	Formality status	Roles
Local government authorities	Sub Ward government (Mtaa) leaders	Formal	Give warning information, mobilise residents to construct and maintain storm water drains, organise campaigns for environmental cleanness, provide accommodation to flood victims
	Ten cell leaders	Formal	Convey early warning information, mobilise their members to clean storm water drains, enforce bylaws, e.g., penalties to households releasing waste water to the settlement
CBOs	KIWAWAM	Informal	More inclined to fishing and ocean related disasters, have not been involved in flood disasters because their activities have not been severely affected
	Beach Management Unit	Formal	Focus on managing the beach and coastal resources
	Federation (Upendo group)	Informal	Environmental cleanness campaigns, including cleaning of storm water drains, help group members in difficult moments, such as illness or loss of life, and also in cheerful times, such as weddings
	SAFINA	Informal	Community cleanness campaigns, small community projects, saving and credit, cleaning of storm water drains, well equipped with necessary tools, formally registered
Religious	Moslem religious groups	Formal	Provides religious and education facilities, especially to the youth
	Galilaya Foundation	Formal	Education services to orphans plus religious/spiritual services, spiritual and material support, as well as accommodation for victims
Political Organisations	CCM, CHADEMA, CUF	Formal	Campaigns on cleaning of storm water drains, encouraging people to relocate to safe, not flood-prone areas
NGOs	Centre for Community Initiatives	Formal	Awareness raising on environmental and sanitation issues, mentors of the two community groups Upendo and SAFINA which engage in community works, including cleaning of storm drains but also operate saving and credit facilities

storm water drains, Ten cell leaders are responsible for enforcing the law prohibiting haphazard dumping of solid waste and discharging of liquid waste. Table 6.1 shows various institutions and their roles in relation to flood reduction.

Not all institutions listed in Table 6.1 are similarly important in dealing with the flood problem. Figure 6.5 portrays the perceived importance of such institutions

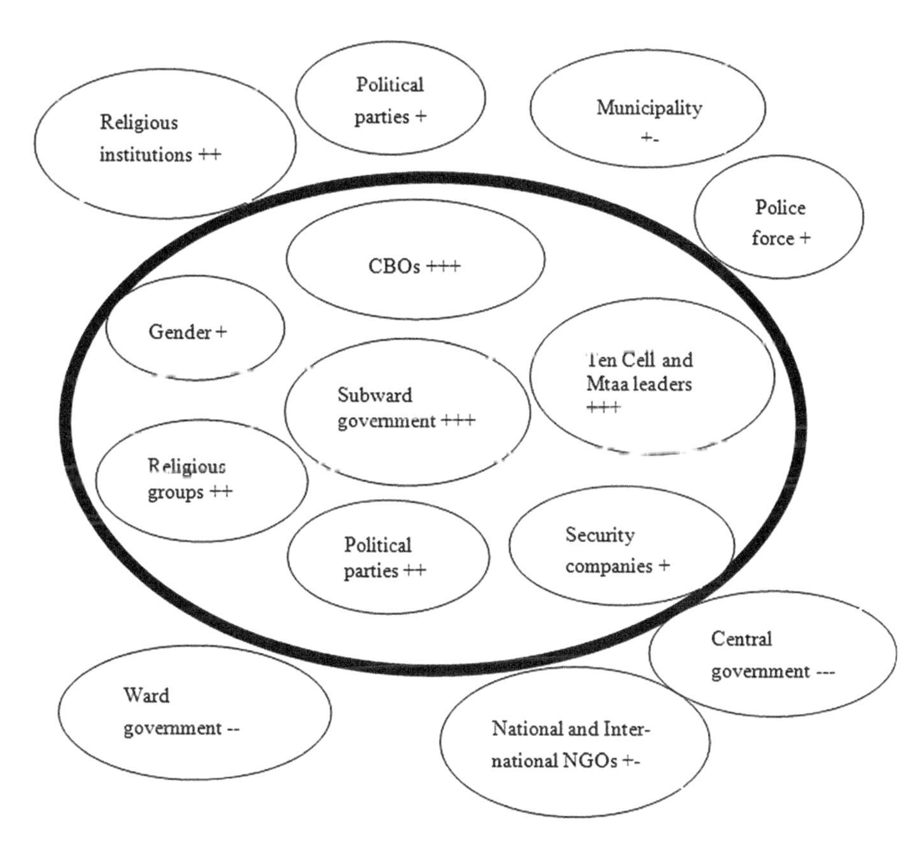

Fig. 6.5 Internal institutions (within the *circle*) and external institutions with a role in flood management in Bonde la Mpunga and the residents' view on the quality of their assistance. Legend: +++ = very helpful; --- = very unhelpful; +- = neither helpful nor unhelpful

from the perspective of the residents. Two types exist: (i) institutions within the settlement (internal) illustrated by the inner circle, and (ii) institutions operating outside the settlement (external), i.e. established outside Bonde la Mpunga but operating and/or offering more or less services to the settlement. Our focus group discussions reveal that residents have more trust in the internal institutions, such as Sub Ward government, Mtaa-leaders and Ten cell leaders, on their efforts to reduce flooding. Other internal key institutions are CBOs through their environmental cleanliness campaigns on a voluntary basis (Table 6.1). Furthermore, religious and gender related institutions were considered important in terms of getting concrete and immediate support in case of a flooding issue. In addition, private security companies offer surveillance and protection to residents during the flooding and after the event. Political parties are relevant in campaigning for environmental cleanliness; however, their motives are often associated with political gains.

In contrast to those institutions belonging to the internal circle, there is less trust in higher government institutions (external) such as the Ward, Municipal or Central Government. Rather, the local residents are convinced that these institutions pursue the developer's targets, and not the interests of the poor local people. The quotation below by a resident of 63 years living in the settlement his whole life illustrates this conviction:

> For almost four years in consecutive now, when it rains residents are agitated finding places to store their material assets. Houses get inundated, people cannot sleep. However, these floods are created by the people who have constructed high rise buildings. The Mtaa government is struggling a lot, for instance they once demolished a wall which was constructed on a water way and blocked the rainy water. The problem is the lands department that issues the construction permit. The rich individuals are given permits to construct and block the water ways (case no 148).

The study found little to no support from higher government institutions, such as the Ward, Municipal or Central Government, to adapt to or cope with flood hazards. The Municipal Government assisted once in constructing a few storm water drains. However, the drains only worked for a short time before failing. Due to technical problems as well as blockages by real estate developers, the drainage system failed. From the perspective of the residents the Municipal Government can be blamed for allowing new developments and – in the experience of the residents – instigating the main cause of the more serious flood impacts in recent years.

The empirical results from the focus group discussions provide information about a variety of formal and informal institutions at local level. But the residents perceive those institutions as most important in terms of dealing with flooding events which act within the community in close collaboration with the residents. These are the Sub Ward government, the Mtaa-leaders and Ten cell leaders and the CBOs. The residents experienced concrete support and activities. Particular importance of institutional activities gains cleaning, maintenance and extension of the drainage system which belongs to the features of the physical vulnerability dimension.

Assessing Attitudinal Vulnerability

To assess attitudinal vulnerability, the number and the characteristics of social networks with consideration of trust are of particular interest. Besides the investigation of these social capital features, the description of hazard experience provides insights on possible behaviour routines in case of a flood and aspects of risk awareness.

The questionnaire survey results provide evidence that *informal networks* are the most important types of networks among the households. Compositions of network persons vary considerably from relatives or family members to friends, neighbours, religious and ethnic groups, or gender related networks. In several cases, mixed networks, characterised by overlapping features (e.g. a relative can also be a neighbour or belong to the same religious group) have been identified. However, relative or family related networks dominate. Thus, we can confirm the observations by Chatterjee (2011) about the basic contributions by networks of self-help groups and religious organisations to risk management.

The findings suggest that households in the studied area depend on informal networks as their focal point for relief and support. Close kin relationships seem to be of paramount importance for the majority of affected households, in terms of trust, to get flood support. This reveals the fact that households particularly turn to relatives rather than non-kin for critical help and support during the flooding events. Thus, the greater the proportion of kin relations in the household network composition, the greater the influence on household responsiveness to flood hazards.

This attitude has its causes in the *hazard experience* of the respondents. 79 % of the households indicated flooding experience during different seasons of a year. The majority (53 %) had to deal with flooding every year during rainy seasons. 26 % indicated to have flooding experiences only during extreme rainfalls, specifically referring to the extreme floods of 1998 and 2011. To get an impression about the intensity of flooding, the respondents were asked to indicate the height of the water level. Half of the households experienced flooding to the waist level and above. And 35 % reported flooding to the knee level (Fig. 6.6). Because of these hazard experiences the residents have developed attitudinal routines in saving their belongings and shelter, which will be described in the next section dealing with coping and adaptation measures.

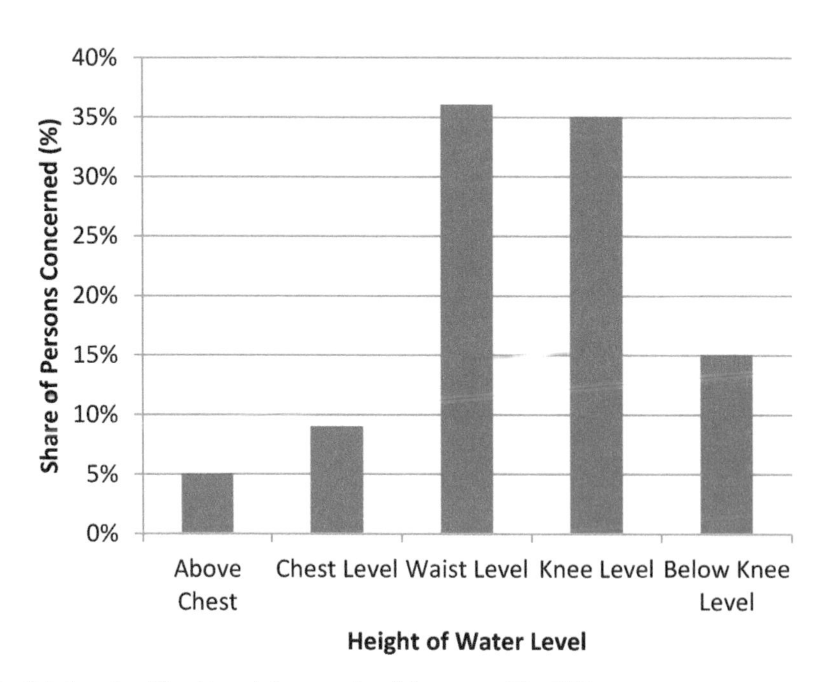

Fig. 6.6 Levels of flood inundation, results of the survey (N = 309)

Concerning the attitudinal characteristics it can be stressed that the informal networks within the community are most important for getting relief and support in case of a flood hazard. In particular, networks among relatives characterised by trustful relations are essential. These findings are based on experience made during the flood events in the past.

Assessing Physical Vulnerability

With the experience of these high inundation levels (Fig. 6.6), it is more likely that households suffer strongly. The swampy nature of the area combined with the frequency of flooding makes the poorer households, which are located in the lower basin, more likely to be affected by flooding. As described earlier in the chapter, the construction of the shopping centre in the upper area of the settlement contributes to surface sealing, and hence, also contributes to surface runoffs and floods in the lower basin. The land use change in this area goes on regardless of the consequences for the poorer residents. They find themselves enclosed by buildings and feel helpless during floods. The 24 year-old respondent who lived in the settlement his entire life, explained:

> The situation is as you can see by yourself; water is throughout coming from down. The construction of shopping centre has made things worse. There is nothing we can do; we just live as it is (case no 14).

Particularly affected is the drainage system which does not function in an appropriate manner. It has not been enlarged, and maintained in the course of the construction activities. More precisely, there is a constant need for drainage clearance. Without this measure the flood causes serious damage (Table 6.2).

The most reported damage in the survey is the destruction of material assets, if water inundates the dwellings. Material assets have been destroyed or washed away by floodwater. This includes clothes, food stores and other equipment. The destruction of dwelling structure itself was reported as the second most experienced damage caused by floods. This includes destruction of parts of the walls and floors.

Table 6.2 Flood hazard experiences

Damage	Frequency
Destruction of household material assets	131
Destruction of dwelling structure	40
Diseases	25
Destruction of green structures	15
Destruction of household mobility	12
Destruction of business	10
Inability of children to attend classes	5
Others	13
Total	251

Multiple answers possible, N = 251, Results of the survey

The respondents reported that sometimes the floodwater causes de-flooring and de-plastering of the walls. Other flood hazard impacts include health risks, like diarrhoea, as well as disruption of access to social services (e.g. schools, health care centres) from destruction of streets.

Summarising, the residents are confronted with the land use change and the lack of infrastructure, especially the drainage system. These are major causes of physical vulnerability. The residents experience serious damages and losses of their physical living conditions. According to the interview results the residents feel exposed and helpless.

Coping and Adaptation Measures to Flooding

The chapter so far presents the empirical findings regarding the four dimensions of social vulnerability on asset, attitudinal, institutional, and physical vulnerability. These findings are related to the exposure and susceptibility of households and the community in the case study area. It hence covers two of the main aspects of the theoretical concept. The following section will now focus on the third main category of social vulnerability that is coping and adaptive capacity. We compile the key findings concerning the most recurrent coping and adaptation measures to flooding.

Households and individuals within the community of Bonde la Mpunga undertake a variety of coping and adaptation measures. Table 6.3 provides an overview of the relevant activities.

The measures are, on the one hand, technical in terms of house and infrastructure repair and re-construction. On the other hand, they include particular behavioural measures to save material assets, such as cleaning the flooded residential houses and the storm water drains. In some cases, the residents leave the location temporarily.

Table 6.3 Coping and adaptation strategies

Measure	Form	Timing
Constructing protection wall	Technical	Anticipatory and/or reactive
Raising foundation/height of dwelling structure		Anticipatory and/or reactive
Rubble filling		Anticipatory and/or reactive
House repair		Reactive
Constructing small storm water drains		Anticipatory and/or reactive
Creating pathways		Concurrent
House cleaning and repair		Reactive
Removal of flood water	Behavioural	Concurrent
Placing material assets on high storage systems		Anticipatory and/or concurrent
Transporting material assets to safe areas		Anticipatory and/or concurrent
Relocation		Anticipatory and/or concurrent
Cleaning storm water drains		Anticipatory and/or reactive

Technical Measures

A typical technical measure is the construction of protection walls to secure main entrances and in some cases the entire dwelling unit to reduce the likelihood of inundation. The height of the walls ranges from 0.5 to 1 m in height. However, households are slowly increasing the height of walls step by step. This is due to new experiences of the immense height of floods in recent years. A flood victim of 42 years who has been living in the residence for the last 20 years, and annually affected by floods, commented:

> The flood affects us each year. There is nothing more we can do other than to fill rubbles and construct protection walls on door entrances to the extent that doors have now become windows (case no 241).

Raising the foundation of dwelling units and toilets (Fig. 6.7) as well as rubble filling to stabilise the ground are common practices. This is done either during the house construction or incrementally after construction.

Elevating the foundations of existing buildings is not only a response to past experienced floods, but also taken as a preventive measure. To raise the foundation, households fill rubble inside the house, sometimes also outside, and make new floors, after which they uncover the roof top and add a few bricks on the top to restore the normal height of the house. Again, it was noted that, due to frequent floods, this strategy is performed more than once by some of the households in their efforts to reduce the possible negative effect of floods. All these activities are connected with house repair measures, which became necessary because of the

Fig. 6.7 Elevated toilets next to a clogged drain (Picture: Sigrun Kabisch 2012)

strong damaging impacts of the flood. The repairs can last weeks to several months, depending on the damages.

Additionally, local infrastructure, such as the drainage system and the pathways, also need to be repaired, extended and cleaned. Some households construct small water drainage channels around their dwellings so as to redirect the water flow away from their property. However, due to the compactness of the settlement this strategy often creates problems for neighbours. Constructed drainage channels release water directly to adjacent houses. Furthermore, floods often cause blockage of access ways. As a response, households place sand bags, stones and/or tree logs on pathways, creating temporary pathways to the dwelling units. Nevertheless, this strategy is only suitable during low depth floods. They are quite ineffective during extreme floods, such as experienced during the flood in 2011.

Behavioural Measures

During a flood, households and individuals attempt to remove the water from their house using buckets or water pumps, depending on their economic resources. Both water removing measures only work in case of low-level inundation. House cleaning and repair follow. Concerning their belongings, some households locate their assets on high storage facilities (e.g. in the roof of the house or temporary trestles inside the house) while others transport them to non-flooded areas (Fig. 6.8). These activities occur before an event strikes, and in some cases, during a flood if the level keeps rising.

Fig. 6.8 People removing their belongings during the flood (Picture: Ardhi University Archive 2011, with permission)

When flooding occurs, some households seek refuge in the house of neighbours, friends or relatives until the water level goes down. Dependent persons, such as children, elders and sick persons are relocated first. If the flood rises and persists, other household members, including men, will also relocate. Our findings reveal that relocation is not a real option. It is taken as one of the last possibilities because of fear of material loss by theft and fear of insecurity at the relocated place. An interview with one house owner who has been living in the settlement for 15 years illustrates this point:

> It is difficult for some of us to relocate unless the house is destructed. How can I relocate before the floods while where I will relocate my house is other people's house.

In the case of Bonde la Mpunga, most residents attempt to remain in their home as long as possible and relocate only if there is a direct threat to their lives.

Besides the household asset related activities and the caring efforts concerning the family members, a major behavioural activity refers to the cleaning of storm water drains. Individual or joint activities by CBOs are essential measures to guarantee a functioning drainage system. Nevertheless, environmental awareness is lacking. Often, the drainage system is filled with rubbish, plastic and organic waste, which causes blockage of water drains and, consequently, intensifies flooding.

Our findings suggest that the coping and adaptation strategies may take various dimensions depending on their timing, intent/goal, resources, motive, spatial scope and form of implementation. These observations have also been described by McCarthy et al. (2001), Smit et al. (2000), Smit and Skinner (2002), and Smit and Wandel (2006). These measures can be categorised as anticipatory (adaptation occurs before the flooding event); concurrent (measures occur during the event) or reactive (measures occur in direct response to a climatic event). The boundaries between anticipatory and reactive measures are blurry due to the nature of incremental construction. Poor households are permanently repairing and adjusting, as their home and material belongings represent their most important assets. Furthermore, the measures we identified are largely technical in involving minor alterations over a long period of time. The behavioural measures which include, for instance, storing material assets in a safe location, often occur concurrently. It can be concluded that the overlapping of measures during different timescale is inescapable in case of a flood in Bonde la Mpunga.

In addition, it was found that some measures are either planned or autonomous. The planned measures occur as a consequence of deliberate policy decisions based on the awareness that conditions have changed or are expected to change and that some form of strategic action is required to maintain a desired state (e.g., cleaning campaigns for drainage). Autonomous adaptation involves changes that will be taken in response to subjective experienced negative events irrespective of any policy, plan or decision, e.g., construction of a protection wall on the door by the house owner. It is worth noting that all strategies listed in Table 6.3 are merely local and autonomous in nature. That means that the support of coping and adaptation strategies by institutions outside of the immediate community is perceived as weak by the local residents (for detailed information concerning the governance structures see Chap. 9).

Conclusion

The focus of this chapter is on the social vulnerability of individuals, households and communities towards flooding. As a conceptual approach for assessing social vulnerability of the flood prone settlement Bonde la Mpunga, we developed a framework consisting of four interlinked dimensions which are named asset, institutional, attitudinal and physical. These dimensions and their indicators are the prerequisites for an integrative appraisal of exposure and susceptibility, as well as coping and adaptation capacity (Fig. 6.1).

Exposure

Exposure is related to the physical condition that increases the likelihood of a negative impact when floods occur. In the case of Bonde la Mpunga, households are clearly exposed based on the low-lying and swampy nature of the area making it more likely to experience floods. This is combined with a backlog in urban infrastructure, most evident in poor waste management and inefficient storm water drainage. This makes a compelling case for the physical vulnerability of the community as a whole as well as the risk of hazards building up over the years. It was mentioned before that Bonde la Mpunga is a dense, saturated settlement with blocked water ways in several points and less infiltration areas. This therefore increases the likelihood of surface runoffs and/or stagnation. In addition to the physical aspects, households are less likely to cope and respond effectively to flooding events, if their source of income is unstable. The vast majority of interviewees derive their income from informal economic activities and operate in proximity to their home. This again increases their vulnerability, given that limited and irregular monetary resources restrain many households from taking the necessary coping measures to resist recurrent floods. Yet, it must be said that exposure does not necessarily translate to vulnerability. This is clear when observing how some wealthier households pump water out of their property and the measures that some of them take to mitigate potential damages. In this particular case, land use control and urban planning could play a key role in reducing the risk of flood and addressing the existing inequalities between exposed households whether they are more and/or less affluent.

Susceptibility

Seen as social and physical preconditions to suffer harm, the susceptibility of households in Bonde la Mpunga is firstly reflected through its location, as well as the quality of the residential buildings consolidated in the low income parts of the settlement. These houses were found to have limited flood clearance heights which would allow the water to flow and pass through the settlement. Furthermore, the intrinsic mixed asset conditions of low education level, irregular income, household size and tenure arrangements are common causes of the increasing adverse effects of recurrent flooding. In the case of Bonde la Mpunga, 41 % of households host more than five members, this implies that more people per unit of area are not only exposed but also sensitive to flooding. This raises the question of density and how it explains the vulnerability of low income households and communities. It shows also that damages reported in Bonde la Mpunga do not occur in a vacuum. Flooding in saturated settlements will translate into greater harm than in less dense communities.

Also, clearly interlinked with susceptibility is the pre-existing governance structure and institutions that operate to prevent, adapt and mitigate flooding events. In Bonde la Mpunga, there is a clear disparity between local governments and community based organisations in terms of their role and responsibilities. Interestingly, the depth and diversity of informal networks is reported to act as a catalyst for immediate relief and support.

Coping and Adaptive Capacity

This component refers mainly to the ability to anticipate, resist and recover from hazardous situations. In Bonde la Mpunga households have developed different skills and resources to cope and adapt to the effects of floods, the number of measures identified implies that residents and stakeholders are not helpless and can step away from a vulnerable state by increasing their rebound and response processes. Here again, aspects of education, access to information and economic resources play a key role in the effectiveness of coping and adaptation measures. Securing belongings and relocating temporarily are among common behavioural measures undertaken during a flood. These choices are not surprising as coping and adaptive alternatives ultimately depend on the source of income, stability and household savings or cash flow. Another important aspect is the number of dependent members of a household (i.e. children, elderly) who might need additional or special support when flooding occurs. Thus, the composition of the households needs particular attention in assessing their lack of capacity. Equally important are physical aspects, such as the resistance of built structures in particular dwelling units when hit by water. In Bonde la Mpunga, damages of dwelling structures are reported to be the second most important effects of flooding which include de-flooring and de-plastering. This explains the different structural efforts made to increase the capacity to recover and adapt to recurrent floods. One aspect that has been proven effective is the diversity of social networks through close kin relationships, gender and religious groups, which remain an important source of mutual support. These CBOs are considered as reliable. Further formal networks such as Sub Ward government, Ten cells and Mtaas are active in the provision of warning information or mobilising drainage cleaning campaigns. But in general, the role of formal institutions in disaster prevention appears to be weak.

Bonde la Mpunga exemplifies the case of several other flood prone settlements in Dar es Salaam. Many residents experience flooding as a normal seasonal occurrence, but also recognise an increase in the impact of recent floods (2009, 2011, 2014). Our investigation reveals that flooding remains in the collective memory of exposed urban settlers and it is part of their yearly cycle. This may well act as a stimulus for strengthening preventive actions to reduce social vulnerability. A perspective for the future lies in joint and concerted actions of both formal and informal institutions in particular on the individual, household and community level. The recognition of these complex connections within African settlements can lead to better prevention measures to reduce the negative effects of flooding in terms of social vulnerability.

References

Adger WN, Brooks N, Bentham G, Agnew M, Eriksen S (2004) New indicators of vulnerability and adaptive capacity. Technical report no. 7. Tyndall Centre for Climate Research, University of East Anglia, Norwich

Badjeck M-C, Mendo J, Wolff M, Lange H (2009) Climate variability and the Peruvian scallop fishery: the role of formal institutions in resilience building. Clim Chang 94:211–232

Birkmann J (2006) Measuring vulnerability to promote disaster-resilient societies: conceptual frameworks and definitions. In: Birkmann J (ed) Measuring vulnerability to natural hazards – towards disaster resilient societies. UNU-Press, Tokyo/New York/Paris, pp 9–54

Blaikie P, Cannon T, Davis I, Wisner B (1994) At risk: natural hazards, people's vulnerability, and disaster. Routledge, London/New York

Bohle HG, Glade T (2008) Vulnerabilitätskonzepte in Sozial-und Naturwissenschaften. In: Felgentreff C, Glade T (eds) Naturrisiken und Sozialkatastrophen. Springer, Berlin, pp 99–119

Brauch HG, Oswald Spring Ú, Mesjasz C, Grin J, Kameri-Mbote P, Chourou B, Dunay P, Birkmann J (eds) (2011) Coping with global environmental change, disasters and security. Springer, Heidelberg/New York

Cannon T (1994) Vulnerability analysis and the explanation of "natural" disasters. In: Varley A (ed) Disasters, development and the environment. Wiley, Chichester, pp 13–30

Cardona OD, van Aalst MK, Birkmann J, Fordham M, McGregor G, Perez R, Pulwarty RS, Schipper ELF, Sinh BT (2012) Determinants of risk: exposure and vulnerability. In: Field CB, Barros V, Stocker TF, Qin D, Dokken DJ, Ebi KL, Mastrandrea MD, Mach KJ, Plattner G-K, Allen SK, Tignor M, Midgley PM (eds) Managing the risks of extreme events and disasters to advance climate change adaptation. A special report of Working Groups I and II of the Intergovernmental Panel on Climate Change (IPCC). Cambridge University Press, Cambridge/New York, pp 65–108

Chambers R (1989) Editorial introduction: vulnerability, coping and policy. IDS Bull 20(2):1–7

Chambers R (1995) Poverty and livelihoods: whose reality counts? Discussion paper no. 347. Institute of Development Studies, Brighton

Chatterjee M (2011) Flood loss redistribution in a third world megacity: the case of Mumbai. In: Brauch HG, Oswald Spring Ú, Mesjasz C, Grin J, Kameri-Mbote P, Chourou B, Dunay P, Birkmann J (eds) Coping with global environmental change, disasters and security. Springer, Heidelberg/New York, pp 603–612

Cutter S, Boruff B, Shirley W (2003) Social vulnerability to environmental hazards. Soc Sci Q 84 (2):242–261

Cutter SL, Emrich CT, Webb JJ, Morath D (2009) Social vulnerability to climate variability hazards: a review of the literature. Final report to Oxfam America. Available via http://adapt.oxfamamerica.org/resources/Literature_Review.pdf. Accessed 28 Nov 2013

De Risi R, Jalayer F, De Paola F, Iervolino I, Giugni M, Topa ME, Mbuya E, Kyessi A, Manfredi G, Gasparini P (2013) Flood risk assessment for informal settlements. Nat Hazards 69(1):1003–1032

Dodman D (2013) The challenge of adaptation that meets the needs of low-income urban dwellers. In: Palutikof J, Boulter SL, Ash J, Stafford Smith M, Parry M, Waschka W, Guitart D (eds) Climate adaptation futures. Wiley-Blackwell, Oxford, pp 227–234

Dodman D, Kibona E, Kiluma L (2011) Tomorrow is too late: responding to social and climate vulnerability in Dar es Salaam, Tanzania. Case study prepared for the Global Report on Human Settlements. Available via www.unhabitat.org/grhs/2011. Accessed 21 Nov 2013

Fuchs S, Kuhlicke C, Meyer V (2011) Editorial for the special issue: vulnerability to natural hazards – the challenge of integration. Nat Hazards 58:609–619

Füssel H-M, Klein R (2006) Climate change vulnerability assessments: an evolution of conceptual thinking. Clim Chang 75:301–329

Green C (2004) The evaluation of vulnerability to flooding. Disaster Prev Manag Int J 13(4):323–329

Heinrichs D, Krellenberg K, Hansjürgens B, Martínez F (2012) Risk habitat megacity. Springer, Heidelberg

Hewitt H (1998) Excluded perspectives in the social conception of disaster. In: Quarentelli EL (ed) What is a disaster? A dozen perspectives on the question. Routledge, London, pp 75–91

Jean-Baptiste N, Kuhlicke C, Kunath A, Kabisch S (2011) Review and evaluation of existing vulnerability indicators for assessing climate related vulnerability in Africa. UFZ-Bericht 07/2011, Helmholtz-Zentrum für Umweltforschung-UFZ, Leipzig

Jean-Baptiste N, Kabisch S, Kuhlicke C (2013) Urban vulnerability assessment in flood-prone areas in West and East Africa. In: Rauch S, Morrison GM, Schleicher N, Norra S (eds) Urban environment. Springer, Heidelberg/New York, pp 203–218

Jha AK, Bloch R, Lamond J (2012) Cities and flooding: a guide to integrated flood risk management for the 21st century. The World Bank, Washington, DC. Available via https://www.gfdrr.org/urbanfloods. Accessed 30 Nov 2013

John R, Jean-Baptiste N, Kabisch S (2014) Vulnerability assessment of urban populations in Africa: the case of Dar es Salaam, Tanzania. In: Edgerton E, Romice O, Thwaites K (eds) Bridging the boundaries. Human experience in the natural and built environment and implications for research, policy, and practice, vol 5, Advances in people and environment studies. Hogrefe Publishing, Göttingen, pp 233–244

Kabisch S, Kunath A, Schweizer-Ries P, Steinführer A (eds) (2012) Vulnerability, risk, and complexity. Impacts of global change on human habitats, vol 3, Advances in people and environment studies. Hogrefe Publishing, Göttingen

Kithiia J (2011) Climate change risk responses in East African cities: need, barriers and opportunities. Curr Opin Environ Sustain 3:176–180

Kiunsi R, Kassenga G, Lupala J, Lerise F, Meshack M, Malele B, Namangaya A, Mchome E (2009a) Building disaster resilient communities: Dar es Salaam, Tanzania. In: Pelling M, Wisner B (eds) Disaster and risk reduction. Cases from urban Africa. Earthscan, London/Sterling, pp 127–146

Kiunsi R, Kassenga G, Lupala J, Malele B, Uhinga G, Rugai D (2009b) Mainstreaming disaster risk reduction in urban planning practice in Tanzania, AURAN Phase II, Project final report

Kombe W (2005) Land use dynamics in peri-urban areas and their implications on the urban growth and form: the case of Dar es Salaam, Tanzania. Habitat Int 29(1):113–135

Kuhlicke C, Kabisch S, Krellenberg K, Steinführer A (2012) Urban vulnerability under conditions of global environmental change: conceptual reflections and empirical examples from growing and shrinking cities. In: Kabisch S, Kunath A, Schweizer-Ries P, Steinführer A (eds) Vulnerability, risks and complexity: impacts of global change on human habitats, vol 3, Advances in people and environment studies. Hogrefe Publishing, Göttingen, pp 27–38
McCarthy J, Canzian F, Leary A, Dokken J, White S (2001) Climate change 2001: impacts, adaptation and vulnerability. Contribution of Working Group II to the third assessment report of the Intergovernmental Panel on Climate Change. Cambridge University Press, Cambridge
Moser C (1998) The asset vulnerability framework. Reassessing urban poverty reduction strategies. World Dev 26(1):1–9
Moser C, Stein A (2011) Implementing urban participatory climate change adaptation appraisals: a methodological guideline. Environ Urban 23(2):463–485
NEPAD The New Partnership for Africa's Development (ed) (2006) Water in Africa: management options to enhance survival and growth. Addis Ababa, United Nations Economic Commission for Africa (UNECA). Available via http://www.uneca.org/awich/nepadwater.pdf. Accessed 4 Dec 2013
O'Brien K, Eriksen S, Nygaard LP, Schjolden A (2007) Why different interpretations of vulnerability matter in climate change discourses. Clim Pol 7(1):73–88
Pelling M (2006) Measuring urban vulnerability to natural disaster risk: benchmarks for sustainability. Open House Int 31:125–132
Pelling M, Blackburn S (2013) Megacities and the coast. Risk, resilience and transformation. Routledge, London/New York
Pelling M, Wisner B (eds) (2009) Disaster and risk reduction. Cases from urban Africa. Earthscan, London/Sterling
Smit B, Skinner M (2002) Adaptation options in agriculture to climate change: a typology. Mitig Adapt Strateg Glob Chang 7:85–114
Smit B, Wandel J (2006) Adaptation, adaptive capacity and vulnerability. Glob Environ Chang 16:282–292
Smit B, Burton I, Klein R, Wandel J (2000) An anatomy of adaptation to climate change and variability. Clim Chang 45:223–251
START (2011) Urban poverty & climate change in Dar es Salaam, Tanzania: a case study, Final report prepared by the Pan-African START-secretariat, Tanzania
Steinführer A, Kuhlicke C (2007) Social vulnerability and the 2002 flood: country report Germany (Mulde River). Floodsite report of T 11-07-08. Available via http://www.floodsite.net/html/partner_area/project_docs/Task_11_M11.3_p44_final.pdf. Accessed 4 Dec 2013
Turner BL, Kasperson RE, Matson PA, Mc Carthy JJ, Corell RW, Christensen L, Eckley N, Kasperson JX, Luers A, Martello ML, Polsky C, Pulsipher A, Schiller A (2003) A framework for vulnerability analysis in sustainability science. Proc Natl Acad Sci 100:8074–8079
UN-Habitat (2011) Cities and climate change. Global report on human settlements 2011. United Nations Human Settlement/Earthscan, London/Washington, DC
United Nations World Water Assessment Programme (2012) Facts and figures: managing water under uncertainty and risk. United Nations World Water Assessment Programme. UNESCO-WWAP 2012. Available via http://unesdoc.unesco.org/images/0021/ 002154/215492e.pdf. Accessed 21 Sept 2013
URT United Republic of Tanzania, National Bureau of Statistics (2005) 2002 population and housing census. Village and street statistics-age and sex distribution, Dar es Salaam Region, vol 7. Dar es Salaam
URT United Republic of Tanzania, National Bureau of Statistics (2013) The United Republic of Tanzania 2012 population and housing census: population distribution by administrative areas. Dar es Salaam
Vincent K (2004) Creating an index of social vulnerability to climate change in Africa. Working paper no. 56. Tyndall Centre for Climate Research, University of East Anglia, Norwich

Vogel C, O'Brien K (2004) Vulnerability and global environmental change: rhetoric and reality. Aviso 13:1–8

Wachinger G, Renn O, Begg C, Kuhlicke C (2013) The risk perception paradox: implications for governance and communication of natural hazards. Risk Anal 33(6):1049–65

Wamsler C (2014) Cities, disaster risk and adaptation. Series critical introductions to urbanism and the city. Routledge, London/New York

Watts MJ, Bohle HG (1993) The space of vulnerability: the causal structure of hunger and famine. Prog Hum Geogr 17:43–67

Wisner B (2004) Assessment of capability and vulnerability. In: Bankoff G, Frerks G, Hilhorst D (eds) Mapping vulnerability: disasters, development and people. Earthscan, London, pp 183–193

Wisner B, Luce HR (1993) Disaster vulnerability: scale, power and daily life. GeoJournal 30(2):127–140

Wisner B, Blaikie P, Cannon T, Davis I (2004) At risk: natural hazards, people's vulnerability and disasters. Routledge, London

Chapter 7
Multi-risk Assessment as a Tool for Decision-Making

Alexander Garcia-Aristizabal, Paolo Gasparini, and Guido Uhinga

Abstract The multi-risk concept refers to a complex variety of combinations of risk (i.e. various combinations of hazards and various combinations of vulnerabilities). For this reason, it requires a review of existing concepts of risk, hazard, exposure and vulnerability, within a multi-risk perspective. The main purpose of the multi-risk assessment is to harmonise the methodologies employed and the results obtained for different risk sources, taking into account possible risk interactions. Given the complexity of processes that the multi-risk problem poses, the framework presented here entails three levels of analysis: the first-level analysis, in which the evaluation of the potential physical damages is performed; the second-level analysis, in which tangible indirect losses are assessed considering the socio-economic context; and the third-level analysis, where a set of specific social context conditions is considered in an indicator-based approach. One of the most challenging elements of the multi-risk assessment is the translation of the quantitative output into useful information for decision-making under uncertainty. This is a critical step to consolidate the importance of the multi-risk analyses and to define their ultimate importance and usefulness for the resolution of critical societal problems. To illustrate the general methodology presented, an example application has been developed in Dar es Salaam (Tanzania), which is one of the case study cities in the CLUVA project.

Keywords Multi-hazard • Multi-risk • Risk harmonisation • Mitigation options • Decision-making • Tanzania

---

A. Garcia-Aristizabal (✉)
Center for the Analysis and Monitoring of Environmental Risk (AMRA),
Via Nuova Agnano 11, 80125 Naples, Italy
e-mail: alexander.garcia@amracenter.com

P. Gasparini
University of Naples Federico II, AMRA S.c.a.r.l., Naples, Italy
e-mail: paolo.gasparini@amracenter.com; paolo.gasparini@na.infn.it

G. Uhinga
Centre for Information and Communication Technology (CICT), Ardhi University,
P.O. Box 35176, Dar es Salaam, Tanzania
e-mail: guido@aru.ac.tz

S. Pauleit et al. (eds.), *Urban Vulnerability and Climate Change in Africa*, Future City 4,
DOI 10.1007/978-3-319-03982-4_7

Introduction

Risk assessment and the quantification of risk are two fundamental components of risk management. The fundamental objective of quantitative risk assessments is to determine the probability distribution of the losses that may be expected in a given area and time interval, and because of the occurrence of a kind of hazardous event. Risk can be considered as the combination of the consequences of an event and the associated probability of its occurrence (e.g. ISO/IEC guide 73 2009; UN-ISDR 2009). The risk assessment practice intends to integrate in a logical and rational way the uncertainties arising at each step of the process and, for this reason, it is generally based on probabilistic analyses.

In the past, the risk assessment for natural hazards was carried out separately for the different hazards of interest, with few attempts to combine the studies in a holistic risk assessment (Marzocchi et al. 2012). However, from the complex relationship between the human activities and the surrounding environment, different kinds of adverse events coming from different sources can produce damages in the same set of exposed elements; furthermore, eventual interactions and cascading effects may increase the total losses with respect to those produced by the initial triggering event. For this reason, an increasing number of scientists interested in natural hazards and risk are turning their interest towards the multi-hazard risk and multi-risk assessment practice. However, an integrated framework for multi-risk assessment is still a major challenge since it implies adopting quite a different perspective with respect to classical single-risk analysis.

Manifold studies addressing multi-hazard and multi-risk problems have been developed at different scales and resolutions. The United Nations Development Program (UNDP 2004) and the World Bank (Dilley et al. 2005) provide examples of low-resolution analyses at global scale. Likewise, regional to local scale applications have been produced in a variety of areas of the world, including Europe (e.g. Grunthal et al. 2006; Del Monaco et al. 2007; Carpignano et al. 2009; Lari et al. 2009; Bovolo et al. 2009; Marzocchi et al. 2012; Selva 2013), Central America (van Westen et al. 2002), North America (e.g. FEMA 2004), and Australia (e.g. Blong 2003; Durham 2003; Schmidt et al. 2011). Reviews of the main initiatives in multi-risk assessment can be found in Del Monaco et al. (2006), Garcia-Aristizabal et al. (2012b), and Kappes et al. (2012). In Europe, in particular, the interest in developing a framework for multi-risk assessment has substantially increased in the last 15 years, as demonstrated by numerous funded projects, such as the TIGRA project ('The Integrated Geological Risk Assessment', e.g. Del Monaco et al. 1999), the TEMRAP project ('The European Multi-Hazard Risk Assessment Project', European Commission 2000), the ESPON project ('The Spatial Effects and Management of Natural and Technological Hazards in Europe', Schmidt-Tomé et al. 2006), the NARAS project ('Natural Risk Assessment', Marzocchi et al. 2009, 2012), the ARMONIA project ('Applied multi-risk mapping of natural hazards for impact assessment', Del Monaco et al. 2007), the MATRIX project ('New methodologies for multi-hazard and multi-risk assessment methods for Europe', e.g. Marzocchi et al. 2012; Nadim et al. 2013), and the CLUVA project

('Climate change and urban vulnerability in Africa', Garcia-Aristizabal and Marzocchi 2012a). The multi-risk framework presented in this chapter was developed in the context of the CLUVA project. The objective was to perform multi-risk assessments in African urban areas considering climate-related hazards taking into account different scenarios of climate change. To consider the possible effects of climate change, a procedure for the analysis of non-stationary climate-related extreme events was also implemented (for details see Garcia-Aristizabal et al. 2015).

Following the definition provided in the working paper on risk assessment and mapping guidelines for disaster management of the European Commission (European Commission 2010: 23), the multi-risk assessment may be understood as the process "to determine the whole risk from several hazards, taking into account possible hazards and vulnerability interactions". An important element to highlight at this point is that the multi-risk assessment adopts a territorial perspective. This perspective is closer to the needs of planners and decision-makers because they may be interested in getting information on the causes of potential damages and how often they may occur in the area of concern, regardless of the source of the damages.

This chapter is structured as follows: first, the transition from single- to multi-risk assessment practice is described. Second, the fundamental concepts and quantitative tools of the multi-risk assessment are introduced. Then, a discussion about the possible advantages of the multi-risk analyses for decision making is presented. Finally, an example of multi-risk assessment implementation is developed in a pilot area located in Dar es Salaam, Tanzania, which is one of the case study cities of the CLUVA project.

From Single- to Multi-risk Assessment

The transition from single- to multi-risk assessment implies a shift from a *hazard-centred* perspective that characterises the single-risk assessment, to a *territorial-centred* one (Carpignano et al. 2009). Therefore, from a multi-risk perspective, the target area of interest is the first element to be defined. The evolution from single-risk to multi-risk analyses consists of three different and complementary elements (Fig. 7.1a):

1. *The Single-Risk (SR) assessment*, generally characterised by a hazard-centred perspective in which the analysis starts from the source of the hazard and concentrates mainly on the propagation in space of the effects on a set of exposed elements (Fig. 7.1b);
2. *The multi-hazard risk (MHR) assessment*, in which different independent hazard sources affecting a given common area of interest are considered. This is the first case in which the change towards a territorial perspective is evident (Fig. 7.1c);
3. *The multi-risk (MR) assessment*, which is a generalisation of the MHR assessment in the sense that it also considers possible interactions and cascading effects (Fig. 7.1c, d).

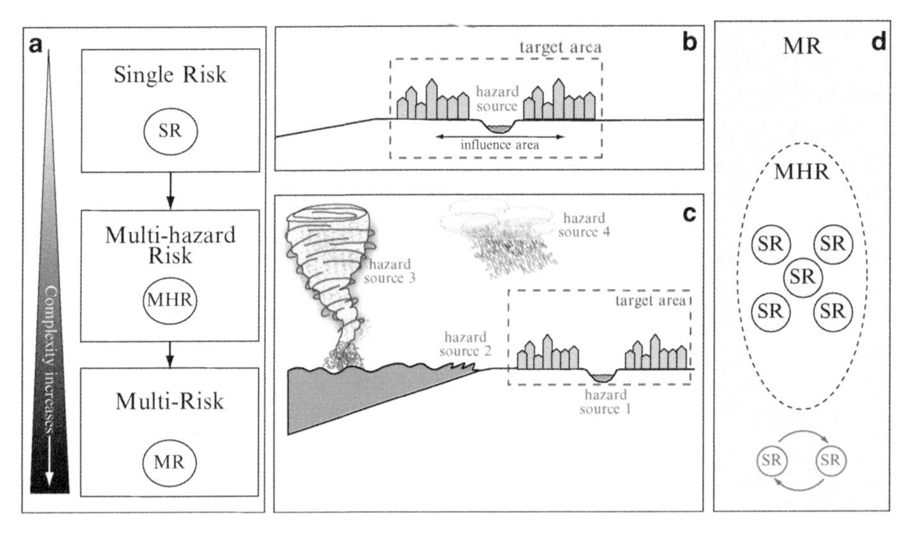

Fig. 7.1 Representation of the transition from single- to multi-risk assessment: (**a**) elements of the transition from the single- to the multi-risk assessment; (**b**) hazard-centred perspective of the single-risk assessment; (**c**) independent hazard sources affecting a given common area of interest as the main characteristic of the multi-hazard risk assessment; (**d**) the multi-risk assessment, in which interactions and cascading effects are considered

Figure 7.1a shows the transition from the SR to the MR assessment. The hazard-centred perspective of the SR is represented in Fig. 7.1b; in this case, the first element of the analysis is normally the hazard source identification followed by the subsequent definition of the impact area and the assessment of the potential effects. The change towards a territorial perspective is reflected in Fig. 7.1c; in this case, the first element of the analysis is the definition of the target area for the analysis. The target area must contain the elements at risk that are exposed to potential damage caused by adverse events and that are of interest for the loss assessment. Note that this change in perspective is common to both the MHR and the MR assessments.

Figure 7.1d illustrates how each element in this transition is a subset in the higher levels of the sequence. In fact, a set of SR analyses compose a MHR analysis, and when the MHR assessment is complemented with the analysis of interactions among events, then the MR level is reached. This sequence is, of course, valid if the SR and MHR analyses are done coherently in a multi-risk framework, following some basic (and often non-trivial) elements of harmonisation. Details of the general setting of a multi-risk framework are presented in the following section.

Reaching the multi-risk level of the analysis may require huge technical efforts and large datasets (for example, data from hazard and vulnerability assessments for each hazard source, and information on conditional probabilities to assess interaction scenarios). In theory, a full multi-risk assessment should consider all the possible hazard sources and the effects on all the exposed elements, and should identify all the possible interaction scenarios (cascading effects). Reaching this level of detail is a difficult task since the complexity and the number of potential scenarios can be overwhelming and discouraging; several of them may be extremely difficult to quantify. In practice, it is necessary to set the problem in a way that the transition SR → MHR → MR is feasible, and the MR level can be reached when it is possible and necessary (e.g. when there is strong evidence for amplifications of the expected loss in some interaction scenarios, and when data and/or models are available for doing that).

The following list introduces a first set of non-exhaustive recommendations to start a multi-risk analysis:

- Set the problem in a multi-risk context, adopting a territorial perspective from the beginning;
- Explore the hazard sources that potentially can cause harm in the defined target area, and perform the MHR assessment producing harmonised results (see section "The multi-risk assessment");
- Explore the possible scenarios of interactions and try to identify those that are more likely to produce significant amplification of the effects;
- Check for the availability of data, physical and empirical models in order to quantify the most important scenarios of interactions.

The Multi-risk Assessment

Considering the general definition of multi-risk provided above, the types of events considered in a multi-risk analysis may include events threatening the same elements at risk (exposed elements, such as population, buildings, or infrastructure) without chronological coincidence, or events occurring at the same time or shortly following each other, because they are dependent on one another or because they are caused by the same triggering event (European Commission 2010). The first case represents what in this chapter is denominated 'multi-hazard risk', whereas the second case represents the possible interactions or cascading effects that are one of the main characteristic elements of multi-risk assessment.

Under certain harmonisation conditions, there is a logical transition from the single- to the multi-risk assessment; therefore, the basic concepts used in single-risk quantification are also part of the core of a multi-risk assessment. Risk is conveniently assessed as a function of the probability that a certain event will occur and of the extent of the damage caused to humans, environment and objects. Conceptually, it is the result of the operation:

$$\begin{aligned} \text{Risk} &= \text{Hazard} \times \text{Damage} \\ &= \text{Hazard} \times \text{Vunerability} \times \text{Value at Risk} \end{aligned} \tag{7.1}$$

Note that in this context both *Hazard* and *Vulnerability* are generally defined in probabilistic terms, while *Value at Risk* has measure units (e.g. monetary cost, number of human lives, etc.). To avoid possible misunderstandings caused by different terminology used by different disciplines, this chapter defines these general terms as follows:

- The *hazard* is understood as the probability that a given intensity value characterising an adverse event can be reached or exceeded in a certain area and within a defined time interval.
- The *vulnerability*, which is a generic concept interpreted and applied in various ways, is considered here as the probable damage to an element at risk given a level of intensity of the hazard. This concept can be regarded as a physical point of view for vulnerability.
- The *values-at-risk* term is considered as a measure of the total exposed values (and therefore the potential losses) in a given area. This parameter can be expressed for example in terms of number of people, economic value or any other metric meaningful for loss assessment.

In more quantitative terms, risk is often quantified using an expression of the form (e.g. Cornell and Krawinkler 2000):

$$\lambda(\ell) = \int_D \int_{I_m} \underbrace{F(\ell|D) dG(D|I_m)}_{\textbf{damage \& loss}} \underbrace{d\lambda(I_m)}_{\textbf{hazard}} \tag{7.2}$$

Where $\lambda(\ell)$ is a measure of the exceedance rate of a given loss value (ℓ, in a specific metric); D is a given damage state (for example, total collapse); $F(\ell|D)$ is the probability that a given loss is reached given the damage state D; $G(D|I_m)$ is the vulnerability (or fragility) term and represents the probability that a damage state D is reached given the occurrence of a certain intensity measure I_m characterising the hazard; and $\lambda(I_m)$ is the hazard term measuring the exceedance rates of a given intensity value of the hazard. Equation 7.2 is commonly used for seismic risk assessment, but it may be used as a general formulation for any kind of natural risk, as it simply describes how the hazard analysis, $\lambda(I_m)$, and the damage analysis, $G(\ell|D)dG(D|I_m)$, are interconnected (Marzocchi et al. 2012).

General Setting

The main purpose of the multi-risk assessment is to harmonise both the methodologies employed and the results obtained for different risk sources, taking into account possible risk interactions. Following the basic principles defined in Marzocchi et al. (2012), the setting of a multi-risk procedure needs to address the following points (once the target area of the analysis has been clearly identified):

1. Definition of the space-time window for the risk assessment and the metric for evaluating the risks. This procedure is part of the harmonisation required for MHR and MR analyses;
2. Identification of the hazard sources relevant for the selected area;

3. Identification of selected hazard scenarios covering all possible intensities and relevant hazard interactions;
4. Probabilistic assessment of each scenario, and propagation of uncertainties arising at each step;
5. Vulnerability and exposure assessment, taking into account the vulnerability to combined hazards; and
6. Loss estimation and multi-risk assessment (including uncertainties).

Multi-risk Framework

Given the complexity of processes that the multi-risk problem poses, the multi-risk framework proposed here contemplates three levels of analysis:

1. **The first-level analysis**, in which the evaluation of the potential physical damages is performed. It results from considering different hazards and physical vulnerabilities. Examples of the kinds of exposed elements considered at this level are buildings, infrastructure, and urban ecosystems.
2. **The second-level analysis**, where a set of *socio-economic factors* representing indirect losses is considered. It implies a (generally) hazard-dependent assessment of tangible indirect losses and is oriented towards the identification of the tangible indirect effects caused by the adverse event.
3. **The third-level analysis**, which is an indicator-based assessment with the objective of identifying hotspots of critical areas where the social context may amplify the losses, or areas in which the post-event and recovery phases may be more critical.

Figure 7.2 is the conceptual representation of the multi-risk framework proposed in this chapter. The first row in the figure represents the first-level physical risk. At this level, probabilistic hazard assessment (Fig. 7.2a) is combined with the physical vulnerability of the exposed elements (e.g. using fragility functions, dose-response or exposure-response functions, as shown in Fig. 7.2b), to produce a risk quantification in terms of "units of loss" per "units of time" (Fig. 7.2c). The loss units (e.g. economic, fatalities, etc.) come from the metric used to define the values at risk, whereas the units of time (e.g. years) come from the hazard assessment. The second and third levels (second and third rows in Fig. 7.2) aim to integrate information from the socio-economic context (Fig. 7.2d) and may produce two complementary kinds of information (Fig. 7.2e): On the one hand, the assessment of indirect effects (level 2) intends to calculate tangible indirect losses that can be added to the results of the direct physical losses estimated in the first level of the analysis (Fig. 7.2f). On the other hand, socio-economic indicators of social fragility and resilience (third level) may be used to calculate a socio-economic index. This index can be used to map hotspots of areas in which the social context may aggravate the loss scenario derived from the physical risk and the indirect loss assessments (Fig. 7.2g).

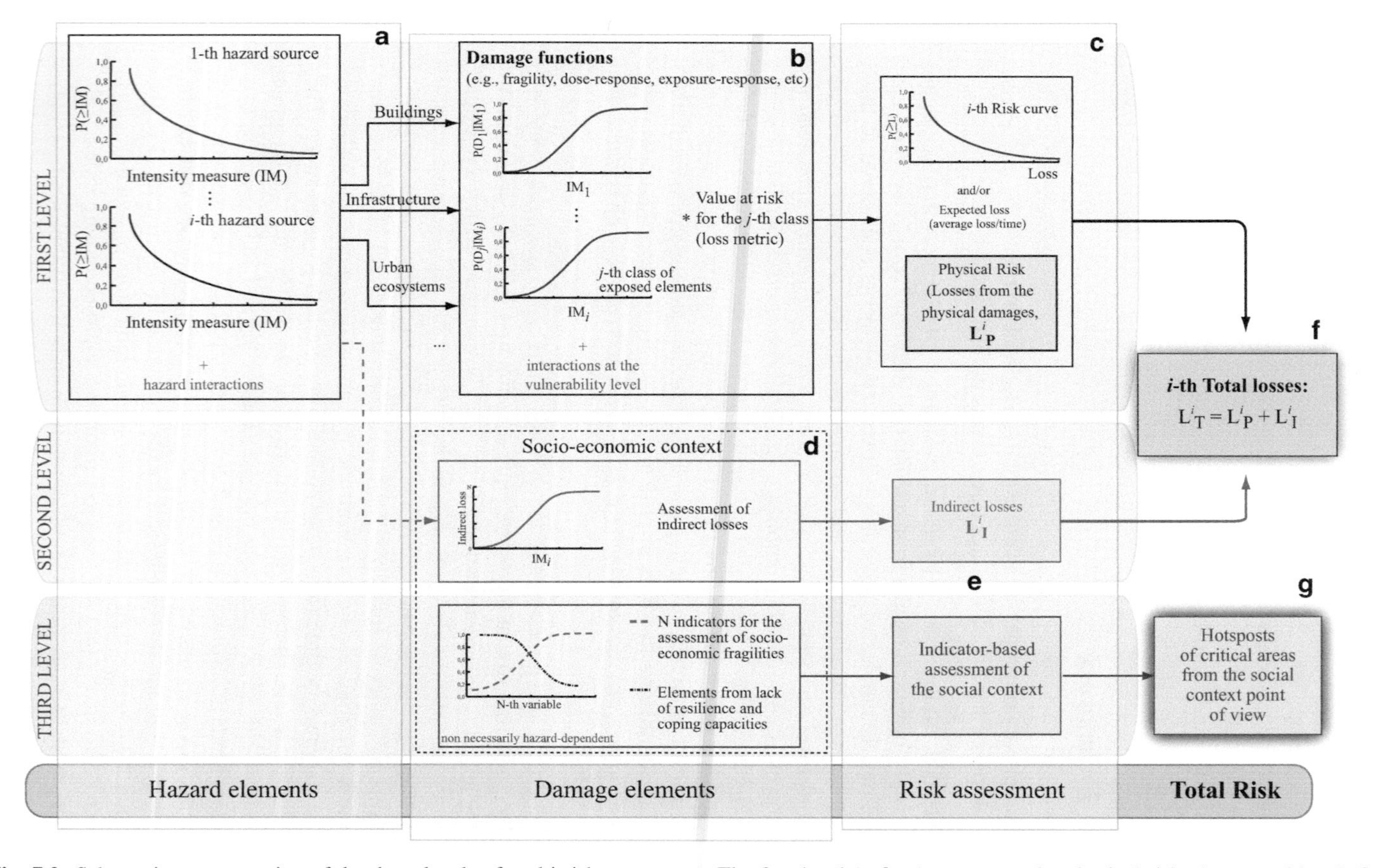

Fig. 7.2 Schematic representation of the three levels of multi-risk assessment. The first level (**a**, **b**, **c**) represents the physical risk; the second level (**d**, **e**) represents the assessment of tangible indirect losses assessed considering the socio-economic context. A total risk factor can be calculated as the total losses resulting from integrating the physical risk with the indirect losses (**f**). In the third-level analysis (**e**, **g**), a set of specific social context conditions is considered in an indicator-based approach

First Level: Multi-risk Assessment for Physical Damages

This level represents the evaluation of the direct effects caused by the adverse events through the quantitative assessment of the *physical* risk. At this level, the formulation for the multi-risk quantification follows the quantitative formulation for single risk assessments. Starting from the defined expression for single risk (Eq. 7.2), the quantitative form of the expected losses for the *physical risk*, L_P^i, associated with the ith hazardous event may be defined as:

$$L_P^i \equiv \lambda_i(\ell) = \int_D \int_{I_m^i} F(\ell|D) dG(D|I_m^i) d\lambda(I_m^i, \Delta t) \tag{7.3}$$

where L_P^i is referred to as the loss assessment considering the ith hazard and after integrating the contributions of all the different types of exposed elements (multi-vulnerability). Note that although rates (λ) are used in the formulation, it is straightforward to use probabilities instead of rates in Eq. 7.3. At this level, the elements described below need to be considered.

Harmonisation of the Risk Assessment

This first consideration implies that the procedures used for risk quantification must be harmonised in time and space, and a common loss metric must be used. The temporal and spatial harmonisation means that each of the different scenarios considered should be assessed, respectively, for the same time frame and the same exposed area. The temporal harmonisation comes fundamentally from the hazard assessment through the definition of the Δt value used to estimate the rates of occurrence (see e.g. Eq. 7.3), whereas the spatial harmonisation is related to both the spatial resolution in the hazard assessment and the resolution at which the inventory of exposed elements is collected.

The final element of harmonisation is the choice of a common metric for the loss assessment. This is a strongly case-dependent parameter; in most of the cases, the risk is quantified in terms of economic loss, fatalities, or number of people requiring assistance, among others. The consequence of the harmonisation process is that the results of different risk assessments become comparable. Note that the multi-hazard analysis following this strategy is the core of the MHR.

Assessing Interactions

Assessing interactions, or cascading effects, consists of the identification of the possible interactions that are likely to happen and that may result in an amplification of the expected losses. This concept is the fundamental part of the MR level of analysis, and what distinguishes it from the MHR.

Assessing interactions or cascading effects is an element poorly developed in the MR practice. Recent important attempts to quantify interactions within a multi-risk framework are those presented in Marzocchi et al. (2012), Selva (2013), and Garcia-Aristizabal et al. (2013a), and Gasparini and Garcia-Aristizabal (2014). The relatively slow development of this field is probably a consequence of the increase in complexity of the problem and the generally limited availability of data to perform reliable calculations. In fact, when the number of scenarios of cascading effects increases, the complexity of the problem becomes overwhelming. Nevertheless, what is necessary in practice is to identify the most important interaction scenarios, i.e. those in which there is strong evidence for possible amplifications of the expected losses.

An important initial step towards the assessment of cascading effects is therefore the identification of the possible scenarios. The term *scenario* is used in a wide range of fields and therefore different interpretations can be found in literature. In general, a scenario may be considered as a synoptical, plausible and consistent representation of an event or series of actions and events; it must be plausible because it must fall within the limits of what might conceivably happen, and must be consistent in the sense that the combined logic used to construct a scenario must not have any built-in inconsistencies (Haimes 2009 and references therein).

Different event-tree-like or fault-tree-like strategies can be used in order to exhaustively identify a complete set of scenarios. In some cases, an adaptive method combining both kinds of approaches is applied in order to ensure an exhaustive exploration of scenarios. A wide description of practical methods for scenario identification and structuring can be found, for example, in Haimes (2009) and Yoe (2012). From the multi-risk assessment point of view, the interaction scenarios of main interest are those that produce an amplification of the effects when compared to the effects produced by the single events.

Taking as reference the works of Marzocchi et al. (2012), Selva (2013), and Garcia-Aristizabal et al. (2013a), it is possible to describe two kinds of interactions, namely: (1) interactions *at the hazard level*, in which the occurrence of a given initial triggering event entails a modification of the probability of occurrence of a secondary event; and (2) *interactions at the vulnerability (or damage) level*, in which the main interest is to assess the effects that the occurrence of one event (the first one occurring in time) may have on the response of the exposed elements against another event. Implicitly, a combination of both kinds of interactions is another possible case; therefore, in the discussion of the interactions at the vulnerability level, both dependent and independent hazards can be considered as possible cases.

Interactions at the Hazard Level

In this case, the interaction problem is understood as the assessment of possible chains of adverse events in which the occurrence of a given initial triggering event entails a modification of the probability of occurrence of a secondary event. Even if this typology of problems can be assessed on a long-term basis, their utility can be highlighted in short-term assessments, in which specific scenarios conditioned to the occurrence of specific events can be assessed and compared among them.

In probabilistic terms, the occurrence of the triggered event (event 2), given the occurrence of the triggering event (event 1), can be assessed as (Marzocchi et al. 2012; Garcia-Aristizabal et al. 2013a; Selva 2013; Gasparini and Garcia-Aristizabal 2014):

$$p(\geq \mathrm{IM}_2^i) = \sum_j p\left(\geq \mathrm{IM}_2^i | \mathrm{IM}_1^j\right) p\left(\mathrm{IM}_1^j\right) \tag{7.4}$$

for $j = 1, 2, 3, \ldots, n$, for the n (exhaustive and mutually exclusive) classes of the intensity measure IM defined for the triggering event. IM_1 is the intensity measure defined for the triggering event, and IM_2 is the intensity measure defined for the triggered event (e.g., Gasparini and Garcia-Aristizabal 2014). This kind of interaction is of central importance in the CLUVA context because it is intrinsic to many climate-related hazards.

Interactions at the Vulnerability Level

This case of cascading effects fundamentally intends to assess the effects that the simultaneous occurrence of two or more events (not necessarily linked among them) may have for the final risk assessment. In this case, the action of different hazards is considered and combined at the vulnerability (or damage) level, and the main interest is to assess the effects that the occurrence of one event (the first one occurring in time) may have on the response of the exposed elements against another event (which may be of the same type as the former, but also possibly a different kind of hazard). Examples of this kind of interactions have been presented in literature, for example, in Lee and Rosowsky (2006), Zuccaro et al. (2008), Garcia-Aristizabal et al. (2013a), and Selva (2013).

In practice, this kind of interaction intends to quantify how the expected damages in the target area (associated with a given hazard) are modified if another hazardous event acts on the same exposed elements simultaneously or within a short time window (in general short enough that the system cannot be repaired). In the case of two hazards having additive load effects (i.e. they act simultaneously over the exposed element), the fragility function will depend on the intensities IM_1 and IM_2 of the two hazards, and therefore, it will represent a fragility surface. The probability that a given damage state (D_k) is reached given the occurrence of any ith value of IM_1 and any jth value of IM_2 can be represented as (detailed descriptions of this case can be found in Lee and Rosowsky 2006; Garcia-Aristizabal et al. 2013a; Gasparini and Garcia-Aristizabal 2014):

$$P(D_k) = \sum_i \sum_j \left[p\left(D_k | \mathrm{IM}_1^i \cap \mathrm{IM}_2^j\right) p\left(\mathrm{IM}_1^i \cap \mathrm{IM}_2^j\right) \right] \tag{7.5}$$

and it may be recursively adapted for considering both dependent and independent events through handling the $p\left(IM_1^i \cap IM_2^j\right)$ term (for details see Gasparini and Garcia-Aristizabal 2014).

Integration of Social Context Variables in the Multi-risk Framework

The risk assessment for physical losses (first level) constitutes the fundamental basis for the quantitative approach of multi-risk. However, *contextual* conditions such as the socio-economic factors and poorly resilient systems may worsen the initial physical loss scenario. In the risk literature, these factors have been generally considered using indicator-based assessments such as those performed in *holistic risk assessment* procedures (Cardona 2001; Barbat and Cardona 2003; Carreño et al. 2007).

The integration of physical and socio-economic vulnerabilities into a single methodology is not an easy task. Nevertheless, it is an interesting challenge for quantitative developments and should be considered in complex social systems, like those found in the African urban areas, where the loss scenario may considerably change because of the tangible and intangible indirect losses caused by the socio-economic context in which an adverse event occurs.

The second and third levels of the multi-risk framework presented here intend to take into account the characteristics of the social context in order to identify elements that may worsen or increase the expected losses estimated in the first-level physical risk.

Second Level: Assessing Tangible Indirect Losses Considering the Socio-economic Context: Extended Quantitative Assessment

In this chapter, the indirect losses are referred to as the tangible indirect effects (in the short- and the long-term) occurring as a consequence of the direct impacts. A full assessment of the tangible indirect losses is a complex task; nevertheless, the most relevant observable elements of indirect impacts can be measured to produce first-order quantifications. Therefore, it is important to note that the results obtained at this level are generally underestimations of the *total* tangible indirect losses. The process of assessing tangible indirect losses in this framework can be summarised as follows:

1. Identification of the most relevant indicators necessary to quantify tangible indirect losses;
2. Definition of the strategy to quantify the indicator using the same metric defined for the quantification of the direct impacts;
3. For each hazard scenario (e.g. for events with different return periods, probabilities or rates), use of the intensity footprint of the event to delineate the impacted areas, and sum up the contributions from all the indicators defined to quantify the total indirect losses; and
4. With the data generated in point 3, a curve is constructed by plotting the indirect losses (L_I) against their respective occurrence or exceedance probabilities.

Note that this approach produces a loss exceedance curve similar to that elaborated in the first-level analysis for the physical risk. In this way, information from the socio-economic context can be straightforwardly integrated into the quantitative multi-risk framework. Examples of socio-economic indicators that can be used for the assessment of tangible indirect losses are, among others:

- Losses in work productivity, which can measure the amount of missing income by workers in a disaster area;
- Losses due to missing income for the commercial activities affected by the occurrence of a disaster;
- Costs of evacuation from areas unfit for use;
- Costs of medical assistance to the people directly and indirectly affected;
- Effects of the loss of functionality of systems and networks, that can also be measured in different terms, as for example, missing income or disruption to productivity and the means of production.

Third Level: Indicator-Based Assessment for Identifying Hotspots of Critical Socio-economic Contexts

The objective of the indicator-based assessment is to calculate an *ad hoc* index based on a predefined set of indicators. The spatial distribution of this index may highlight the areas with particular social context conditions in which the effects of a disaster can be aggravated or where the post-event and recovery phases of the risk management process may show particular difficulties. Contrary to the assessment of tangible indirect losses presented in the previous paragraph, this indicator-based measurement is fully built on a subjective basis and has no specific loss units. Therefore, it summarises in an index a logical combination of subjective information, defined as representative of the socio-economic context. In general, the information required at this level is related to intrinsic parameters limiting the response in an emergency or worsening an adverse situation.

Given the subjective character of the information, the quantification procedure is a rather arbitrary process since there is not a standard process to identify the most appropriate set of indicators. The important point is that it has to be logical and transparent enough so that the output information can be easily interpreted. In general, the process can be summarised as follows:

1. Identification of the most relevant indicators to represent the socio-economic context; as stated before, the indicators should measure intrinsic parameters that may play an important role for limiting the response or that may worsen an adverse situation in that area;
2. Identification of a metric to measure the parameter (e.g. a percentage, a number of units or people available playing a specific role, etc.; for details see the examples of indicators);
3. Definition of a mechanism for measurement normalisation. In general, it implies (1) to define critical (low and high) thresholds and use them to map the measurement from an unbounded arbitrary scale to a normalised value (e.g. in the range 0–1); and

4. If necessary, application of weights to the different indicators (to ensure that the indicators considered as the most important are the most informative), and define the strategy to combine the results of the measurements in order to obtain the final index.

A simple way to quantify social context indicators is to measure quantities in percentage terms. This kind of information provides the opportunity to explore a wide range of potential parameters representing intrinsic characteristics of the population, as for example:

- *Demographic characteristics*: for example, percentages of more vulnerable categories (e.g. younger and older people) with respect to the total population; percentage of people with any kind of disability, etc.
- *Risk perception characteristics*: for example, risk awareness and acceptance, education level, communication capacities (see e.g. Chap. 6 in this volume).
- *Resilience-related characteristics*: for example, availability of resources, human capital/livelihood, insurance coverage (health, property), access to informal support networks, etc.

Finally, it is necessary to define a strategy to produce an indicator-based index and a mechanism of normalisation of the measurements. An interesting example of this kind of strategy is the methodology presented by Cardona (2001) and used by different authors in indicator-based risk assessments (e.g. Barbat and Cardona 2003; Carreño et al. 2007). These works apply an interesting framework to quantify an index (denominated as *aggravating factor*, F_S) using a set of input descriptor variables that represent elements contributing to aggravate the risk. The descriptors should permit identifying the characteristics of the urban area that increase the level of risk, and therefore, can be used to highlight critical areas (Cardona 2001). The kinds of indicators of that approach are grouped in two categories: (1) indicators representing social fragilities; and (2) indicators representing lack of resilience. The F_S is calculated as the weighted sum of the different indices calculated from the identified indicators (for details see Cardona 2001; Carreño et al. 2007). Tables 7.1

Table 7.1 Examples of social fragility indicators

Social fragility indicator	Measure (example)
More vulnerable categories	Percentages of children, elderly people, and people with disabilities, with respect to total population
Risk awareness, education	Number of people trained or informed about risks, percentage with respect to total population
Slums-squatter neighbourhoods	Slum-squatter neighbourhoods area/Total area
Delinquency rate	Number of crimes per 100,000 inhabitants
Population density	Inhabitants/ha of built area

Modified from Carreño et al. (2007)

Table 7.2 Examples of lack of resilience indicators

Lack of resilience indicator	Measure (example)
Insurance coverage	Percentage of population without insurance coverage (health, property)
Hospital beds	Number of hospital beds per 1,000 inhabitants
Health human resources	Medical professionals per 1,000 inhabitants
Rescue and firemen manpower	Number of units per 1,000 inhabitants

Modified from Carreño et al. (2007)

and 7.2 show examples of indicators used employing this approach (modified from Carreño et al. 2007).

To estimate the social fragility and lack of resilience indices, it is necessary to convert the index measurements from an unbounded arbitrary scale to a normalised value. A possible approach for this operation is using fuzzy sets theory (Zadeh 1965). Fuzzy sets are often applied to model and deal with imprecise information such as inexact measurements, imprecise concepts, imprecise dependencies, modelling of expert knowledge, or representation of information extracted from inherently imprecise data (Berthold and Hand 2003).

Output of the Multi-risk Assessment

The potential output of the multi-risk assessment can be summarised in the following two categories:

- Multi-hazard loss assessment output: Characterised by harmonised loss exceedance curves and expected annual losses; and
- Indicator-based output: Characterised by the identification of hotspots highlighting areas with critical socio-economic context.

Considering the loss assessment output, a standard concept for the representation of disaster risk results is the loss exceedance curve, which indicates the probability of exceeding a certain loss level. Likewise, the expected annual loss is an average value that allows us to define a kind of risk index. In the multi-risk framework presented in this chapter, the loss assessment is harmonised and considers the additive contribution of direct and indirect losses quantified using a common metric. It is worth noting, however, that when assessing interactions it is important to avoid duplicating the contribution of the indirect losses.

Conversely, the output provided by the indicator-based assessment of socio-economic context variables provides a spatial distribution of the calculated index. This kind of map provides complementary information, highlighting areas of particular social context conditions.

Applicative Example: Multi-hazard Risk Assessment in Dar es Salaam, Tanzania

This section illustrates the practical implementation of a multi-risk analysis in a pilot area located in the case study city of Dar es Salaam, Tanzania. The example is presented and developed step by step, starting from selecting the target area, identifying the exposed elements of interest for the analysis, and the spatial and temporal settings for harmonisation. The multi-risk analysis level is defined after identifying the available input data. Finally, the results obtained are used to link the concepts of multi-risk assessment and decision-making.

Selecting the Target Area and Identification of Exposed Elements

The pilot area for the quantitative analysis in Dar es Salaam is located in the Msimbazi river basin. The interest of the study is to perform a multi-risk analysis in areas occupied by or mixed with informal settlements; with this scope we selected a set of mixed areas located in the Sub Wards of Suna, Magomeni Idrisa, Dosi, Hanna Nassif, Kigogo Mbuyani, Mtambani, Mwnyimkuu, Mzimuni Idrisa, Mzimuni Makumbusho, and Mzimuni Mtambani. The zones identified as mixed/informal settlements in the pilot area are shown in Fig. 7.3 (represented in red); the footprints of the buildings in these areas are illustrated as yellow polygons underneath the blue shaded plot that aggregates the results.

Harmonisation Setting: Spatial Resolution, Timeframe and Loss Metric

In order to obtain comparable results, the spatial resolution, the temporal window of reference and the loss metric need to be harmonised. In this study, the time window used for all the analyses is one year (Δt=1 year). For the loss assessment, we have considered direct physical damages quantified, when possible, using two kinds of loss metrics: (1) economic losses and (2) number of people requiring assistance (PRA). PRA measures the number of people directly affected by the adverse event and that may require some sort of assistance . Note that PRA can straightforwardly be converted to monetary loss units if necessary. Finally, the results for the different risks analysed are aggregated in equal surface units that are represented by the blue cells in Fig. 7.3.

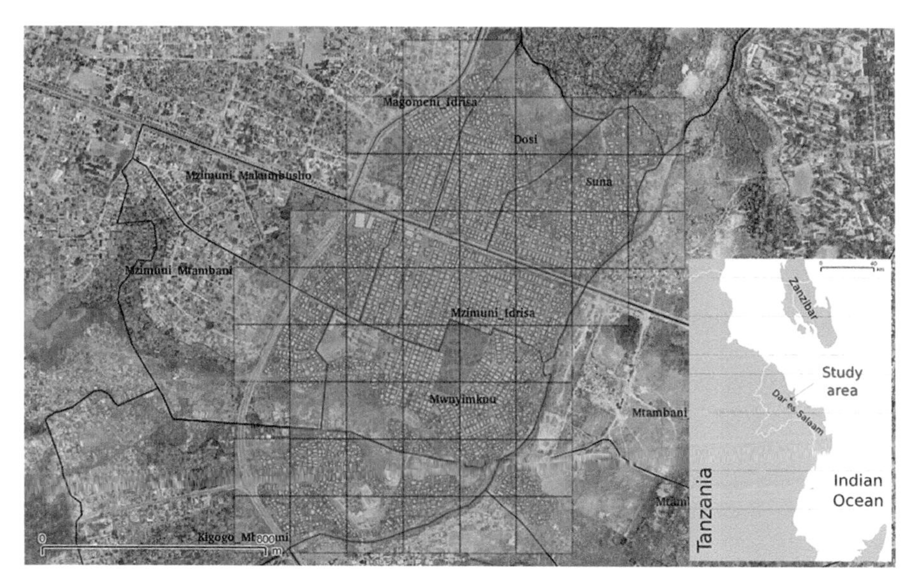

Fig. 7.3 Target area for the pilot study; the *red* areas represent the identified mixed/informal settlements in the study area. The *yellow* polygons are the residential structures defining the set of exposed elements. The *blue* shaded plot shows the regular grid (200 × 200 m) used to aggregate the results. Coordinates of the left-lower corner of the grid: Latitude −6.813°; Longitude 39.253°

Table 7.3 Climate-related triggering effects for Dar es Salaam

	Triggering event				
Triggered event	Extreme temperature and heat waves	Wind	Intense rain	Consecutive dry days	Sea level rise
Flooding			x		x
Drought	x			x	
Desertification	x			x	

Identifying Hazard Sources, Scenarios of Cascading Effects and Setting the Level of Analysis

The events of relevance in this study are climate-related hazards. As possible hazard sources, we have identified extreme meteorological events (temperature/heat waves, wind, precipitation, and consecutive dry days), flooding (fluvial or sea level rise), droughts, and desertification. Many of these events may be classified as triggering and triggered events, as shown in Table 7.3.

The main scenarios of cascading effects are well known cases of interactions at the hazard level, and most of them are generally already considered by the specific procedures for the (single) hazard assessment. For instance, the main chains of events are represented by:

- Intense rain → Fluvial flooding;
- Sea level rise → Flooding;
- Extreme temperature/heat waves → Drought;
- Consecutive dry days → Drought;
- Extreme temperature/heat waves → Desertification;
- Consecutive dry days → Desertification.

The Dar es Salaam pilot test is characterised by the presence of different exposed elements diversely vulnerable to different climate-related hazards. Considering the available information from the hazards and the vulnerabilities, in this pilot example, we present a comparative analysis taking into account fluvial floods and heat waves. Therefore, because the cascading effects component in this example is represented by sequences already considered by the standard routines used in their assessment (e.g. rain → floods), the kind of analysis can be framed as a 'multi-hazard risk assessment' (see e.g. Fig. 7.1).

Quantitative Analysis: A Level 1, Multi-hazard Risk Assessment

The available information in this pilot case allows us to perform a level 1 multi-hazard risk assessment. No information of indirect losses or social context indicators is used. Therefore, combining the available hazard, vulnerability and exposure information we conduct a comparative analysis of floods and heat wave risks. Detailed information about the floods hazard assessment can be found in Chap. 2 of this volume; descriptions of the fragility functions used for the flood risk assessment are available in Chap. 3; and the exposure-response functions used for the heat-waves risk assessment are from Garcia-Aristizabal et al. (2013b).

Using the available flood hazard information, the flood vulnerability (fragility functions) of structures in informal settlements, and the exposure information collected by the authors, the flood risk calculations were performed, aggregating the results in sub-areas as shown in the grid in Fig. 7.3. The resulting flood risk map presents the annual expected losses as PRA/year (Fig. 7.4). Figure 7.4a shows the area of interest for the analysis and an example of a flood hazard map (water depth for a return period of 300 years) for reference. The 50th percentile values of the expected loss at each cell of the grid are given in Fig. 7.4b. Risk curves are calculated at each cell and an example is presented in Fig. 7.4c. It is worth noting that, when possible, uncertainties were considered and propagated; therefore, different percentiles are used to express the results: the 50th percentile is used as the *best value* of the displayed parameter, whereas the 16th and 84th percentiles are used to represent the uncertainty bounds.

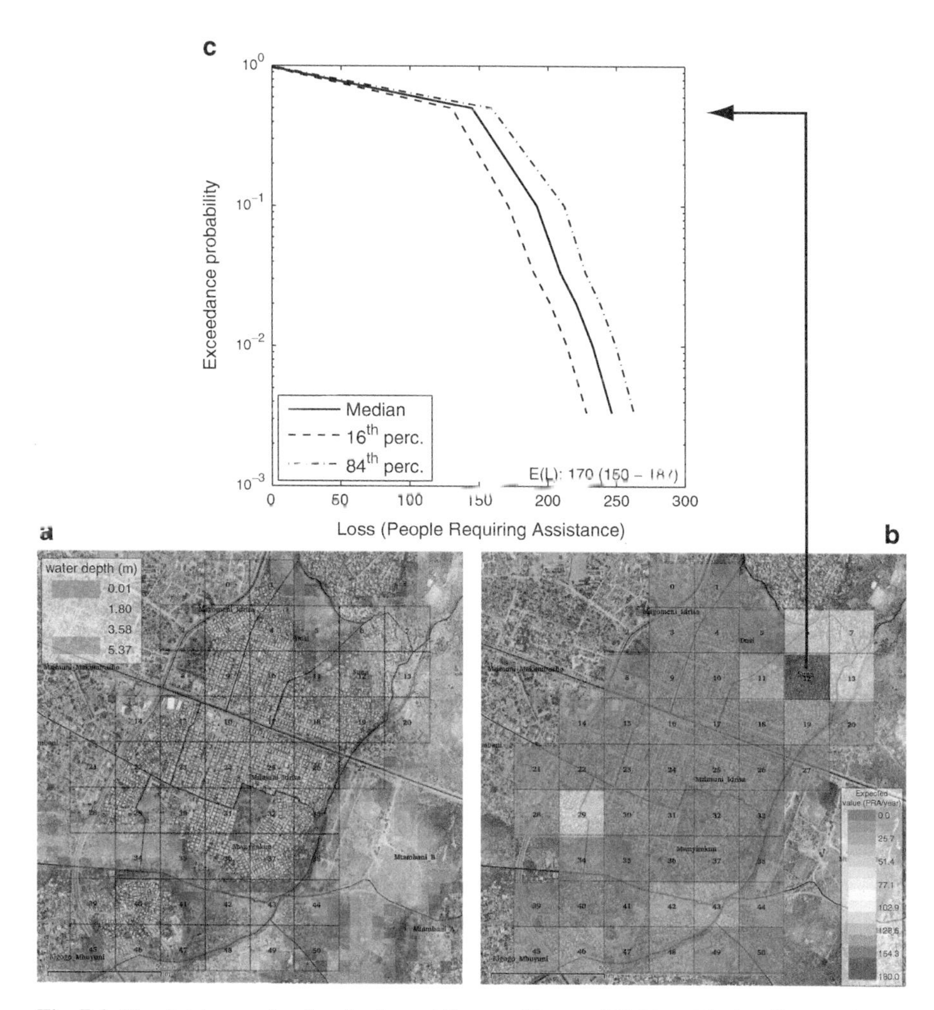

Fig. 7.4 Flood risk map showing the Annual Expected Losses (PRA/year) in the pilot area in Dar es Salaam. (**a**) Flood hazard map (return period of 300 years) and footprint of the considered exposed elements; (**b**) 50th percentile of the expected losses at each grid element; (**c**) Example of the risk curve at Cell 12: the *solid line* is the 50th percentile, and the *dashed lines* are the 16th and 84th percentiles of the exceedance probabilities, representing the uncertainty bounds

For the quantification of the heat wave risk, the hazard information was obtained from Garcia-Aristizabal et al. (2013b) considering the maximum temperature in heat waves. The exposure-response functions were obtained from literature (see Garcia-Aristizabal et al. 2013b; Bell et al. 2008; McMichael et al. 2008). Following the harmonisation procedure, the heat wave risk calculation was performed using the same calculation grid used for the flood risk (Fig. 7.3); likewise, the harmonisation requirement implies that also in this case the metric for the loss assessment has to be in PRA terms.

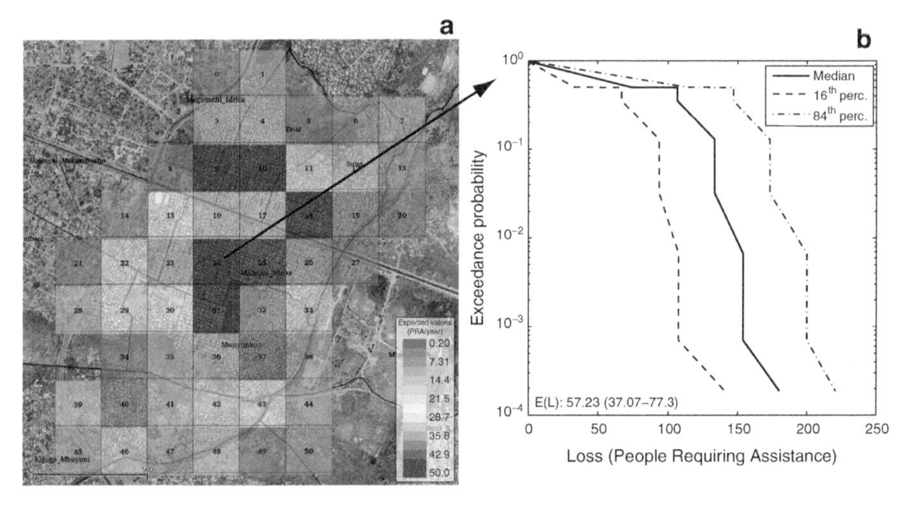

Fig. 7.5 Heat wave risk map showing the annual expected losses (PRA/year) in the pilot area in Dar es Salaam. (**a**) 50th percentile of the expected losses at each grid element; (**b**) Example of the risk curve in the Cell 24: the *solid line* is the 50th percentile, and the *dashed lines* are the 16th and 84th percentiles of the exceedance probabilities, representing the uncertainty bounds

Figure 7.5 shows a heat wave risk map with the annual expected losses (PRA/year) for the pilot test area in Dar es Salaam. Figure 7.5a represents the 50th percentile values of the expected loss at each cell of the grid. As in the previous case, a risk curve is available at each cell for the 16th, 50th, and 84th percentiles (Fig. 7.5b).

Given that the hazard assessment procedures related to the scenarios considered in this example intrinsically consider the cascading effects (hazard level interactions), the main effort in the multi-risk analysis was oriented towards the harmonisation of the results. The harmonised results for the worked example are shown in Figs. 7.4 and 7.5. These results, along with the implicit annual risk curves calculated for each cell of the domain, are harmonised and then can be ranked and compared. Figure 7.6 is a comparative example in which the resulting risk curves for floods and heat waves are shown for different areas of the analysis domain. Figure 7.6a displays the calculation grid and the cells selected for the example, while Fig. 7.6b compares the plots of the flood (black) and heat wave (red) risk curves for Cell 12; in this area (located in the Suna Sub Ward), the flood risk indicates higher values relative to the heat wave risk. For example, if we take as reference the risk values with 10 % exceedance probability in both cases, the flood risk ranges between ~180 and ~240 PRA/year, whereas the heat wave risk ranges between ~50 and ~110 PRA/year. Considering the results of Fig. 7.6c (Cell 29), the curves are closer with respect to the case presented in Fig. 7.6b, indicating that the differences in this area are less marked with respect to the previous case (in fact, the median of the losses for floods is located around the 16th percentile of the heat

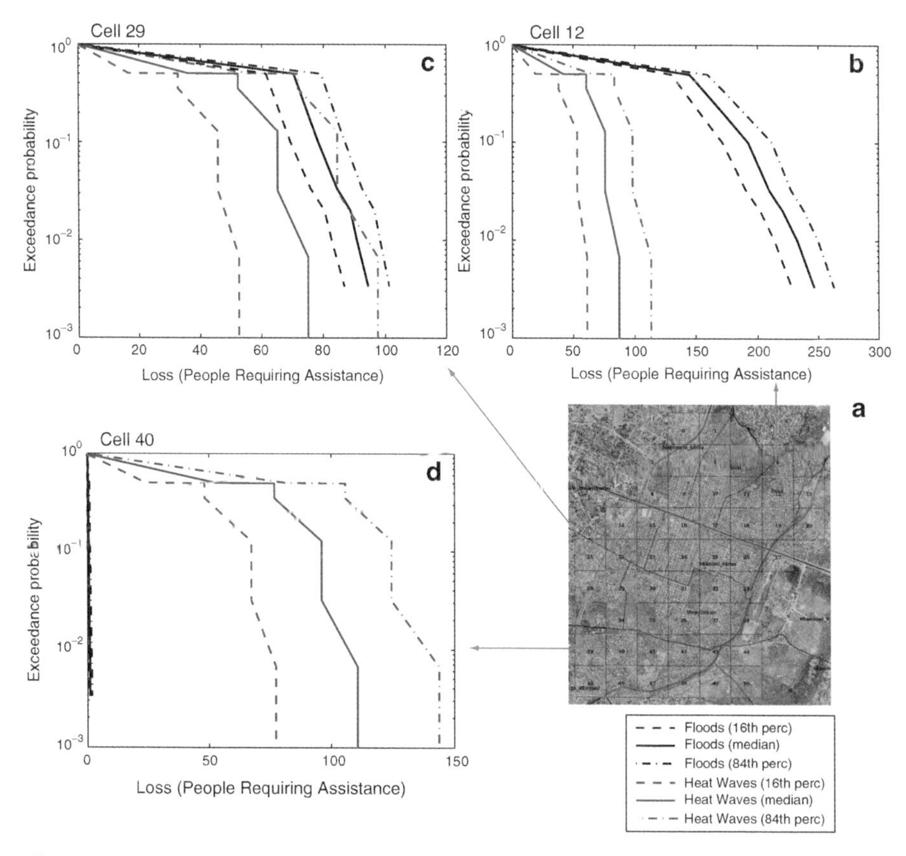

Fig. 7.6 Comparative plot of the flood and heat wave risk curves (exceedance probabilities and annual losses): (**a**) grid domain for the analysis; (**b**) risk curves in the Cell 12; (**c**) risk curves in the Cell 29; (**d**) risk curves in the Cell 40. For details, see the text

wave risk). Opposite to the case in Fig. 7.6a, d shows an example in which the heat wave risk is clearly higher than the flood risk (Cell 40).

Finally, to quantify total risks for the whole area of the pilot test, the analyses were also performed aggregating the results in a unique risk curve for the whole area. The aggregated results for the whole area are presented in Fig. 7.7. The floods and heat wave risk curves are shown, respectively, in black and red colours ($\Delta t = 1$ year). The solid lines are the 50th percentiles, whereas the dashed lines are the uncertainty bounds (16th and 84th percentiles). The results obtained show interesting features: first, considering the lower loss values (the more probable and frequent), for example PRA/year < ~500, the curves show comparable results in both cases. There is a range of PRA values (roughly between 500 and 2,000) in which the curve for floods decreases with a higher slope respect to the heat waves curve; therefore, in this range of losses the heat wave risk is significantly higher than the flood risk.

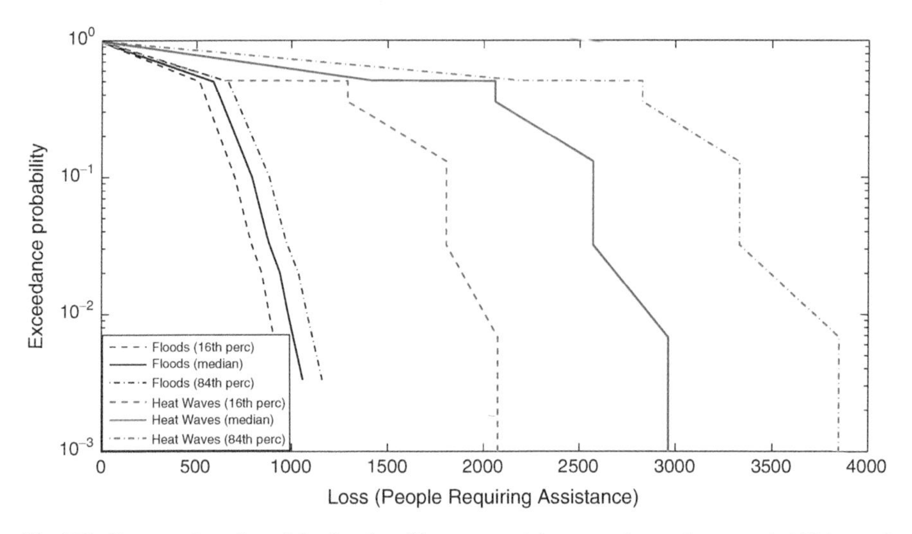

Fig. 7.7 Comparative plot of the flood and heat wave risk curves (exceedance probabilities and annual losses) for the whole domain

Multi-risk Assessment and Decision-Making

The risk assessment and management process represents a set of logical, systemic, and well-defined activities that provide the decision-maker with a sound identification, measurement, quantification, and evaluation of the risk associated with certain natural phenomena or man-made activities (Haimes 2009). The entire process of risk assessment and management is a circular loop involving feedback and updating of information. The multi-risk concept enters this loop as an additional element complementary to the single-risk assessment practice and, therefore, a clear definition of the evaluation mechanism of multi-risk results is required. For this reason, one of the most challenging issues of multi-risk assessment is the translation of the output of the quantification and analysis processes into useful information for decision-making under uncertainty. This is a critical step to consolidate the value of multi-risk analyses and to define their ultimate importance and usefulness in the resolution of critical societal problems. In this section, the results of the worked example are applied to illustrate the application of the multi-risk output in decision-making activities.

Ranking and Comparing Risks of Different Nature

In a multi-risk framework, the introduction of the concept of probability in the risk assessment is useful for many important reasons, as for example: (1) with a common metric for the loss assessment, it allows an immediate ranking of the risks and facilitates the comparison and management of events of different nature; (2) it may serve as a supporting tool during the decision-making phase, for example, to perform cost/benefits analyses of management strategies; and (3) it allows the definition of the fundamental concepts of societally *acceptable* and *tolerable* risk. In this way, the multi-risk assessment establishes an objective reference both for land use planning and for the selection of risk mitigation actions.

For the results of the multi-risk quantification to be supportive and effective, they must capture not only the *average* risk value (i.e. the expected losses) but also the extreme values; therefore, it is strongly desirable to present the output of the analysis using both the risk curves and the expected values (see e.g. section "Output of the multi-risk assessment") because these can be used together to compare and rank the existing risks, as illustrated in Figs. 7.6 and 7.7.

The expected annual losses are calculated by averaging events of high probabilities and low consequences with events of low probabilities and high consequences; therefore, this parameter must be used with caution for risk-based decision-making given that, by definition, major disasters are generally caused by the low probability extreme events. Nevertheless, the expected value is a resource that, when used together with the full risk curve, becomes a useful tool in the risk analysis process. Conversely, the risk curve provides a more complete picture of the potential consequences and their likelihood. A detailed analysis of the risk curve, in fact, may provide important information that the average does not capture; for example, it makes it possible to compare the effects of high probability/low consequence events with the effects of low probability/high consequence ones.

From Multi-risk Assessment to Decision-Making

The results of the multi-risk analysis can be an essential tool for decision-making based on a comprehensive appreciation of all the risks threatening a particular target to be protected. To build the link between multi-risk assessment and decision-making, it is necessary to define proper multi-risk evaluation strategies. Decisions may include *yes* or *no* choices on a single action, may rank a series of alternative options, or may select the best option among a set of alternatives; the harmonisation strategy required by multi-risk assessments provides appropriate results with objective elements to support decision-making activities.

Ranking and comparing risks, as discussed in the previous paragraph, is the most direct tool that the multi-risk assessment produces for decision-making. Another significant decision often faced by the decision-maker is to answer the question: Is the risk acceptable? The probabilistic approach adopted by quantitative risk and multi-risk assessments provides evidence to answer this question. It is important to distinguish two important concepts: *acceptable* and *tolerable* risk. The risk is *acceptable* when the probability of occurrence is so small or the consequences are so slight that individuals or the society are willing to take or be subject to the risk that the event might occur. A risk that is not acceptable must be managed in order to reduce it –if possible– to an acceptable level, or at least to a tolerable level. A *tolerable* risk is then a non-negligible risk that has not yet been reduced to an acceptable level (because it is not possible to reduce the risk further, or the costs of doing so are excessive) and then is tolerated (Yoe 2012). A tolerable risk is still unacceptable, but its severity has been reduced to a point where it can be tolerated.

Deciding whether an assessed risk is acceptable or not and determining a tolerable level is a subjective matter, and decision-makers must take into account different elements, such as the scientific evidence, the uncertainty, the defined objectives on the analysis, and the available resources, among others. Results of multi-risk assessments may provide objective data with all of these required characteristics. Even though no particular set of rules exists for establishing acceptable and tolerable levels of risk, decision-makers use several principles, such as the precautionary principle, the weight of evidence, the ALARA (as low as reasonably achieved), the ALOP (appropriate level of protection), or by using *reasonable relationships* between the cost of the mitigation actions and the reduction in risk. A review of the different approaches can be found in Yoe (2012).

Beyond the direct comparative analyses, decision-makers can also use the multi-risk results to evaluate the effects of different risk management options (RMOs). When different RMOs are formulated, the decision-maker must assess and compare a number of options to find the best one. In this process, the results of the multi-risk assessment may be also of great help. To show an illustrative example we consider the risk curves for floods and heat wave risks presented in Fig. 7.7 for the pilot test in Dar es Salaam, Tanzania. To assess different possible mitigation options (MOs), a decision tree can be used to represent alternative MOs, as shown in Fig. 7.8 for the floods case. The decision node in Fig. 7.8a considers different possible MOs oriented to reduce the flood risk. The first path (Fig. 7.8b) shows the case without implementing any MO, which represents the existing risk. The other two paths

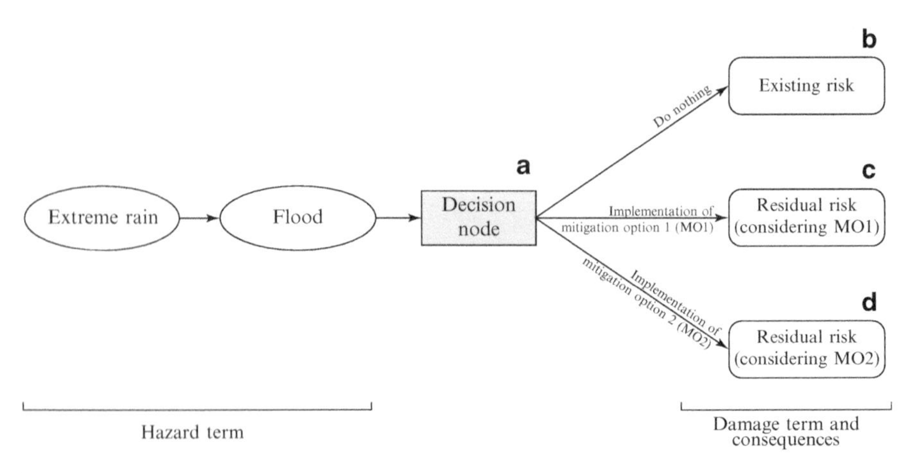

Fig. 7.8 Simplified example of a decision tree in which a decision node is used to assess the consequences of different scenarios defined by the implementation of alternative mitigation options

are two possible mitigation options (MO1 and MO2) that might be proposed (Fig. 7.8c, d, respectively).

If one or more risks are judged unacceptable, it is necessary to reduce them through risk management options. Considering the results of a multi-risk analysis, the expected values of risk and the risk curves can be used to estimate the effectiveness of the RMOs. Building a levee, for example, is one option (among many others) for managing flood risk and this is the case that we use here for the illustration. Figure 7.9a shows the risk curves derived from the assessment of

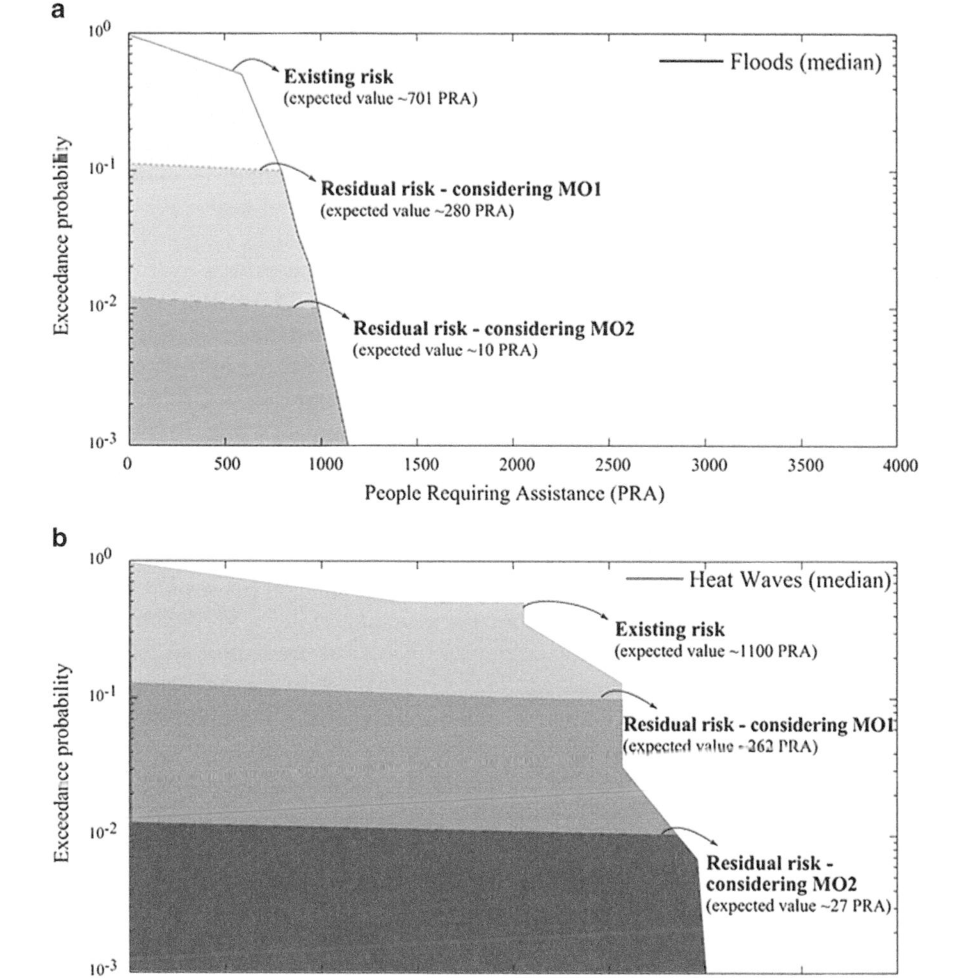

Fig. 7.9 Risk curves representing the existing risk and the residual risk considering different mitigation options: (**a**) Example for the flood risk; (**b**) Example for the heat waves risk

the three different paths after the decision node shown in Fig. 7.8. The "Existing risk" curve in Fig. 7.9a corresponds with the *do nothing* path in Fig. 7.8b, and the expected annual consequences for this scenario is ~701 PRA/year.

Suppose that two MOs have been suggested: the first flood risk mitigation option, MO1, is an option in which a levee to contain the water level of the flood with 10 years of return period is proposed. The second mitigation option (MO2) consists of a more expensive levee to contain the water level corresponding with the flood with a return period of 100 years. For simplicity, we ignore the risk of levee failure. The residual risks after considering the MO1 and MO2 are represented in Fig. 7.9a. The expected annual consequences for the scenarios defined by MO1 and MO2 are, respectively, 280 and 10 PRA/year. The cheaper MO1 reduces the expected annual consequences by 60 %, whereas the more expensive MO2 reduces them by 98 %. It is worth noting that different kinds of MOs (for example, relocation of people settled in floodplains) can also be assessed; nevertheless, we limit our discussion here to these two MO examples for simplicity.

A similar analysis is presented in Fig. 7.9b for the case of heat waves. In this case, the MOs can be the activation of an emergency plan to support the more vulnerable people in the area of interest in the case of temperatures overcoming a given threshold (for example, those with return periods of 10 and 100 years).

In a multi-risk analysis, similar exercises can be developed considering scenarios of cascading effects. After assessing a set of MOs, a set of data composed by (1) the cost of implementing the MO, (2) the residual risk, and (3) the reduced (or mitigated) risk, is available to help the decision-maker. For example, the decision-maker can use this information to perform cost/benefit analyses (Marzocchi and Woo 2007), and to decide where and how to operate more effectively, considering the available resources.

This section contains a basic example for using the multi-risk results to support decision-making activities. However, this field represents an interesting area for further and more detailed developments. For example, in the worked example we have mainly used the expected annual consequences to measure risk reduction; nevertheless, more specific analyses may be required to perform assessments focused on the more extreme events, which in many cases are the scenarios of primary importance for the risk management. Examples of such kinds of analyses are the risk partitioning tools (Haimes 2009). Another field requiring further research is on the handling and interpretation of the epistemic and aleatory uncertainties considered in the assessments.

Conclusions

The harmonised quantitative assessment of (a) all the types of hazards threatening a specific target, (b) their interactions, and (c) the expected losses, is the result provided by an ideal multi-risk approach. These harmonised assessments are necessary in order to allow the decision-makers to make objective decisions. The theory and examples discussed in this chapter aimed to show, in general terms, how the approach can be implemented and how the results can be used to help decision-makers. However, the potential applications span a wider range of cases according to the specific needs of the problem at hand.

For example, a long-term multi-risk assessment, covering several decades, is useful for city and land use planning purposes. The assessment of the occurrence of each hazard with a given intensity, the probabilistic assessment of cascading events, and the quantitative assessment of the impacts on the target area is a fundamental tool for deciding how a city should develop in time or for developing the regulations for land use. This is particularly important for areas prone to the effects of time-varying agents, such as those related to climate change. Conversely, short-term multi-risk assessments (days to months) are useful for analysing the relative consequences of several possible scenarios, allowing several disaster mitigation options to be analysed and compared, and the residual risks calculated.

In order to perform multi-risk calculations, comprehensive databases of past events, fragility curves for all types of vulnerable assets, and the relevant socio-economic information must be available. Although these data are generally difficult to be retrieved or may not be available at all (especially in many developing countries), multi-risk estimates with a poor data set are nonetheless a valuable way to approach decision-making in support of emergency planning and response, and even to assess the effects of adaptation measures (see e.g. Chap. 9). In this case, Bayesian methods can be a useful approach for integrating different sources of information such as past data, results of deterministic modelling and/or data derived from expert opinions.

Acknowledgements This chapter substantially enhanced by constructive comments from Farrokh Nadim, Sarah Lindley and Ingo Simonis. Credits also go to Sandra Fohlmeister and Cynthia Skelhorn for helping to improve the original manuscript. The work presented in this chapter was developed in the framework of the FP7 European project CLUVA, grant no. 265137.

References

Barbat A, Cardona O (2003) Vulnerability and disaster risk indices from engineering perspective and holistic approach to consider hard and soft variables at urban level. Technical report, IDB/IDEA Program on Indicators for Disaster Risk Management, Universidad Nacional de Colombia, Manizales, Colombia. http://idea.unalmzl.edu.co/. Accessed 12 Dec 2011

Bell ML, O'Neill MS, Ranjit N, Borja-Aburto VH, Cifuentes LA, Gouveia NC (2008) Vulnerability to heat-related mortality in Latin America: a case-crossover study in Sao Paulo, Brazil, Santiago, Chile and Mexico City, Mexico. Int J Epidemiol 37:796–804

Berthold MR, Hand DJ (eds) (2003) Intelligent data analysis: an introduction, 2nd edn. Springer, Berlin/Heidelberg

Blong R (2003) A new damage index. Nat Hazards 30(1):1–23

Bovolo CI, Abele SJ, Bathurst JC, Caballero D, Ciglan M, Eftichidis G, Simo B (2009) A distributed framework for multi-risk assessment of natural hazards used to model the effects of forest fire on hydrology and sediment yield. Comput Geosci 35(5):924–945. doi:10.1016/j.cageo.2007.10.010

Cardona O (2001) Holistic evaluation of the seismic risk using complex dynamic systems (in Spanish). PhD thesis, Technical University of Catalonia, Barcelona, Spain

Carpignano A, Golia E, Di Mauro C, Bouchon S, Nordvik J (2009) A methodological approach for the definition of multi-risk maps at regional level: first application. J Risk Res 12(3–4):513–534. doi:10.1080/13669870903050269

Carreño ML, Cardona O, Barbat A (2007) Urban seismic risk evaluation: a holistic approach. Nat Hazards 40:137–172. http://dx.doi.org/10.1007/s11069-006-0008-8. 10.1007/s11069-006-0008-8

Cornell CA, Krawinkler H (2000) Progress and challenges in seismic performance assessment cornell. PEER Center News 3(2). http://peer.berkeley.edu/news/2000spring/index.html

Del Monaco G, Margottini C, Serafini S (1999) Multi-hazard risk assessment and zoning: an integrated approach for incorporating natural disaster reduction into sustainable development. Technical report, TIGRA project (The Integrated Geological Risk Assessment), grant No. Env4-CT96- 0262. European Commission DG XII, Environment and Climate Program, Brussels

Del Monaco G, Margottini C, Spizzichino D (2006) Report on new methodology for multi-risk assessment and the harmonisation of different natural risk maps. Technical report D3.1, ARMONIA project (Applied Multi-Risk Mapping of Natural Hazards for Impact Assessment), grant No. 511208. Rome, January 2006

Del Monaco G, Margottini C, Spizzichino D (2007) ARMONIA methodology for multi-risk assessment and the harmonisation of different natural risk map. Technical report D3.1.1, ARMONIA project (Applied Multi-Risk Mapping of Natural Hazards for Impact Assessment), grant No. 511208. Rome, 31 January 2007

Dilley M, Chen RS, Deichmann U, Lerner-Lam AL, Arnold M, Agwe J, Buys P, Kjiekstad O, Lyon B, Yetman G, Dilley M, Chen RS, Deichmann U, Lerner-Lam AL, Arnold M, Agwe J, Buys P, Kjiekstad O, Lyon B, Yetman G (2005) Natural disaster hotspots: a global risk analysis. The World Bank Hazard Management Unit, Washington, DC

Durham K (2003) Treating the risks in Cairns. Nat Hazards 30(2):251–261

European Commission (2000) TEMRAP: the European multi-hazard risk assessment project. Technical report, European Commission DG XII, Environment and Climate Program, grant ENV4-CT97-0589, Brussels

European Commission (2010) Commission staff working paper: risk assessment and mapping guidelines for disaster management. Technical report, European Commission, Brussels

FEMA (2004) Using HAZUS-MH for risk assessment. In: HAZUS-MH risk assessment and user group series, Federal Emergency Management Agency, FEMA 433

Garcia-Aristizabal A, Marzocchi W (2012a) Bayesian multi-risk model: demonstration for test city researchers. Technical report D2.13, CLimate change and Urban Vulnerability in Africa (CLUVA project), grant No. 265137. URL http://www.cluva.eu/deliverables/CLUVA_D2.13.pdf

Garcia-Aristizabal A, Marzocchi W, Woo G, Reveillere A, Douglas J, Le Cozannet G, Rego F, Colaco C, Fleming K, Vorogushyn S, Nadim F, Vangelsten BV (2012b) State-of-the-art in multi-risk assessment. Technical report D5.1, MATRIX project (New methodologies for multi-hazard and multi-risk assessment methods for Europe), grant No. 265138. URL http://matrix.gpi.kit.edu/Deliverables.php

Garcia-Aristizabal A, Marzocchi W, Di Ruocco A (2013a) Probabilistic framework for multi-hazard assessment. Technical report D3.4, MATRIX project (New methodologies for multi-hazard and multi-risk assessment methods for Europe), grant No. 265138. URL http://matrix.gpi.kit.edu/Deliverables.php

Garcia-Aristizabal A, Marzocchi W, Ambara G, Uhinga G (2013b) Reports and map on multi-risk Bayesian scenarios on one selected city (Dar es Salaam, Tanzania). Technical report D2.14, CLimate change and Urban Vulnerability in Africa (CLUVA project), grant No. 265137 URL http://www.cluva.eu/deliverables/CLUVA_D2.14.pdf

Gasparini P, Garcia-Aristizabal A (2014) Seismic risk assessment, cascading effects. In: Beer M, Patelli E, Kougioumtzoglou I, Au I (ed) Encyclopedia of earthquake engineering, Springer Reference. Springer, Berlin/Heidelberg, pp 1–20. ISBN: 978-3-642-36197-5, doi:10.1007/978-3-642-36197-5_260-1, url http://link.springer.com/referenceworkentry/10.1007/978-3-642-36197-5_260-1

Garcia-Aristizabal A, Bucchignani E, Palazzi E, D'Onofrio D, Gasparini P, Marzocchi W (2015) Analysis of non-stationary climate-related extreme events considering climate-change scenarios: an application for multi-hazard assessment in the Dar es Salaam region, Tanzania. Nat Hazards 75(1):289–320. doi:10.1007/s11069-014-1324-z, url http://link.springer.com/article/10.1007/s11069-014-1324-z

Grunthal G, Thieken H, Schwarz J, Radtke KS, Smolka A, Merz B (2006) Comparative risk assessments for the city of Cologne – storms, floods, earthquakes. Nat Hazards 38(1–2):21–44

Haimes Y (2009) Risk modeling, assessment, and management, 3rd edn. Wiley, Hoboken

ISO/IEC guide 73 (2009) Risk management – vocabulary (and ISO 31010: Risk management – Risk assessment techniques)

Kappes MS, Keiler M, von Elverfeldt K, Glade T (2012) Challenges of analyzing multi-hazard risk: a review. Nat Hazards 64(2):1925–1958. doi:10.1007/s11069-012-0294-2

Lari S, Frattini P, Crosta GB (2009) Integration of natural and technological risks in Lombardy, Italy. Nat Hazard Earth Syst 9(6):2085–2106. doi:10.5194/nhess-9-2085-2009

Lee KH, Rosowsky DV (2006) Fragility analysis of woodframe buildings considering combined snow and earthquake loading. Struct Saf 28(3):289–303. doi:10.1016/j.strusafe.2005.08.002

Marzocchi W, Woo G (2007) Probabilistic eruption forecasting and the call for an evacuation. J Geophys Res 34:L22310

Marzocchi W, Mastellone ML, Di Ruocco A, Novelli P, Romeo E, Gasparini P (2009) Principles of multi-risk assessment: interactions amongst natural and man-induced risks, Project report. European Commission, Directorate-General Research – Environment, grant No. 511264

Marzocchi W, Garcia-Aristizabal A, Gasparini P, Mastellone ML, Di Ruocco A (2012) Basic principles of multi-risk assessment: a case study in Italy. Nat Hazards 62(2):551–573

McMichael AJ, Wilkinson P, Kovats RS, Pattenden S, Hajat S, Armstrong B, Vajanapoom N, Niciu EM, Mahomed H, Kingkeow C, Kosnik M, O'Neill MS, Romieu I, Ramirez-Aguilar M, Barreto ML, Gouveia N, Nikiforov B (2008) International study of temperature, heat and urban mortality: the 'ISOTHURM' project. Int J Epidemiol 37:1121–1131

Nadim F, Liu Z, Garcia-Aristizabal A, Woo G, Aspinall W, Fleming K, Vangelsten BV, van Gelder P (2013) Framework for multi-risk assessment. Technical report D5.2, MATRIX project (New methodologies for multi-hazard and multi-risk assessment methods for Europe), grant No. 265138. URL http://matrix.gpi.kit.edu/Deliverables.php

Schmidt J, Matcham I, Reese S, King A, Bell R, Henderson R, Smart G, Cousins J, Smith W, Heron D (2011) Quantitative multi-risk analysis for natural hazards: a framework for multi-risk modelling. Nat Hazards 58:1169–1192. doi:10.1007/s11069-011-9721-z

Schmidt-Tomé P, Callio H, Jarva J, Tarvainen T, Greiving S, Fleischhauer M, Peltonen L, Kumpulainen S, Olfert A, Schanze J, Barring L, Persson G, Relvao AM, Batista M (2006) The spatial effects and management of natural and technological hazards in Europe (ESPON). Geological Survey of Finland

Selva J (2013) Long-term multi-risk assessment: statistical treatment of interaction among risks. Nat Hazards 67(2):701–722

UN-ISDR (2009) UN international strategy for disaster reduction, terminology section, UNISDR, Geneva, Switzerland. http://www.unisdr.org/we/inform/terminology. Accessed 25 Sept 2013

UNDP (2004) Reducing disaster risk: a challenge for development. United Nations Development Programme Bureau for Crisis Prevention and Recovery, New York

van Westen CJ, Montoya AL, Boerboom LGJ, Badilla Coto E (2002) Multi -hazard risk assessment using GIS in urban areas: a case study for the city of Turrialba, Costa Rica. In: Proceedings of the regional workshop on best practices in disaster mitigation: lessons learned from the Asian urban disaster mitigation: lessons learned from the Asian urban disaster mitigation program and other initiatives, 24–26 September 2002, Bali, Indonesia, pp 120–136

Yoe C (2012) Principles of risk analysis. Decision making under uncertainty. Taylor & Francis group, LLC, Boca Raton

Zadeh LA (1965) Fuzzy sets. Inf Control 8:338–353. Academic. URL http://www-bisc.cs.berkeley.edu/Zadeh-1965.pdf. Accessed 25 Sept 2013

Zuccaro G, Cacace F, Spence R, Baxter P (2008) Impact of explosive eruption scenarios at Vesuvius. J Volcanol Geotherm Res 178(3):416–453

Chapter 8
USSDM – Urban Spatial Scenario Design Modelling

Andreas Printz, Hany Abo-El-Wafa, Katja Buchta, and Stephan Pauleit

Abstract Climate change, population growth and expansion of settlement areas are among the major challenges that African cities are facing. The aim of the study was to model growth of settlement areas under different scenarios of population density and flood protection to support development of strategies for urban development for risk reduction and protection of green areas which deliver vital ecosystem services for the future city.

The GIS-based USSDM – Urban Spatial Scenario Design Modelling approach – was developed and applied to simulate the likely patterns of settlement area increase in Dar es Salaam. The model is based on cellular automata principles and further developed as a participatory design tool for urban planners with strong involvement of stakeholders and experts. Specific features are transparency of model settings, a user friendly interface and easy adjustment of model parameters to respond to different assumptions on urban growth conditions as well as fast simulation runs.

According to USSDM scenario simulations, higher population density settings could minimise the future settlement expansion of Dar es Salaam by up to 5,000 ha until 2025, corresponding to approximately 8 % of the farmland area in 2008. Preventing future expansion of settlement areas into the flood prone areas would significantly reduce exposure of the human population to flooding. Recommendations for strategic development of a multifunctional urban green infrastructure planning in African cities are made.

Keywords Urban growth • Urban planning • Urban green • Scenario modelling • Dar es Salaam

A. Printz (✉) • H. Abo-El-Wafa • K. Buchta • S. Pauleit
Chair for Strategic Landscape Planning and Management, Technical University of Munich (TUM), Emil-Ramann-Str. 6, 85354 Freising, Germany
e-mail: printz@wzw.tum.de; hanywafa@mytum.de; katja.buchta@gmx.net; pauleit@wzw.tum.de

S. Pauleit et al. (eds.), *Urban Vulnerability and Climate Change in Africa*, Future City 4,
DOI 10.1007/978-3-319-03982-4_8

Introduction

Africa is well known for the highest indices of vulnerability related to climate change. Urban areas will be particularly at risk due to their concentration of population, assets and functions, and often, due to their location in areas exposed to natural hazards (Pelling and Wisner 2009).

The African continent adds another superlative – it has the world's highest rates of population growth and urbanisation (Fig. 8.1): by 2050, African cities are expected to accommodate three times as many people as today (UN-Habitat 2014). On the background of extreme poverty, low education, low institutional capacities, limited availability of resources and loss of ecosystem services, African cities have to address huge and highly complex challenges. In this context, climate change may be "just" an additional stressor, which amplifies the already high baseline vulnerability of African cities.

African cities suffer from uncontrolled growth of their settlement areas, mainly as informal settlements with insufficient access to infrastructure, health, education services and income (UN-Habitat 2008). In most cities, a great proportion of new urban settlers are rural migrants. Many of them can be considered as environmental refugees as they were forced to leave their marginal agricultural areas (Barrios et al. 2006; IPCC 2007).

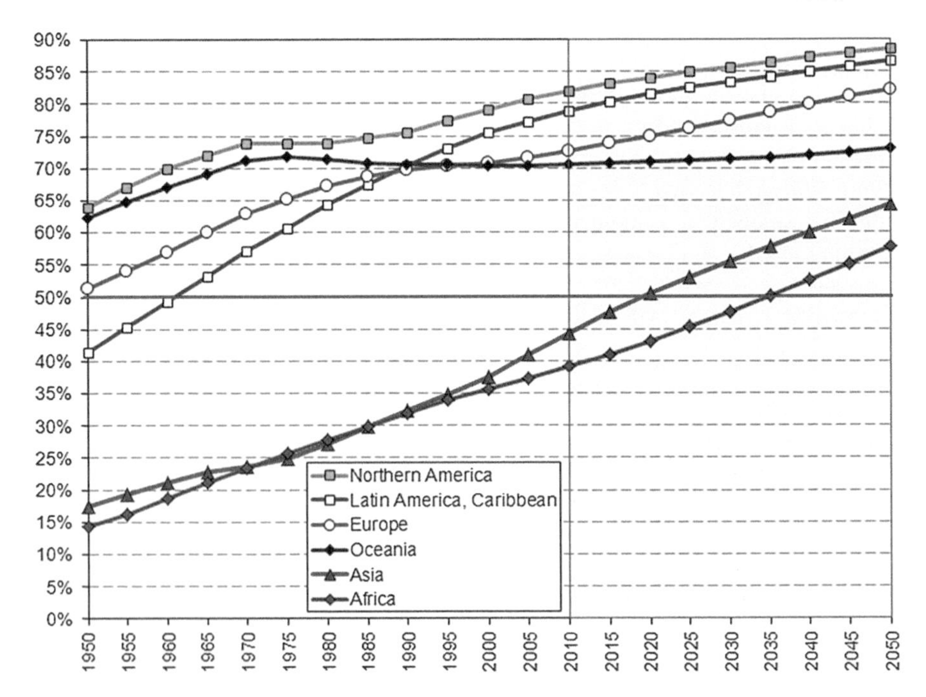

Fig. 8.1 Urban population by major geographical area (in % of total population) (Based on UNDESA 2011)

Climate change projections show strong evidence that this movement in many African regions is expected to be further aggravated in the future due to deterioration by climate change and/or desertification (Chap. 2; Niang et al. 2014). Although African urbanisation trends have been overestimated in some cases and are more complex than just being unidirectional (Potts 2012), the strong population growth rates were also confirmed in the recent population census for the CLUVA case study cities Dar es Salaam and Addis Ababa.

New city dwellers often arrive with few resources and settle in precarious areas within the urban periphery or in remaining open areas within the existing built-up area. In many cases these areas are threatened by natural hazards such as flooding. Climate change projections show a trend for an increase of extreme weather events, which will result in increased risks of flooding or lead to more frequent and intense heat waves (see Chaps. 1 and 2 of this volume).

Besides the increased risks of direct threats in certain areas, the need for vital ecosystem services will rise as they enable mitigating some of the threats, such as moderating heat waves or providing flood protection. Ecosystem services can also contribute to the survival of people in critical situations of extreme hazards, e.g. by providing safe spaces for resettlement and supply of food and fuel wood.

However, to be able to derive strategic knowledge for urban planning, there is a need to provide spatially explicit information on areas at risk as well as the distribution of ecosystem services. For planning the future city, information is also required on growth patterns of urban areas across rapidly developing African cities. Better knowledge is required about how expected population growth rates translate into spatial patterns of expanding settlement areas for a city; how this will increase the vulnerability of the city as more people will live in hazard prone areas; and how ecosystem services will decline due to loss of green areas. Importantly, the impact of different urban development strategies on urban vulnerability and its adaptive capacity needs to be known so that well informed decisions can be made and ensure that the capacity of urban planning is strengthened to respond to the combined challenges of climate change and urban growth.

Therefore, the aim of the study was to develop a spatial scenario modelling approach for supporting strategic planning in African cities. The objectives were to:

(a) Develop a land demand driven GIS-based urban scenario model which can be used in participatory planning contexts;
(b) Determine and visualise the expected spatial expansion of settlement areas using the model according to the different scenarios in Dar es Salaam;
(c) Explore the impact of urban growth on urban vulnerability to climate change related natural hazards.

The approach was developed and applied for the case study cities of Addis Ababa and Dar es Salaam. For the chapter's comprehension, this paper will focus on the Dar es Salaam case.

Urban Development History in Dar es Salaam

The formative period of the city is linked to the shift of the administration seat and capital of the former German colonial regime from Bagamoyo in 1890. During this time period, town planning introduced racial segregation: Different settlement zones for Germans/Europeans, Asians and Africans were established with decreasing size of plots, proportion of green areas and services standards. In 1918, for approximately 50,000 inhabitants, principles of a garden city represented the top level of urban living standards. Britain took over in 1918 without modifying the German urban regime order. Even though Dar es Salaam has not been the official state capital since 1974, it continued to attract -as the biggest Tanzanian city- a large number of new settlers to reach 1.5 million inhabitants in 1992. The attempts of the Master Plans of 1968 and 1979 to regain a planning initiative are considered to be surpassed by the reality of strong informal settlement activities. The impact of other development plans for the city has also remained limited as well as that of slum upgrading programmes (Mng'ong'o 2004; Kiunsi 2013). Dar es Salaam currently has an annual growth rate of more than 4 % which places it among the ten fastest growing cities of the world (City Mayors Statistics 2011). With a background of strong population pressure and weak planning, about 80 % of the settlement area of Dar es Salaam in 2013 was informal, severely lacking infrastructure and services. Around 50–70 % of the total population is estimated to have a connection to the piped water supply, while only 10 % are connected to a sewer system (UN-Habitat 2010).

Informal settlements are often found in areas with a high exposure to natural hazards, such as floodplain areas. An example of the disastrous consequences of such development was the December 2011 flood which led to 20 casualties, 200 persons injured, and approximately 10,000 people displaced (IFRC 2011).

In the CLUVA project, the consequences of urbanisation on land use change, and in particular the city's green structure, were analysed for the years 2002 and 2008. The comparison was based on the Urban Morphology Type (UMT) approach (see also Chap. 4).

The different pathways of transformation of green areas (left side) to settlement areas (right side) and the further transformation of settlement areas were analysed (Fig. 8.2).

The main chain of transformation was from remaining/degraded "Bushland" to "Mixed Farming", while "Mixed Farming" at the periphery in turn was 'invaded' by the transitional settlement type "Scattered Settlement" which further developed into the "Villa/Single Storey" settlement type. Only relatively small areas further changed to settlement types with a higher population density such as "Condominiums". Also, the land use in "Riverine" (and flood prone) areas was strongly altered into "Mixed Farming" as well as into "Scattered Settlement".

Even if the comparison is based on only two dates, it shows that "Mixed Farming" seems to be an important precursor of the settlement, and the settlement development itself, at least over a longer period, is characterised by settlements with a low population density.

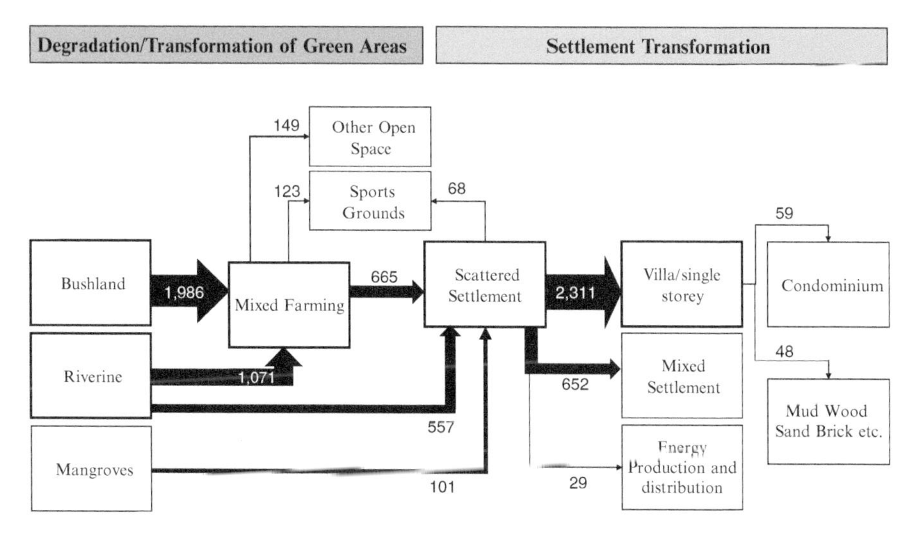

Fig. 8.2 Change detection of land use (UMT units) 2002–2008: degradation of green areas and settlement transformation in Dar es Salaam; *numbers* indicate area changed (ha) (Reproduced from Lindley et al. 2013)

During the 6-year period (2002–2008) "Green Area" UMTs lost more than 30 % of their total area. A more detailed land cover study by Renner (in Lindley et al. 2013) found strong evidence for a general loss of wood cover of 9–12 % during this time period.

Theoretical Background

A literature review identified several factors that drive urban growth patterns. Barredo et al. (2003) classified them into the following categories:

1. Environmental – there could be environmental constraints present which prevent specific areas from development and in turn favour other areas, such as steep slope and flood areas.
2. Neighbourhood – related to the dynamics of land use and its relation to other land uses in the surrounding area. For example, new residential areas tend to grow close to existent residential areas, while in some other cases certain land uses have a repulsive effect on other land uses. It is considered as one of the most influential factors that drives patterns of urban growth.
3. Spatial characteristics, such as the accessibility or proximity to transport networks.
4. Urban and regional planning policies that affect urban dynamics, notably land use zoning.

5. Other categories, such as economic development, socio-economic and political systems, and individual preferences. Modelling of categories is rather challenging because it involves human decision-making processes and should be integrated into econometric models.

The overall urban land use demand is generated exogenously to the cellular models (White et al. 1997) as it is not reflected by the urban local dynamics but by the growth of a city (Barredo et al. 2003).

There are two main different approaches in modelling spatial dynamics of urban areas. The traditional top-down approach tends to break down the system into manageable elements and components using deterministic models. The other approach that provides a higher capability in modelling complexities of spatial systems is a bottom-up approach that considers urban development as a self-organising system in which specific constraints and controls affect local decision-making processes producing macroscopic patterns of urban form (Ward et al. 2000). Examples of such an approach are cellular automata models (CAs) (Batty et al. 1999; Ward et al. 2000; Barredo et al. 2003).

The cellular structure used in CA models represents the adjacency or proximity characteristic similar to that of land parcels in urban systems. The structure is usually a lattice of a uniform gridded space. The value of the cell can represent one state of the cell (e.g. residential or commercial) from a group of possible states defined by the system being modelled (Ward et al. 2000).

Adjacent cells change their states based on the application of simple transition rules that can be considered as the generators of growth or decline. The cell can change from undeveloped to developed or other attributes of the cell can change (Batty and Xie 1997). The change of the cell state depends on functions that react to the cell's attributes as well as what is happening in its neighbourhood (Batty et al. 1999).

Urban dynamics models have been introduced in different cities in developed countries, such as the models produced for Cincinnati (White and Engelen 1993), San Francisco (Clarke et al. 1997), Marseilles (Meaille and Wald 1990) and the Washington/Baltimore corridor (Clarke and Gaydos 1998). Such cities' growth is usually characterised by planned settlements, which is not the case in the majority of African cities. The models used in the developed countries' context are unsuitable to capture the urban growth dynamics within a developing country's context. Typically, these kinds of models are expensive, data intensive, customised and limited to specific countries. Vector based models require complex programming skills, or the source code may not be accessible (Sietchiping 2004).

In most developing countries, and particularly in Africa, the availability of spatial information for cities is rather limited and in some cases non-existent. In most cases, the usability of existing spatial data for urban dynamics modelling is challenging, due to issues such as the lack of a consistent classification methodology, low resolution, and because data is outdated. These issues are considered as obstacles in sharing spatial information among different sectors and also across different cities in Africa (Barredo et al. 2004). Despite these data limitations, there has been considerable work on modelling the spatial expansion of African cities.

In Lagos, a cellular automata approach was applied (Barredo et al. 2004). A probabilistic approach was adopted in which several factors drive land-use dynamics and the city's growth that represent the accessibility and suitability. Main determinants were a land-use zoning factor, which represents the legal regulations for future land uses, in addition to a neighbourhood factor and a stochastic parameter that simulates the degree of uncertainty that is characteristic for most social and economic processes. Another example is the modelling of informal settlement dynamics and the dynamic visualisation of the simulation results for the case of Yaoundé, Cameroon (Sietchiping 2004).

In Dar es Salaam, a model based on standard GIS software and designed with the principles of Cellular Automata (CA) was developed by Hill and Lindner (2010a, b). In this model, the transformation process of non-urban cells into informal urban areas was based on a so-called 'transition potential' that is calculated using certain suitability factors.

The modelling approach developed and applied in the CLUVA project has its similarities with Hill and Lindner's (2010a, b) model as they both are raster based GIS models that are designed using the principle of cellular automata. However, Hill and Lindner's model was based on a binomial logistic regression model which defines transition potential through an extensive process of model calibration and validation. Moreover, the spatial extent of this model was confined to a limited area within the city of Dar es Salaam and the particular focus of the application was to explore the potential of transport infrastructure to guide future informal residential development decisions.

In contrast to the prediction oriented machine-like CA-based models, in CLUVA, a scenario-discourse based design model approach was applied. Therefore, the SSDM (Spatial Scenario Design Model) approach for a land use and water management in Benin (Schwarz von Raumer and Printz 2007) was transferred and further developed towards specific urban planning situations of fast growing settlement areas of the CLUVA reference cities Dar es Salaam (Buchta 2013a, b) and Addis Ababa (Abo-El-Wafa 2013a, b).

The main aim was to develop a flexible model particularly suitable to support scenario development for urban planning and assess the impacts of different pathways of urban development and urban growth impacts on land use types that provide several ecosystem services to the city. Therefore, the model should allow the user to quickly change settings and be able to add or change the weights of influencing factors according to outcomes from planning workshops.

Methodology

Urban Spatial Scenario Design Modelling Approach

The Urban Spatial Scenario Design Modelling (USSDM) approach was developed in two stages. The initial core model was first developed and then discussed and customised to the case of Dar es Salaam with the contribution of stakeholders in several workshops. The stakeholders were involved in the determination of modelling parameters that included the expected settlement land demand, exclusion criteria, influencing factors and the scenarios to be modelled.

USSDM Design

USSDM is a raster-based model developed in a GIS environment on the main principles of cellular automata (CA) where the study area is represented in a GIS environment by a grid of raster cells. The size of the cells was 100 m by 100 m, corresponding to the minimum area of one ha in the urban morphology type survey (Cavan et al. 2012). The entire study area of Dar es Salaam was therefore divided into 149,896 raster cells.

The overall logic of the USSDM approach is depicted in Fig. 8.3. Land uses which were not expected to undergo change from non-settlement to settlement within the temporal scope of the model, were excluded beforehand ("excluded areas"). The likelihood of land use change from non-settlement to settlement and the potential for an increase of population density in already built areas had to be determined for the cells. For this purpose, the model calculates a transformability index for each cell using a weighted overlay operation of specific influencing factors which have been determined as factors in the transformation of cells into settlement cells. Among these factors is the neighbourhood influencing factor which is calculated based on the assumption that cells that lie close to settlement cells have a higher likelihood to transform than other cells that lie far away from settlement cells. The future spatial demand for urban land is estimated separately based on the projected population growth rate and assumptions of population densities according to local urban planning experts. After each modelling step, exclusion areas and the neighbourhood influencing factor are updated to include the new settlement areas that were transformed in the previous modelling step.

Scenario Building Process

The scenario building process involved local urban planning experts (academic and administrative) and decision-makers. It included three steps:

- Scoping and defining the scenarios;
- Defining exclusion criteria, expected settlement land demand, influencing factors;
- Discussion of results and model-handover.

At the beginning of the scenario building process a scoping study was carried out to identify model inputs, useful output parameters, scales, time horizons and expected drivers of the model. In this study, population growth is assumed to be

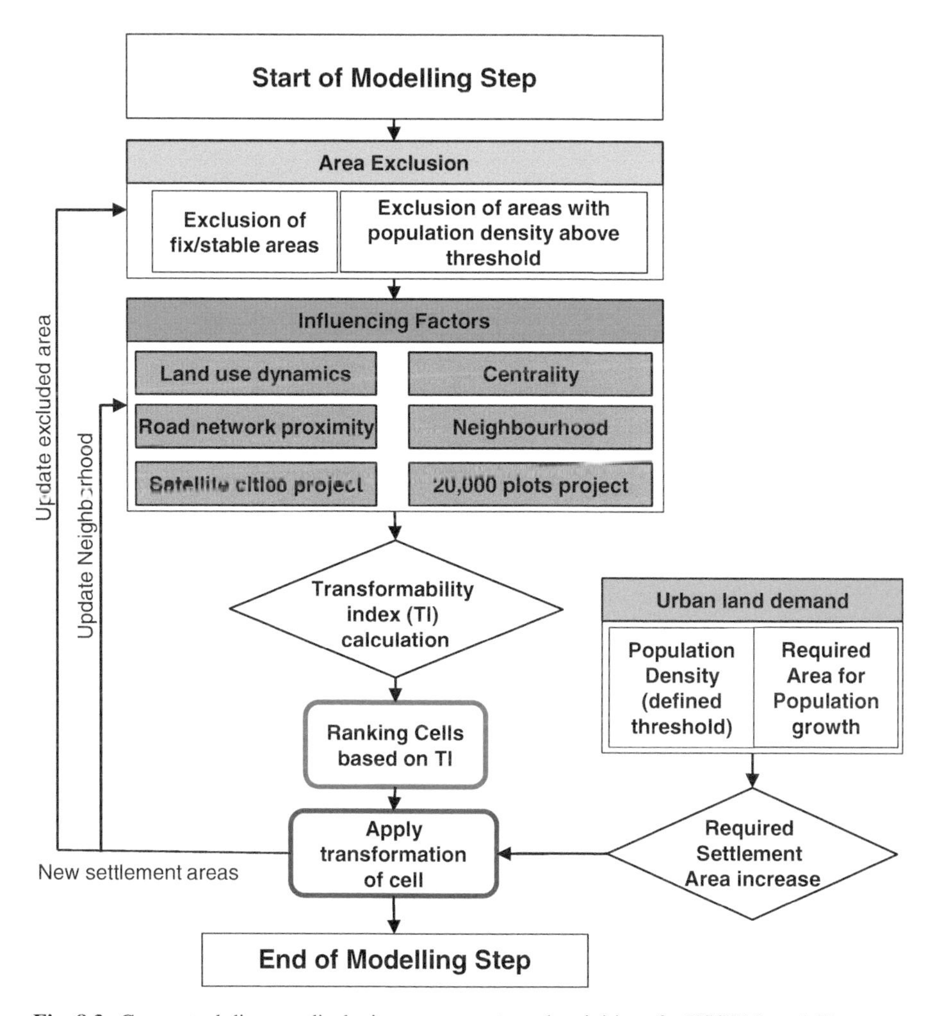

Fig. 8.3 Conceptual diagram displaying components and activities of a USSDM modelling step

the main driver for urbanisation. Moreover, different threshold values for maximum population densities and excluded areas were defined for the scenarios.

The scenario building and the definition of the influencing factors were discussed during workshops and several meetings with local experts. This process was informed by the draft new Master plan as stakeholders from the municipal administration and researchers from Ardhi University were directly or indirectly involved in its preparation (Chap. 10). Intermediary and final outputs were also presented and discussed in workshops with research partners, urban planners and other stakeholders. This helped to check and improve the plausibility of scenario outputs.

In addition to a presentation of the output, administrative urban planning staff members were "trained on the job". The model, together with a background information document and an operating manual were provided to the local administrative staff as well as to the academic partners of the CLUVA project.

Transformability Index

Within the scope of this research, the urban morphology types (UMTs) that are likely to transform within the temporal scope of the model into settlement are: farmland, green areas (other vegetated areas) and selected recreation and settlement UMTs. These areas are regarded as the dynamic UMT classes and are subject to transformation while all other areas are considered as stable areas and excluded from modelling which included parks, military areas, and government/institutional buildings, such as schools and hospitals (Fig. 8.4).

Transformability scores for a number of "influencing factors" were determined depending on the properties and location of each grid cell. All influencing factor maps were normalised to a score range from 0 to 100 with 0 being the least probable and 100 as the most probable to transform to settlements (Fig. 8.5). The overall transformability index was then derived by the overlay and combination of the different influencing factors, which are briefly described hereafter. A detailed description of the entire methodology can be found in Buchta (2013a, b) and Abo-El-Wafa (2013a, b).

Land Use Dynamics Geospatial change detection analysis of Dar es Salaam's development in the period from 2002 to 2008 and interviews with experts were performed to assess the likelihood of transformation of other UMTs to residential UMTs (settlements) in the period from 2002 to 2008.

Centrality/Sub-centres Proximity The influencing factor represents proximity to planned areas with central urban functions in Dar es Salaam: "New metropolitan central areas" and "Local central areas" according to the Draft Master Plan. Using the Euclidean distance tool in the spatial analyst toolbox in ArcGIS 10.0, a proximity analysis was applied which generated a raster file where grid-values were equal to the distance to the nearest sub-centre.

Road Network Proximity The influencing factor represents the proximity to the road network of Dar es Salaam by considering the different importance of roads (small roads, major roads, district roads). Again, proximity analysis was based on the Euclidean distance tool in the spatial analyst toolbox. Although it can be assumed that the road network will develop dynamically in conjunction with settlement growth, the model did not take future changes of the road network into account due to a lack of data.

Neighbourhood The neighbourhood influencing factor was calculated based on the assumption that cells close to existing settlement cells have a higher likelihood to transform than other cells that are located far away from settlement cells, even if

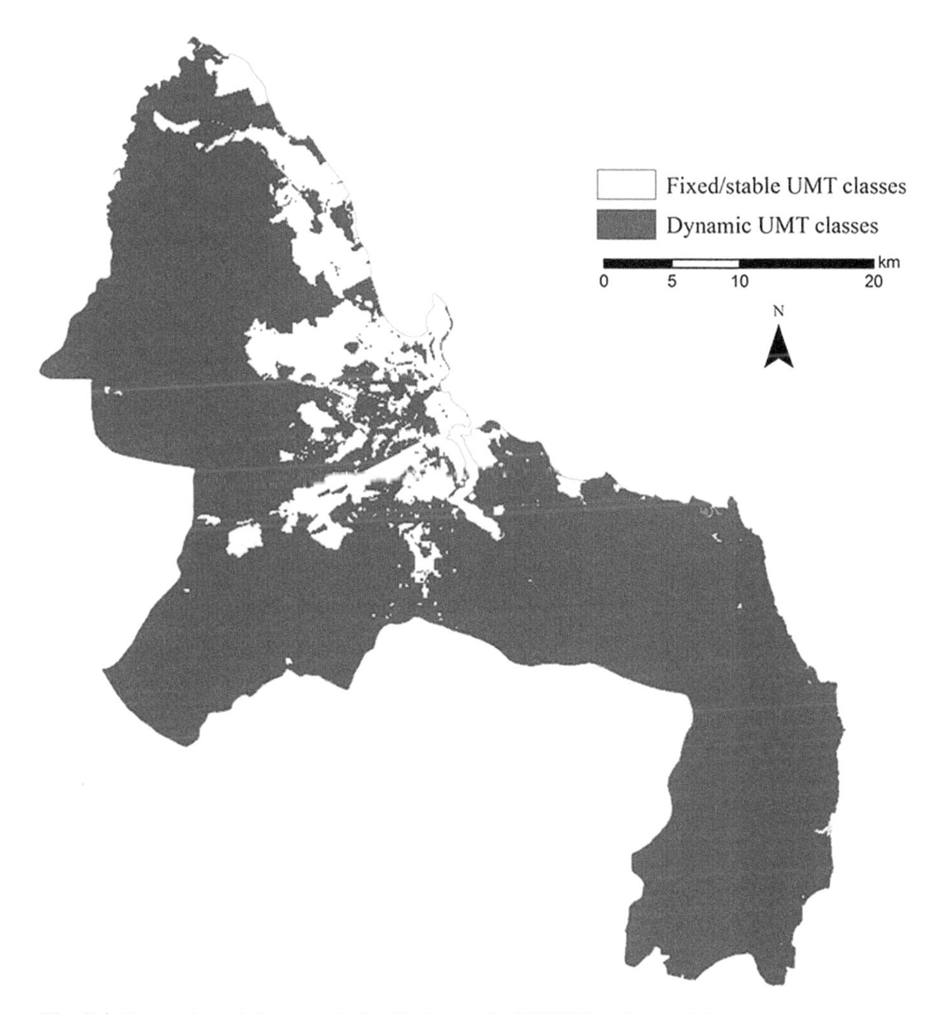

Fig. 8.4 Dynamic and fixed/stable UMT classes for USSDM in Dar es Salaam

they have the same attractiveness for urban development according to other influencing factors. The neighbourhood factor was automatically updated after each time step by recalculating the score of each cell to include the new settlement cells that were developed in the former modelling step providing the nature of urban dynamics.

Urban City Development Projects The influencing factors represent the designated areas of the "Satellite cities project" and the "20,000 plots project" according to the UN-Habitat Citywide Action Plan for Upgrading Unplanned and Unserviced Settlements in Dar es Salaam (2010).

The change of areas from non-settled into settlement was modelled in addition to densification of existing settlements up to a threshold value. Settlement areas that already exceeded the respective threshold values were not changed.

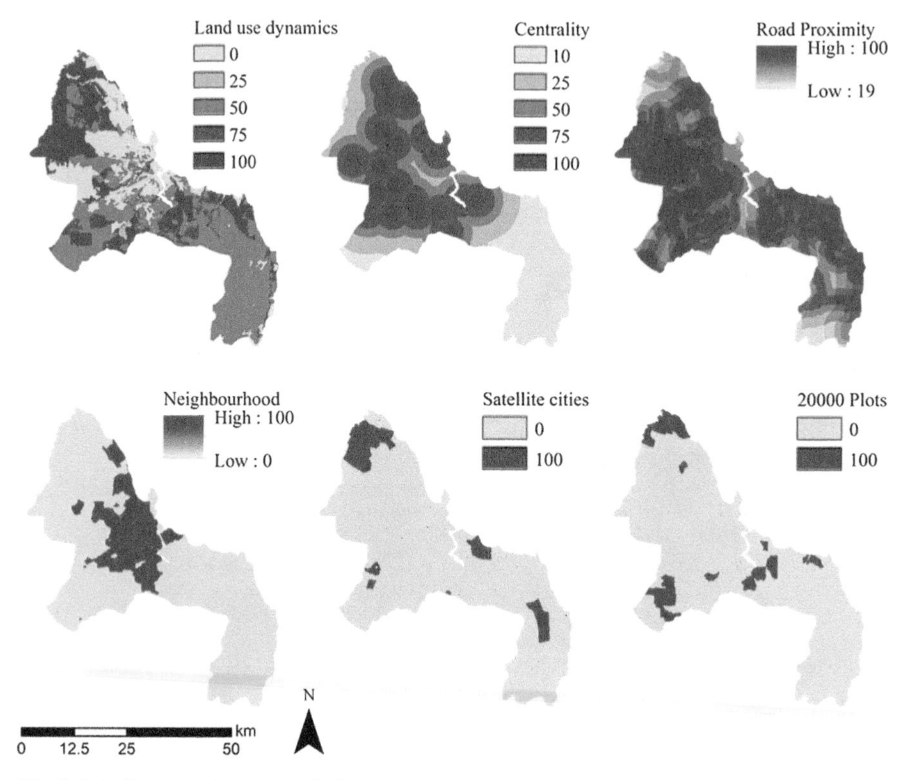

Fig. 8.5 Influencing factors for USSDM in Dar es Salaam (Reproduced from Lindley et al. 2013)

Definition of the Scenarios

The temporal scope for the scenarios was set to the modelling period 2008–2025. Main results of the model were produced for 2012, 2015, 2020, and 2025.

The modelling scenarios for USSDM were based on two planning parameters: Population density (high vs. low) and whether settlements in flood prone areas were allowed or not, resulting in four final modelling scenarios (Fig. 8.6). While the definition of these scenarios was initially suggested by the researchers, stakeholders agreed in the workshops that understanding the implications of different population densities and whether measures were taken to restrict settlement of flood prone areas was of particular relevance.

Two types of population density scenarios were created: scenarios 1 and 3 as low density scenarios, corresponding to business as usual, and scenarios 2 and 4 as high density scenarios. Achieving a higher population density by implementation of a compact urban development model was considered to be a key issue for urban development in Dar es Salaam, given the rapid expansion of the urban area and its associated losses of farmland and natural land with the associated decline of important ecosystem functions (Lindley et al. 2013).

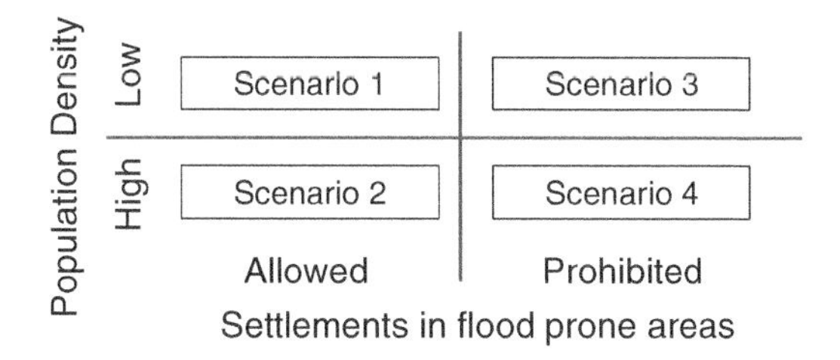

Fig. 8.6 Summary of criteria for USSDM modelling scenarios (Reproduced from Lindley et al. 2013)

For Dar es Salaam the settlement density was set for the "business as usual" scenarios 1 and 3 to 150 inhabitants/ha; according to experts this corresponds to a mean density in Dar es Salaam for low storey settlement zones. For the high density scenarios 2 and 4, the density was set to 350 inhabitants/ha, which was considered to be a feasible value for multi-storey settlement zones.

In the high flood risk scenarios (scenarios 3 and 4) the flood prone areas were excluded from the areas that could be transformed to settlements. The Topographic Wetness Index served in determining flood prone areas using the threshold (19.27) defined by the CLUVA partner team working on flood risk area studies (Chap. 3).

Settlement Land Demand

The number of cells which were needed to accommodate population growth in each time step was based on the estimated future settlement land demand. This land demand in turn was derived from the annual population growth rate and assumptions of maximum population density for new settlements in the different scenarios. As the population and urbanisation dynamics of Dar es Salaam cannot be considered in isolation from the neighbouring areas, it was agreed with local experts to assume that 10 % of the new population would settle outside the boundaries of the study area.

Population was set to be 3.1 million inhabitants in 2008 and estimated to be 7.1 million in 2025, corresponding to a population increase of four million inhabitants within 17 years (UNDESA Population Division 2011). This corresponds to a mean annual growth rate of 5 % (Fig. 8.7). It can be assumed that this population growth rate is a rather conservative estimate as the latest population census data for 2012 was already close to the projected 2015 data with a population of 4,364,541 (National Bureau of Statistics 2013). This new census data was not available during the modelling process; hence the UNDESA data were taken as the basis for the modelling.

The projected population growth rate of approximately 5 % (5.05 % for 2010–2015, 5.12 % for 2015–2020 and 4.96 % for 2020–2025) was considered in all scenarios (UNDESA 2011). The population increase was then translated into settlement land demand based on the respective density values of the different scenarios.

Table 8.1 shows the final values for model parameters that were used in the case of Dar es Salaam.

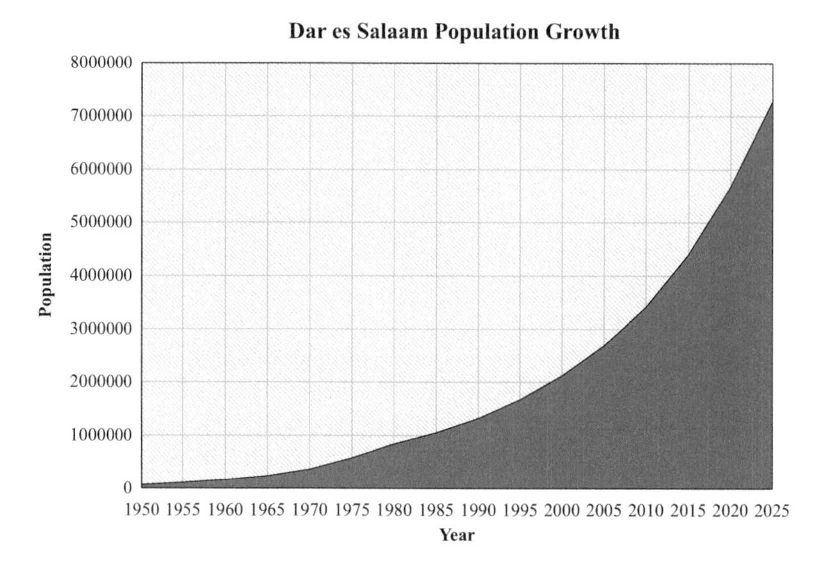

Fig. 8.7 Historical population development and future projection in 2011 in Dar es Salaam (Reproduced from UNDESA Population Division 2011)

Table 8.1 Final USSDM settings for Dar es Salaam

Model parameter	Description		
Population 2008	3.1 million		
Population 2025	7.1 million		
Population increase	4.0 million		
Temporal scope	2008–2025		
Spatial resolution	100 m × 100 m		
Total number of cells	149,896		
Geospatial change detection	2002–2008		
Influencing factors	Land use dynamics		
	Centrality/sub-centres proximity		
	Road network proximity		
	Neighbourhood		
	Known urban city development project: satellite cities project		
	Known urban city development project: 20,000 plots project		
Modelling scenarios (Scenario x = Sx)		Population density threshold	Settlement development in flood prone areas
	S 1	150 persons/ha	Allowed
	S 2	350 persons/ha	Allowed
	S 3	150 persons/ha	Prohibited
	S 4	350 persons/ha	Prohibited

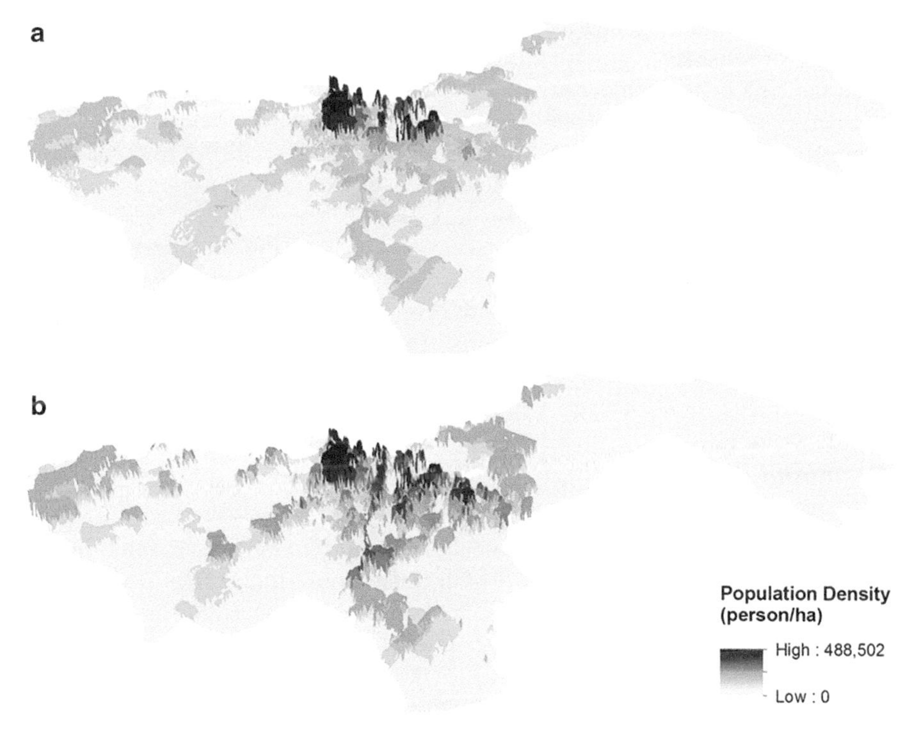

Fig. 8.8 3D visualisation of settlement development USSDM output (**a**) Low population density (scenario 1) (**b**) High population density (scenario 4) (Reproduced from Pauleit et al. 2013)

Results

Settlement Area Development

In 2008 the settlement area covered 69,847 ha (47 % of the total administrative area). According to the modelled scenarios, the settlement area in 2025 will have increased to 79,798 ha (53 %), 75,289 ha, (50.1 %), 79,647 ha (53.0 %) and 74,702 ha (49.7 %) for scenarios 1, 2, 3, and 4, respectively. This means that approximately 5,000 ha more land would have been converted to settlements in the low density scenarios when compared with the high density scenarios.

In the low population density scenarios (Fig. 8.8a), clusters in the Northern, North-Western and Western part of the city will have developed up to 2025. Major roads, new development projects, the proximity to central locations and neighbouring settlements will have increased the attractiveness, and accordingly the settlement development, in these areas. Little settlement development would occur at the Southern part of the city in spite of the existence of district roads and one major road in this area; this is due to the lack of central urban functions (centrality) and no neighbouring settlements which lowered the attractiveness of this area for settlers.

In the high population density scenarios (Fig. 8.8b), the Northern, Central and Western parts of the city areas would have higher population densities when compared to low population density scenarios. The figure also shows that larger areas in the North-West would have lower density settlements, saving more space for 'green areas' or 'farming' UMTs.

In comparison to the low population density scenarios (Fig. 8.8a), in which there will be one main centre of high density and vast areas of low density surrounding this, the high population density scenarios (Fig. 8.8b) predict that several clusters of high density will evolve in different parts of the city.

Farmland

In 2008 the farmland (field crops, horticulture, and mixed farming) covered an area of 60,711 ha, corresponding to 40.4 % of the total administrative area. According to the modelling scenarios, loss of the farmland will reach 6,886 ha, 2,993 ha, 7,255 ha, and 3,085 ha for scenarios 1, 2, 3, and 4, respectively. Thus, implementing the high density development would reduce the loss of farmland by more than 50 % when compared with 'business as usual' (Fig. 8.9).

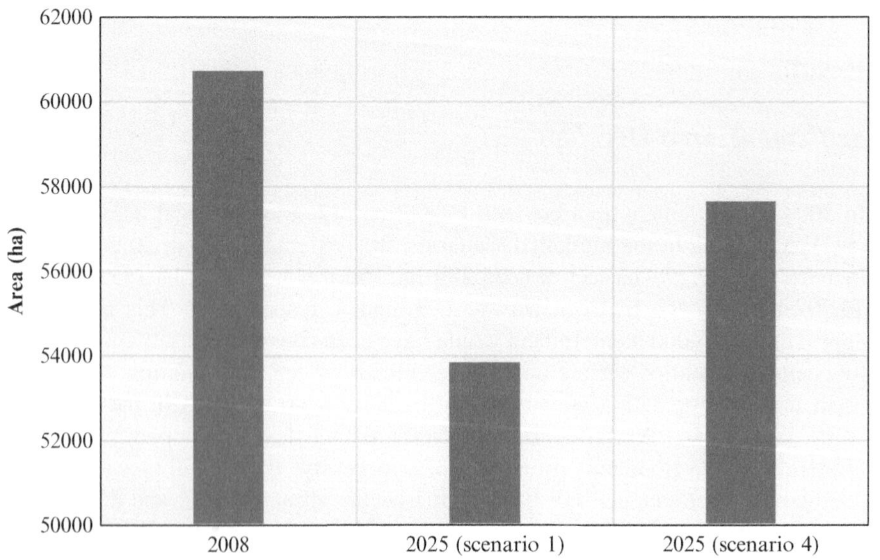

Fig. 8.9 Estimated loss of farmland: scenario comparison (scenarios 1 and 4)

Green Areas

In 2008, green areas (mixed forest, riverine, bush land, mangrove, marsh/swamp, parks, other open space) covered an area of 8,123 ha or 5.4 % of the city's surface area. According to the modelling scenarios, by 2025 loss of green area will reach:

- 2,352 ha for scenario 1;
- 1,717 ha for scenario 2;
- 2,082 ha for scenario 3,
- and 1,319 ha for scenario 4.

Considerably less green areas would be converted to settlement if high population density and prohibiting settlements in flood prone areas were applied (scenario 4). In comparison to 'business as usual' (scenario 1), an area of 1,033 ha would be saved (Fig. 8.10).

Susceptibility to Flooding

In 2008 settlements that are located in flood prone areas covered an area of 3,049 ha corresponding to 2 % of the city's surface area. Mean population density in the flood prone areas was 26 inhabitants/ha. This figure remained the same in the scenarios

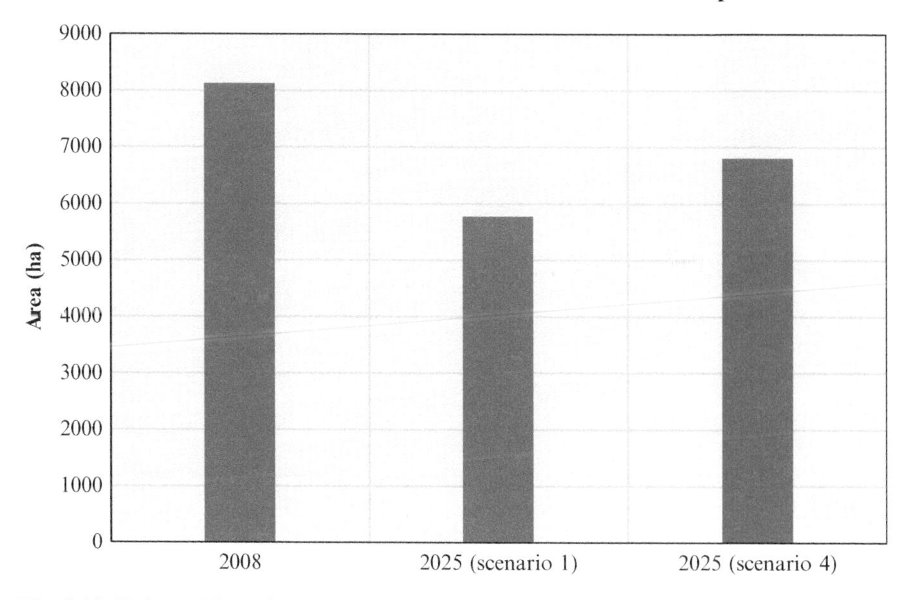

Fig. 8.10 Estimated loss of green areas: scenario comparison (scenarios 1 and 4)

where further settlement development in flood prone areas was prohibited, i.e. excluded from the modelling (scenarios 3 and 4). In the other scenarios (scenarios 1 and 2) the settlements located in flood prone areas covered a total area of 4,177 ha (mean population density 56 inhabitants/ha) in the case of low population density (scenario 1), and 4,031 ha (mean population density 64 inhabitants/ha) in the case of high population density (scenario 2) in 2025. There is a small difference between the surface area of settlements located in flood prone areas in both scenarios (146 ha) due to the fact that these areas are of high attractiveness to settlers and if not properly protected, would be among the first areas to be settled regardless of the set population densities.

An analysis of the settlement location in flood prone areas in 2008 and the modelling output of scenario 1 for the year 2025 were conducted (Fig. 8.11).

According to the output of the low density scenario (scenario 1), there will be a high concentration of settlement development in flood prone areas, and 1,025 ha of riverine and 238 ha of mangroves will be transformed into settlements by 2025.

The high density scenario (scenario 2) will have led to less settlement development in riverine areas but the overall population density will have increased in flood prone areas as development has taken place in the form of dense housing. Implementation of this scenario may reduce the loss of regulating ecosystem services; however, people's susceptibility to flooding has increased as dense settlements were allowed to develop in the river corridors.

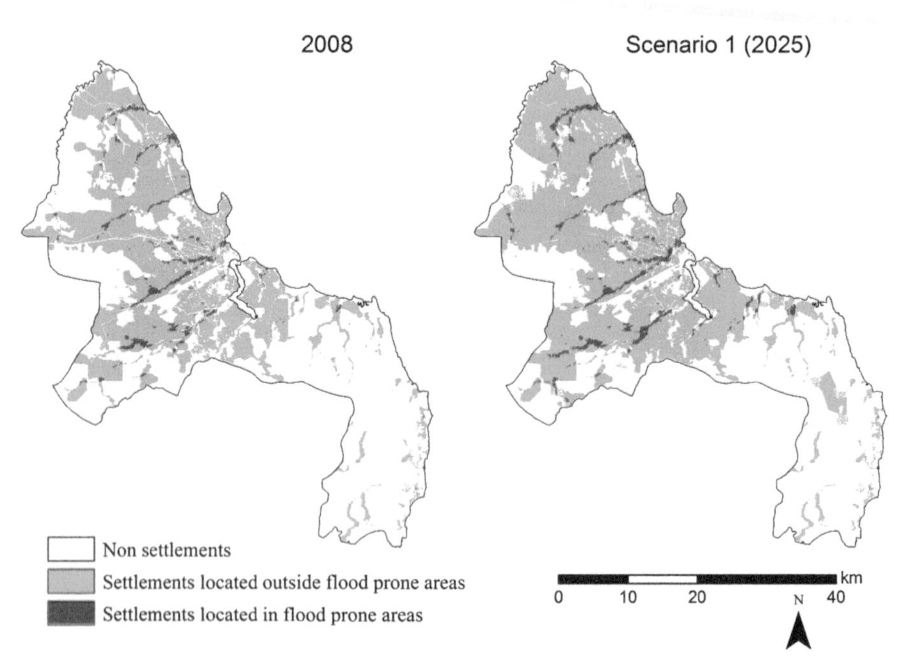

Fig. 8.11 Settlement location in relation to flood prone areas in Dar es Salaam 2008 and 2025 (scenario 1)

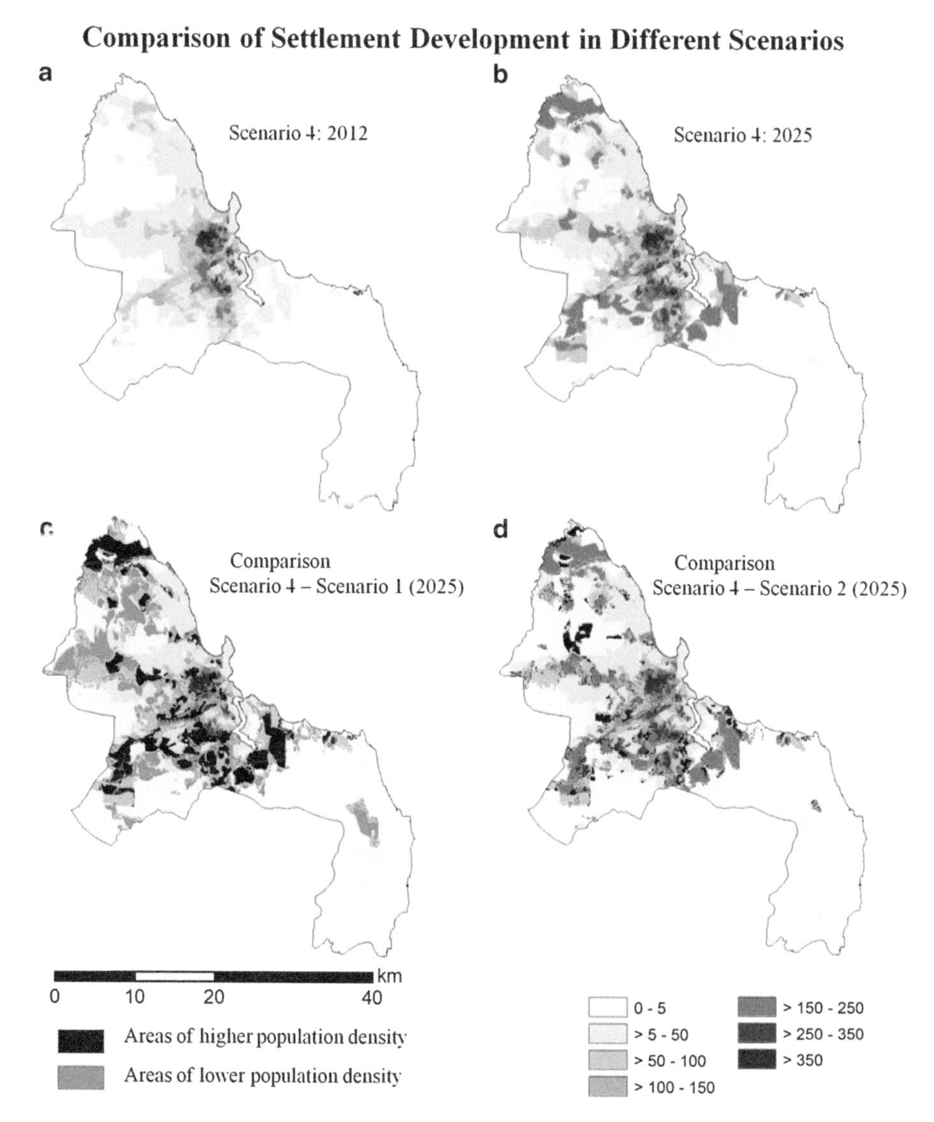

Fig. 8.12 Comparison of settlement development in different scenarios. (**a**) Scenario 4 population density in 2012, (**b**) Scenario 4 population density in 2025, (**c**) Comparison of the population density between scenario 4 and scenario 1 (2025), (**d**) Comparison of the population density between scenario 4 and scenario 2 (2025) (Reproduced from Pauleit et al. 2013)

Figures 8.12a, b show the effects of population increase of scenario 4 (high density, no development in flood prone areas) between 2012 and 2025 – darker areas show higher population density.

Figure 8.12c shows the difference between this high density scenario and the low density, business as usual scenario (scenario 1). Due to the higher population density and the prohibition of further settlements in flood prone areas, the areas marked in green could be "saved" from settlement, whereas areas with an increased population density are highlighted in black.

Figure 8.12d shows the impact of prohibiting further settlements in flood prone areas under the high density scenarios. Thus, if the settlement strategy aims for high population densities in addition to prohibiting further settlements in flood prone areas, population density would be increased under scenario 4 in the areas marked in black; in contrast, the areas marked in green (which actually correspond to the flood prone areas) would be "saved" from further development.

As expected, different settlement densities could lead to large differences in the conversion of farmland and woodlands into settlements. Prohibiting development of settlements in the flood risk areas, however, would not lead to a large increase of settled areas in other parts of the city.

Discussion

Settlement Development

The scenarios showed that Northern, North-West, South-West and Central parts of Dar es Salaam would be the most attractive areas for settlement development along the four main roads (Bagamoyo Road, Morogoro Road, Pugu Road, and Kilwa Road). The road network, however, is not the only factor which determines settlement expansion; rather, it is the combination of several factors that influence the attractiveness of an area to become a settlement area. Overall, areas that are most likely to develop into new settlements were serviced by roads, close to existing settlements, and moreover, close to one of the planned central locations according to the draft new Master Plan for Dar es Salaam. The Southern part of the city would have low attractiveness due to the fact that central urban functions and already existing settlement areas are distant (at least as long as there is no bridging of the bay).

By promoting high density settlements within the designs of all new development projects, such as the satellite sites project and the 20,000 plots project (Mwiga 2011), the efficient use of attractive areas for settlement development will be achieved, which in turn will lead to fewer losses of farmland and green areas that are considered as important provisioning and regulating ecosystem services.

A polycentric development was proposed in the draft of the new Master Plan for Dar es Salaam in 2013. A density simulation of this proposed specific structure could show to which extent this might contribute to relieve the pressure from the old city centre and main roads. Furthermore, pressure on flood prone areas near the city centre would be reduced. These areas are occupied mainly by people with low incomes due to the short distance to job opportunities in the city centre. The new clusters could result in new job opportunities, and hence people may prefer to live close to these areas.

Farmland and Green Areas

Farmland (field crops, horticulture, and mixed farming) and green areas (mixed forest, riverine, bush land, mangrove, marsh/swamp, parks, and other open space UMTs) are especially vulnerable to urban settlement development.

Densification of settlement areas and the prohibition of settlement development in flood prone areas would have a big impact on the saving potential of the farmland and green areas. In comparison to the business as usual scenario, the losses in these areas would be greatly decreased. Implementing such a scenario would strongly contribute to the city's food security which is an important issue in a city with increasing malnutrition problems (Kinabo 2003; Minot et al. 2006).

With the loss of farmland and green areas, important regulating ecosystem services would also be lost that are needed to reduce vulnerability to climate change. For instance, trees in green areas reduce air and surface temperatures, sequester and store carbon, provide firewood, promote biodiversity and are of cultural importance (see also Chap. 4). In CLUVA, effects of the different scenarios on the stock of merchantable wood in different UMT types were estimated (Renner in Lindley et al. 2013). Corresponding to the respective loss of area of "Farmland" and "Green Areas" in each scenario, the loss of relative growing stock of wood (large stem in m^3 merchantable stem wood) in scenario 1 would be 30 % in "Farmland" and 62 % in "Green areas", while for scenario 4, the losses would be "only" 25 % and 55 %, respectively. Scenario 4 could thus save 52,000 m^3 of merchantable stem wood. The most significant losses within the scenarios' results would be in the UMT type 'Riverine areas'.

Specific types of green areas (riverine and mangrove) are located in flood prone areas. This resulted in more losses in these UMTs in scenarios 1 and 2 (where settlement development in flood prone areas is allowed). Conversely, in scenarios 3 and 4, other green types (mixed forest and bushland) would suffer higher losses as more settlement development would be directed into these UMTs. The modelling output of scenario 2 (high population density, settlement development in flood prone areas allowed) has saved some of the mangroves and riverine areas from being lost in comparison to scenario 1 (low population density) even though the settlement development in flood prone areas is still allowed. This indicates that applying high population density alone would result in a more efficient use of areas outside the flood prone areas.

Scenario 4 would have the best results in terms of preserving ecosystem functions of all urban development scenarios modelled in this research as loss of mixed farming areas would be reduced (from −13 % in scenario 3 to −5 %), and losses of other types of green areas (mixed forests, riverine, bush land, etc.) would be even more strongly reduced.

Susceptibility to Flooding

The business as usual scenario 1 (low population density and the allowance of settlement development in flood prone areas) would result in vast losses of farmland and green areas, not only in the floodplains. Decreased surface permeability would lead to increasing flood risks combined with decreasing potentials of provisioning and regulating ecosystem services, while at the same time a higher proportion/part of the population will be exposed.

Population density also plays a role in the susceptibility of the population to flooding. In the first two scenarios where settlement development in flood prone areas is allowed, the new settlement development in flood prone areas in the case of high population density (scenario 2) will be much less than with low population density (scenario 1). However, the increase in population density might lead to larger numbers of people on a smaller area of land that are susceptible to floods in certain areas.

In scenarios 3 and 4, where settlement development in flood prone areas is prohibited, areas outside the flood prone areas would gain in their attractiveness. These areas are particularly located in the North-West and Western parts of Dar es Salaam. As the population density threshold is set low in scenario 3, these attractive areas would be rapidly settled.

Conclusion

USSDM – Lessons Learned

USSDM was designed as a modelling tool to be readily applied in a real world context in order to support strategic urban planning. The feasibility of running a spatial scenario model in the context of African cities was demonstrated. Feedback from the stakeholders during the planning workshops was very positive. In the parallel case of Addis Ababa (not presented in this chapter), results from the USSDM were even discussed in the current process of developing the new Master Plan.

Considering the current debate on climate change and as suggested by Hill and Lindner (2010a, b), USSDM is aimed at simulating the impacts of urban growth and planning measures on natural resources, such as loss in green infrastructure due to settlement land demand.

One of the main strengths of the USSDM is its flexibility in changing the parameter settings and incorporating different influencing factors as inputs to the model. Therefore, it will be well suited to determine the sensitivity of urban development to the different weighting of influencing factors. Ease of use, transparency of the modelled urban development processes, ability to quickly change the scenario settings, and ability to add or remove influencing factors were important criteria for the model's development.

While the experience with developing and applying the USSDM have thus been positive, some limitations should be highlighted. Importantly, assumptions concerning current average population density and the possible increase of density made for Dar es Salaam were rather coarse inputs into the model. A further analysis and more specific knowledge about the population densities for different urban morphology types and the speed of change over time in the process of densification could lead to considerable improvement of the model by, e.g. differentiating not only between settled and non-settled areas but also distinguishing urban zones, such as city centre, formal housing areas, or informal settlements at the periphery.

USSDM also did not explicitly differentiate between informal and formal settlements due to the fact that it models only a single type and cannot analyse more than one type at the same time. USSDM could also be used to model specific types of settlements; however, influencing factors, exclusion areas and land demand calculations should be changed to be suitable for this scope.

Using the ArcGIS v10.0 newly added time feature, the future expansion of settlement areas was visualised during the workshops in front of stakeholders of different professional backgrounds which incited lively discussions and debates about the different trends and patterns of development that are expected to occur and the differences between the scenarios. These discussions have helped in working together to identify possible measures that could be put in place to control and direct this development, and to minimise the impact on ecosystem services. This could enhance the communication process between scientists and planners from one side and policy makers and the general public from the other side, providing new possibilities for participatory planning and decision-making.

In future applications, special emphasis should be placed on testing the effect of further policy measures such as exclusion of areas by protecting them from development. Also, the model allows testing the effect of further influencing factors that favour specific areas in terms of development.

The city is a complex system that has several interlinked components and relations that impact its growth. There are several social, behavioural and cultural factors that influence the urban settlement growth. The model could be coupled with social science research where spatial indicators or empirical research outputs could serve as inputs to the model. Moreover, the USSDM approach introduced in this research could be integrated with other models (e.g. transportation models) to enhance the incorporation of factors and agents that influence the urban settlement dynamics within a single modelling package.

Not all of the influencing factors that are found in literature were incorporated into the model's current design, such as the economic development, the socio-economic and political system. Integrating both ecological and socioeconomic aspects could boost the model to become an even more powerful planning tool, e.g. to assess the impacts of different urban scenarios on the flow of ecosystem services. The example of change of the stock of timber wood (Renner in Lindley et al. 2013) would be just one example, but further ecosystem services, such as modification of air temperatures, could be added. The UMT approach (Chap. 4) would enable linking assessment of the provision of ecosystem services with socio-demographic and economic data, e.g. in order to estimate the demand of ecosystem services, such as food provision by urban agriculture in the different neighbourhoods of Dar es Salaam. Modelling the spatial and temporal dynamics of ecosystem service supply and demand could provide a powerful tool for urban planning.

However, when expanding the scope and the complexity of the model, care should be taken that its logic still remains transparent for its users. The "design" oriented approach of USSDM is not intended to deliver deterministic forecasts but to explore different spatial patterns and trends according to planners' and stakeholders' assumptions. This is in contrast to many stakeholders' expectations as well as to some CA-based land use change models which attempt to integrate the full range of urban complexity within their model. Doing this, modellers may run into the dilemma of seemingly being more accurate and holistic, but at the same time generating results in a "black box" by including highly sophisticated mathematical functions. Mathematical functions are often based on extensive historical analyses – the correct implementation into the model is shown by being "proven" through the model's calibration of historical situations.

Further development of the model's interface would increase the user friendliness of the model for its use by multiple expert users.

Strategic Development Towards Reducing Vulnerability in Dar es Salaam

USSDM application in the case study city of Dar es Salaam has shown where and to what extent future settlements are likely to develop according to the estimated population growth. Results clearly demonstrate that the "business as usual" scenario of low density development would rapidly lead to the consumption of huge areas of farmland, forests and shrubland, as well as floodplains that are important for ecosystem services, such as flood water retention, and thus would greatly increase the vulnerability of Dar es Salaam to the effects of climate change.

Potential impacts of urban development on the cities' green structure and its ecosystem services were of main interest for USSDM results. Other dimensions of urban sustainability could not be taken into account. Recommendations for spatial urban planning can be derived from the results of this study (Fig. 8.13):

- Planning should promote urban development at considerably higher densities in order to better protect green structure and its ecosystem services such as food provision, flood protection, stormwater management and moderation of the urban heat island (Chap. 4). Therefore, high population density settlements should be developed outside flood prone areas and where remnants of natural vegetation will not be affected. This could be achieved through implementing higher population densities in the designs of development projects, such as the satellite cities project or the 20,000 plots project. Another means is through

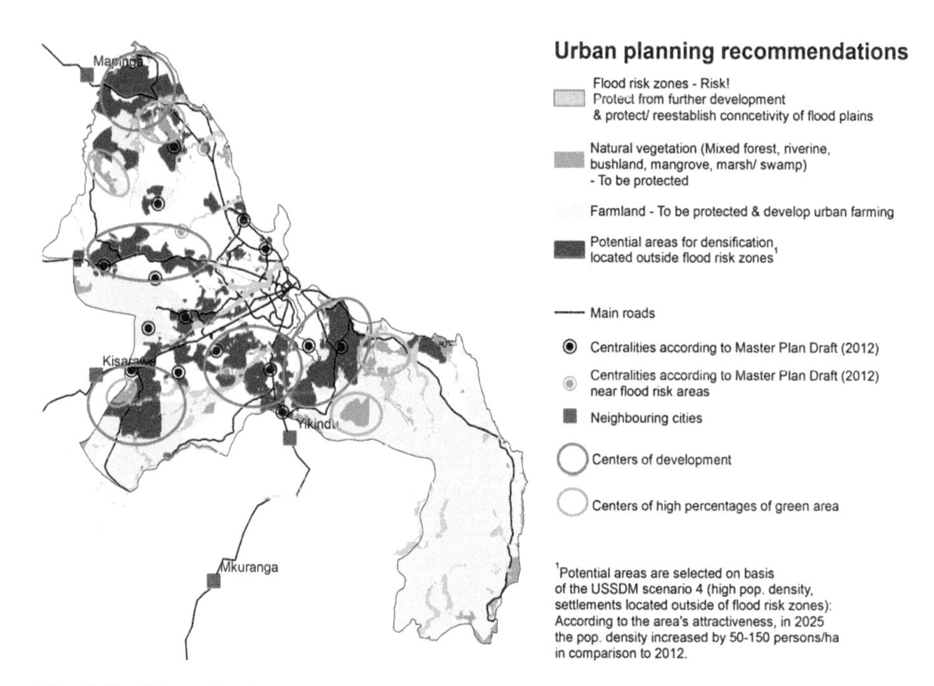

Fig. 8.13 Urban planning recommendations for Dar es Salaam based on USSDM (Reproduced from Pauleit et al. 2013)

adopting new policies and land use zoning regulations which set specific building heights and population densities in the newly developed areas.

- Densification should be combined with development of public mass transport systems. The USSDM model showed that effective transport routes and sub-centres with supply functions are an important means for directing urban growth towards a polycentric urban structure. Promoting such polycentric development is already intended in the new Master Plan as it would help to relieve the pressure from the core city and reduce unchecked sprawl at the fringe. However, the planning of these future poles of urban development should be carefully linked to the protection of green structure. From Fig. 8.13, it may be noticed that some of the locations with central urban functions in the West of the core city are placed in the middle of farmland where they are not connected to the existing network of major roads. Development around these nodes would lead to large losses of farmland. Therefore, it is suggested that the location of these central nodes should be reconsidered and balanced against the needs for protecting and developing a coherent green infrastructure network. The latter should be based on a strategic spatial plan building on scientific evidence (Chap. 4).
- Riverine areas have a major role within a multifunctional green infrastructure, in particular for flood protection. Therefore, further settlement development should be prohibited in flood prone areas and the population already living in these risk zones should be relocated in socially acceptable ways. Providing local options

for livelihood diversification, with the presence of investment driven development is also an important aspect for the resettled people to adjust to their new surroundings and not to return to their place of origin (Arnall et al. 2013). The results from the USSDM clearly support adoption of taking such an approach. While strong government action may be needed to protect the floodplains, linking conservation with use values, e.g. by promoting site adapted urban farming in these areas, may increase the success of such efforts as residents would benefit from it.

- Securing productive farming areas in the periphery as well as in central urban areas must be a major goal of future urban development as a large part of the urban population is at least partly depending on self-supply of food. Given expected growth rates of Dar es Salaam and Sub-Saharan cities more widely, integrating urban farming into urban development should be a top priority. Urban farming also would act as an added-value land use in hazard areas, providing employment for the city inhabitants instead of supplying them with high risk housing (Howorth et al. 2001). City farming can also protect land against pests, thieves, squatters, garbage dumping, and vandals; it can also raise use and rent value of land (Mougeot 1994).
- In USSDM, it was assumed that 10 % of the additional urban population would settle in the neighbouring municipalities as it appeared to be unrealistic that projected population growth would only be accommodated within Dar es Salaam's administrative area. The linkages between the city and its neighbouring municipalities will certainly continue to strengthen with ongoing urban development. Therefore, it seems important to widen the scope of planning to the regional level. Strategic spatial planning at the regional level would help to alleviate the pressure on the urban core, promote development of an effective settlement pattern and a functional green infrastructure. The USSDM presented in this chapter could be easily expanded to support planning and adequate governance arrangements at regional level (see Chaps. 9 and 10 for further information).

References

Abo-El-Wafa H (2013a) Urban spatial scenario design modelling (USSDM) in Addis Ababa: background information. Available: http://www.cluva.eu/CLUVA_publications/CLUVA-Papers/USSDM_Addis Ababa_Background Information_JULY2013a.pdf. Accessed 24 Feb 2014

Abo-El-Wafa H (2013b) Urban spatial scenario design modelling (USSDM) in Addis Ababa: technical user guide. Available: http://www.cluva.eu/CLUVA_publications/CLUVA-Papers/USSDM_Addis Ababa_Technical User Guide_JULY2013b.pdf. Accessed 24 Feb 2014

Arnall A, Thomas DS, Twyman C, Liverman D (2013) Flooding, resettlement, and change in livelihoods: evidence from rural Mozambique. Disasters 37(3):468–488

Barredo JI, Kasanko M, McCormick N, Lavalle C (2003) Modeling dynamic spatial processes: simulation of urban future scenarios through cellular automata. Landsc Urban Plan 64(3):145–160

Barredo JI, Demicheli L, Lavalle C, Kasanko M, McCormick N (2004) Modelling future urban scenarios in developing countries: an application case study in Lagos Nigeria. Environ Plan B 31(1):65–84

Barrios S, Bertinelli L, Strobl E (2006) Climatic change and rural-urban migration: the case of sub-Saharan Africa. http://webdoc.sub.gwdg.de/ebook/serien/e/CORE/dp2006_46.pdf. Accessed 15 May 2014

Batty M, Xie Y (1997) Possible urban automata. Environ Plan B 24:175–192

Batty M, Xie Y, Sun Z (1999) Modeling urban dynamics through GIS-based cellular automata. Comput Environ Urban Syst 23(3):205–233

Buchta K (2013a) Urban spatial scenario design modelling (USSDM) in Dar es Salaam – background information. Available: http://www.cluva.eu/CLUVA_publications/CLUVA-Papers/USSDM_Dar es Salaam_Background Information_JULY2013.pdf. Accessed 24 Feb 2014

Buchta K (2013b) Urban spatial scenario design modelling (USSDM) in Dar es Salaam – technical user guide. Available: http://www.cluva.eu/CLUVA_publications/CLUVA-Papers/USSDM_Dar es Salaam_Technical User Guide_JULY2013.pdf. Accessed 24 Feb 2014

Cavan G, Lindley S, Yeshitela K, Nebebe A, Woldegerima T, Shemdoe R, Kibassa D, Pauleit S, Renner F, Printz A, Buchta K, Coly A, Sall F, Ndour NM, Ouédraogo Y, Samari BS, Sankara BT, Feumba RA, Ngapgue JN, Ngoumo MT, Tsalefac M, Tonye E (2012) Green infrastructure maps for selected case studies and a report with an urban green infrastructure mapping methodology adapted to African cities. CLUVA deliverable D2.7. Available: http://www.cluva.eu/deliverables/CLUVA_D2.7.pdf. Accessed 15 Jan 2014

City Mayors Statistics (2011) The world's fastest growing cities and urban areas from 2006 to 2020. http://www.citymayors.com/statistics/urban_growth1.html. Accessed 21 July 2014

Clarke KC, Gaydos L (1998) Long term urban growth prediction using a cellular automaton model and GIS. Int J Geogr Inf Sci 12(7):699–714

Clarke KC, Gaydos L, Hoppen S (1997) A self-modifying cellular automaton model of historical urbanization in the San Francisco Bay area. Environ Plan B 24:247–261

Hill A, Lindner C (2010a) Land-use modelling to support strategic urban planning – the case of Dar es Salaam, Tanzania. Paper presented at the 45th ISOCARP congress, Porto, 18–22 Oct 2009

Hill A, Lindner C (2010b) Modelling informal urban growth under rapid urbanisation: a CA-based land-use simulation model for the city of Dar es Salaam, Tanzania. Dissertation, TU Dortmund

Howorth C, Convery I, O'Keefe P (2001) Gardening to reduce hazard: urban agriculture in Tanzania. Land Degrad Dev 12(3):285–291

IFRC (2011) Disaster relief emergency fund (DREF) Tanzania: floods. http://www.ifrc.org/docs/appeals/11/MDRTZ013.pdf. Accessed 15 May 2014

IPCC (2007) Boko M, Niang I, Nyong A, Vogel C, Githeko A, Medany M, Osman-Elasha B, Tabo R, Yanda P (2007) Africa. In: Parry ML, Canziani OF, Palutikof JP, van der Linden PJ, Hanson CE (eds) Climate change 2007: impacts, adaptation and vulnerability. Contribution of Working Group II to the fourth assessment report of the Intergovernmental Panel on Climate Change. Cambridge University Press, Cambridge, pp 433–467

Kinabo J (2003) A case study of Dar es Salaam City, Tanzania. Paper prepared for the FAO technical workshop on "Globalization of food systems: impacts on food security and nutrition", Rome, 8–10 Oct 2003

Kiunsi R (2013) The constraints on climate change adaptation in a city with a large development deficit: the case of Dar es Salaam. Environ Urban 25(2):321–337

Lindley S, Gill SE, Cavan G, Yeshitela K, Nebebe A, Woldegerima T, Shemdoe R, Kibassa D, Pauleit S, Renner F, Printz A, Buchta K, Abo El Wafa H, Coly A, Sall F, Ndour NM, Ouédraogo Y, Samari BS, Sankara BT, Feumba RA, Ambara G, Kandé L, Zogning MOM, Tonye E, Pavlou A, Koome DK, Lyakurwa RJ, Garcia A (2013) A GIS based assessment of the urban green structure of selected case study areas and their ecosystem services. CLUVA deliverable D2.8. Available: http://www.cluva.eu/deliverables/CLUVA_D2.8.pdf. Accessed 15 Jan 2014

Meaille R, Wald L (1990) Using geographical information systems and satellite imagery within a numerical simulation of regional growth. Int J Geogr Inf Syst 4(4):445–456

Minot N, Simler K, Benson T, Kilama B, Luvanda E, Makbel A (2006) Poverty and malnutrition in Tanzania; new approaches for examining trends and spatial patterns. International Food Policy Research Institute, Washington, DC

Mng'ong'o OS (2004) A browning process: the case of Dar es Salaam City. Dissertation, Royal Institute of Technology, Department of Infrastructure, Division of Urban Studies, Stockholm

Mougeot LJ (1994) African city farming from a world perspective. In: Egziabher AG, Lee-Smith D, Maxwell DG, Memon PA, Mougeot LJA, Sawio CJ (eds) Cities feeding people: an examination of urban agriculture in east Africa. International Development Research Center, Ottawa, pp 1–24

Mwiga BG (2011) Evaluating the effectiveness of the regulatory framework in providing planned land in urban areas (The case of Dar es Salaam city 20,000 plots project, Tanzania). Master thesis, Faculty of Geo-Information and Earth Observation of the University of Twente, Enschede

National Bureau of Statistics, Ministry of Finance (2013) 2012 Tanzania population and housing census, Dar es Salaam. http://www.nbs.go.tz. Accessed 15 Jan 2014

Niang I, Ruppel OC, Abdrabo MA, Essel A, Lennard C, Padgham J, Urquhart P (2014) Africa. In: Barros VR, Field CB, Dokken DJ, Mastrandrea MD, Mach KJ, Bilir TE, Chatterjee M, Ebi KL, Estrada YO, Genova RC, Girma B, Kissel ES, Levy AN, MacCracken S, Mastrandrea PR, White LL (eds) Climate change 2014: impacts, adaptation, and vulnerability. Part B: Regional aspects. Contribution of Working Group II to the fifth assessment report of the Intergovernmental Panel on Climate Change. Cambridge University Press, Cambridge, United Kingdom and New York, NY, USA, pp. 1199–1265

Pauleit S, Buchta K, Abo El Wafa H, Renner F, Printz A, Kumelachew Y, Kibassa D, Shemdoe R, Kombe W (2013) Recommendations for green infrastructure planning in selected case study cities. CLUVA deliverable D2.9. Available: http://www.cluva.eu/deliverables/CLUVA_D2.9.pdf. Accessed 15 Jan 2014

Pelling M, Wisner B (eds) (2009) Disaster risk reduction. Cases from urban Africa. Earthscan, London

Potts D (2012) Whatever happened to Africa's rapid urbanisation? Africa Research Institute, London

Schwarz von Raumer H-G, Printz A (2007) Ein "Spatial Scenario Design Model" zur strategischen Unterstützung der Landnutzungspolitik im Ouémé-Einzugsgebiet (Benin). In: Strobl J, Blaschke T, Griesebner G (eds) Beiträge zum 19. AGIT-Symposium, Salzburg, pp 725–730

Sietchiping R (2004) A geographic information systems and cellular automata-based model of informal settlement growth. PhD dissertation, University of Melbourne

UNDESA Population Division (2011) World Urbanization Prospects: The 2011 Revision. UNDESA. Available: http://www.un.org/en/development/desa/publications/world-urbanization-prospects-the-2011-revision.html. Accessed 24 March 2014

UN-Habitat United Nations Human Settlements Programme (2008) The state of African cities 2008. A framework for addressing urban challenges in Africa. United Nations Human Settlements Programme, Nairobi

UN-Habitat United Nations Human Settlements Programme (2010) Citywide action plan for upgrading unplanned and unserviced settlements in Dar Es Salaam. United Nations Human Settlements Programme, Nairobi

UN-Habitat United Nations Human Settlements Programme (2014) The state of African cities 2014. Re-imagining sustainable urban transitions. United Nations Human Settlements Programme, Nairobi

Ward DP, Murray AT, Phinn SR (2000) A stochastically constrained cellular model of urban growth. Comput Environ Urban Syst 24(6):539–558

White R, Engelen G (1993) Cellular automata and fractal urban form: a cellular modeling approach to the evolution of urban land-use patterns. Environ Plan A 25:1175–1199

White R, Engelen G, Uljee I (1997) The use of constrained cellular automata for high-resolution modelling of urban land-use dynamics. Environ Plan B 24:323–344

Chapter 9
Multi-level Governance, Resilience to Flood Risks and Coproduction in Urban Africa

Trond Vedeld, Wilbard J. Kombe, Clara Kweka-Msale, Ndèye Marème Ndour, Adrien Coly, and Siri Hellevik

Abstract This chapter examines how climate change adaptation becomes integrated as a policy field within multi-level governance in the two coastal cities of Dar es Salaam, Tanzania and Saint Louis, Senegal. We explore the ways in which this policy sector works towards *resilient cities* as it interfaces with the governance of flood risks.

In Dar es Salaam, we find that adaptation and flood risk management have no substantive organisational home at municipal level. These policy fields are not well integrated into the city's land use planning and development at a local level. Public officials, to a limited degree, encourage citizen participation in flood risk management and land use planning. In Saint Louis, public officials, especially municipal planners, actively encourage citizen participation in flood risk management and local development planning. We suggest that it is not the size and adaptive capacity of the municipality, *per se*, that matter for the integration and functioning of climate risk management at local level. Rather, it is the way multi-level governance enables or constrains the ability of public officials to enhance the responsiveness of citizens and their input into the coproduction of services and water resources management. The chapter refers to theories of coproduction within multi-level governance to explain drivers and barriers to adaptation and resilience.

T. Vedeld (✉)
Department of International Studies, Norwegian Institute for Urban and Regional Research (NIBR), Gaustadalleén 21, 0349 Oslo, Norway
e-mail: trond.vedeld@nibr.no

W.J. Kombe • C. Kweka-Msale
Institute of Human Settlements Studies, Ardhi University, P.O. Box 35176, Dar es Salaam, Tanzania
e-mail: kombewilbard@yahoo.com; nicasclara@yahoo.com

N.M. Ndour • A. Coly
Department of Geography, University of Gaston Berger, B.P. 234, Saint Louis, Senegal
e-mail: sodamareme2007@gmail.com; adrien.coly@ugb.edu.sn

S. Hellevik
Sund Energy, Meltzers gt 4, 0257 Oslo, Norway
e-mail: siribbo@hotmail.com

S. Pauleit et al. (eds.), *Urban Vulnerability and Climate Change in Africa*, Future City 4,
DOI 10.1007/978-3-319-03982-4_9

Keywords Climate resilience in urban Africa • Multi-level governance • Floods • Coproduction

Introduction

Cities are increasingly acknowledged as important strategic actors and places for governing climate risks and enhancing resilience (Betsill and Bulkeley 2007). Many cities do take a high degree of leadership in preparing adaptation programmes and protection against flood risks even in the absence of guidance from national policies and enabling frameworks. The concept of *resilient cities* has, to this end, been widely promoted as a desirable goal within the policy context of climate change (UN-Habitat 2011; UNISDR 2012, 2014; Satterthwaite and Dodman 2013). Consequently, many observers claim that it is the city government that needs to bring coherence to agendas that have previously been addressed in uncoordinated manners, e.g. climate change adaptation, disaster risk management and sustainable urban development (Satterthwaite et al. 2007, 2009; Satterthwaite 2011; Pelling 2011; Bulkeley 2010, 2013; O'Brien 2012; Satterthwaite and Dodman 2013).

It is also widely recognised, however, that many cities in both developing countries and developed countries are confronted by important barriers to the integration and institutionalisation of the climate risk agenda that require better understanding (Bulkeley 2013; Hanssen et al. 2013; Roberts and O'Donoghue 2013). Municipal authorities' capacity to act on adaptation is often severely constrained by multi-level governance that restricts mandates and resources at city, sub-city and local/community levels (Satterthwaite 2011; Vedeld et al. 2012; Satterthwaite and Dodman 2013).

This study is designed as a comparative analysis of how urban flood risk management is integrated in multi-level governance within two cities – in one larger city and one smaller city – as it interfaces with the climate change adaptation agenda. Flood risks are considered as local manifestations of (extreme) climate risks and extreme weather (Douglas et al. 2009).

Several observers in this regard suggest that larger municipalities will have greater capacity and an advantage in integrating adaptation in governance compared to smaller ones (Rauken et al. 2014). Analysing the two city cases from a perspective of multi-level governance, we put forward a different hypothesis, however. We argue that it is not the size and scale of the municipality which is decisive for the functioning of local climate risk management. Rather, it is the way the arrangement of coproduction operates within the system of multi-level governance that best explains difference in approaches. Coproduction is defined as "the process through which inputs used to produce a good or service are contributed by individuals who are not 'in' the same organisation" (Ostrom 1996: 86).

We gave priority to analysis of the vertical steering and policy integration issues within and across key climate risk-relevant sectors, and how coordination and cooperation is enabled at two local levels; (i) municipal level (city and sub-city level); and (ii) local community level. The focus on the local level is motivated by the fact that climate change will impact differently across urban neighbourhoods. This means that a significant share of climate risk governance needs to take place at the local level and, thus, be analysed at this level, i.e. as close as possible to the scene of events of those that are potentially impacted and need to act upon or manage an extreme risk.

So far, considerable research on urban vulnerability and climate change adaptation in developing countries has been undertaken (Adger et al. 2009; Ziervogel and Parnell 2012; Carmin et al. 2012; Satterthwaite and Dodman 2013; Roberts and O'Donoghue 2013). However, less research has been done on institutional drivers and barriers to establishing a resilient city – especially in Africa – and its relation to governance at different levels and scales (Vedeld et al. 2012). Research on this topic is also limited for comparative city cases across socio-political and climatic settings (Kern and Alber 2009; Bulkeley 2013; Hanssen et al. 2013). Our study contributes to this debate and builds upon earlier research from the growing international literature on climate change, adaptation, and urban governance (Betsill and Bulkeley 2007; OECD 2009; Bicknell et al. 2009; Pelling 2011; Cartwright et al. 2012; Hanssen et al. 2013; Bulkeley 2013; Satterthwaite and Dodman 2013; ICLEI 2014).

Analytical Framework

The integration of adaptation into government and governance across levels and scales is considered critical to long term climate resilience (Bulkeley 2013). The focus on resilience and *resilient cities* is of relevance to all urban settings and concerns "capacities to withstand or recover from all direct and indirect impacts of climate change" (Pelling 2011; Satterthwaite and Dodman 2013: 292). Resilience can be defined more precisely as:

> The ability of a system and its component parts to anticipate, absorb, accommodate, or recover from the effects of a hazardous event in a timely and efficient manner, including through ensuring the preservation, restoration, or improvement of its essential basic structures and functions. (IPCC 2012: 3)

The resilience discourse within the climate adaptation literature is allied with contemporary governmental and governance discourses about the sharing of responsibilities between state and non-state actors for risk management, and conditions for innovation and change towards a new, more sustainable system state. This inherently normative stand relates to discussions about transitional adaptation and transformation (Pelling 2011). Resilient cities in this chapter is operationalised with inspiration from the *Ten Essentials for Making Cities Resilient* (UNISDR 2014); e.g. existence of organisation and coordination mechanisms to reduce risks in responsive manners; budget for (flood) risk management; risk and vulnerability assessments; early warning systems and emergency management capacities; critical stormwater infrastructure (to address underlying vulnerability); land use principles in place and enforced; climate-action strategy in place; and sound environmental management.

Consequently, we bring together theories from (i) multi-level governance, (ii) coproduction (and related network governance theory), and (iii) resilient cities. The governance analysis explains the interplay between actors, levels and sectors of

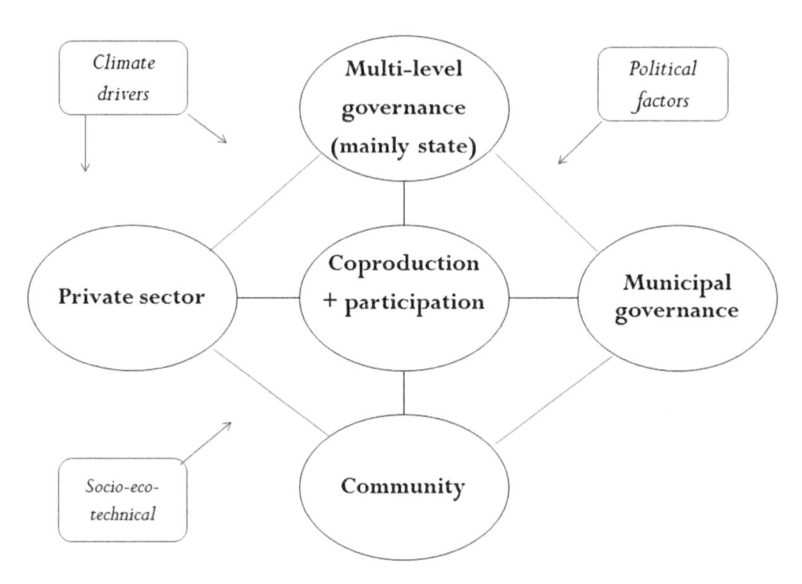

Fig. 9.1 Multi-level governance and coproduction (Authors' own design; inspired by Ostrom 1996 and Bulkeley 2013)

government in addressing complex public policy challenges related to climate risk management – as a *wicked problem* (Bache and Flinders 2004; Betsill and Bulkeley 2007; Weber and Khademian 2008; Peters 2008; Sorensen and Torfing 2009; Osborne 2010; Bulkeley 2013; Sorensen and Torfing 2014). The analysis of coproduction within multi-level governance focuses particularly on the encounters at the interface between public officials and citizen groups (or private sector) related to forms of collaboration (engagement/disengagement) and forms and degrees of participation in service delivery and possible coproduction in flood risk management (Ostrom 1996, 2005; Bulkeley 2013). Figure 9.1 suggests the institutional focus of the analysis. Multi-level governance, which is typically state and hierarchy dominated, constrains the capacity for municipal governance (through processes of decentralisation, deconcentration and delegation), including through the enabling of the private sector and community (see below) (Betsill and Bulkeley 2007). Municipal governance typically involves four modes of governance: self-governance, provisioning, regulation, and enabling (Bulkeley 2013).

Methodology

The research focus led to a case study approach with a mix of research methods; policy and planning reviews, institutional analysis, in-depth interviews (planners, officials), focus group interviews with citizens, local observations of floods and transect walks, and validation workshops and meetings with local officials, civil society and academia. Altogether we conducted 49 semi-structured interviews in the two cities, allowing for a variety of topics to be addressed. Each interview lasted about 1.5–2 h, and was conducted mostly in the office of the interviewee by a lead interviewer and 2–3 team members.

We first explored which institutions and policy sectors were involved and relevant for "climate adaptation policies at city and community levels" which is our main research "case". Given the "triple" governance challenges faced by the municipality in both cities of (i) planning new settlements for a rapidly increasing population; (ii) meeting existing gaps in providing quality service delivery and infrastructure for all citizens and addressing underlying vulnerabilities, and (iii) addressing "new" or future climate risks such as floods, the empirical data collection focused on the integration of climate risks/adaptation in three key sectors and related actors/agencies: (i) urban planning (strategic, land use and development planning); (ii) water resources management (flood risk and storm water management, sewerage, water supply); and (iii) disaster risk management. In concrete terms, we looked for actions considered critical to promote resilience at the level of the local communities, including, *inter alia*: whether or not climate action plans had been established through participatory approaches; institutional homes and mandates for understanding and addressing key issues; functioning coordination mechanisms; assigned staff, budgets and resources; planning and zoning principles that were enforced and respected; and whether or not broader underlying vulnerabilities were addressed in ways responsive to community-based demands and needs for adaptation and coproduction (ref. the criteria identified in the *Ten Essentials for Making Cities Resilient*, UNISDR 2014).

We considered both incremental adaptation measures that mirror resilience more narrowly defined, while also determining if there was evidence of transitional adaptation or transformation, with reference to the second and third "levels of adaptation" as recognised by Pelling (2011). Transitional adaptation, according to Pelling, involves changes in governance; while transformation requires more radical or deeper changes in political and cultural values and structures. For our purpose the cities and urban communities then needed to be assessed with regard to two key dimensions; resilience and the capacity to act and deliberately change or adjust urban development through different forms of adaptation at different levels. A wide range of complementary adaptation and disaster risk management approaches are available to enhance resilience to flood risks, including: to reduce risk and exposure; reduce vulnerability; share risks; and prepare, respond and recover from floods (IPCC 2012).

We carried out field-work in two case study locations in each city in order to understand how sub-city levels of governance operate and how local community groups are enabled or constrained in climate change adaptation and (flood) risk management, particularly in encounters with government officials and agencies. The two locations were chosen as informal settlements that are highly exposed and vulnerable to different kinds of floods, with a known history of engagement in flood risk management at the level of the community and Sub Wards. The areas included 8,000–12,000 people. In Dar es Salaam, the two settlements were both located within Kinondoni Municipality. Bonde la Mpunga is an informal, underserviced and unplanned settlement and Sub Ward close to the coast and subject to flash flood and floods from smaller rivers due to inadequate drainage. Suna is located further inland and is exposed to both severe river flooding and stagnant water, as it is located in a river valley bottom. We utilised Bonde la Mpunga as the main case. In Saint Louis, Diaminar is a low-lying area and Sub Ward that is also exposed to flash floods due to inadequate drainage, similar to Bonde la Mpunga. It is also affected by salinisation of housing from sea water intrusion. The second case, Goxxumbacc is exposed both to coastal flooding and flash floods, while in the past it was also subject to floods from the River Senegal. Diaminar is utilised as the main case for comparison. The four cases illustrate different kinds of flood risks observed, and how these engage different kinds of responses, interactional relationships, and variation in public agency responses.

Drivers and Barriers

We utilised a multi-level governance framework to systematically analyse and compare the integration of climate adaptation policies and flood risk management across sectors, actors and levels (Fig. 9.1) (Osborne 2010; Pelling 2011; Bulkeley 2013). The extent to which different modes of governance are actually deployed at different levels and have been successful are the result of different factors that act as both drivers and barriers to the cooperation or coproduction of knowledge and action between state, municipal and non-state actors. We distinguished three sets of factors; *institutional*, *political* and *socio-technical* (Bulkeley 2013), in addition to climate drivers (Fig. 9.1).

Institutional factors in our study are those that shape the capacity of urban institutions at all levels – both formal and informal – to respond, withstand or improve systems' resilience. The main institutional factors and actors and related interactional processes are captured in the circles of Fig. 9.1. Our key focus is on institutional factors related to the *multi-level governance* context within which urban adaptive capacity is enabled and capacity to act and respond to climate risks at local levels largely determined. The institutional factors are conditioned by political factors, socio-technical factors and climate drivers. From an adaptation perspective, we need to understand who and what need to become more resilient; i.e. individual citizens and their assets in local settlements that are exposed and vulnerable to floods. They need support from resilient services and governance systems.

Regarding the *state level* and state-municipal interactions within the broader institutional arrangements, we investigated how the state, at national and regional levels, works through different governance modes in vertical (and horizontal) coordination in the selected sectors to balance powers and resources and enable or constrain the adaptation of city and sub-city level actors. This concerns processes related to (Vedeld 2003; Peters 2008; Manor 2011): (i) Decentralisation and devolution of key functions and resources of the public sector; (ii) Delegation of public authority to e.g. civil society or private market actors; and, (iii) Deconcentration of public authority and tasks to regional state bodies or to a variety of (semi-) autonomous corporations or agencies (e.g. in water supply or sewerage provisioning).

In both cities, the state engages directly within the urban territory, in e.g. land management, service provision and infrastructure. In this regard, we are also interested in how the state makes available powers, mandates, and *financial and other resources* for the municipality.

At the *municipal level*, we analysed organisational arrangements and operations of the municipality in relation to different ideal modes of municipal governance. These modes of governance relate to a specific set of processes and techniques that municipal authorities deploy and related resources and knowledge. Bulkeley distinguishes between four modes of municipal led governance: municipal self-governing, provisioning (by the municipality itself of services and infrastructure), regulation (through hierarchical steering and control), and enabling (for detailed explanation cf. Kern and Alber 2009; Bulkeley 2013). We focus mostly on the capability of the municipality for enabling, i.e. capacity to facilitate, coordinate and encourage actions of non-state actors through coproduction or forms of networks, partnerships or mobilisation (Ostrom 1996; Sorensen and Torfing 2014).

At *community, neighbourhood or sub-city levels*, civil society actors or private business may interact and condition municipal actions from above and from below, or engage in non-state voluntary or community self-regulation modes of governance (De Sardan 2011).

Political factors are also important, although not central to our analysis. Political factors relate to the attention to adaptation (risks and vulnerabilities) given by the political leadership at different levels (e.g. municipal councillors) within the political-economic context. It also relates to the role of specific economic and political interests and actors that shape resilience and urban development locally (Bulkeley 2013: 102).

Socio-ecological or socio-technical factors relate to the ecosystem context, social and demographic conditions, including urbanisation trends and forms and settlement patterns (e.g. in flood-exposed areas), and urban landscapes and morphologies (Ostrom 2005; Bulkeley 2013; see also Chaps. 4 and 6).

Criteria for Choice of the Cities

As a consequence of the research focus on how adaptation becomes integrated in multi-level governance and drivers/barriers to resilient cities in relation to flood risks, we selected one larger (Dar es Salaam) and one smaller city (Saint Louis) faced with fairly similar types of flood risks. The cities were situated in similar eco-system contexts (coastal/river delta) and conditioned by broadly similar governance challenges. In particular, we wanted to explore the importance of size

Table 9.1 Size, vulnerability and climate drivers in the case study cities

	Dar es Salaam	Saint Louis	Comments
Demography and area			
Population size (2012)	4,000,000	200,000	Dar es Salaam much larger
Population density	1,500 citizens per km^2	5,000 citizens per km^2	Higher density in Saint Louis
Size of urban territory	1,590 km^2	40 km^2	Dar es Salaam much larger
Underlying vulnerability			
Rate of population growth	4.4 %	2.4 %	Urbanisation is fastest in Dar and promotes informality and settlement in flood-risk areas
Share of population in informal areas	80 %	29 %	"Informal areas" not defined fully the same way
Access to critical services – % of total urban population			
Access to piped water	25 %	97 %	Indicates much better water supply services in Saint Louis
Access to sewerage	11 %	15 %	
Reliance on pit latrines	92 %	90 %	
Exposure; geo-physical pattern			
Exposure of people, places, assets – percentage of territory flood exposed	8 % of territory below 10 m above sea level	All of city territory below 4 m above s.l.	Dar has a low-lying coastal strip; Saint Louis is all low-lying
Climate drivers and exposure			
Annual rainfall	1,100 mm	300 mm	Heavy cloudbursts common
Flood risk types	River, flash floods, coastal	Flash floods, coastal	Flood risk from River Senegal lowered due to new canal made in 2003
Climate/flood risk projections	No major change in rainfall	More extremes expected	Flood risk expected to increase in both cities due to changes in social conditions more than change in rainfall

and scale for urban flood risk governance. The two cities were chosen as two coastal cities among the five cities studied under CLUVA (see also Chap. 1 and Table 9.1).[1] Both are former capital cities and regional economic centres with several similarities in institutional structures and presence of most of the sector agencies relevant for addressing flood risks. As such, the two cities represent institutionally "rich" case studies and are interesting for our comparison. Selecting two cases that vary in terms of past experiences of extreme flood events, size and adaptive capacity, political factors, and multi-level governance (structures and processes) enabled us to compare and analyse the implications of governance for coproduction,

[1] The five cities being researched under CLUVA are Addis Ababa, Dar es Salaam, Douala, Ouagadougou, and Saint Louis (CLUVA 2014a, b).

coordination and integration of adaptation strategies in the municipalities' and local communities' approaches to urban development.

Some obvious differences between the two municipalities complicated a rigid comparison, however. For example, the two cities faced rather different rates of population growth, and also quite different rates of urbanisation and risk exposure and vulnerabilities. Saint Louis is faced with greater magnitude and frequency of extreme flood events. Moreover, the climate scenarios for Saint Louis suggest changes towards greater climate variability than those for Dar es Salaam (see Chap. 2 and CLUVA 2014a, b). The two cities are also located in rather different cultural and institutional settings; one in Francophone West Africa and one in Anglophone East Africa. Moreover, while Senegal was among the pioneers in introducing democratic elections and a multi-party system in the 1970s, and decentralisation in the mid-1990s, Tanzania introduced a multi-party system only in 1992, and politics are still dominated by one main socialist party (CCM).

The precise criteria chosen for the comparison of the two cities were the following. First, the most important institutional factors to be compared related to the multi-level governance context within which the actual capacity for urban response and action to climate change and flood risks evolve. We assumed that the vertical relationships of government and governance were critical for framing the relationships between local actors, e.g. determining the responsibilities and powers of the municipality in different sectors, and enabling or constraining coproduction between different government, civic and private actors (UN-Habitat 2011; Bulkeley 2013). From the outset we assumed (based on existing literature) that the city of Dar es Salaam was likely to be less resilient compared to Saint Louis (Diagne 2007; Ndour 2010; Coly et al. 2011b; Vedeld et al. 2012; Kiunsi 2013). However, while this assumption proved correct, we also made this an object of our investigation and comparative analysis.

Second, size and scale of a city and the capacity of the municipality to adapt have been considered important factors by some observers for the choice of adaptation approaches (Rauken et al. 2014). The argument is that a large municipality will typically have more resources and capacity for specialisation and allocation of resources to a new policy domain such as adaptation (Rauken et al. 2014). Our approach included a test of this hypothesis.

Third, both cities have recently experienced severe floods and related extreme weather events. In this regard, new knowledge and awareness about climate change and its risks and also experiences of extreme impacts were considered a potentially important driver for explaining urban municipal responses.

Finally, we also considered the degree to which political factors appeared to influence the framing of climate policies at city level, and thus, approaches to adaptation among planners and other actors.

Both cases contributed to an understanding of how adaptation is established (or not) through coproduction in multi-level governance in cities in coastal Africa operating under *extreme financial constraints and relatively centralised hierarchical steering*. Salient features about the size, scale, socio-economic vulnerability and climate exposure of the two cities are provided in Table 9.1.

Findings from the Case Study Cities

The presentation of the findings from the research is organised as follows; first, we present a comparative analysis of the multi-level governance structure in both cities as reflected in the vertical state-municipal relationships, indicating similarities and differences in structures and capacity for coordination of the different levels. Second, we present each of the city cases, initially through a brief background, and subsequently, through a detailed presentation of the municipal organisational structure and the municipality's capacity to operate within the key sectors of concern (planning, adaptation, disaster risk management, and water resources management). Finally, we analyse the (vertical) interactional relationships between public officials (state and municipal) and community level actors and institutions in relation to coproduction of relevant flood risk responses and services at the local level.

Multi-level Governance Structure – Centralised and Hierarchical Government

The multi-level governance structure critically constrains the capability of each of the municipalities for autonomous self-governing and provisioning of key public services and infrastructure. The degree and form of centralisation keeps the municipalities from accessing important mandates, powers and financial resources. Key state ministries and agencies in both countries are interventionist in their dealing with the municipality related to urban planning and development functions. This concerns ministries dealing with functions such as urban planning, adaptation/ environment, disaster risk management and local administration.

Both countries basically have a three-tier government structure characterised by a relatively strong unitary state, a relatively autonomous, yet weak, municipal level, and an even weaker regional elected level (with elected regional councils). Dar es Salaam city is located within the region of Dar es Salaam, while Saint Louis is similarly located within the region of Saint Louis.

The regional elected councils in both countries are weak in terms of administrative capacity and resources for coordinating services and development. Most importantly, the regional state level, as well as the district state level, represents the 'strong' levels in terms of oversight and coordination and the provision of key relevant services related to land use planning and control, water infrastructure, flood risk management, and resettlement of flood victims. The deconcentrated state services at this level, considered part and parcel of the decentralisation reform, are in both cities located 'next-door' to the municipal services and to some degree overlap and integrate with those of the municipality in local operations.

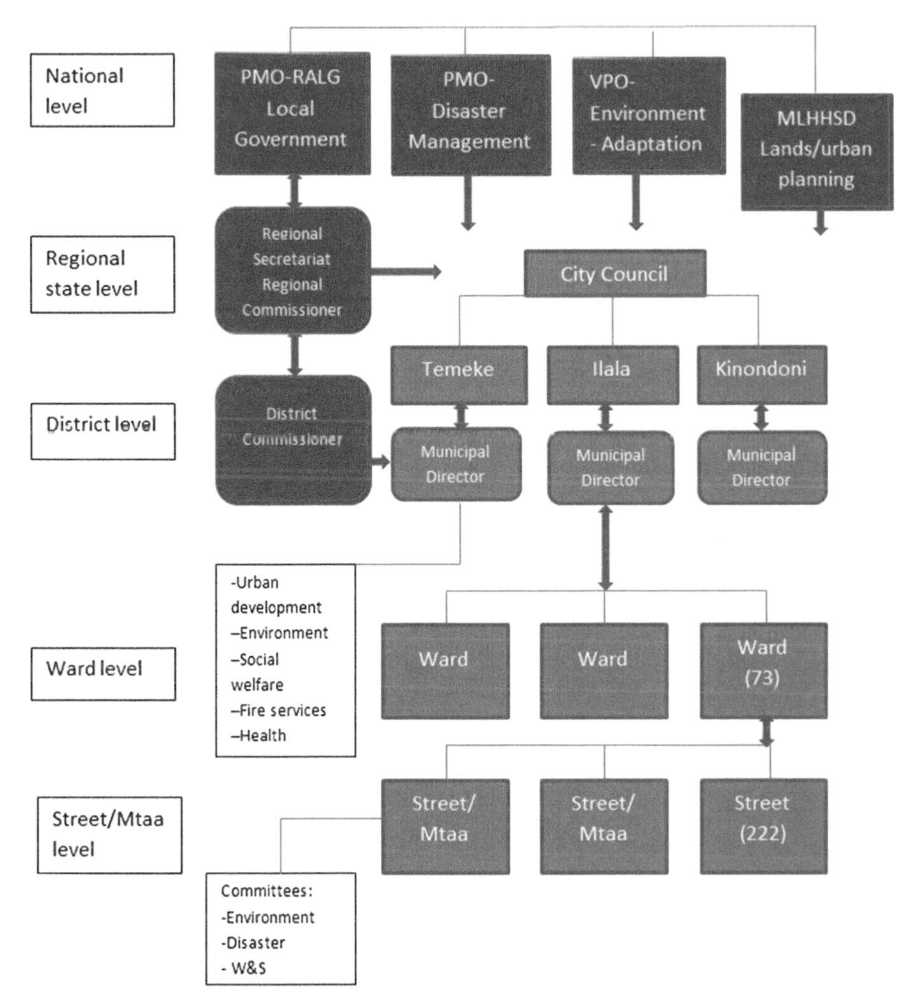

Fig. 9.2 The organisational structure of Kinondoni Municipality and Dar es Salaam City Council (*blue*) and key state agencies (*red*) engaged in urban development (abbreviations explained in the text)

The state services are largely coordinated by the Governor at regional level and Prefect at district level in Saint Louis, and, similarly, by the Regional Commissioner and Regional Secretariat at regional level and the District Commissioner at district level in Dar es Salaam. The Prefect and District Commission are on par with the Mayors in the authority hierarchy (see Figs. 9.2 and 9.4).

The regional state agencies thus play the key role in coordination with important implications for adaptation and disaster risk management services within the city territory. Only Saint Louis has a Climate Advisor in the city planning office and this office enhances municipal governance on these issues. Coordination in flood risk management is mainly done by the Governor/Regional Commissioner at regional level and by the Prefect/District Commissioner at city level. The coordinating committees at this level tend to function through hierarchical instruments, for example, by providing information and guidance in environment and/or disaster risk management committees, while also instructing service agencies on what to do in general development (development committees). The regional and national level state coordinating committees for adaptation and disaster risk management have limited administrative, technical and financial capacity to actually perform their duties within the city territory. Also, important semi-state and private corporations are involved in sewerage and water supply in both cities. But they are mandated to work mostly in formal areas and, thus, in middle-class neighbourhoods and the city centres.

While the municipality in each city has some degree of autonomy, and is supposed to be overseen by the regional state level only in their legal handling of budgets and functions, in reality this autonomy is lacking in the sense that the city has very limited powers and legal mandates, finances, staff, and technical competence. This undermines their capacity to perform their planning and service functions, including in climate risk management.

Dar es Salaam – Short Background

Dar es Salaam is the largest city and the main economic centre of Tanzania. The city has experienced a series of floods arising from smaller rivers and lack of drainage as well as coastal flooding over the last few years (Ardhi 2011). In 2011, flash floods after heavy rains killed over 40 people and displaced thousands of people and destroyed houses and assets across the city. CLUVA's analysis of extreme rainfall events, based on climate projected data until 2050, suggests an increase in the frequency of extreme events, but reduction in intensity and limited increase in rainfall (Chap. 2). Flood impacts will increase, however, but mainly due to likely increased concentration of people in flood prone areas (CLUVA 2014a).

It has been in the context of these different climate change risks, while being confronted with the multiple challenges of providing development to a rising population, that the urban authorities have started to respond to climate change adaptation and flood risks more systematically. We exemplify these issues through the two local case studies; Bonde la Mpunga (our main case) and Suna.

Multi-level Governance and Barriers to Municipal Adaptive Capacity

The municipality of Kinondoni is responsible for a rather limited set of services, including development planning, strategic planning, fire services, solid waste collection, health/education, social welfare, and environmental management. Hence, it lacks critical capacity for self-governing and provisioning, as well as in enabling coordination. The lack of institutional capacity is illustrated by the lack of financial and other resources for operations. The overall annual budget for the municipality represents only about 31 Euro per capita. Adaptation and disaster risk management have no allocated budget (beyond the general budget for fire and rescue services). The budget at Sub Ward levels is only about 1–2 Euro per capita per year and covers only minor running costs of a small office (PMO-RALG 2012). A local planner confirmed that the "finance was very limited" for the list of priority development activities prepared by the Sub Wards (Interview MLHHSD planner on 4th June 2012). Key urban development programs are mostly funded directly by central government agencies, public enterprises or by external partners (donors).

Municipal Government Structure

Beyond the constraints set by the state and multi-level governance, the organisational structure of Dar es Salaam City Municipality is complex and multi-layered with clear structural deficiencies in terms of vertical coordination and horizontal communication. The municipality is composed of a four-layered structure (Fig. 9.2), headed by the City Council (DCC). It is sub-divided into three autonomous municipal councils; Kinondoni (531 km^2), Ilala (210 km^2) and Temeke (652 km^2). The Kinondoni Municipality, which is the focus of our study, includes close to 50 % of the city population and most of its high-income residential areas. Each municipality is divided into Wards (total 73 Wards) and Sub Wards or street level (222 Mtaas), for further explanation of organisation of the Mtaas, see Chap. 6. However, the three municipalities tend to govern with limited regard for the City Council. On many accounts they act as if they are more accountable upwards to some of the state ministries which are heavily involved in the strategic and daily governing of the city, such as the Prime Minister's Office for Regional Administration and Local Government (PMO-RALG) and Ministry of Lands, Housing and Human Settlement (MLHHSD). To a lesser degree, municipal officials show genuine downward accountability to the Wards/Sub Wards or the citizens they are meant to serve (Mng'ongo 2005; Kombe and Kreibich 2006; START 2011).

PMO-RALG is key to the governance of the city, including through its control of employment of human resources. MLHHSD is the "custodian" of all land in Tanzania, and central to land management and enforcement in Dar es Salaam, land being defined as state owned. Moreover, MLHHSD has taken charge of developing the new Dar es Salaam Master Plan (2010–2030).

Centralised Adaptation and Disaster Risk Management

There are no substantive institutional "homes" mandated for either adaptation or disaster risk management with the City Council or the municipality that can build knowledge and coherence between these two agendas and key planning and service operations (as suggested to be critical by Birkmann and von Teichmann 2010).

Most actions on adaptation and disaster risk management are centralised in national offices (Vice President's Office and Prime Minister's Office, respectively) and there are no staff and resources at the level of the Ward and Sub Wards to promote these service sectors. National policies and strategies exist, but provide limited guidance for urban adaptation. The National Climate Change Strategy (2012) provides some general suggestions for what is required for improving urban settlements and enforcement of land use zoning, and possible relocation of flood-affected communities (Kiunsi 2013, see also NAPA 2007). Awareness about climate change adaptation is low.

The most recent national strategy on DRM was approved in 2004 (building on an Act from 1991). The lack of coordination mechanisms between the Commissioner level and Municipal Directors in emergency operations was evident during the 2011 extreme flood that strongly impacted the Suna settlement (ref. Interview 15.9.2012 Ward representative). The emergency response rested heavily on assistance from the Red Cross country office located in Dar es Salaam. While "the military and the police were called upon and were present during the event, they did not have the required operational equipment such as life jackets, vehicles, blankets, first aid. They only contributed human resources", and, "officials at the Ward level during the flood were mostly confused and did not help much" (Interview Red Cross official 5.6.2012).

Urban Planning – Unplanned Expansion of Settlements

Flood risk issues have not been well integrated in urban planning in the case study areas, and land use planning is done with limited active involvement of the citizens, even if this is mandated (Kyessi 2002; Kombe and Kreibich 2006; Vedeld et al. 2012). Expansion of housing in the two informal areas we studied has mostly happened in an unplanned manner. Close to 80 % of the citizens of Dar es Salaam live in unplanned or informal settlements, most of which lack basic infrastructure and services, such as storm drains and basic sanitation facilities. Some 20 out of the total 150 informal settlements in the city have been identified by CLUVA researchers as extremely vulnerable to floods with current rainfall and social patterns (Kombe and Kreibich 2006; John et al. 2012; Herslund et al. 2012; Vedeld et al. 2012; Kiunsi 2013). The failure of the planning and governance arrangement to meet basic service needs is illustrated by the fact that only 11 % of the population have access to sewerage, 25 % to piped water, and more than 90 % of the citizens of Dar es Salaam continue to rely on pit latrines and autonomous systems of sewerage. Moreover, the outlook is gloomy. A water engineer from the key water corporation (DAWASA) claimed that they "have no plans to cover the informal areas with sewerage" (ref. Interview 5.6.2012).

Few, if any, specific legal covenants are in place or enforced regarding risk exposed zones (for example, in the NAPA). The new draft Master Plan identifies flood risk zones, but has no substantive mention of climate adaptation (Interview 4.6.2012 MLHHSD decision maker/planner; see also URT 2011; Moss and Happold 2012; Kiunsi 2013).

Integration of Flood Risk Management into Systems of Coproduction

The case of Bonde la Mpunga illustrates that public agencies and officials are not capable of encouraging or mobilising high levels of citizen participation and input into the local management of flood risks and related land conflicts (Fig. 9.3).

A key governance issue arose some years ago as the central government allowed private developers to construct middle-class houses and new commercial buildings in the settlement in violation of its protection in the original Master Plan from 1979. The new buildings were also constructed such that they blocked natural drainage, and forced flood water into the houses of many of the low-income inhabitants. Local people have protested this development through media, local demonstrations and by mobilising local politicians. Although the central government has sent urban

Fig. 9.3 Drains blocked by solid waste in Bonde la Mpunga (Picture: Trond Vedeld 2012)

planners and water engineers to engage in dialogue, no shared solution acceptable to the local population has been found, presumably because a coalition of private interests and central government interest collaborate to exploit the high value land. The interventions of various state and municipal agencies in local affairs have resulted in conflicts and mutual disengagements, more so than coproduction. A MLHHSD planner reflected on the ambiguous role of the urban authorities; "wrong decisions have been made on transfer of hazardous land to commercial area and residential area" (Interview 5.6.2012 MLHHSD planner).

The weak enforcement of land use zoning also allows for continuous expansion of the settlement, because poor people have few other areas in which to settle (Mng'ongo 2005; Kombe and Kweka 2012; Vedeld et al. 2012). To this end, several of the respondents suggested that, due to the lack of enforcement capacity of higher level authorities, powers should be given to the Sub Ward/Mtaa and Ward levels in land development planning and control; a statutory role they do not have currently (Interview MLHHSD planner 5.6.2012). They are now mostly involved in local development planning with minor development projects (Kombe and Kreibich 2006).

The problem of enforcing land use zoning is compounded by political factors. For example, following the 2011 flood in the Suna settlement, the government decided on a plan to resettle many of the flood affected households (with support of UN-Habitat). However, local residents came together to form a local political bureau of CCM, the main political party, in order to resist what was perceived as "forced" resettlement. This reflects on how local people in informal settlements establish protection through ties to political representatives and local Big men to secure tenure and houses, against the aims of the central government administration (Interview Ward representative 15.9.2012; Interview Official in Disaster Management Office, PMO 5.6.2012). Such "political patronage" illustrates the ambiguity of the government in addressing informality; the administrators seek to move people; politicians engage to protect them.

Saint Louis – Short Background

The city of Saint Louis is a smaller city, yet a regional economic centre in Northern Senegal. It is located on four low-lying islands near the mouth of the River Senegal on the shore of the Atlantic Ocean (fifth largest city in Senegal). CLUVA's analysis of extreme rainfall events, based on climate projection data until 2050, suggests that intensity and frequency of extreme events will significantly increase and enhance flood risks (see Chap. 2 and CLUVA 2014a, b). The city has experienced several recent extreme flood events from the River Senegal, for example in 1994, 1999, and 2003. In 2003, the city was under threat of a major flood which prompted dramatic interventions on the part of the central government. In 2010, more than 80,000 people or 40 % of the 200,000 citizens were affected by floods following extreme rainfall.

It is on the background of significant climate-change risks, increasing population and settlement of people in flood-exposed informal areas, that the urban authorities from the mid-1990s started to take flood risks increasingly more seriously and from around 2005 responded more systematically to the climate change adaptation agenda. We illustrate these governance issues in Diaminar (our main case) and Goxxumbacc, which are two flood-exposed and vulnerable settlements with a large share of the dwellers living in unplanned areas.

Multi-level Governance and Barriers to Municipal Adaptive Capacity

The degree of centralisation in Senegal is significant, and there are important multi-level governance constraints on the capacity of the municipality of Saint Louis for self-governing and provision of key services and infrastructure (Pinto 2004). However, these constraints do not hinder the combined state-municipality system to provide relatively good services and water infrastructure for flood risk management by African standards. Regional state agencies are actively engaged in urban development and urban investments, and interact directly with municipal and local councillors and civil society actors (Fig. 9.4).

The city of Saint Louis is managed by a City Council with 70 councillors, and is administratively divided in 22 local districts managed by local District Councils. These Councils were introduced in 1994 as an institutional reform to strengthen planning and implementation capacity at local level, and ensure greater participation of local people and civil society. The reform was inspired by the decentralisation trends of the time, and the program was supported by several donor agencies (see below).

Even so, the mandate of the municipality is relatively limited to functions such as economic development planning, environmental management, strategic city planning and development, solid waste management, and development programming and prioritising at the level of the District Councils (in much the same way as in Dar es Salaam). Moreover, the municipality might be at least as constrained financially as the municipality in Dar es Salaam. As suggested by the Prefect, "the Mayor has no administrative, financial or technical services" (Interview 10.5.2012). The total annual budget is only about 3 million Euro per year, or about 15 Euro per capita per year. However, a small budget line item (about 1 % of the budget) is allocated to flood risk management and staff within the city planning office working on adaptation. Moreover, the combined efforts of direct state investments and municipal planning and investments, have resulted in Saint Louis having embarked upon several major investments in protective infrastructure against flooding in recent years, including a system of dykes around the city and expansion of sewerage into some of the informal areas (e.g. Diaminar).

Centralised Adaptation and Disaster Risk Management

Reflecting central state dominance in multi-level governance, the municipality of Saint Louis has no statutory mandate to govern in the areas of adaptation and disaster risk management, even if the role of the city in disaster risk management is

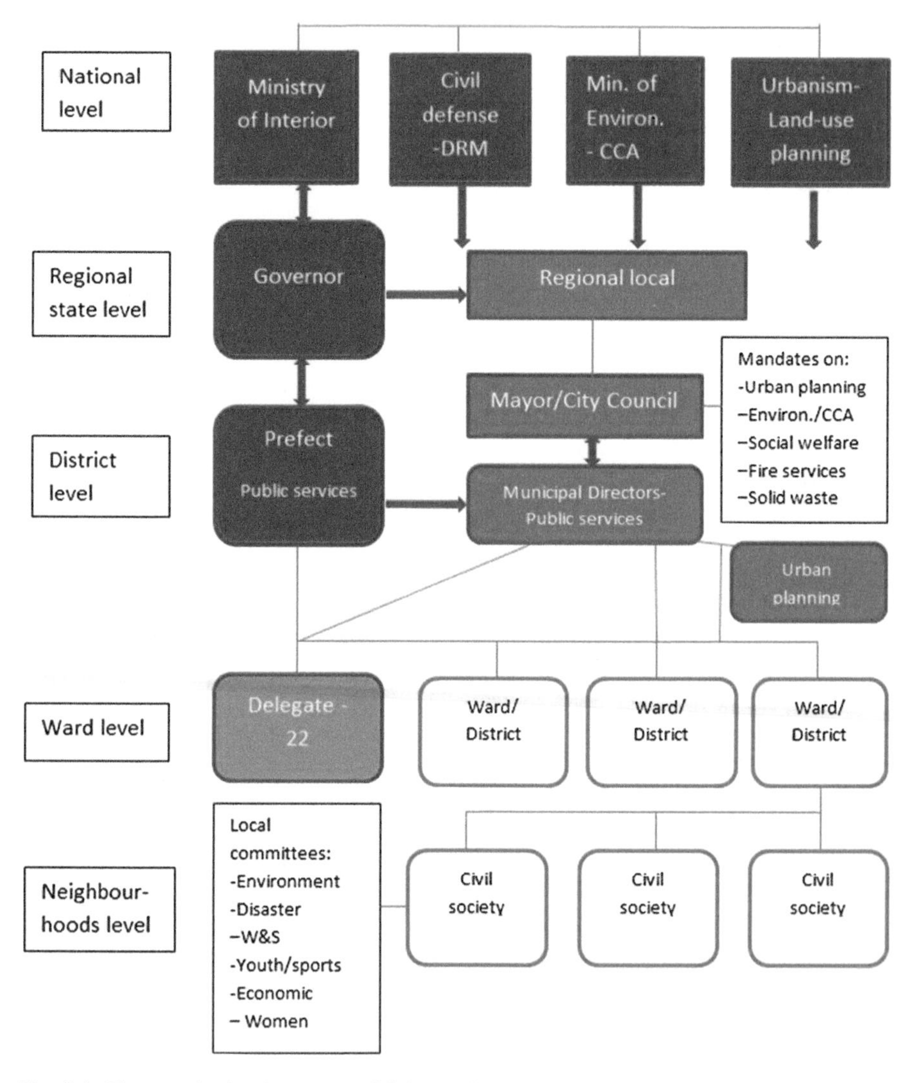

Fig. 9.4 The organisational structure of Saint Louis Municipality and key state agencies engaged in urban development

recognised as important in the national disaster risk management strategy. The National Action Program on Climate Change Adaptation from 2006 has limited guidance on urban adaptation. The agenda is centrally vested with the Ministry of Environment and Protection of Nature (at the time). Disaster risk management is a central state domain and placed with the Civil Protection Agency of the Ministry of Interior. Emergency operations are headed by the Governor at Regional level or the Prefect at departmental level (at the level of the Municipality). A program to enhance coordination of public and civic actors in flood risk management with the Mayor was launched in the period 2006–2008 and supported by an NGO (Enda-Tiers Monde) and local state authorities (Water department and Senegal National Sanitation Office (ONAS) (Diagne 2007)). However, these governance structures faded as the external support stopped, and few traces of this initiative were found today, beyond the capability left in the city planning and development office (ADC).

Urban Planning and Water Governance

While the city municipality is firmly in control of strategic city planning through its planning office (ADC), land use planning and management is largely in the hands of the regional state body of the national planning ministry. Except for the latest city strategic development plan, Horizon 2030, neither general city plans nor sector plans have directly addressed climate change issues, according to our review of these plans (Horizon 2013). However, given that close to 70 % of the population live in formal and planned areas, the planning system is performing relatively well according to African standards. Moreover, water and flood risk management have been substantially treated in the new strategic plan, reflecting both awareness and knowledge and organisational commitment to these issues on the part of the city planners and the leadership of ADC, as well as with the Mayor (Diagne 2007; Niang 2007; Ndour 2010; Jørgensen et al. 2012). All key agencies involved in water resources, storm water and flood risk management are directly organised under the state, as state bodies or semi-state (sewerage) or private corporations (water supply) under state oversight. Water governance in Saint Louis performs well on some accounts, but not all. For example, a significant achievement is that 97 % of the population have access to piped water. However, only 15 % of the population have access to sewerage, and close to 90 % rely on pit latrines and various autonomous systems of sewerage. ONAS works not only in the formal areas, as the new sewerage program in Diaminar demonstrates. Finally, a key issue is that the Mayor is not involved in the governance structure for the management of the River Senegal.

Our interviews suggest that knowledge and awareness about the linkage between climate change, extreme rainfall and flood risks are present at different levels among urban authorities, including City Council members, water resources engineers (ONAS), environmental officers (located with the Deputy Mayor) and urban planners (ADC), and the Governor, especially within the city planning office (ADC) (Interview 8.5.2012 Environmental officer). ADC is perceived as well-positioned and capacitated to "take on a stronger role in coordinating climate risk responses" (Interview 8.5.2012 NGO representative).

However, covenants in the land use act about illegal settlements in zones close to the river and the coast are mostly not strictly enforced due in part to political factors (e.g. political patronage). "A key problem is the interventions by politicians in critical land management decisions, which makes the state soft [on enforcing land management] ... Often the local state Delegate takes part in political conflicts and enhances local conflicts" (Interview 8.5.2012 City planners).

Integration of Flood Risk Management in Systems of Coproduction

The most interesting and unique aspect of Saint Louis' municipal arrangement is the way a system of coproduction has been institutionalised within the municipal government and governance structure. Through the decentralisation reform that started in 1994, with the establishment of District Councils (DCs) and a City Planning and Development Office (ADC), a political-administrative system developed to support the DCs through preparation of District Development Plans (DDP). This created administrative capacity and systems of local governance and coproduction at local neighbourhood levels. The reform received considerable donor support initially, but the level of support has fallen over the last decade or so (Niang 2007).

The ADC planners actively encourage citizen participation in local development planning through continuous local dialogue and meetings. They ensure local input into the DDPs and related development programs with a bearing on flood risk management. The DCs rally around the preparation of the DDPs and smaller development projects, including flood risk issues. The unique feature of the DCs is that they are constituted from elected representatives of the most important local civil society associations within the local administrative territory. These may be local business groups, youth and sports groups, women's groups or religious groups. We found 23 registered local associations in Diaminar (while Goxxumbacc had 94 such associations), reflecting a particularly strong presence of local civil society organisations. The District Councils are not part of the formal decentralised structure, but in practice work through a coproduction arrangement with city planners. Hence, they stand out as institutionalised coproduction arrangements that bridge the public-private divide (Ostrom 1996, 2005). The Youth and Sports Clubs, interestingly, are the most active in terms of facilitating the mobilisation of people for actions in flood risk preparedness, management and risk reduction in flood exposed communities (Ndour 2010; Vedeld et al. 2012). A problematic side of this municipal organisation is that it raises issues of political accountability because elections are not through secret ballots and the councillors do not report formally to the City Council. However, the structure has proved to have real strength in governance practice regarding the mobilisation of local people for planning, flood risk management and development. ADC planners involve the councillors which subsequently mobilise own citizen members for discussions. Many genuine and broad needs and demands are discussed in these local encounters and subsequently included in the DDPs (Diagne 2007; Ndour 2010; Coly et al. 2011a, b, c). The success of the Saint Louis local governance model has made it a country-wide approach to city and sub-city level governance in Senegal.

For example, in Diaminar, constructive interaction between ADC and the District Council has resulted in enhanced local flood risk awareness (through local training assisted by a local NGO) and the preparation of a local flood action plan and mapping of flood risk sites and a new drainage/sewerage system being established (Fig. 9.5). The local District Development Plan (DDP) has been predominantly oriented towards finding solutions to flooding. Diaminar is typical of an

Fig. 9.5 Construction of new drains and sewerage in Diaminar (Picture: Trond Vedeld 2012)

informal settlement allowed to evolve on low lying wasteland over the last few decades.

The government recently responded to local demands and embarked upon a sewerage and urban restructuring program in parts of Diaminar (with donor support). This has prompted the need for a resettlement program, with UN-Habitat support and aided by ADC. The people of Diaminar, spearheaded by the Youth groups, would typically engage in cleaning of drains prior to the rainy season, placing sand bags or waste dumps to protect their own houses, and raise door steps of houses and latrines. Local people in 2010 were also observed providing cash as co-funding of fuel or repairs of local pumps operated by the fire brigade. Such civic engagement is also a reflection of a rich local culture and dense networks of community organisation (Coly et al. 2011c).

Despite such local engagement and community capacity, the Diaminar neighbourhood proved to be highly dependent on external assistance to reduce the impacts of flash floods during the extreme event of 2010, e.g. from the pumping station operated by the fire brigade and the local Red Cross. As observed by the Governor, "The District level is still very weak" and lacks financial resources (Interview 8.5.2012 with Governor's office).

However, Saint Louis has also experienced how centralised political factors can overrule municipal government and lead to maladaptation. In 2003 the city was threatened by a new major flood from River Senegal due to heavy upstream rainfall (cloudbursts). Reflecting in part the experience from the devastating 1999 flood that impacted the whole of Saint Louis, including Goxxumbac which is situated at the spit of "Langue de Barbarie" at the mouth of the river, the Governor, in consultation with the central government, decided to have a canal created across the spit to open up a new outlet for River Senegal. This decision was taken without consulting the Mayor of the city or local citizens (except through a rapidly organised public meeting including people from Goxxumbac). This canal has since 2003 reduced considerably the risk of river flooding in the city of Saint Louis (Kane 2010; Vedeld et al. 2012). But the canal has grown from 4 to 2 km in 10 years, and it has greatly increased other environmental issues such as coastal erosion and salinisation. Several local planners and practitioners considered this an ecological catastrophe.

Comparison Between City Cases in Integration of Adaptation

The subsequent section provided a comparative overview of how the two cities have integrated adaptation into existing service sectors and into the multi-level governance system with reference back to our definition of resilience. Tables 9.2 and 9.3 outline some deliberate actions taken by the government at state and/or municipal levels on adaptation *per se* (as a distinct policy area) as well as actions on the integration of adaptation in governance and selected service sectors (planning, water/storm water, disaster risk management) in Dar es Salaam and Saint Louis. The role of coproduction in multi-level governance is discussed separately.

Discussion of Findings

The comparison suggests that the two cities have chosen very different approaches to the integration of climate change adaptation and flood risk management in the municipal government and multi-level governance system. Climate change adaptation – narrowly defined – remains a fairly nascent agenda in both cities and is not

Table 9.2 Comparison between the case study cities in integration of adaptation and building of '*resilient cities*' in multi-level governance and coproduction

Resilience according to adaptation actions		
Multi-level governance	**Dar es Salaam**	**Saint Louis**
Strategies/policies for adaptation and disaster risk management	Exists only at national level with limited policy guidance	Exists only at national level with limited policy guidance
Organisational homes for adaptation	Only at national state level (Vice President's Office)	Only at national or regional state level
Organisational homes for disaster risk management	At national state, regional state and ward levels	At national, regional state levels, ward, civil society level
Organisational home for flood risk management	At national level	At regional state level
Coordination mechanisms for adaptation and disaster risk management	Only at national level	At regional state level (Governor) for both policy areas
Participation in governance/planning	Weak	Medium
Engagement of local groups	Weak	Strong
Resources, budgets and mandates for adaptation and flood risk management	Weak at municipal level; no budget for flood risk management	Medium at municipal level; small budget for flood risk management
Coproduction	**Dar es Salaam**	**Saint Louis**
Capacity for coproduction	Weak	Strong; institutionalised

well integrated in planning and governance at municipal level, with some exception regarding the way climate risks are treated by the city planning office in Saint Louis and in local development planning.

Disaster risk management is basically a national and regional state anchored agenda, and the municipalities are not much involved in disaster preparedness or direct coordination of emergency responses. Flood risk management interfaces with all sectors discussed in the study, and reflecting the many public and private actors involved remains a fragmented agenda as regards municipal provision and regulation.

As a policy area, it has several institutional homes and is not well defined and coordinated at municipal levels in either of the cities. The notable exception is regarding the inclusion of flood issues in the District Development Plans and strategic plan for Saint Louis. Adaptation is only indirectly addressed in the draft Master Plan for Dar es Salaam; however, this may become a key element if the municipality engages with CLUVA researchers in a new city Climate Action Plan (Kombe et al. 2013).

Table 9.3 Comparison between the case study cities in integration of adaptation and urban planning and flood risk management

Urban planning	Dar es Salaam	Saint Louis
Strategic plan and integration of adaptation	New draft Master plan (2013): No direct mention of adaptation; flood risk zones mapped	Horizon 2030 (2013): Flood risk/ water resources management integrated. Limited mention of adaptation
Land use planning	Not done for 80 % of city territory; flood risk zones known but not enforced	Done for 70 % of city territory; flood risk zones known but to limited degree
Local development planning	Do not include reference to flood risks	District Development Plans have flood risk management at their core
Disaster risk management (DRM)	**Dar es Salaam**	**Saint Louis**
Early warning system and national-local DRM system	Weakly established DRM system within Prime Minister's office; coordinated with Regional and District Commissioners; no home for DRM with Mayor's office; limited informal early warning at local levels	Reasonably well DRM capacity at national level; DRM coordination committee with Governor; no home for DRM with Mayor's office; only informal early warning system through cell phones
Resettlement of flood victims	Small resettlement program (UN-Habitat); large implementation problems	Small resettlement program (UN-Habitat); large implementation problems
Flood risk/storm water management	**Dar es Salaam**	**Saint Louis**
Organisational home for flood risk and storm water management	No real organisational home with municipality or state; many state and private actors involved	New ministry established for flood risk management (2012); many state and private actors involved; no home with municipality
Sewerage	Sewerage only provided in formal areas	Sewerage mostly provided in formal areas
Storm water infrastructure	Storm drains along main roads only	Dykes, drainage, opened new outlet for River Senegal
Adaptation	**Dar es Salaam**	**Saint Louis**
Adaptation projects (incremental approach)	A few adaptation projects; no major investments in storm water/flood risk management	A few adaptation projects; substantive investments in flood risk management

Multi-level Governance and Coproduction

Regarding the multi-level governance context in Saint Louis the state agencies and semi-state/private corporations (e.g. in water supply, sewerage) contributed directly to substantive urban planning and infrastructural development (drains, dykes) in protection of informal areas. The Governor and the Prefect, including street level bureaucrats such as the local state Delegate, actively engaged with local Ward officials and civil society groups.

This was less the case in Dar es Salaam. In Dar es Salaam, adaptation and flood risk management had no substantive organisational homes at municipal level and these policy fields were not well integrated at the local level. Public officials acted ambiguously in local contexts and to limited degree encouraged participation of citizen input to management of flood risks, water resources and land use planning.

It seems that the smaller size and particular organisational structure in Saint Louis facilitated closer day-to-day interactions between public officials and local people and an institutionalisation of coproduction in services. There existed relatively close social networks between state level officials and urban planners. This ensured responsiveness and citizen participation in flood risk management. Scale probably contributed to this finding (see also Chap. 5). Processes of deconcentration in Saint Louis appeared coordinated with the decentralised and devolved municipal arrangements to ensure vertical integration by the municipality and urban state planners in collaborative development planning. In turn, better coverage and efficiency in service provision and infrastructure was supported by direct state and donor funding in the city.

A lack of financial and other resources within the two municipalities proved a critical barrier to their capacities to provide services and enable coordinated flood risk management. While both city municipalities operate under extreme financial constraints, Dar es Salaam had a larger annual municipal budget per capita (31 Euro against 15 Euro for Saint Louis).[2] Even so, Saint Louis performs better in terms of investments in flood risk management, sewerage and water supply (according to the indicators chosen, see Table 9.2). This reflects upon Senegal being a higher-income per capita country than Tanzania.

Community Factors

We suggest that the relative efficiency in participation of citizens in coproduction in Saint Louis is also a likely effect of the high community capacity (or social capital) represented by the high density of local civic associations in the local settlements. To this end, there are many types of community groups and local "Big men" involved in governance in both the cities, which is also recognised in the literature on local governance in Africa (Crook 2010; De Sardan 2011; Crook and Booth 2011).

[2] A municipality in Norway would have about 5,000 Euro per capita at its disposal. This is an astonishing difference.

Size, Scale, and Adaptive Capacity of the Municipality

Regarding size and adaptive capacity of the municipality, we find that the larger city Dar es Salaam has executed a set of smaller or incremental adaptation activities, but has done less well in coordination, provisioning, and enabling of flood risk management at the local level than the smaller city Saint Louis. In fact, the large size and complex multi-layered structure may have discouraged such participation. This seemingly contradicts the findings of Rauken et al. (2014), who compared municipalities in Norway, and found that the largest municipalities revealed the most advanced adaptation approaches. We agree, however, that large size potentially might contribute to explain the presence of early awareness among individual officials on adaptation and possibly the adoption of incremental adaptation activities (as in Dar es Salaam). Moreover, if municipalities are extremely small, such as in some Scandinavian countries (with below 10,000 people), resource constraints within the municipality might hinder their ability to pay attention to adaptation (such as lack of specialised staff, knowledge, or finance).

Prior Experience with Extreme Floods

Prior experience with extreme weather events and floods may have been a contributing factor, but probably not a major factor in explaining why climate risks over time have prompted attention by municipal planners and politicians in Saint Louis – more so than we observed in Dar es Salaam. Saint Louis is contextually placed in a more extreme eco-region (Sahel), subject to more frequent floods and droughts, and the city will be expecting more extreme events in the future. Moreover, a greater share of the city population and local economy is affected by floods. However, prior experience with extreme events interfaces with other institutional and political factors, and it is hard to assign relative causal value to either. For example, in Saint Louis the lingering controversy around the opening of the new outlet of the River Senegal in 2003 has maintained flood risks and impacts of flood on the local political agenda (e.g. Interview 8.5.2012 Governor and City Planning Office).

Political Factors

Comparing the two case study cities, we argue that political factors were not decisive for the approaches taken by municipal officials to the integration of adaptation activities in municipal government and governance in the two cities. However, the personal engagement of the city Mayor in Saint Louis likely facilitated the engagement of planners and sector officials in addressing flood risks. This finds support in evidence from the case of Durban, South Africa (Roberts

2010; Roberts and O'Donoghue 2013).[3] Moreover, political factors probably become decisive once important decisions are on the agenda about transitions in the governance or service delivery system, or about whether infrastructural investments should address vulnerability and inequality (Pelling 2011; Bulkeley 2013). Political factors are important for explaining why and how people are allowed to settle in unplanned and risky areas. The decentralisation reform in Saint Louis in 1994/1995 can serve as an example of a transition in urban government which, in turn, influenced local governance and resilience.

Conclusion

The experiences from Dar es Salaam and Saint Louis suggest that resilience as a concept was useful for pointing to important factors that define a resilient city and what and who need to become resilient, i.e. individual citizens and their assets and places and local institutions. None of the cities are in this regard 'resilient'. As conceptualised by UNISDR (2014) resilient cities points to the scope for defined improvements of basic structures and functions of the system at risk (see also IPCC's definition of resilience (2012: 3)). However, given the major barriers revealed in multi-level governance in both cities, we need concepts that are more practical and provide a better understanding on how to act and move beyond resilience and incremental adaptation to transitional adaptation (change in governance) and transformation (more progressive changes within political systems or power structures and value systems). This is in line with Pelling (2011: 69) and Bulkeley (2013).

To this end the concept of coproduction in multi-level governance was fruitful for understanding drivers and barriers to municipal governance. The concept helped focus on the multitude of actors involved, interactional relationships, the high degree of centralisation, financial constraints, and the diverse levels of motivation and levels of responsiveness of public officials to local citizens.

For Saint Louis we argued that public officials actively encouraged citizen participation in planning and their input into the production of public goods (flood risk management), while in Dar es Salaam the system rather discouraged high levels of citizen contributions. Hence, we propose that the arrangement of coproduction within the system of multi-level governance – and the way governance was practised – were the factors that best explained the differences in

[3] The municipality of Durban is recognised internationally as innovative and active in promoting climate change adaptation. Adaptation has been encouraged by municipal officials and integrated in urban development despite limited guidance from the state. The case illustrates the importance of local entrepreneurs – and agency – among the city planners in experimenting with climate strategies and incremental adaptation. It suggests that political factors and limited funding can be overcome. It also suggests that incremental adaptation and learning-by-doing lead to positive results provided there is cooperation across sectors (Roberts and O'Donoghue 2013).

adaptation approaches between the two cities. Large size and scale did not help furthering adaptation. In this regard, we concur with Ostrom (1996, 2005) in that good agency performance resulted not from the capacity or strength of public sector agencies *per se*, but rather from increasing their responsiveness to customers and from cooperation across the public-private divide.[4]

We argue that size and adaptive capacity of the municipality *per se* were not particularly important in the extent to which adaptation and climate risk management were integrated, coordinated and functioned at local level. We suggest that smaller size and simpler organisational structure (Saint Louis), combined with committed leadership and organisational cultures, facilitated active citizen input into service delivery and coproduction in flood risk management.

In conclusion, we propose several additional findings with relevance for theory. First, extreme financial constraints at municipal level did not constitute a decisive hindrance against adoption of incremental adaptation (in any of the two cities). The lack of political guidance and resources could be overcome by engaged municipal officials and local champions (Saint Louis). Second, benign political factors and additional resources were required to promote more progressive transition in governance and larger (infrastructural) investments aimed at addressing risks and vulnerability and inequality on a broad scale (as in Saint Louis). But political factors were not decisive for getting incremental or small-scale adaptation activities going. Third, weak political systems led to maladaptation (opening of the new river mouth in Saint Louis) and/or lack of substantive adaptive actions (Dar es Salaam).

Finally, we argue that transition towards *climate resilient cities* for Dar es Salaam and Saint Louis will require significant strengthening of multi-level governance and capabilities to act and coordinate at city and community levels (vertically and horizontally). Some key politico-institutional measures, given that coordination is foremost required at the local level (according to subsidiarity principles), are;

- Decentralise more powers and resources to city and sub-city levels;
- Enhance greater responsiveness to informality and inequality among public officials;
- Provide the Wards and Sub Wards with a formal mandate, resources and powers for (community-led) land development control and flood risk management;
- Enable greater commitment to participation of citizens in coproduction of relevant services and local planning; and
- Put in place organisational homes and structures for coordination of the joint disaster risk management and climate change agendas, including for storm water management, supported by city-level climate action plans.

---

[4]That being said, even in Saint Louis the coordination of urban adaptation and flood risk management by hierarchical instruments was relatively weak and fragmented. This is also not surprising given that this has also been observed in many other African countries (Cartwright et al. 2012; Roberts and O'Donoghue 2013; CLUVA 2014b) as well as in many European countries (OECD 2009; Bulkeley 2010; Hanssen et al. 2013; Rauken et al. 2014).

Business as usual is clearly not enough in either of the cities if the aim is to address flood risks in a more resilient and sustainable manner.

References

Adger N, Lorenzoni I, O'Brien K (2009) Adapting to climate change. Thresholds, values, governance. Cambridge University Press, New York

Ardhi (2011) Dar es Salaam case study. Climate change, disaster risk and the urban poor: cities building resilience for a changing world. Ardhi University, Dar es Salaam

Bache I, Flinders M (2004) Multi-level governance. Oxford University Press, Oxford

Betsill M, Bulkeley H (2007) Looking back and thinking ahead: a decade of cities and climate change research. Local Environ 12(5):447–456

Bicknell J, Dodman D, Satterthwaite D (eds) (2009) Adapting cities to climate change. Understanding and addressing the development challenges. Earthscan Climate, London

Birkmann J, von Teichmann K (2010) Integrating disaster risk reduction and climate change adaptation: key challenges – scales, knowledge, and norms. Sustain Sci 5:171–184

Bulkeley H (2010) Cities and the governing of climate change. Annu Rev Environ Resour 35:229–253

Bulkeley H (2013) Cities and climate change, critical introductions to urbanism and the city. Routledge, Oxon/New York

Carmin J, Anguelovski I, Roberts D (2012) Urban climate adaptation in the global south: planning in an emerging policy domain. J Plan Educ Res 32(I):18–32

Cartwright A, Parnell S, Oelofse G, Ward S (eds) (2012) Climate change and the city scale. Impacts, mitigation and adaptation in Cape Town. Routledge/Earthscan, Oxon

CLUVA (2014a) Main page. http://www.cluva.eu/deliverables/CLUVA_ D1.7.pdf/. Accessed 10 Mar 2014

CLUVA (2014b) Main page. Available at http://www.cluva.eu/. Accessed 10 Mar 2014

Coly A, D'Almeida A, Diakhaté MM, Lo M, Sy BA, Ndour NM, Gueye S, Sall F, Sy AA (2011a) Report on climate related hazard in the city of Saint-Louis. Internal CLUVA report (D5.2), University of Gaston Berger

Coly A, Ndour NM, Gueye S (2011b) Assessment of the institutional capacity of the city of Saint-Louis. Internal CLUVA report (D5.3), University of Gaston Berger

Coly A, Ndour NM, Gueye S (2011c) Social vulnerability assessment in the city of Saint-Louis. Internal CLUVA report (D5.3), University of Gaston Berger

Crook RC (2010) Rethinking civil service reform in Africa: 'islands of effectiveness' and organisational commitment. Commonw Comp Polit 48:479–504

Crook C, Booth D (eds) (2011) Working with the grain? Rethinking African governance. IDS Bulletin 42(2):iii–iv, 1–101

De Sardan J-PO (2011) The eight modes of local governance in west Africa. In: Crook RC, Booth D (eds) Working with the grain? Rethinking African governance. IDS Bull 42(2):22–31

Diagne K (2007) Governance and natural disasters: addressing flooding in Saint Louis, Senegal. Environ Urban 19(2):552–562

Douglas I, Alma K, Magenta M, McDonnell Y, McLean L, Campbell J (2009) Unjust waters. Climate change, flooding and the urban poor in Africa. In: Bicknell J, Dodman D, Satterthwaite D (eds) Adapting cities to climate change. Understanding and addressing the development challenges. Earthscan Climate, London, pp 201–223

Hanssen G, Mydske P, Dahle E (2013) Multilevel coordination of climate change adaptation: by national hierarchical steering or by regional network governance. Local Environ 18(8):869–887

Herslund L, Mguni P, Lund DH, Souleymane G, Workneh A, Ouedraogo JB (2012) Exemplary policies, strategies and measures. CLUVA deliverable D3.6. Available: http://www.cluva.eu/deliverables/CLUVA_D3.6.pdf. Accessed 10 Mar 2014

Horizon (2013) Horizon 2030, Nouvelle Métropole Africaine, City Planning Office (ADC), Saint Louis

ICLEI (2014) Main page. Available at http://www.iclei.org. Accessed 10 Mar 2014

IPCC (2012) Summary for policy makers. In: Barros V, Barros V, Stocker TF, Qin D, Dokken DJ, Ebi KL, Mastrandrea MD, Mach KJ, Plattner G-K, Allen SK, Tignor M, Midgley PM (eds) Managing the risks of extreme events and disasters to advance climate change adaptation, A special report of Working Groups I and II of IPCC. Cambridge University Press, Cambridge/New York, pp 1–19

John R. Mayunga J, Kombe W, Fekade R, Ouedraogo JB, Tchangang R, Ndour NM, Ngom T, Gueye S, Sall F (2012) Preliminary findings on social vulnerability assessment in the selected cities. CLUVA deliverable D5.4. Ardhi-ARU, Dar es Salaam. Available: http://www.cluva.eu/deliverables/CLUVA_D5.4.pdf. Accessed 10 Mar 2014

Jørgensen G, Herslund L, Sarr C, Nakouye N, Sine A, Workneh A, Workalemahu L, Bekele E (2012) Base line scenarios for urban development of selected case study areas. CLUVA deliverable D3.5. Available: http://www.cluva.eu/deliverables/CLUVA_D3.5.pdf. Accessed 10 Mar 2014

Kane C (2010) Vulnérabilité du système socio-environnemental en domaine sahélien: l'exemple de l'estuaire du fleuve Sénégal. De la perception à la gestion des risques naturels. Thèse de doctorat de 3e cycle, Université de Strasbourg

Kiunsi R (2013) The constraints on climate change adaptation in a city with large development deficit: the case of Dar es Salaam. Environ Urban 25:321–333

Kyessi AG (2002) Community participation in urban infrastructure provision, servicing informal settlements in Dar es Salaam. Published PhD thesis. SPRING publication series no. 33, University of Dortmund, Germany

Kern K, Alber G (2009) Governing climate change in cities: modes of urban climate governance in multilevel systems. In: OECD competitive cities and climate change, OECD conference proceedings, Milan, pp 171–196

Kombe WJ, Kreibich V (2006) Governance of informal urbanization in Tanzania. Mkuki Na Nyota, Dar es Salaam

Kombe WJ, Kweka C (2012) Institutional analysis for climate change in Dar es Salaam City, Internal CLUVA report (D5.1). Ardhi, Dar es Salaam

Kombe WJ, Kyessi A, Kassenga G, Shemdoe R (2013) City local climate action plan 2014–2020 of African cities facing climate change. CLUVA (D5.7). ARDHI University

Manor J (2011) Perspective on decentralisation, Working paper no. 3, ICLD. Swedish International Centre for Local Democracy, Visby

Mng'ongo O (2005) Browning process: the case of Dar es Salaam. Published PhD thesis, Royal Institute of Technology, School of Architecture and Built Environment

Moss D, Happold B (2012) Afri Arch, BuroQ. Dar es Salaam Master plan 2030. Preliminary Draft, Dar es Salaam

National Adaptation Programme of Action (NAPA) (2007) Vice President's Office, Division of Environment, Dar es Salaam

National Climate Change Strategy (2012) United Republic of Tanzania, Vice President's Office, Environment Division, Dar es Salaam

Ndour NM (2010) Intégration de la question de l'eau dans la gouvernance urbaine: analyse des systèmes de décision dans la Commune de Saint-Louis. Mémoire de Master 2, UGB, Géographie

Niang D (2007) Gouvernance Locale, maitrise d'ouvrage communale et stratégies de développement local au Sénégal: l'expérience de la Ville de Saint-Louis. Thèse de doctorat de 3^{ieme} cycle

O'Brien K (2012) Global environmental change II: from adaptation to deliberate transformation. Prog Hum Geogr 36(5):667–676
OECD (2009) Competitive cities and climate change. In: OECD conference proceedings, Milan, Italy, 9–10 Oct 2008
Osborne S (2010) Introduction: the (new) public governance: a suitable case for treatment? In: Osborne S (ed) The new public governance. Emerging perspectives on the theory and practice of public governance. Routledge, Oxon
Ostrom E (1996) Crossing the great divide: coproduction, synergy, and development. World Dev 24(6):1073–1087
Ostrom E (2005) Understanding institutional diversity. Princeton University Press, Princeton/Oxford
Pelling M (2011) Adaptation to climate change: from resilience to transformation. Taylor & Francis Books, London
Peters BG (2008) The two futures of governing. Decentering and recentering processes in governing, 114 Reihe Politikwissenschaft/Political science series. Institut für Höhere Studien, Wien
Pinto RF (2004) Service delivery in francophone West Africa: the challenge of balancing deconcentration and decentralisation. Public Adm Dev 24:263–275
PMO-RALG (2012) Annual budgets for Tanzanian municipal councils. Online. http://beta.pmoralg.go/tz/lginformation/monotor1a.php. Accessed 15 Sept 2012
Rauken T, Mydske PK, Winsvold M (2014) Mainstreaming climate change adaptation at the local level. Local Environ. doi:10.1080/13549839.2014.880412
Roberts D (2010) Prioritizing climate change adaptation and local level resilience in Durban, South Africa. Environ Urban 22(2):397–413
Roberts D, O'Donoghue S (2013) Urban environmental challenges and climate change action in Durban, South Africa. Environ Urban 25(2):299–319
Satterthwaite D (2011) Editorial. Why community action is needed for disaster risk reduction and climate change adaptation? Environ Urban 23(2):339–350
Satterthwaite D, Dodman D (2013) Towards resilience and transformation for cities within a finite planet. Environ Urban 25(2):291–298
Satterthwaite D, Huq S, Pelling M, Reid H, Lankao PR (2007) Adapting to climate change in urban areas. The possibilities in low-and middle-income countries, Human settlements discussion paper series. IIED, London
Satterthwaite D, Huq S, Pelling M, Reid H, Pelling M, Lankao PR (2009) Adapting to climate change in urban areas: the possibilities and constraints in low- and middle-income nations. In: Bicknell J, Dodman D, Satterthwaite D (eds) Adapting cities to climate change. Understanding and addressing the development challenges. Earthscan Climate, IIED, London, pp 3–47
Sorensen E, Torfing J (2009) Making governance networks effective and democratic through metagovernance. Public Adm 87:234–258
Sorensen E, Torfing J (2014) Enhancing social innovation by rethinking collaboration, leadership and public governance, social frontiers. The next edge of social innovation research. Nesta, Glasgow
START (2011) Urban poverty & climate change in Dar es Salaam, Tanzania: a case study. Ardhi, Dar es Salaam
UN-HABITAT (2011) Global report on human settlements 2011. Cities and climate change. Nairobi, UN-HABITAT
UNISDR (2012) 2012 making cities resilient report. UNISDR, Geneva
UNISDR (2014) Online page. http://www.unisdr.org/campaign/resilientcities/toolkit/essentials. Accessed 8 Mar 2014
URT (2011) New draft urban development and management policy. United Republic of Tanzania (URT), Unpublished. Dar es Salaam

Vedeld T (2003) Democratic decentralisation and poverty reduction: exploring the linkages. Forum for Development Studies No 2, Norwegian Institute for Foreign Affairs (NUPI), Oslo

Vedeld T, Hellevik S, Kombe W, Kweka-Msale C, Coly A, Ndour NM, Guéye S, Klausen JE, Saglie I-L (2012) Report on planning system and government structure in 2 case cities. CLUVA deliverable D3.1. Available: http://www.cluva.eu/deliverables/CLUVA_D3.1.pdf. Accessed 10 Mar 2014

Weber EP, Khademian AM (2008) Wicked problems, knowledge challenges, and collaborative capacity building in network settings. Public Admin Rev March/April 2008

Ziervogel G, Parnell S (2012) South African coastal cities: governance responses to climate change adaptation. In: Cartwright A, Parnell S, Oelofse G, Ward S (eds) Climate change and the city scale. Impacts, mitigation and adaptation in Cape Town. Routledge/Earthscan, Oxon, pp 223–243

Chapter 10
Towards Climate Change Resilient Cities in Africa – Initiating Adaptation in Dar es Salaam and Addis Ababa

Lise Herslund, Dorthe Hedensted Lund, Gertrud Jørgensen, Patience Mguni, Wilbard J. Kombe, and Kumelachew Yeshitela

Abstract The chapter explores the possibilities and barriers for integrating climate change adaptation into urban development and planning in the case study cities of Dar es Salaam and Addis Ababa. It identifies adaptation measures in collaboration with stakeholders that are meaningful and need urgent attention across various parts of the urban governance system and discusses possible pathways towards increased city resilience.

The study shows that only limited momentum exists among city stakeholders for a broad citywide adaptation strategy addressing the complexity of climate change impacts in both case study cities. This is – for the time being – due to inability to address all measures relevant for making the cities resilient as several more urgent issues, such as rapid urbanisation and poverty, take precedence among stakeholders. Therefore, a more incremental approach of addressing the most pressing matters that can mobilise a range of stakeholders and create synergistic effects with other critical urban problems needs to be prioritised. Such projects can furthermore create knowledge and relations between important actors and institutions.

In Addis Ababa, the project that stakeholders could support and that could address the immediate flooding and drought problems of the poor was 'integrated water management', a citywide approach based on the common interests and possible synergies across city sectors and levels for better water management. In Dar es Salaam, the governance system appears to be too fragmented to drive a

L. Herslund (✉) • D.H. Lund • G. Jørgensen • P. Mguni
Department of Geosciences and Natural Resource Management, University of Copenhagen, Rolighedsvej 23, 1958 Frederiksberg C, Copenhagen, Denmark
e-mail: lihe@ign.ku.dk; dhl@ign.ku.dk; gej@ign.ku.dk; pamg@ign.ku.dk

W.J. Kombe
Institute of Human Settlements Studies, Ardhi University, P.O. Box 35176, Dar es Salaam, Tanzania
e-mail: kombewilbard@yahoo.com

K. Yeshitela
Ethiopian Institute of Architecture, Building Construction and City Development (EiABC), Addis Ababa University, Addis Ababa, Ethiopia
e-mail: kumeyesh@googlemail.com

S. Pauleit et al. (eds.), *Urban Vulnerability and Climate Change in Africa*, Future City 4,
DOI 10.1007/978-3-319-03982-4_10

similar institutionally-led adaptation effort. Here 'integrated local projects' in the most vulnerable areas addressing land management, upgrading and livelihood projects generate most resonance among stakeholders.

Keywords City resilience • Incremental approach • Integrated projects • Vulnerable areas • Water management

Introduction

Climate change is expected to present serious challenges for cities all over the world, and in particular in the least developed countries. The five CLUVA case study cities Dar es Salaam, Addis Ababa, Saint Louis, Douala and Ouagadougou are already confronted with severe effects such as flooding and drought (Chap. 1). The cities need to become more resilient[1] and adapt to present and future effects of climate change despite the many challenges. The question is how to do so?

This chapter initially presents a theoretical framework for spatial strategy-making (Healey 2009) and identifies a number of measures which are commonly considered important for adaptation in African cities (Parkinson et al. 2007; Douglas et al. 2008; Satterthwaite et al. 2009; de Sherbinin et al. 2011; UN-Habitat 2011; Jha et al. 2012; Roberts et al. 2012) in order to guide us in a strategic process towards more climate change resilient cities in terms of possible actions. Due to existing deficits in governance systems it is not realistic that these measures can be implemented without mobilising broad support from a wide variety of actors (Roberts 2008; Healey 2009) and without having to deal with a diversity of barriers (Chap. 9). It is therefore important to focus on the ability of measures to *act in practice* rather than to be mere 'desk-strategies' listing the right things to do without receiving the necessary commitment from key stakeholders (Healey 2007, 2009).

It is important to recognise that we refer to *a particular understanding* of strategy-making which differs from the strategy-making of the 1960s in European spatial planning, which aimed for comprehensive master plans prepared by experts based on the modernist faith in the power of scientific analysis to predict and shape the future. Such rational planning approaches focused mainly on physical patterns of spatial development. As the comprehensive master plans failed to deliver the desired outcomes, planning theory evolved towards more collaborative approaches acknowledging the importance of gaining knowledge and support from a variety of actors as well as the futility of making detailed plans and strategies for the long term in a highly dynamic and unpredictable environment. Consequently the idea of strategy-making and strategic planning processes changed accordingly

[1] By resilience we refer to the literature on evolutionary resilience, i.e. 'the ability of complex socio-ecological systems to change, adapt, and, crucially, transform in response to stresses and strains' (Davoudi 2012: 302) in order to maintain important functions (Chap. 9).

(Salet and Faludi 2000; Albrechts 2004; Healey 2007). Thus when we talk about strategy-making and taking the first steps in a strategic process towards more resilient cities we understand strategy-making as an interactive and collaborative process of identifying the most important issues for a broad range of actors in the cities, a process of scoping and creating momentum for action, a process that builds on both local, context-specific knowledge and expert knowledge and a process of selecting and framing short and long term actions that resonates with stakeholders (Healey 2007, 2009). This process may or may not result in a written strategy, but has the explicit aim to create transformations in practice.

With this in mind, the aim of the chapter is to explore the possibilities and barriers for integrating climate change adaptation into urban development and planning by means of initiating a strategy-making process searching for momentum and possible ways to frame adaptation to create action in practice. We seek to answer the question of how to approach making cities more resilient with a particular focus on adaptation towards urban flooding. Which adaptation measures have the most realistic chance of being implemented in the two case study cities? And where can support and momentum be attained across various parts of the urban governance system (Healey 2009)? The CLUVA case study cities have not yet come far in adapting to climate change, and therefore flooding and drought are rapidly becoming pressing problems (Herslund et al. 2012). We have chosen the two CLUVA cities: Dar es Salaam and Addis Ababa as case studies as these cities showed most interest in the topic, and because both cities were in a process of developing city-wide master plans, making the identification of possible adaptation measures salient. Furthermore, while the cities share some of the same challenges they are widely different in terms of geography as well as governance systems (see e.g. Chap. 9 of this volume). In Dar es Salaam lives have been lost and hundreds of people have been resettled after floods in 2011. In Addis Ababa floods each year force a number of families to leave their homes, occasionally destroy the houses completely, and have severe impacts on urban farming and livelihoods (Jørgensen et al. 2012).

In the two case study cities climate change adaptation is rarely mentioned explicitly in city plans or sector plans, and furthermore when there are plans and planned activities which might improve resilience, there are severe implementation deficits (Herslund et al. 2012; Vedeld et al. 2013). Thus an integrated urban climate change adaptation strategy seems far off. At the moment, in both countries climate change is being addressed at the national level (see section "Who is in charge?"), and coping with the effects by local communities or individuals is ongoing (Herslund et al. 2012). The city level is rather weak or missing, but possibly an important nexus for a more strategic urban climate change adaptation by being the point at which community-based adaptation options may be linked to the funds and skills of sectors and ministries at the city and national levels (Jørgensen et al. 2014), and thus the entry point for our attempts to initiate a strategy-making process. Urban authorities have a key role to play in making cities more resilient to climate changes (UN-Habitat 2011), but the question is where to start when climate change adaptation tends to drown in more urgent urban development problems? In the case study cities massive annual population increase and the resultant increase in demand for urban services and land have set a rapid pace of urbanisation that the city planners are struggling to address. Consequently, the cities already suffer from an 'adaptation deficit' due to housing backlogs, lack of sanitation and drainage, etc., leaving the urban poor living in informal settlements in a vulnerable position (Dodman et al. 2011; START 2011). In Dar es Salaam 70 % of the city are informal areas[2] (Kombe 2005). In Addis Ababa the situation is similar as a large part of the city consists of slums and informal settlements with low or no service provision (UN-Habitat Ethiopia 2008; Jørgensen et al. 2012).

The main contribution of this chapter is – in collaboration with city stakeholders – to explore the possibilities and barriers for getting climate change adaptation on the agenda and into African city development by taking the first steps in a strategy-making process. It searches for momentum in order to extract the most realistic measures and possible entry points from where to proceed with making cities more resilient. The work is mainly based on workshops with multiple stakeholders conducted in the case study cities Dar es Salaam and Addis Ababa in March 2013 but also on interaction with stakeholders throughout the CLUVA project 2010–2013 (Chap. 1 and www.cluva.eu), interviews with resource persons and people living in vulnerable areas (see section "Methods").

Theoretical Framework

Our approach is, as mentioned, strongly inspired by Healey's version of spatial urban strategy-making, which we relate to adaptation planning. Thus we introduce this approach and demonstrate why it is relevant in the African context. Then we move on to adaptation and measures that literature identifies as important in an African context, as well as important conditions for implementation.

Strategic Planning Relevant to the African Context

Adaptation of a city to climate change is complex and will involve different sectors and levels of governance (Chap. 9). To obtain multi-benefits and harvest synergies between measures, coordination and a strategic prioritisation of efforts are key (Leck and Simon 2013; Lund et al. 2012; Roberts et al. 2012). According to Albrechts (2004), for spatial planning to be 'strategic' it must at least be characterised by prioritisation focusing on a limited number of key issues that are

[2] By informal areas we mean areas that are outside the formal urban structure, i.e. there are no formal plans governing development, and no formal provisioning of infrastructure (housing, roads, sewerage etc.).

considered most urgent. "As it is impossible to do everything that needs to be done, 'strategic' implies that some decisions and actions are considered more important than others and that much of the process lies in making the tough decisions about what is most important" (Albrechts 2004: 752). Healey's (2009) idea of 'strategic planning' relies on a collaborative process where learning through more pragmatic and incremental measures and activities characterised by a broad and diverse involvement of local stakeholders, forms a basis for finding out what 'really matters' and what is most important to do (Albrechts 2004; Healey 2009). In these approaches, the linking of efforts and learnings throughout the process is necessary to build up knowledge, networks and collaboration. According to Healey (2009), for a plan or strategy to be 'transformative', special attention must be paid to four contextual dimensions for it to have an effect in practice.

The first of these contextual dimensions, *mobilising attention,* is about finding the momentum for a specific strategy. What forces and actors are driving or supporting a climate change adaptation agenda? What are the opportunities for an agenda 'to take on' and to be supported? The second dimension, *scoping the situation,* includes finding out what is at stake, and in which arenas strategic ideas and visions can be promoted. Who are the critical stakeholders active now and what is the relative power of a particular strategy-making initiative in relation to other ongoing urban dynamics? The third dimension, *enlarging intelligence,* involves mobilising knowledge resources. Urban development dynamics as a field is too complex to be covered by a single discipline – multi-disciplinarity is needed both from academics and professional practitioners from different fields. Also, the people who live or work in the areas that the strategy aims to influence must be involved. What is going on locally? The fourth dimension, *creating frames and selecting actions,* is about developing the projects and ideas that have power to create transformations. Frames arise through collective processes of 'sense-making'. The process allows those involved to position their activities in a wider context and create a way of valuing and justifying what they do (Healey 2009) (Fig. 10.1).

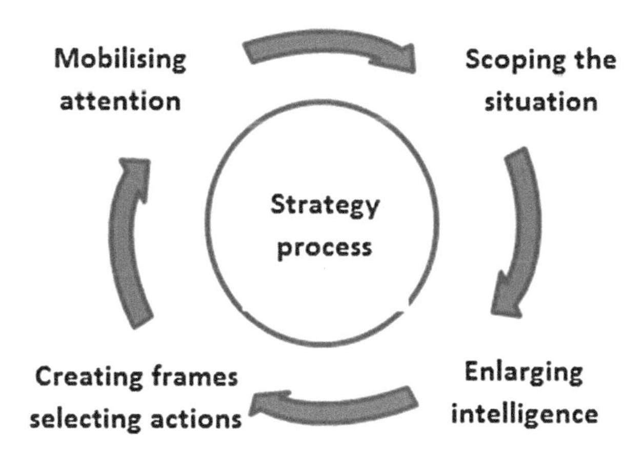

Fig. 10.1 Spatial strategy-making consists of four dimensions: mobilising attention, scoping the situation, enlarging intelligence and creating frames/selecting actions (Adapted from Healey 2009: 442)

Healey's approach is to be understood as a reaction to the strategic planning of the 1960s following a rational planning approach and made up of logic and progressive stages in planning, beginning with clear goals and comprehensive assessments giving exact and reliable knowledge of present conditions and projections of the future (Allmendinger 2009). This was to be followed by detailed, all-encompassing planning and implementation carried out by professionals. In the rational approach, scientific, expert knowledge was seen as the most reliable and legitimate type of knowledge as it claims objectivity (and is the basis for prioritising and deciding what is most important to do) (Bryson 1995). In Europe, this approach largely failed (Albrechts 2004; Healey 2007). Nonetheless, in African countries and cities it is often the rational planning approach that is taken as a heritage from colonialism (Watson 2009). Master plans are being prepared by ministry officials and foreign experts, without much involvement of stakeholders and affected residents. And also in the African context these plans largely fail due to a variety of reasons (Watson 2009). Consequently, more attention must be paid to the strategy-making process and to involving local knowledge (Albrechts 2004; Healey 2009; Watson 2009). Perhaps it is even more pertinent in an African context due to the fragmented governance system and the widespread lack of resources which makes stakeholder mobilisation and support more crucial.

With a starting point in Healey's dimensions we will initiate a search for measures where momentum, support and resonance with the experienced problems of a varied set of stakeholders in the case study cities can be found. Furthermore, we will discuss the barriers and opportunities as well as the possible road towards the fourth dimension – *selecting actions and frames*. In the following we identify a number of adaptation measures that might be relevant for the selected case study cities.

Adaptation Measures

City adaptation to climate change is a relatively new field and according to leading researchers and main development agencies in the field there is no exact tool-kit for how to do it mainly because of uncertainty of climate change impacts, their complexity and context dependency (Bicknell et al. 2009; UN-Habitat 2011). Surveying contemporary literature on planning for climate change adaptation and urban flood management in developing cities, recent guidelines for city climate change adaptation and urban flood management by UN-Habitat, the World Bank, World Meteorological Organisation, UNIDSR and experiences from climate change adaptation activities in Durban, we have identified several groups of important measures for adaptation (see Fig. 10.2 for a summary).

The main measures relate directly to spatial initiatives to reduce the vulnerability to one of the most urgent climate change related challenges; that of urban flooding. *Land use management* is extremely important because poorly planned and managed urbanisation strongly contributes to the growing flood hazard (Tucci 2007; UN-Habitat 2011; Jha et al. 2012; UNIDSR 2012). As cities and towns sprawl to

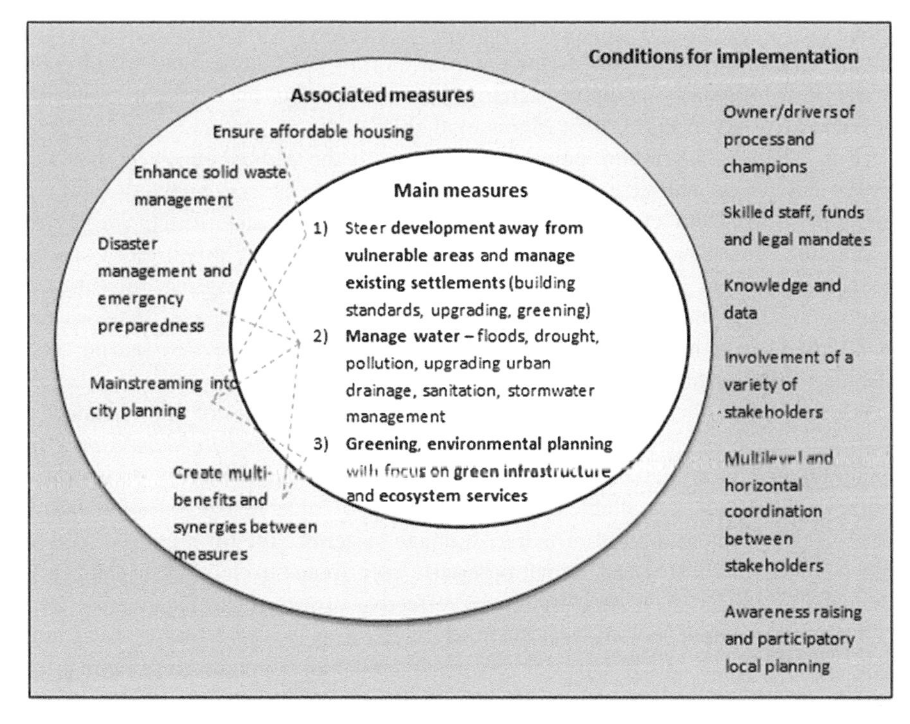

Fig. 10.2 Main and associated adaptation measures to increase resilience to climate change as well as the main conditions for success

accommodate population increase, large-scale urban expansion often occurs in the form of unplanned development in floodplains as well as in other flood-prone areas. Furthermore, a very high proportion of urban population growth and spatial expansion takes place in the dense, lower-quality informal settlements where sanitation and drainage infrastructure are lacking, and public services, such as solid waste collection, are highly insufficient (Parkinson and Mark 2005; Douglas et al. 2008; Satterthwaite et al. 2009). Improved land use and higher building standards (Chap. 3) can directly reduce vulnerability by directing settlements away from flood prone areas and by creating buildings that can withstand occasional flooding.

Better handling of stormwater and *improvement of urban drainage* are also essential to adaptation (Tucci 2007; UN-Habitat 2011; Jha et al. 2012). Many urban dwellers in developing cities experience flooding and environmental health problems because of lacking or ineffective drainage systems and ill-planned construction that blocks natural watercourses (Parkinson et al. 2007; Douglas et al. 2008; Satterthwaite et al. 2009). Recent thinking towards more sustainable drainage practices is encouraging the use of landscape-based, natural drainage arrangements rather than upgrading the sewer system of an area. Landscape based methods are often referred to as Sustainable Urban Drainage Systems (SUDS) (Armitage et al. 2012). In theory, SUDS offer several advantages also for developing cities such as the possible creation of multiple synergies like improved conditions for urban agriculture, provision of structures for stormwater management in informal areas and freshwater aquifers recharge also addressing the drought problems (Fryd et al. 2010; WRGC 2013; Mguni et al. 2014).

Related to SUDS is the increasing attention to the *ecosystem services* of the urban landscape and green spaces. Urban green spaces and wetlands provide services in the form of infiltration and storage of stormwater, which are key to adaptation. In addition other climate benefits can be realised: urban green spaces reduce urban heat island effects, reduce air pollution, and improve recreational opportunities, while urban wetlands provide important hydrological functions for flood alleviation and maintain river/stream flows during the dry season, etc. (Bolund and Hunhammar 1999; Biggs et al. 2010; Mafuta et al. 2011; Balaban 2012; Roberts et al. 2012; Nickel et al. 2013) (see also Chap. 4 of this volume).

While these main measures are essential, they need *associated measures* to be implemented (Fig. 10.2). The ability to direct settlement away from flood prone areas requires available alternatives for affordable housing (Satterthwaite et al. 2009). The creation of effective drainage systems calls for a more effective management of solid waste, which otherwise contributes to flooding by blocking drainage systems. When flooding occurs, effective disaster risk management procedures are needed, including early flood forecasting to avoid loss of lives and property (Jha et al. 2012). Disaster risk management also includes resettlement of people from risk areas, which must include economically feasible reconstruction of productive activities like jobs and education with sufficient income generation and restoration of livelihoods (de Sherbinin et al. 2011).

As climate change thus affects land use planning, drainage, waste management, green area management, and housing, it is important to mainstream adaptation measures into all the relevant sector plans (Roberts 2008). Furthermore, a focus on multi-benefits should be set, as climate change effects are complex and might not occur exactly as predicted. Therefore, it is meaningful that measures taken are also beneficial for other purposes (Roberts 2008, 2010; Lewis 2009), such as the mentioned added values of green areas and ecosystem management for biodiversity, mitigation, recreation, health, etc. Moreover, as 'economic development' is the key concern for most city governments in emerging countries, climate change adaptation needs to go hand-in-hand with development aspects (UN-Habitat 2011; Simon 2013). Kok et al. (2008) argue that there are considerable synergies to be harvested if development is pursued simultaneously with an effort to reduce vulnerabilities. These include reduced poverty, improved health, energy and food security and infrastructure, for instance.

Conditions for Success

As the complexity of adaptation (Fig. 10.2) and Healey's dimensions (Fig. 10.1) indicate, multi-disciplinarity and collaboration between multiple scales of governance and across sectors are essential for collaborative strategic planning.

This is especially valid for climate change adaptation, as the socio-ecological systems which are affected by climate change cut across scales and sectors.

Skilled staff and access to funding are a common precondition to be in place for new planning themes and adaptation measures. Carmin et al. (2012) concludes that the successful adaptation in the two resource restrained cities of Durban and Quito has been driven by internal incentives, knowledge and awareness. Furthermore, short-term action has been essential. The process of developing headline strategies and visions must be accompanied by on-going concrete activities and mainstreaming into sector plans in order to generate knowledge (*ibid.*).

An *institutional owner* and one or more 'champions' are also needed to take on responsibility for putting climate change on the agenda, and to drive the process and mobilise stakeholders towards implementation by means of their intimate understanding of the power dynamics within a city (Healey 2009). Champions can be politicians, public managers and staff, local leaders and people involved in NGOs who have specific knowledge and engagement in the matter (Taylor et al. 2012). Capacity building efforts are a possible way to further support "champions" in dealing with the manifold challenges of adaptation to climate change (Roberts et al. 2012).

But champions exist only in relation to their *'followers'*; while the champions are the catalysts for change, change is not the act of one individual but a result of an alignment of interests and resources (Brown et al. 2013). Hence, another precondition for transformative adaptation strategies is the involvement of a variety of stakeholders (e.g. public, private, NGO, local groups, university, utility companies) both in order to create ownership and to gain resources of different kinds (UN-Habitat 2011).

Awareness-raising at all levels of government and among people in affected areas on the projected and experienced impacts of climate change is also an important condition for climate change adaptation. Education and awareness-raising can be done by linking practising planners, engineers and local community leaders in arenas where knowledge and experiences of impacts and possible solutions can be exchanged (Roberts 2008; Herslund et al. 2012). A tool proposed in the UN-Habitat guidelines (2011) for regulating land use and building structures within designated vulnerable areas is local land use plans. Developing such local land use plans in collaboration with the local communities in question can create ownership and awareness of the situation. Measures developed in collaboration with local people are then in better compliance with the actual problems in the particular area.

Because of the complexity of climate change many various types of knowledge are needed. Effective adaptation will therefore require that many different sources of knowledge are tapped. This can be done by forming partnerships to knowledge institutions, such as universities, at different levels (Roberts 2010), by observing and including local citizens in activities and meetings and by engaging a wide array of stakeholders. However, producing accurate, reliable data that takes into account the climatic, environmental, technical, as well as social aspects is also a complex task. Furthermore, it is not sufficient to produce data and knowledge, it must also be updated and managed by a responsible institution where it can be accessed and used (Nyed and Herslund 2013).

The CLUVA research has revealed significant challenges to climate change adaptation and disaster risk management in Dar es Salam and Addis Ababa (see also Chap. 9; Herslund et al. 2012; Jørgensen et al. 2014). It is clear that changes in governance and planning systems and institutions are required in order to improve effectiveness and capacity of the city to adapt. This involves improved multi-level and inter-sectoral coordination, increased involvement of local people and local levels of administration and improved data management. However, it also becomes apparent that there is potential in decentralisation and that involvement of affected people from informal settlements can provide vital information, both to local planning authorities and to citizens which may lead to improved practices. In the following we explore the barriers and opportunities for adaptation and identify possible adaptation measures in Dar es Salaam and Addis Ababa which may have a realistic chance to become implemented by initiating a strategy-making process in the two cities.

Methods

Case studies were chosen as a method because the selection of appropriate adaptation measures is expected to be highly context-dependent as resource availability, administrative and political support, institutional anchorage, etc., will have an influence on which measures are likely to be implemented. The two case study cities face many of the same challenges but are quite different both in terms of geographical features and governance systems. As such they offer useful insights for adaptation efforts of other cities. Both cities were in a process of master plan revision. During our surveys, in Addis Ababa city researchers were directly involved in the master plan process, whereas in Dar es Salaam, the planning was undertaken by foreign consultants affecting our choice of interventions even though the basic approach was similar in the two cities. In both cities we had workshops at city and local levels supplemented by interviews with key informants identified during workshops or by our city partners. During our first workshops in both cities we mainly explored vulnerability as experienced by stakeholders at different levels in the cities as well as how the different actors coped and addressed the challenges they faced. In the later workshops we focussed more concretely on identifying measures that would have potential to achieve support from, or at least make sense, to actors at different levels. These later workshops are thus the main empirical basis of this chapter. Figure 10.3 shows a flow chart of our interventions in the two cities.

Local level workshops were held on-site for three main reasons: Firstly, to gain knowledge of local perspectives, insights of experienced problems and possible measures, and to understand how flooding affected peoples' daily lives; secondly, to give participants the opportunity to show the effects and measures taken, e.g. how high water levels had been, locations of small ditches that had been dug to drain the areas, where possible physical measures could be undertaken, etc.; and thirdly, simply to give people an opportunity to tell their stories in their own context.

	Exploring the consequences of climate change	Workshops: Exploring vulnerability	Workshops: Identifying measures	Interviews
Dar es Salaam	• June 2011: Visit to 3 local sites: Suna, Kigogo and Bonde de la Mpunga	• July 2012: City level workshop (Ministry staff and municipal officers (planning, environment, agriculture, fishery), NGO, utility companies, local area representatives, meteorologist, and university staff) • July 2012: Local workshop. Environmental committee.	• March 2013: City level workshop (Ministry staff, municipal officers, local area representatives and university staff) • March 2013: Local workshops in Suna and Kigogo with local environmental comittee	• July 2012: Senior urban planner, Ministry of Lands, Housing and Human settlements • July 2012: Urban Planner, city council and municipality • July 2012: Resettled woman, now back in Suna, street vendor, restaurant owner • July 2012: Local leader • March 2013: City Engineer • March 2013: Municipal head of solid wastet management • March 2013: Ward leader, Kinindoni
Addis Ababa		• November 2012: City level workshop	• March 2013: City level worksop, master plan staff from four teams (Environmental planning, housing, Industry, regional planning) • March 2013: Local workshop with participants from two areas: Akaki and Mekanissa (farmers, industry owners, seminarium)	• March 2013: Urban planner • March 2013: Head of Environmental Protection Authority • March 2013: Informal interviews with farmers, Akaki site • March 2013: Informal interview with dam manager, upper catchment

Fig. 10.3 Flowchart of workshops and interviews in Dar es Salaam and Addis Ababa

City level workshops aimed to engage stakeholders from different institutions in common discussions of both the causes of problems and possible solutions and measures. In Dar es Salaam, city level workshops included ministry staff, municipal officers, local area representatives, utility companies and university staff. In Addis Ababa, we had a unique opportunity to engage with the sector teams preparing proposals for the upcoming revision of the city structure plan, where CLUVA university partners were participating. This made it possible to discuss which adaptation efforts could be meaningfully integrated in the city level plan. Here the city level workshop did not include local area representatives or ministries. The advantage of limiting the city-level workshop to plan revision teams was that participants in the middle of an actual planning process could focus on concrete measures to improve the future of the city. One limitation was that these specialists did not sufficiently represent the complexity of the governance situation in the city. The actual opportunities to move forward are probably much more complex than the sessions indicated.

The main limitation of the stakeholder interactions in both cities was the difficulty in getting all invited to attend and achieving representation from all relevant, important sectors and levels of governance. This was partly due to an apparent lack of interest from these actors, lack of contacts among local facilitators and to practical barriers such as severe congestion problems in Dar es Salaam making it difficult to get to the workshops' venues. In particular, the actual decision-makers and leaders were hard to involve as well as politicians who are in a strong position to set the agenda for adaptation and need to be on board to create momentum. Likewise, the number of NGOs that took part was limited, although they play a crucial part in adaptation because they can mobilise resources. In consequence, we were only able to extract a partial understanding of the situation. Thus, our processes, results and recommendations should be seen as a first step of a strategic process in the cities to adapt, where further involvement and stakeholder interaction is needed.

Interviews aimed to gain more depth with particular perspectives which had been identified at the workshops as particularly important. In Dar es Salaam, the areas identified were: urban planning, which resides both at Ministry and City Municipality levels; resettlement practices, as these were perceived as ineffective; and solid waste management, as solid waste was considered a main contributor to flooding. In Addis Ababa, the interviews were more related to institutional anchorage as this had been identified as being key for implementation, which had been severely challenged in previous planning efforts.

In the following we first explore the situation in Dar es Salaam and then in Addis Ababa, identifying the current efforts and barriers to adaptation, measures needed, and detected entry points that could hold potential as a starting point for a wider adaptation effort in the cities.

Results

Integrated Local Projects in Dar es Salaam

Who Is in Charge?

In Tanzania a National Adaptation Policy was prepared in 2007 by the Environmental Ministry that is mandated to request other government departments to include actions on climate change adaptation and to ensure that environmental committees and officers are in place at all governance levels. However, these are in fact only in place at national level. Concerning disaster risk, it is the Disaster Management Unit in the Prime Minister's Office that oversees risk reduction in the whole country (see Chap. 9). It is also at the national level where there has been an on-going initiative by the Ministry of Lands and Human Settlements (MLHS) to prepare a new master plan for Dar es Salaam City. The last urban plan for Dar es Salaam is from 1979. In the inception report made to initiate the new master plan process, no reference has been made to climate change as such or to the challenges the city is likely to face in the future (Vedeld et al. 2013). Moreover, the MLHS has outsourced the preparation of the new master plan to foreign consultants without incorporating a participation process with involvement of local stakeholders.

In Dar es Salaam there are three municipalities: Kinondoni, Ilala and Temeke and a City Council which in principle coordinates trans-municipal matters. In the municipalities there are two further levels of governance: Wards and Mtaas (Sub Wards) (see Chap. 9, Fig. 9.2 for organogram). The municipalities are responsible for a number of the elements important in climate change adaptation including waste management, drainage, detailed land use planning, and the enforcement of land use and building regulations. The common land use planning practice includes zoning of hazardous or unbuildable areas, but presently there is no guidance on management of such areas and no mechanisms for enforcing land use development control in informal settlements (Vedeld et al. 2013).

Similar to land use control, urban services like drainage and solid waste management are not offered in informal areas. Concerning solid waste, it is entirely left to the local communities in informal areas to transport the waste to municipal waste collection points and often these points are far away. Throughout Temeke municipality[3] there are four major collection points. A third of the local communities in the municipality have formed an organisation around waste collection. In the rest of the communities the collection is not formalised and is either done by informal collectors or not at all. The solid waste manager in Temeke municipality estimates that only between 20 and 40 % of the waste produced in informal areas end up at the collection points (Mbarouk and Mamuya 2013). From the collection points it is then the responsibility of the municipality to bring the waste to the landfills. However, due to insufficient funds for fuel, maintenance of trucks and wages for the drivers, less than half of the waste at the collection points is transported to the city landfill by the municipality. In total, this means that it is only 10–20 % of the waste produced in the informal areas that end up in the landfill (*ibid.*). This results in considerable amounts of waste clogging up rivers, open areas and drainage structures, leading to increased flood risk.

Local Adaptation

Flooding is affecting the everyday lives of people in the informal areas around rivers and in low lying areas. It destroys houses, and affects livelihoods and mobility. According to inhabitants of the vulnerable sites, problems with flooding are increasing, and local sites visited were all seriously affected in 2011.

The vulnerability to flooding has many faces in the local communities as low quality services and poor physical structures intensify the problem. Moreover, poverty and lack of assets limit what the people can do themselves (see also Chap. 6 of this volume). On top of this, mistrust, conflicts and deficient participation end up making people stop taking action (Fig. 10.4).

At the first city session (Fig. 10.3) stakeholders discussed what made the city and local areas vulnerable. Particularly, the lack of basic services (e.g. poor drainage, inefficient waste management and lack of sanitation) was rated as important

[3] The total population of Temeke municipality in 2012 was 1,368,881 according to the Population and Housing Census 2012 (National Bureau of Statistics 2013).

Fig. 10.4 Abandoned houses in local areas destroyed by floods (Picture: Dorthe Hedensted Lund 2013)

together with high population density. Also, how well local communities were prepared and the degree of participatory decision-making were mentioned to be key items for the vulnerability to flooding.

Coping with vulnerability seems to be a very local concern, however. It is mainly the community and individuals who have taken action up to now, such as cleaning of river areas, organising an excavator to dig out waste and silt from waterways, constructing drainage channels, raising the ground level, as well as conducting door-to-door waste collection. However, lack of coordination among coping activities can also intensify the problem (e.g. the raising of the ground level or construction of drainage by individual households may negatively impact other households). Moreover, the development of new, larger-scale middle-class housing may redirect water to the poorer households, which end up with even more water in their houses (John et al. 2012).

The actual measure taken by authorities to deal with the problems of flooding in 2011 was resettlement of people from affected areas. In one case study Ward, 370 households were affected.[4] A third of them were not entitled to resettlement as they were tenants, but the rest were moved to tents in the outskirts of the city.

[4] The population of the Ward was approximately 37,000 according to the 2012 Population and Housing Census General Report (National Bureau of Statistics 2013).

However, half of them are back again today. People returned mainly because they could not make a living far away from the main markets and public transportation. According to the local Ward leader, he cannot prevent people from coming back. He and the municipality have the right to tell people to leave, but the issue of political interference and patronage is strong. In his Ward, a local spokesman of the resettled people that returned, came back, contacted a municipal councillor of his own party and got him to stop the municipal public officers from taking action. However, as the Ward leader also said, people have no other option of where to go, so he will not try to intervene. The resettlement has caused anger among the inhabitants who have stopped the local organisation of waste collection, and also the monthly cleaning session of the river areas is on stand-by for the time being.

Adaptation Measures Needed

The city sessions (Fig. 10.3) in Dar es Salaam revealed that stakeholders across different governance levels and sectors were aware of the problems of flooding and easily could list several factors contributing to vulnerability. They all felt that flooding had become a pressing problem for the city and shared both personal and professional experiences where they had come across flooding. In spite of that they lacked both tools and resources, but also support from their leadership in actually addressing it. The measures listed by stakeholders were better land use management (of vulnerable and also open/green areas), improved urban services and better disaster management (resettlement). On top of this, the stakeholders also pointed out that social matters needed high priority, both in vulnerable areas and in the resettlement of people (Table 10.1). However, a discussion of the possible

Table 10.1 Adaptation measures listed by stakeholders of Dar es Salaam city

Focus	Measures	Responsible	Barriers
Land use development control	Stop building or minimise bad effects from building in vulnerable areas	Municipalities/Ministry of Lands and Human Settlements	Lack of enforcement
			No plans and maps
	Protect green zones		Political interference
Disaster management	Early warning systems	Disaster management unit	Lack of resources and competences
	Resettlement programs		
Social improvement	Help people in vulnerable areas	No actor feels responsible	
	Participatory rehabilitation		
Physical services	Improve storm water drainage systems	Municipalities/Ministry of Local Government and Regional Administrative	Scarce resources
	Improve solid waste collection		Not involved in informal areas

responsible actors and institutions for taking action on the suggested measures created debate.

Most of the measures, such as land use management and provision of services are already the responsibilities of the municipalities and of the Disaster Management Unit (resettlement). However, the Disaster Management Unit does not have resources for providing compensation to all resettled and to actually prepare the areas where people were resettled to and deal with the social and livelihood matters of the resettled. The municipal engineers and planners also expressed frustration with their ability to actually improve on waste management and stop the building activities that are taking place in vulnerable areas. The main reasons mentioned by the stakeholders were lack of the following: resources; proper plans and maps on vulnerable areas; and guidance on managing such areas. Furthermore, the municipality does not have resources to work in the informal areas. Additionally, several cases were mentioned where politicians had overruled their decisions and let developers build on vulnerable lands. The municipality representatives pointed to the Ministry of Lands and Human Settlements (which is responsible for the new master plan) to take on a larger responsibility. The municipal actors felt they had not been properly involved in the master plan process and hoped that better maps, designations and guidance on managing vulnerable areas would be available (Table 10.1). Also, the Ministry of Local Government and Regional Administration (under the Prime Minister's Office), were pointed at to play a stronger role in coordinating local projects and making sure that flooding was tackled and coordinated with surrounding areas.

The stakeholders investigated ways to proceed in providing better urban services and land management, while resources, plans or guidelines are scarce. Suggestions were to make a 'river plan', 'focus on the most vulnerable areas' and to 'help local communities'. Discussions centred on combining upgrading efforts of the physical structures with livelihood projects and land use management plans. Among the stakeholders it was acknowledged that the NGO-run upgrading programmes may be the most important adaptation efforts in the city.[5] Thus, collaboration with NGOs would be of great importance for the future.

Entry Points for Adaptation Measures

Stakeholders generally agreed on the listed general measures (Table 10.1), even though, for some individuals, it was the first time they had actually discussed climate change and flooding with colleagues and the other stakeholders. However, even for those who regarded climate change as a plausible cause of flooding, they felt that the issue of climate change was too abstract to let real action follow. One comment was that the city should not 'wait around' for projections and the

[5] Projects such as 'Citywide Strategy for Upgrading Unplanned and Unserviced Settlements in Dar es Salaam' and 'Community Infrastructural Upgrading Programme (CIUP)'.

making of grand plans when they actually had many current problems with flooding to tackle. Another comment was that talking of climate change would make it too easy for the politicians to delay action on the problem because of the long time-frame.

The stakeholder sessions did not reveal any adaptation efforts or ready-to-go initiatives to start from, but reinforced that something must be done, particularly in the vulnerable informal areas, and both in terms of improved physical structures and land management. Furthermore, it became obvious that any interventions need to be combined with livelihood projects and involvement in order to make any difference. It seems the local level may provide the only promising opportunity for starting efforts. As community groups organise themselves to fill the infrastructural and service gaps left by centralised institutions, it is easier to anticipate that adaptation could start here. It is difficult, however, to envisage the adaptation of Dar es Salaam being the result of institutionally-driven (top-down) decision-making, as the institutional set-up and governance system currently appear to be preoccupied with more pressing urbanisation problems to allow for a meaningful adaptation. Thus, instead of 'waiting around' for yet another master plan, adaptation could initially take a starting point in 'integrated projects' on a local level to support and build on what is already taking place in the vulnerable informal areas. Such integrated projects would need to combine upgrading, livelihood activities and land management.

Another necessary starting point would be to integrate climate change adaptation and flooding into the master plan and sector plans, although this does currently not seem likely.

Integrated Efforts in Addis Ababa

Who Is in Charge?

In Ethiopia the Federal Environmental Protection Authority (FEPA) has taken the lead in elaborating a climate change adaptation plan. The city of Addis Ababa also drafted a "climate change adaptation action plan" in 2011 (Jørgensen et al. 2012). However, no sector or institution has taken ownership or been given the authority to follow up with integrating climate change adaptation in plans, sectors and projects so far (Herslund et al. 2013). As the city structure plan was being revised during our field work (2013) and climate change was being addressed directly (mitigation) and indirectly (environmental sector) our main focus was to explore the opportunities for adaptation at city level in this revision process.

The previous structure plan from 2002 addressed a number of issues of relevance to create resilience to climate change, such as protecting and rehabilitating the green structure, the provision of sanitation services and waste management to a larger proportion of the citizens, provision of formal housing opportunities, rehabilitation of settlements, and resettlement of people living in vulnerable informal settlements despite that climate change adaptation was not mentioned. However, most of these intentions were never implemented because of insufficient resources, lack of coordination, legal frameworks and lack of stakeholder involvement imposing major barriers to do so (ORAAMP 2002; Desalegn 2013).

Addis Ababa is divided into ten Sub-Cities and 116 Woredas (districts). The legislative body of Addis Ababa is the City Council, and the administrative functions are located in a number of sector-based bureaus and authorities, which have different institutes under them (Hailu 2013). The structure plan revision is organised as a project office under one of these institutes; the Urban Plan and Information Institute (Desalegn 2013), with 18 teams related to different sectors consisting of staff from in- and outside of the city administration. The structure plan project office was established in 2012, with the intention of being dissolved as soon as the city structure plan would be in place. The structure plan has now been prepared and is waiting for political approval (Master plan staff workshop 2013).

After the approval, the Urban Planning Institute is responsible for down-scaling the structure plan to local development and action plans. There are no set guidelines and procedures as to the involvement of local communities in the development of plans and no specific involvement sessions are foreseen. The structure plan stipulates the desired strategic development, whereas the local development plans serve to guide the implementation. The local development plans, however, only offer guidelines to the different implementing departments in the city administration and not specific regulations (Desalegn 2013).

Local Adaptation

The local workshop participants (Fig. 10.3), being representatives of different farming associations, an Evangelical seminar and the metal industry had experienced severe flooding on several occasions over the last years. Flooding reduces the number of harvests during the year and causes erosion which affects farming lands. It has destroyed production material and forced factories to close down for long periods. Flood risk also led to psychological strains as floods occurred as flash floods from which people have barely escaped. Moreover, the workshop participants were concerned with the pollution of the rivers having negative health effects.

The participants did not, however, attribute the increasing flood problems to climate change but to urbanisation and construction, as well as poorly communicated and managed releases from the up-river dams. In response to this the dam manager claimed that releases are properly announced. Because of uncertainty over rainfall, the dam management staff is unable to release smaller amounts of water before the rainy season, because if rainfall is low, Addis Ababa cannot be supplied with drinking water.

The coping strategies to flooding consist of making constructions to protect properties from water, e.g. building small dikes, dumping soil and planting trees as buffers. But individual efforts in some cases have adverse effects on other residents as local dikes increase flooding in other areas, for instance. A coordinated effort

Fig. 10.5 A member from the Akaki urban farmers association is sharing his experiences (Picture: Dorthe Hedensted Lund 2013)

is needed to reduce siltation, minimise blockage of the rivers and make sure there are areas where the river can flow over in a controlled way along the river. Furthermore, there is a need to make more controlled discharges from the dams, or to have buffer capacity in the upper areas to obtain a more steady flow of water, and not to cause flash floods. Also, pollution control needs to be enforced to reduce the negative health effects on people and livestock (Fig. 10.5).

According to the local participants flood protection, such as the establishment of buffers, is the responsibility of the local Environmental Protection Authority, but they do not find that they receive sufficient help. Some have requested to be relocated, but did not receive any response from government. The local participants are convinced that they should have a say in the structure plan revision process as they are the victims of the present development.

Adaptation Measures Needed

From the workshops and interviews (Fig. 10.3) a number of concrete adaptation measures were identified such as conservation of catchment areas, river rehabilitation, provision of formal, dense housing leaving space for green areas for

Table 10.2 Adaptation measures listed by structure plan teams in Addis Ababa

Focus	Measures	Barriers
Water management	Controlled and well communicated dam releases, coordinated effort to avoid siltation of rivers, upstream buffer areas to avoid flash floods	Lack of resources, coordination difficulties, unpredictable amounts of rain
Green areas and rivers	Designate green corridors along rivers (for flooding buffer zones, SUDS, reduce pollution and recreation)	Inhabitants must be resettled
		Poor waste management
		Green area initiatives in last structure plan still not implemented
Housing	Provide formal housing (dense in order to preserve more land for green areas)	Lack of resources
Pollution	Improve pollution regulations	Lack of enforcement

stormwater infiltration, enforcing pollution regulations, etc. Climate change and adaptation were issues considered by the structure plan teams on regional planning, housing, industry and environment who participated in the city session (Fig. 10.3). The participants identified the main challenges of climate change to be urban flooding as well as drought.

In relation to hinting at important measures for climate change adaptation the most progressive sector was the environmental sector which focused on how green areas could contribute to adaptation. A further concern was pollution, which aggravates the negative effects of flooding. The environmental sector presented a number of proposals for the structure plan revision in order to maintain and rehabilitate green areas in the city, focusing on their importance for multiple functions. In that respect sustainable urban drainage systems (SUDS) were also considered to be relevant, as infiltration of rainwater could improve the groundwater supply. A second proposal was rehabilitation of the six major rivers[6] in the city. Green corridors along the rivers would secure green recreational areas near all households, reduce pollution and create flood buffer zones. This, however, would require that informal settlers along the river are resettled, and that waste management is improved to avoid direct release of industrial and domestic pollutants in the rivers. The team thought that leading politicians have an interest in green areas, and that this interest can be further developed. The director of the Addis Ababa Environmental Protection Agency (EPA) agreed with this (Hailu 2013) (Table 10.2).

The housing sector also mentioned some proposals indirectly related to adaptation such as the provision of formal housing opportunities. A reason for informal settlements is the large backlog of formal housing. Furthermore, new housing should increase the density in order to conserve space for open, green areas. Most of the housing in Addis Ababa is presently not connected to a sewerage

[6] Akaki, Bulbula, Bantyketu, Kebena, Tinishu Akaki, Ras Mekonen rivers.

system, but has septic tanks with insufficient capacity to properly maintain and empty. Therefore, huge investments are needed in housing and infrastructure such as sewerage and water supply, both to redirect stormwater and to minimise pollution of receiving water bodies.

The regional planning team and the industry team were mainly concerned with drought and water scarcity, particularly with how the city could preserve water. All in all, the participating revision teams were all very concerned about water management: ensuring water availability and minimising floods, as well as coping with the negative effects of climate change that could be expected in the future.

Entry Points for Adaptation Measures

There is awareness among the Addis Ababa city structure plan team that flooding and drought can be expected to get worse as climate change progresses. To the extent that we were able to scope the situation, there appears to be commitment to deal with this issue, particularly when framed as integrated water management. The local citizens are severely affected by water issues. Also, the different teams in the structure plan revision are aware of water management problems and have a number of proposals. Flooding and adaptation is becoming more important at the Addis Ababa Environmental Protection Agency (AAEPA) as funds are earmarked according to the director, and some of the conservation strategies related to the upper catchments and the river buffers are important activities creating more resilience in Addis Ababa.

Framing adaptation as *water management* is promising as it is a relevant issue in many sectors and among the local people. Benefits from an integrated water management plan would likely improve the resilience of Addis Ababa to climate change, and the sectors present at our workshop would have a role to play in the creation and implementation of such a proposal. An integrated water management plan was thus considered a useful frame by pointing to concrete actions as well as synergies and benefits, such as conservation of water for periods of drought, increased infiltration to reduce the runoff to rivers, decreasing pollution of water resources by means of watershed protection and creating buffer zones along the rivers. Among the different stakeholders, the idea of integrated water management was more relevant than a strategy to deal with climate change, because water management was directly related to problems already being felt. In order to make an integrated water management plan it is important that the relevant and necessary sectors in the structure plan as well as different local stakeholders are working together to coordinate efforts, which may prove to be a challenge. The structure plan revision process was very pressed for time, and so far the teams had only focused on their own areas and no integration had occurred. Furthermore, the lack of human and financial resources, fast human resources turnover, lack of commitment, and insufficient legal frameworks, which have caused implementation failure before, are still present. These barriers will need to be addressed with respect to implementing an integrated water management plan, and framing adaptation as water management does not reduce the complexities of harvesting the possible synergies.

Integrated water management cannot be achieved by any one institution but requires coordination and collaboration among a number of different authorities, which is doubtlessly challenging in the current institutional set-up. The structure plan teams stressed the necessity of *institutionalising* climate change adaptation and water management at both federal and city levels and to create commitment in government to properly address the issue. The city-level participants especially argued that it will be meaningful to form a new institution or to transfer authority to anyone that either promotes collaboration between existing institutions or combines the authority necessary for regulation and implementation. Furthermore, actions would have to be prioritised as capacity is insufficient to take on all of the identified measures. The stakeholders all identified the EPA at different levels to be the obvious institution to take on ownership of integrated water management and thereby adaptation. This, however, calls for some restructuring of the EPA and probably a new institution to support collaboration and integration would be needed. As such, there may be an opportunity to address adaptation as an institutionally led process, anchored in the EPA. But this needs to be combined with local efforts to address the serious problems being experienced.

Discussion

The case studies illustrate that the challenges of climate change adaptation for developing cities are many and severe. As the cities grow quickly, large urban service and management deficits arise. Insufficient solid waste management, drainage and sanitation combined with limited land use management and enforcement make up much of the present problem with flooding, drought and pollution. Thus, whether it is the effect of climate change or it is climate variability, the cities are not resilient.

Searching for Momentum

Climate change adaptation has been dealt with at the national level in the environmental sector of both cities. In neither city has the agenda of climate change adaptation been taken on explicitly. In both cases local vulnerable communities live and deal with flooding, drought and pollution in their everyday life and cope in many of the same ways by digging drainage, raising the ground level, cleaning up waste, etc. The local people manage to cope but find themselves very alone in these efforts. They also feel they should be more involved in planning and decision-making processes being relevant for adaptation to climate change.

At the moment, only limited momentum has developed among stakeholders and in the political system for a broad city-level climate change adaptation strategy. For many stakeholders, it was the first time at the workshops they were confronted with the issue of climate change and had to reflect on adaptation.

In the cities, stakeholders were initially sceptical towards large plans such as an overriding climate change adaptation plan, as it is too abstract and not sufficiently focused on the pressing problems experienced in everyday life of citizens and administrators. In both cities the implementation of large plans like flooding plans, master plans, and similar, have been lacking. However, while cities might not have come so far in the steps towards making their city resilient, they are already experiencing problems that most probably will be aggravated by climate change and there is an awareness of flooding and drought being very serious problems for city development across levels and sectors.

In Addis Ababa a strong interest across city sectors and levels was identified for water management, including provision of potable water, flooding and pollution. These are problems that all stakeholders can easily relate to and consider highly important. Combining local plans and integrated water management could be a good match as local waste management, local drainage solutions and social efforts to help people being resettled or find livelihoods can be combined with water protection, creating conditions for the plans to become a reality. A combination of top-down and bottom-up activities could most probably be a necessary step in Addis Ababa to increase the chance for implementation of their plans.

In Dar es Salaam the institutional landscape seems too fragmented for city-level top-down integrated water management planning at the moment. Here a bottom-up approach in local projects seems to be a more realistic starting point. As in Addis Ababa, local plans can be used to address different vulnerabilities while also integrating local needs for new livelihoods. As mentioned before (see section "Conditions for success") UN-Habitat (2011) proposes local plans as a tool for better regulating land use and building structures within designated vulnerable areas. However, local projects would also have to consist of upgrading basic services, social projects and possibly the promotion of 'climate proof' livelihood activities, like urban agriculture, that also work to enhance the ecosystem services of open areas to find resonance among local inhabitants (Chap. 4).

Scoping the Situation

Synergies and coordination are the main strengths of integrated efforts, however, it is also their weakness in the present urban governance situation of the two case study cities where multi-level and cross-sectoral coordination is limited. In such a governance situation, 'champions' are needed to drive efforts forward. Even if there is resonance with experienced problems, if no real commitment is made from an institution or other actors, any measures are less likely to be implemented.

The city and municipality seem like the 'natural' level for coordination of possible integrated projects. The municipality is the main actor in adaptation as also identified by stakeholders, being responsible for urban services and land use management. However, in Dar es Salaam no obvious city stakeholder is in place to initiate or push for an effort. The 'integrated local projects' would need a national level kick-start, either from the Ministry of Environment, as the only institution explicitly addressing climate change adaptation, or the Ministry of Lands and Human Settlements that coordinates the master plan.

In Addis Ababa, main champions are the CLUVA partner (Ethiopian Institute of Architecture, Building Construction and City Development) and the university staff taking part in the revision of the city structure plan. They are pressing to get climate change and water management integrated into different sections of the plan. A challenge here is how to create ownership in the municipality and coordination between offices afterwards when the project staff that has prepared the structure plan steps out and the municipality offices need to take over. Stakeholders point to the EPA to be the institutional anchorage point, but also that it has to be strengthened to take on this task.

'Climate change adaptation' as a main theme in the structure plan in Addis Ababa does not appear to have become a reality so far. In the form of enhanced focus on water management the topic could become a subject being included in several topics in the structure plan, but it may also end up 'only' being integrated into the environmental section where the CLUVA partner (the university) is driving the effort. In Dar es Salaam, integrating climate change adaptation in the master plan in revision does not seem likely and the plan might not be effective as other city institutions and implementing actors have not been much involved in its development; many of them do not even know of its existence.

A major limitation of the stakeholder interaction we have performed is the failure to include top politicians or decision-makers in our sessions. Champions higher up in the system are necessary to push for a more sustainable city development and frame the broader climate change adaptation agenda in the cities. The initial impression is that the 'leadership' in the different institutions represented by stakeholders in Dar es Salaam and the head of the structure plan team in Addis Ababa are not presently tuned into addressing climate change adaptation. More pressing development issues are currently at stake. Furthermore, the stories told about political pressure and patronage do not raise hope for a strong commitment among politicians either.

Creating Frames – The Pathways Ahead

As the stakeholder sessions show, stakeholders are aware of the situation and feel that flooding is a major problem. They see a need for action to be taken which is a first step in the strategy process of creating momentum for adaptation. But as stakeholders and city institutions are short on resources, time and competences, identifying the problems and needed actions (as the sessions started) is one thing. The difficult part is to prioritise measures. In this chapter the prioritisation is based on which actions are likely to have the largest effects, taking resonance and ability to implement into account. Initially this is not easy as all are pressing problems. Therefore, integrated focused approaches, such as 'integrated water management' in Addis Ababa and 'local plans for vulnerable areas' in Dar es Salaam seem to be frames that can address a combination of problems as well as build on resources, stakeholders and activities already in motion. Framing adaptation efforts under the headline of 'integrated river or water management' in Addis Ababa, or 'helping the vulnerable areas' in Dar es Salaam is thus more likely to gain general momentum and support than a headline adaptation strategy, even though these two strategies may contain some of the same elements. Such initiatives seem to be a more realistic starting point for a 'learning-by-doing' process. Carmin et al. (2012) also found in the South African urban adaptation that headline strategies and visions had to be accompanied by on-going concrete activities and mainstreaming into sector plans in order to generate knowledge. The integrated projects suggested for the case study cities can serve as a tactical process, or as Healey (2009) calls them, 'pragmatic and incremental projects', that can initiate the strategic process and create learning and knowledge, but also collaboration among institutions and stakeholders that at present do not work together. Similarly, Vedeld et al., in Chap. 9 of this volume, refer to this as *'incremental adaptation'* where *'transitional adaptation'* or *'transformation'* involves changes in governance and possibly changes in political as well as cultural values and structures.

To create possible changes in governance bringing together main actors and institutions in initial projects can serve as a starting point for more collaboration across governance levels and sectors. During such a process possible champions could be identified or developed which may serve as a key indicator marking readiness for a transition from incremental approaches in the strategic process to more citywide and coordinated activities and plans. Champions higher up in the system are necessary to push for a more sustainable city development and frame the broader climate change adaptation agenda in the cities in the longer term.

For many stakeholders, it was the first time at the workshops they were confronted with the issue of climate change and had to reflect on adaptation. To continue holding sessions and working towards a more comprehensive and coordinated climate adaptation effort and strategy could be an important tool to inform and make the most important stakeholders aware of the wider problems, while encouraging them to share knowledge and coordinate different concrete efforts. This will have to go on at the same time as the integrated projects take off to provide knowledge and learning to the wider process and develop and strengthen the collaboration among stakeholders.

Conclusion

Urban authorities have a key role to play in strategic adaptation and to make cities more resilient to climate change (UN-Habitat 2011). However, the exploration of ongoing climate change adaptation efforts in the two case study cities, Dar es Salaam and Addis Ababa, reveals a rather weak city-level institutional landscape for planning, with limited resources and competences as well as deficient cross-sectoral and multilevel coordination and interaction with government institutions and NGOs at the core. So the conditions for the cities to develop a transformative climate change adaptation strategy are not favourable at present.

At the moment it does not seem realistic that all the insufficiencies can be dealt with here and now. However, it is more probable that they can be dealt with in a step-wise manner. Thus, an incremental approach is recommended which is addressing the most vulnerable areas first through integrated, focused interventions that address the combination of problems in particular areas or a cross-cutting but 'forgotten' or 'invisible' topic such as water. At the same time efforts to integrate the topics of water management and vulnerability reduction into main city plans like the structure plan in Addis Ababa and the master plan in Dar es Salaam, should be on-going. This, however, requires that there are champions among the higher level politicians and administrations to drive the effort and that the cities are able to coordinate across institutional levels and sectoral boundaries. Following Healey's (2009) notion of strategic planning, scoping for champions and momentum for even parts of the needed measures should be among the first steps for the cities.

The main entry point in the cities of Dar es Salaam and Addis Ababa for getting climate change adaptation into city development is the urgency of flooding problems that both citizens and professionals already face. There is awareness of the problems and of what kind of measures could make a difference. The general recommendations to make the case study cities more resilient to climate change must include both short term integrated projects as well as longer term efforts to integrate climate change adaptation into plans, policies and practices. As the recommendations mainly build on the local and city sessions and the interviews conducted with a selection of stakeholders, they only capture a partial image of the situation in the cities. Therefore, it should be noted that the measures suggested are just a first step, and a further and on-going process of stakeholder involvement is definitely necessary.

Short-Term Integrated projects and plans addressing pressing and multiple problems currently identified in the cities should be commenced. Other relevant on-going projects like upgrading efforts, housing developments, waste management, and green area development should be coordinated and 'climate proofed'.

Focus should be placed on building networks between the stakeholders of such projects, including local people, NGOs, municipality staff, and national level coordinators to create a basis for coordination, awareness raising, learning-by-doing and the development of champions.

In Addis Ababa

- Develop an 'integrated water management' plan for Addis Ababa that addresses flooding, drought and pollution by watershed protection, designation and management of buffer zones and different 'integrated local projects' in the most vulnerable areas that take into account local waste management, drainage solutions and social issues.

In Dar es Salaam

- Initiate 'integrated local projects' in the most vulnerable areas (and resettled areas) that combine urban service upgrading, livelihood projects and local land use management and regulation.

Middle- and Long-Term Climate change adaptation and measures addressing flooding and drought must be dealt with in main city plans and integrated into policies and practices in the core fields of land use planning, management and mapping, urban drainage and water (waste and stormwater), and environmental planning. In addition adaptation must also be integrated into the sectors and activities of solid waste management, housing, informal area upgrading and regularisation, disaster management and resettlement.

Focus should be placed on involving stakeholders across levels. Local people, municipalities and national bodies, and also relevant NGOs and utility companies, should be involved in the integration and strategy-making of adaptation into the different fields and sectors in order to develop knowledge, create awareness, foster ownership and to improve coordination.

On top of this, network building between relevant stakeholders across the listed fields, sectors and projects should be on-going, including politicians and decision-makers, to enhance coordination and finding synergies across fields and working towards a longer-term city adaptation effort or a strategy closely connected to the problems that the cities face.

In order to move from an incremental approach – the short-term projects – towards a more citywide adaptation effort politicians as well as decision-makers must become part of the process to push towards a more sustainable city. However, a city-level adaptation plan does not need to start with an all-encompassing plan to be effective. It can also be started by coordinating the variety of local projects and activities as well as integrating efforts and sector plans. While such a pathway may not capture all conceivable effects which may result from climate change in the long term, it could more likely foster quick action and be transformative rather than an overarching master plan for climate change adaptation. This in turn will generate experiences and learning that can be applied in the continuing process towards adaptation.

The two cases illustrate that there are no one-size-fits-all solutions, when it comes to adaptation. The important measures to take, according to literature and guidelines prepared by development organisations for climate change adaptation, are numerous (Fig. 10.2). This can be overwhelming for a developing city struggling with a multitude of urban problems. Efforts to make a city resilient have to be explicitly and closely connected to the problems the city already faces in order to be relevant for stakeholders and citizens. A developing city is unable to address *all* the issues which are important to create resilience, but it might be able to address some: those that can mobilise a range of stakeholders and their resources, which can create synergy effects, and which resonate with the experienced problems in the cities.

References

Albrechts L (2004) Strategic (spatial) planning reexamined. Environ Plan B 31(5):743–758

Allmendinger P (2009) Planning theory. Palgrave Macmillan/Houndsmills, Bassingstoke

Armitage N, Vice M, Fisher-Jeffes L, Winter K, Spiegel A, Dunstan J (2012) The South African guidelines for sustainable drainage systems. Water Research Commission (WRC), Cape Town. http://www.wrc.org.za/Pages/Preview.aspx?ItemID=10575&FromURL=%2fPages%2fDisplayItem.aspx%3fItemID%3d10575%26FromURL%3d%252fPages%252fDefault.aspx%253fdt%253d%2526ms%253d%2526d%253dThe%2bSouth%2bAfrican%2bGuidelines%2bfor%2bSustainable%2bDrainage%2bSystems%2526start%253d1. Accessed 4 Nov 2013

Balaban O (2012) Climate change and cities: a review on the impacts and policy responses. Metu J Fac Arch 29(1):21–44

Bicknell J, Dodman D, Satterthwaite D (2009) Conclusions: local development and adaptation. In: Bicknell J, Dodman D, Satterthwaite D (eds) Adapting cities to climate change. Understanding and addressing the development challenges. Earthscan, London, pp 359–383

Biggs R, Westley FR, Carpenter SR (2010) Navigating the back loop: fostering social innovation and transformation in ecosystem management. Ecol Soc 15(2):9–33

Bolund P, Hunhammar S (1999) Ecosystem services in urban areas. Ecol Econ 29:293–301

Brown RR, Farrelly MA, Loorbach DA (2013) Actors working the institutions in sustainable transitions: the case of Melbourne's stormwater management. Glob Environ Chang 23:701–718

Bryson JM (1995) Strategic planning for public and nonprofit organizations, revised edn. Jossey-Bass, San Francisco

Carmin J, Roberts D, Anguelovski I (2012) Preparing cities for climate change: early lessons from early adaptors. In: Hoornweg D, Freire M, Lee MJ, Bhada-Tata P, Yuen B (eds) Cities and climate change: responding to an urgent agenda, vol 2. World Bank, Washington, DC, pp 470–501

Davoudi S (2012) Resilience: a bridging concept or a dead end? Plan Theory Pract 13(2):299–307

de Sherbinin A, Castro M, Gemenne F, Cernea MM, Adamo S, Fearnside PM, Krieger G, Lahmani S, Oliver-Smith A, Pankhurst A, Scudder T, Singer B, Tan Y, Wannier G, Boncour P, Ehrhart C, Hugo G, Pandey B, Shi G (2011) Preparing for resettlement associated with climate change. Science 334(6055):456–457

Desalegn W (2013) Interview with urban planner Walelegn Desalegn, Addis Ababa Urban Planning Institute, 7 Mar 2013

Dodman D, Kibona K, Kiluma L (2011) Tomorrow is too late: responding to social and climate vulnerability in Dar es Salaam, Tanzania. Case study prepared for cities and climate change: global report on human settlements. UN-Habitat, Nairobi. Available: http://mirror.unhabitat.org/downloads/docs/GRHS2011/GRHS2011CaseStudyChapter06DaresSalaam.pdf. Accessed 15 Dec 2013

Douglas I, Kurshid A, Maghenda M, McDonnel Y, McLean L, Campbell J (2008) Unjust waters: climate change, flooding and the urban poor in Africa. Environ Urban 20(1):187–205

Fryd O, Bergen Jensen M, Ingvertsen ST, Jeppesen J, Magid J (2010) Doing the first loop of planning for sustainable urban drainage system retrofits: a case study from Odense, Denmark. Urban Water J 7(6):367–378

Hailu HS (2013) Interview with Haileselassie Sebehatu Hailu, General Manager of Addis Ababa City Government Environmental Protection Authority. Addis Ababa Environmental Protection Authority, 8 Mar 2013

Healey P (2007) Urban complexity and spatial strategies – towards a relational planning in our times. Routledge, London

Healey P (2009) In search for the "strategic" in spatial strategy making. Plan Theory Pract 10(4):439–457

Herslund L, Mguni P, Lund DH, Souleymane G, Workneh A, Ouedraogo JB (2012) Exemplary policies, strategies and measures. CLUVA deliverable D3.6. Available: http://www.cluva.eu/deliverables/CLUVA_D3.6.pdf. Accessed 5 Jan 2013

Herslund L, Lund DH, Yeshitela K, Workalemahu L, Kombe W, Kyessi AG (2013) Strategic measures for the two cities. Most important strategies – recommendations. Recommendations for process and prioritisation of data needs. CLUVA deliverable D3.7 and D3.8. Available: http://www.cluva.eu/deliverables/CLUVA_D3.7.pdf. Accessed 15 Dec 2013

Jha A, Bloch R, Lamond J (2012) Cities and flooding. A guide for integrated urban flood risk management for the 21st century. GFDRR/The World Bank, Washington, DC. https://www.gfdrr.org/urbanfloods. Accessed 3 Feb 2013

John R, Mayunga J, Kombe W, Fekade R, Ouedraogo JB, Tchangang R, Ndour NM, Ngom T, Gueye S, Sall F (2012) Preliminary findings on social vulnerability assessment in the selected cities. CLUVA deliverable D5.4. Available: http://www.cluva.eu/deliverables/CLUVA_D5.4.pdf. Accessed 5 Sept 2012

Jørgensen G, Herslund L, Sarr C, Nakouye N, Sine A, Workneh A, Workalemahu L, Bekele E (2012) Base line scenarios for urban development of selected case study areas. CLUVA deliverable D3.5. Available: http://www.cluva.eu/deliverables/CLUVA_D3.5.pdf. Accessed 5 Jan 2013

Jørgensen G, Herslund LB, Lund DH, Workneh A, Kombe W, Gueye S (2014) Climate change adaptation in urban planning in African cities: the CLUVA project. In: Gasparini P, Manfredi G, Asprone D (eds) Resilience and sustainability in relation to natural disasters: a challenge for future cities, SpringerBriefs in Earth sciences. Springer, Cham/Heidelberg/New York/Dordrecht/London, pp 25–37

Kok M, Metz B, Verhagen J, Van Rooijen S (2008) Integrating development and climate policies: national and international benefits. Clim Policy 8(2):103–118

Kombe W (2005) Land use dynamics in peri-urban areas and their implications on the urban growth and form: the case of Dar es Salaam, Tanzania. Habitat Int 29:113–135

Leck H, Simon D (2013) Fostering multiscalar collaboration and co-operation for effective governance of climate change adaptation. Urban Stud, online: 1(18):1221–1238

Lewis M (2009) Climate change adaptation planning for a resilient city. Durban's municipal climate protection programme. Environmental Planning & Climate Protection Department, Durban. http://www.durban.gov.za/City_Services/development_planning_management/environmental_planning_climate_protection/Pages/default.aspx. Accessed 3 Mar 2011

Lund DH, Sehested K, Hellesen T, Nellemann V (2012) Climate change adaptation in Denmark: enhancement through collaboration and metagovernance. Local Environ Int J Justice Sustain 17(6–7):613–628

Mafuta C, Formo R, Nellemann C, Li F (2011) Green hills, blue cities. An ecosystem approach to water resources management for African cities. GRID Arendal/UN-Habitat. http://www.zaragoza.es/contenidos/medioambiente/onu/348-eng.pdf. Accessed 10 July 2012

Mbarouk M, Mamuya L (2013) Interview with head of solid waste management (Mbarouk) and his secretary (Mamuya) in Temeke municipality. Temeke Municipality solid waste management office, 14 Mar 2013

Mguni P, Herslund L, Jensen MB (2014) Green infrastructure for flood risk management in Dar es Salaam and Copenhagen: exploring the potential for transitions towards sustainable urban water management. Water Policy (in press). doi: 10.2166/wp.2014.047

National Bureau of Statistics, Ministry of Finance Dar es Salaam and Office of Chief Government Statistician Zanzibar (2013) 2012 population and housing census, Dar es Salaam, Tanzania. Available online: http://www.meac.go.tz/sites/default/files/Statistics/TanzaniaPopulationCensus2012.pdf. Accessed 7 Sept 2014

Nickel D, Schönfelder W, Medearis D, Dolowitz DP, Keeley M, Shuster W (2013) German experience in managing stormwater with green infrastructure. J Environ Plan Manag 1(21):403–423

Nyed PK, Herslund L (2013) Map of high-risk areas for selected case areas. Towards a holistic perspective on city-level vulnerability assessment. CLUVA deliverable D3.4. Available: http://www.cluva.eu/deliverables/CLUVA_D3.4.pdf. Accessed 30 Aug 2014

ORAAMP (2002) Addis Ababa development plan, executive summary report. Office of the Revision of Addis Ababa Master Plan (ORAAMP)

Parkinson J, Mark O (2005) Urban stormwater in developing countries. IWA Publishing, London/Seattle

Parkinson J, Tayler K, Mark O (2007) Planning and design of urban drainage systems in informal settlements in developing countries. Urban Water J 4(3):137–149

Roberts D (2008) Thinking globally, acting locally – institutionalizing climate change at the local government level in Durban, South Africa. Environ Urban 20(2):521–537

Roberts D (2010) Prioritizing climate change adaptation and local level resilience in Durban, South Africa. Environ Urban 22(2):397–413

Roberts D, Boon R, Diederichs N, Douwes E, Govender N, McInnes A, McLean C, O'Donoghue S, Spires M (2012) Exploring ecosystem-based adaptation in Durban, South Africa: "learning-by-doing" at the local government coal face. Environ Urban 24(1):167–195

Salet W, Faludi A (2000) The revival of strategic spatial planning. Royal Netherlands Academy for Arts and Sciences, Amsterdam

Satterthwaite D, Huq S, Reid H, Pelling M, Lankao PR (2009) Adapting to climate change in urban areas: the possibilities and constraints in low- and middle-income nations. In: Bicknell J, Dodman D, Satterthwaite D (eds) Adapting cities to climate change. Understanding and addressing the development challenges. Earthscan, London, pp 3–47

Simon D (2013) Climate and environmental change and the potential for greening African cities. Local Econ 28(2):203–217

START (2011) Urban poverty & climate change in Dar Es Salaam, Tanzania: a case study. Washington DC, START secretariat (global change SysTem for Analysis, Research and Training). http://start.org/download/2011/dar-case-study.pdf. Accessed 2 Dec 2012

Taylor A, Cocklin C, Brown R (2012) Fostering environmental champions: a process to build their capacity to drive change. J Environ Manag 98:84–97

Tucci CE (2007) Urban flood management. World Meteorological Organization. http://www.gwp.org/Global/GWP-SAm_Files/Publicaciones/GestiondeInundaciones/Gestion-de-inundaciones-urbanas-ing.pdf. Accessed 19 Mar 2012

UN-Habitat (2011) Planning for climate change: a strategic, value-based approach for urban planners. UN-Habitat/Ecoplan International, Inc. http://mirror.unhabitat.org/downloads/docs/GRHS2011/GRHS2011CaseStudyChapter06DaresSalaam.pdf. Accessed 19 Mar 2012

UN-Habitat Ethiopia (2008) Addis Ababa urban profile. United Nations Human Settlement Programme, Nairobi

UNIDSR (2012) The 10 essentials for making cities resilient. http://www.buildresilience.org/2013/index.php/unisdr-making-cities-resilient-campaign. Accessed 5 Aug 2013

Vedeld T, Kombe W, Coly A, Ndour NM, Kweka-Msale C, Hellevik S (2013) Recommendation of how climate change can be better integrated in the planning and government system. CLUVA deliverable D3.2. Available: http://www.cluva.eu/deliverables/CLUVA_D3.2.pdf. Accessed 13 Dec 2013

Watson V (2009) The planned city sweeps the poor away: urban planning and 21st century urbanization. Prog Plan 72:151–193

WRGC (2013) Water resilient green cities in Africa. Danida funded research project in Dar es Salaam and Addis Ababa. http://ign.ku.dk/english/research/landscape-architecture-planning/landscape-technology/water-green-africa/. Accessed 5 Jan 2014

Chapter 11
Fostering Transformative Climate Adaptation and Mitigation in the African City: Opportunities and Constraints of Urban Planning

Susan Parnell

Abstract Mainstreaming climate resilient strategy into the systems of urban design, construction and management has to take seriously both climate adaptation and mitigation in shifting the practices of urban planning if the interests of the urban poor are to be advanced. Because of the way that urban poverty has been conceived, the adaptation agenda tends to focus on small-scale household interventions rather than strategic spatial planning, development controls and enforcement. These city scale planning actions are more closely tied to the climate mitigation agenda, something that has had little traction in African cities where planning is weak and often considered part of the urban problem. Few professions have such a poor reputation or are so badly understood as town planning, and this is nowhere more so than in the fragile African context where illegitimate colonial legacies, weak local government and low levels of professional capacity make embedding the climate agenda into the planning regime especially difficult. However, pro-poor planning and planners cannot be bypassed if a sustainable city, rather than a set of projects, is to be promoted. In an effort to make clear the barriers and opportunities to a transformative climate agenda, this chapter sets out the importance of rethinking poverty in less individualised ways, thus enabling the reform of urban planning practices that are typically found in African cites and through which institutional change might be realised.

Keywords Poverty • Urban planning • City-scale action

S. Parnell (✉)
Department of Environmental and Geographical Sciences and African Centre for Cities, University of Cape Town, UCT, P Bag, 3 Rondebosch, Cape Town 7701, South Africa
e-mail: susan.parnell@uct.ac.za

S. Pauleit et al. (eds.), *Urban Vulnerability and Climate Change in Africa*, Future City 4,
DOI 10.1007/978-3-319-03982-4_11

Introduction

There has been something of a turnaround in attitudes, with growing acceptance of the political and intellectual importance that should be ascribed to cities. Papers that begin by asserting the centrality of the urban question in Africa, that point out that climate adaptation and mitigation are real threats to African urban integrity and vitality and, even that urban planning in Africa needs to be revitalised, talk to the converted as these are now, apparently, accepted points of departure for both scholars and policy makers (Watson and Agbola 2013; UN-Habitat 2014; Parnell and Pieterse 2014). Obviously, giving greater attention to urban questions is good news, but moving on from this small victory opens up the vast frontiers of defining what it will take to change the developmental pathways of cities that are currently dysfunctional and exposed to climate risk as well as those urban places that will come into being over the next part of the twenty-first century (see also Chaps. 1 and 5). It may well be that this kind of reflective exercise requires both normative and historical reflection (Winkler 2009, 2011), as well as the kind of sector specific engagement on new urban challenges, like climate change, that the contributions in this book present.

In the African context, making the case for 'the urban' (Pieterse 2008; Myers 2011), for city scale climate adaptation and mitigation action (Cartwright et al. 2012; Simon and Leck 2014) and, especially for planning (UN-Habitat 2009; Berrisford 2014), has not been easy. Anyone with grounded experience of Africa also knows that formal recognition of the central role of cities in addressing big challenges, such as climate mitigation/adaptation, land management or biodiversity, recognises that this is a fragile moment of acceptance (Napier et al. 2013; Elmqvist et al. 2013). Africa has long been characterised by strong anti-urban sentiments and its planning regimes are widely seen as dysfunctional or illegitimate; it is not uncommon for planning itself to be tagged as the cause of the problem (Watson 2009; Winkler and Duminy 2014; Berrisford 2014). Taking steps, politically and administratively, to implement the new planning processes that are essential in developing robust and resilient cities will require at least the equivalent level of drive as that already taken over the last two decades to make the case for the dramatic policy shifts already seen.

The purpose of this chapter is to affirm that a revival of urban planning lies at the core of the developmental agenda of Africa. Moreover, it proposes that the future of the African city rests on climate appropriate planning.[1] To make progress on this vision, however, I argue that climate mitigation and adaptation must jointly form the foundation of a resurgent planning vision and practice. Furthermore, planning will never achieve resilience in urban African contexts without pragmatic engagement with the realities of the continent – key in this regard are the issues of poverty

[1] An early version of ideas on which this paper draws were prepared for MAPS. The author would like to acknowledge the comments of Emily Tyler and the ongoing support of the National Research Foundation.

and inequality.[2] Thus the core argument of this chapter is that the treatment of climate change (mitigation and adaptation) and poverty cannot be separated in any way, conceptually or operationally, in the reshaping of African cities. The task of the chapter is to explore barriers to this integration.

While the focus in this chapter is on the role of planning in the transformation of the African city, this should not be read as placing undue trust in the ability of one profession or of local government (where planning is typically hosted). Rebuilding planning capacities and the redefining of planning goals and processes must be considered alongside a raft of other improvements to cities, including the allocation of funds and powers from national governments and related impacts on city management, the rights of citizens to mobilise and organise their communities, the reconfiguration of global trade and investment protocols, and the end to violence and conflict. However, in setting out such a utopian course of action we need to be alert to possible barriers that will, intentionally or unintentionally, preclude the effectiveness of the call for planning based interventions for resilient cities. In scoping out likely obstacles, the first task is to remind us why, even under conditions of extreme poverty, such as exist in many African cities, climate mitigation (not adaptation) cannot be ignored or overlooked. Second, it is useful to review the reasons why, to date, climate adaptation has tended to dominate climate mitigation in the articulation of anti-poverty strategies as this distortion is likely to carry over to planning reforms. Third, it is possible to explain the tardiness in taking up climate mitigation within the poverty agenda by demonstrating that the focus on micro-scale dynamics which has dominated poverty thinking over the period of climate change mobilisation has not translated readily into city-wide planning reforms. Implied in this analysis is the view that, for mitigation and adaptation to become a more central feature of anti-poverty thinking in African cites, some theoretical adjustment in the understanding of poverty is needed, allowing more space for the crucial role of government, in general, and planners, in particular. To realise the revival of state-led (if not state controlled or delivered) urban governance and service delivery, specific tools and instruments of urban planning (spatial and land use planning, standards for services and buildings, byelaws, zoning etc.) will need to become a more prominent object of climate scientists' attention.

Taken together, the arguments of the first three parts of the chapter suggest that the failure to adequately link poverty, climate (adaptation and mitigation) and planning, while understandable intellectually and politically, has undermined the impact and effectiveness of building climate resilient cities across Africa. Unpacking the conflicting rationalities of these developmental discourses is key to making practical changes in African cities. The arguments proposed imply that planning should be embraced as a potentially progressive instrument of urban change, but the affirmation of planning must come with caveats that include:

[2] Although the focus of this chapter is on poverty it may well be that the unchecked expansion of wealth in the African city is the more important barrier to climate resilience.

foregrounding the centrality of public good and universal values in professional practice; acknowledging that climate mitigation and adaptation are inseparable priorities for urban planning in Africa; recognition that the initial phase of climate awareness when cities introduced climate adaptation plans must be followed up with the integration/mainstreaming of innovations into everyday urban practice through a robust review of the planning regimes of African cities; commitment to the view that climate resilient cities will necessarily need to extend and adjust every aspect of planning practice, requiring reform of everything, including: the relationship between the administrative and political structures of local government through which planning operates; strategic planning goals; the norms and standards of design; engagement with vulnerable communities and enforcement practices; and the critical issue of urban finance. Accepting the imperative to use planning also implies major reform of professional practice and training. Such substantive reforms to the urban planning system will not, however, be possible without some intellectual ground-clearing in development and poverty thinking. This leads us to focus our attention on the currently confused relationship between climate and poverty.

Why Climate Mitigation Cannot Be Marginalised from a Poverty Agenda

The social and economic costs of climate impacts are most often borne by low-income groups (Moser and Satterthwaite 2008; Tyler and Moench 2012). Thus, a widespread hope is that the imperative to shift the current developmental path and make human settlements more sustainable, for example through the United Nations' Sustainable Development Goals (SDG) process, will gain momentum from the fight for climate change action, with mitigation as a central message to inspire the better management of the built and natural environment (Revi 2009; Hodson and Marvin 2010). Conversely, it is clear that climate action, especially that associated with mitigation, will never gain traction in the global south unless the transformations proposed unambiguously advance a social, not only environmental, agenda. Therefore, the ability to articulate the developmental advantages of climate action is key (Parnell et al. 2007; Simon and Leck 2013).

The potential that would come from making mitigation a more explicit part of an anti-poverty struggle could, if realised, be very powerful politically (Bulkeley and Betsill 2013). But the mobilising potential of linking climate mitigation and poverty in a translational research agenda is not the only reason that understanding the relationship, or interface, between the two is important. Other imperatives for a closer integration of the languages and practices of poverty relief and climate mitigation have to do with the global shift in world population, the urbanisation of poverty and the false dichotomy between adaptation and mitigation in climate-related city government often experienced by poor people.

Population, Poverty and Mitigation

Demography matters in the debate about the relative emphasis on mitigation and adaptation because of the role of population size and the global distribution of poverty. Expanding populations obviously put more people at risk from climate related events, but even more important are shifts in where people live, how people live and how the settlement systems in which they reside are designed, built and managed (Martine et al. 2008). Retrofitting past structures, and better planning and design of current and future human settlement can have huge implications for overall levels of global emissions (Blanco et al. 2009; IPCC 2012). Of central importance for the mitigation agenda is the macro picture of population change – specifically, the growing population concentration in the global south, especially in Asia and Africa, and the shift in population into cities (Fig. 11.1). The net increase in urban land use cover is the standard way that the urban population issue is expressed for mitigation studies (Seto et al. 2012), but the overall trend towards the physical expansion of urban areas masks the importance of the city-specific dynamics of social and spatial differentiation, where intra-city inequality is a crucial element driving mitigation that, once aggregated, also has global impact.

The absolute expansion in the number of people contributing to greenhouse gas emissions is a small, if important, part of the global environmental change story (IPCC 2012). Historically, poor people, because they consume little and have much lower carbon footprints than the international elite and industry, have been dismissed as having any significant role in the feedbacks between human settlement and global warming (Satterthwaite et al. 2009). Although the impact of the poor on overall emissions may be less problematic than that of the rich does not mean their lives, lifestyles and built environments should be discounted in mitigation scenarios, not least because the (expanding) scale of the poor population makes them, and their potentially wealthier offspring, an important driver of change. Mitigation action that is universal, not solely focussed on the current elite, is an obvious precaution in securing lower rates of emission increases.

Urbanisation of Poverty

In the twenty-first century poverty is no longer a predominantly rural phenomenon (Martine et al. 2008; Satterthwaite and Mitlin 2013; Turok and McGranahan 2013). Even noting that income figures horribly underestimate urban poverty, the rates of increase in cash based poverty are higher in urban areas than they are in rural areas (Fig. 11.2). Urbanisation and natural urban growth, especially the high growth rates of poor cities, means that increasingly large numbers of the poor partake in emission generating activities, such as eating food sourced far away from their homes and undertaking long motorised journeys to work. Moreover, predictions of the doubling of the urban population over the next 30 years (United Nations 2012) imply that cities in the global south, which are now only half-built, will become a more significant driver of global change in years to come. The design of these cities and materials used in their construction are key to future emission scenarios. Even under current conditions, much more could be done by way of mitigation in the cities of the south. While it is true that the poorest of the poor live in structures sourced from recycled materials and natural resources, these low carbon buildings

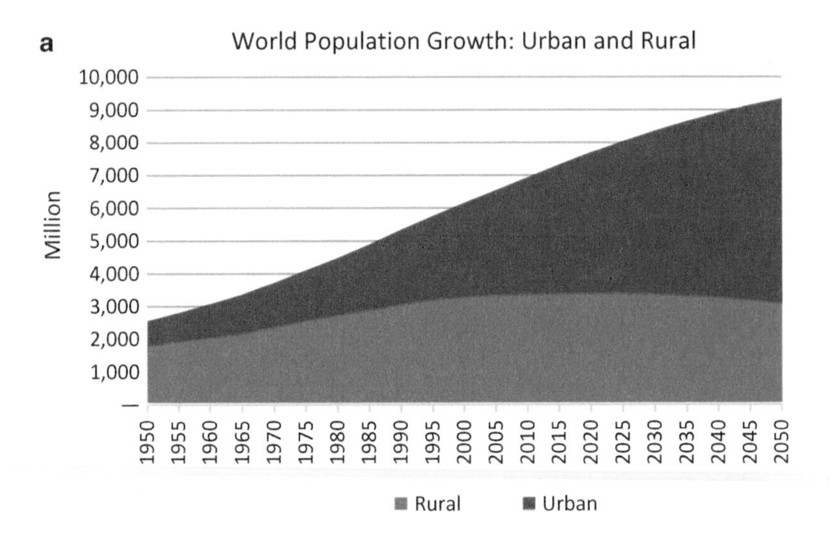

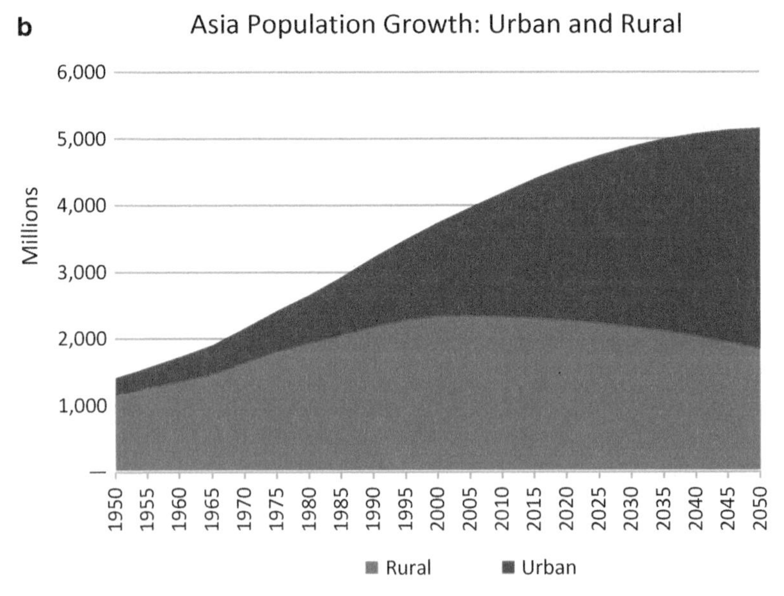

Fig. 11.1 (**a–c**) Urban population growth through 2050 showing the dominance of the projected contributions of Asia and Africa relative to that of the world (Compiled by the author from data in UN 2012)

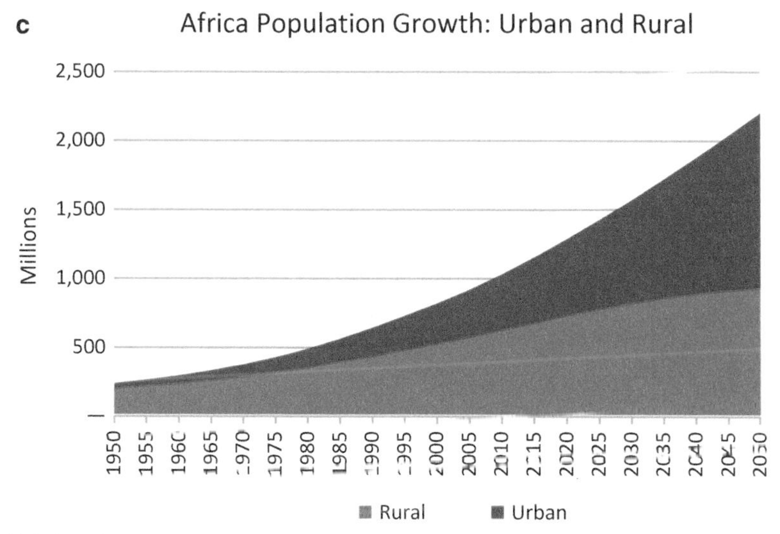

Fig. 11.1 (continued)

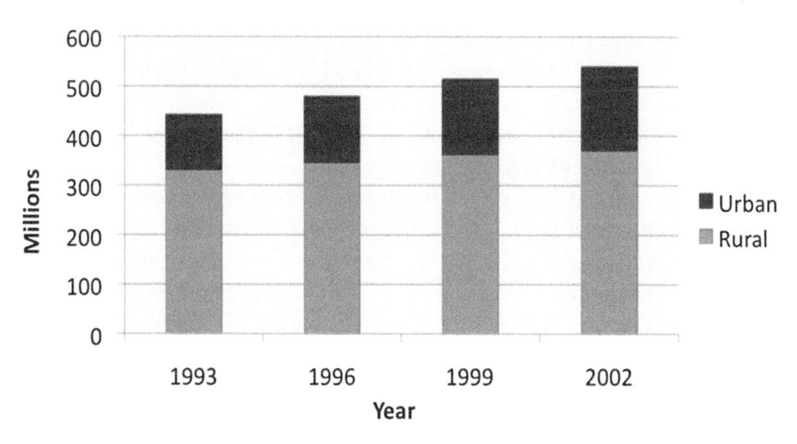

Fig. 11.2 Rates of urban and rural poverty increases (Compiled by the author from data in UN-Habitat 2003)

are not the norm for the urban poor, who are typically renters in low quality, low efficiency buildings developed privately or as low-cost state housing. Energy efficiency is not necessarily the hallmark of the neighbourhoods of the urban poor even when major programmes of retrofitting occur (Lewis and Jooste 2012; Swilling and Annecke 2012; Silver 2014). Without more effective urban planning regimes, the introduction of comprehensive climate mitigation action is a pipedream.

Climate innovative cities of the global north, like London or New York, have demonstrated that the management of cities can enhance or detract from meeting climate mitigation targets (Rosenzweig et al. 2011). However, it is widely acknowledged that effecting low carbon transitions requires effective and innovative governance, including at the city scale (Hunt and Watkiss 2011), something few nations or cities in the global south can deliver – at least in the near future (Anguelovski and Carmin 2011; Carmin et al. 2012; Cartwright et al. 2012). If the majority of city dwellers in poor countries were poor, low-level energy consumers, climate mitigation would be less important than it is. The rise of a new urban middle class in the emerging economies is much lauded by economists. While some debate takes place over the classification of rich or poor for these, generally urban, residents (ADB 2011; Turok 2013; Pieterse 2014), it is clear that there is not only the urbanisation of poverty, with its associated expansion in energy (albeit small amounts), but also the concentration of a new middle class in nations and cities that remain poor overall.

These demographic and spatial transitions imply a dramatic expansion of the absolute and relative proportion of the world's population living in places where mitigation is not a central concern and where cities are not managed well enough to introduce mitigation measures that rest on effective state or organised private sector and civil society capacity. It also highlights that little attention has been given to what happens when, as is widely desired and expressly targeted in the MDGs, millions of people are lifted out of poverty. What will be the appropriate climate mitigation response, given the legitimate increase in resource consumption by those escaping poverty? This is not to imply that a transition from adaptation to mitigation is needed as GDP increases, poverty falls or urbanisation rises. Rather, it implies that the agendas of poverty and mitigation will need to co-exist as part of national and sub-national climate resilient responses forged in response to poverty and inequality. The distribution of poverty, especially urban poverty, is such that, although poor countries and cities in poor countries experience the deepest rates of poverty, the actual numbers of the poor are greater in middle income nations – where the urban middle classes of the BRICS and other nations are flourishing (Martine and McGranahan 2014).

Mitigation and Adaptation Are Not Divisible in Shaping Vulnerability

The theoretical co-existence and interdependence of urban climate mitigation and adaptation action is now widely acknowledged in well-resourced and well-functioning institutions that have money and strong human capacity even when this entails cooperative governance (Rosenzweig et al. 2011). But where planning is weak and governance is fractured (either internally between scales or sectorally between departments, or even between the state and other powerful players such as chiefs, and informal or illegal service providers), operational and institutional tension over the policy direction followed by the city occurs and climate action is held up by the conflicting rationalities of different development imperatives (Parnell et al. 2007; Chap. 9). In cities with large poor populations, these latter conditions are the norm (Bicknell et al. 2009). Hodson and Marvin (2013) warn that, even in well-resourced cities, the fragmentation of urban governance means that mitigation-based ideas like green growth are being used to marginalise issues of justice and promote the interests of the elite, a pattern that is undoubtedly more true in fast growing middle income cities where the private sector drives the bulk of urban expansion, often in edge cities under the rubric of green growth that utterly excludes the poorest (Watson 2013). Clearly, planning failure is as much a part of mitigation failure as it is of adaptation failure. For example, the Asian Development Bank (2010) points out that there is evidence of a systemic failure in the delivery of formal services to many urban poor communities. To some extent this can be attributed to the limited resources of local authorities, but is also due to official aversion to being seen to 'legitimise' settlers' claims by providing utility connections to dwellings that are considered temporary. Under these conditions, neither mitigation nor adaptation can flourish.

Integrating mitigation and adaptation in cities of the global south is hard precisely because these are places where a single or integrated system of urban management rarely exists. Because city governments are unable to deal with large poor populations the governance regimes of rich and poor residents are typically institutionally (legally and fiscally) separated in ways that suit the rich, but expose the poor. This is a point made repeatedly by vulnerability scholars (Pelling 2010; Ziervogel and Parnell 2012), but urban fragmentation is also of consequence to city scale mitigation efforts that must not only synthesise different modes of infrastructure and service provision, but also embed these in law and in the service delivery norms of all residents (Graham and Marvin 2001).

Integrating climate programmes, whether focused on adaptation or mitigation, into the core business of city government requires major institutional reform and a substantial expansion of capacity in order to execute the new reformed agendas. Because different sectors and activities lend themselves to adaptation (usually land use and disaster risk response) and others to mitigation (typically led through energy or transport) it means that internal institutional completion in municipalities (and other spheres of government) prevails. A more holistic view could ensure that good adaptation is also effective mitigation and that both efforts are strongly directed at positively reducing short and long-term poverty through the restructuring of the city form and function.

The lack of co-ordination in climate related mitigation and adaptation activity in cities is also attributable to the fact that, as new issues without obvious professional champions or institutional drivers, they are rarely part of the core business of local government budget allocations (Cartwright et al. 2012). In the poorest contexts, climate mitigation and adaptation programmes may be separate externally funded programmes. This programme format makes it much harder to move from policy level to implementation, especially in ways that are durable over time and scalable across the city or system of cities. The difficulty of integrating climate mitigation into the machine of managing a city more developmentally is compounded by the often-naive assumption that climate mitigation and developmental agendas are easily fused. As we will see in later sections there are conflicting rationalities within urban development that have bearing on the interface between mitigation and development discourses. Before unpacking the underlying assumptions of the poverty literature that play out in climate mitigation efforts, it is useful to briefly reflect on why adaptation rather than mitigation has been the preferred focus of development advocates.

Mitigation, the Cinderella of Poverty Studies

If, as the previous section argued, it is so important that climate mitigation becomes an acknowledged pillar of anti-poverty sustainable development thinking, especially at the urban scale, why is it that poverty related climate action nearly always appears to be couched in the language and focus of adaptation thinking? The reasons mitigation failed to gain traction as an anti-poverty tool can be summarised as the outcome of the political vanguardism of climate adaptation proponents, the rapid emergence of a strong anti-rural lobby within climate science, donors' insistence on focussing climate action only on low income contexts and national governments' reluctance to embrace urban areas where mitigation concerns are concentrated. Together these forces delayed the early articulation of a link between poverty and climate mitigation everywhere, but especially in the global south.

Across the global south (including for urban areas) lobbying, research and funding for climate adaptation overshadows mitigation by a considerable margin (Simon 2010; Simon and Leck 2013). This is no coincidence. Led by powerful voices familiar with the global politics of development (with contributions from, for example, leaders from BRAC, IIED and the Rockefeller Foundation), a strategic choice was made in early phases of climate science and mobilisation to differentiate mitigation and adaptation imperatives and to push for climate adaptation as a priority.[3] In essence the view was that, for the global south, an adaptation (not mitigation) focus would do more to realise poverty reduction under conditions of global climate change (Bicknell et al. 2009). There was also a view that vulnerabilities tended to be focused on the larger extreme events, with smaller 'risk-accumulation processes' often overlooked as developmental rather than climate-related issues. This concern not only fed an adaptation focus but also underscored adoption of the language of global environmental change, not just climate change, in highlighting the connections to social justice (Parnell et al. 2007; Simon and Leck 2013).

[3] By the time an urban climate change meeting was convened at Bellagio in 2007 this position was, if not fixed, clearly well entrenched across donors, scholars and activists.

The (informal but influential) positions were no doubt made with the best intentions and were premised on the view that making mitigation the responsibilities of northern nations allowed for the rightful apportion of blame for human induced climate change. Also, stressing southern adaptation imperatives left southern parties free to focus on mobilising as beneficiaries from the potentially large climate adaptation fund (at one time thought to be larger than all donor assistance). Finally, stressing adaptation and downplaying mitigation facilitated the positioning of poor people to take adaptive action to protect themselves against predicted climate related hazards. This was seen as useful, especially in the context of weak and ineffective states whose capacity to regulate and enforce planning across the city scale in the way that is necessary to drive mitigation (which typically requires strong enforcement or significant incentives from a vigilant authority) was unrealistic.

In the early 1990s, as thought leaders galvanised their communities to respond to the science of climate change that was being assessed and informed by the IPCC, important decisions on priorities were made. The first wave of climate responses, both mitigation and adaptation, were heavily skewed to rural areas (especially agriculture and forestation), with powerful agenda setting organisations, like START, ignoring and even being hostile to urban dimensions of climate change.[4] Big science, the home of mitigation research, was thus very slow to pick up on urban climate research. A simple tracking of the dates of urban and rural focussed climate mitigation research would confirm the belated recognition of the problems of cities. Such a scan of the published literature would also confirm the mismatch in research on poverty and mitigation at the city scale. Even the 2004 launch of the Urban Global Environmental Change (UGEC) Programme failed to fundamentally shift the early tangential trajectories of climate and poverty work, though the UGEC schema of urban feedbacks (Sánchez-Rodríguez et al. 2005) did much to reintegrate global change thinking. This positioning of cities within a planetary frame is now commonplace, even in urban studies circles not usually concerned with environmental resources and flows (Brenner et al. 2011).

So, in poverty circles, mitigation lost out initially because of the overt focus given to adaptation and cities lost out to climate mitigators' emphasis on forestation and rural land use cover issues. Central to the rural subsistence push was the (mistaken then and even more so now) view that poverty was concentrated in rural areas[5] and the also widely challenged view that subsistence held the key to enhanced rural livelihoods.

[4] Evidence of this pro-rural emphasis is found in the programme design of START's climate training throughout the 1990s and 2000s. Only recently have START actively engaged the urban agenda's relevance to climate (see 2013 their Durban meeting).

[5] See extensive discussions of the under-recognition of urban poverty in UN-Habitat 2009; Satterthwaite and Mitlin 2013 etc.

The rapidly growing cities of the emerging economies also failed to attract the attention of global science because of the underlying view, widely pushed by donors, that poverty was primarily a phenomenon of low income nations. As a result the realities of middle income nations, where higher rates of consumption and the availability of greater resources for infrastructure meant that mitigation was a more significant issue than in ultra-low consumption, infrastructure depleted contexts, were largely disregarded in the pro-adaptation design of climate related donor assistance and popular education. A similar pattern of urban mitigation neglect stemmed from the tendency of national governments to drive mitigation through large efforts, such as the South African Long Term Mitigation Scenario (LTMS) process, or major renewable energy programmes. Large, expensive and ambitious programmes such as these are normally the preserve of national departments and the treasury, and the devolution of power to make effective changes at the city scale are rare. The innovative city scale downscaling of the LTMS and other energy based mitigation thinking has been largely donor funded and NGO driven (Lewis and Jooste 2012). South Africa is, in fact, very unusual in the global south in that a major debate is underway over the reallocation of powers and functions relating to the built environment, with devolution to the city scale that might enable significant local government action for mitigation. In poor countries and countries with high levels of urban poverty it is more common for the control over the institutional and fiscal levers of mitigation to be centralised. With that in mind, it is tempting to imagine that the blocks to a better climate/poverty interface lie on the climate science and policy end of the spectrum. In fact, until recently, conceptual barriers within the poverty community have conflicted with the adoption of mitigation as a pro-poor activity area.

How Dominant Perspectives on Poverty Militate Against Introducing a Mitigation Focus to Climate Change Research

Thus far it has been described why the climate community failed to engage a key problem of mitigation in poor cities. What may already be apparent is that the adoption of a new agenda like climate mitigation also depended on having a receptive host. In fact, the particular configuration of the poverty literature at the time of the late 1990s proved hostile to the structural interventions that climate mitigation implied and receptive to the more individualised and community focussed interventions proposed by adaptation science. The overview that follows provides an outline of the shifts in the thinking about urban poverty, highlighting possible reasons why climate mitigation was not readily embraced (for a comprehensive overview of the evolution of poverty studies, see Satterthwaite and Mitlin 2013; Mitlin and Satterthwaite 2013).

Trends in Poverty Research of Relevance to the Adoption of Climate Mitigation

The field of poverty studies is deeply divided with many different ideological and disciplinary traditions of approaching poverty. Poverty specialists are interested in whether poverty is absolute or relative, and if so, at what level. There is also keen interest in understanding the shifting patterns of poverty – leading to a major focus on chronic versus transient poverty. For our purposes of looking at poverty studies as a potential host for climate mitigation, the most important area of debate in poverty studies relates to shifts in the core definition of the problem. In short, before the 1980s the dominant view was that of economists who looked at lack of income. For reasons of data availability and professional influence, these income-based methods of poverty assessment prevail, notwithstanding extensive critique of the narrowness of economistic studies of poverty. Indeed, it could be argued that the tendency to ignore mitigation in the global south comes out of an economistic view that those who spend little money have no significant global impact.

The second most powerful influence on poverty interventions came with engineers and bankers who, working largely with donors and the big multilateral agencies, focussed on access to basic needs and infrastructure. Subsequent rejection of the focus on infrastructure for basic needs had as much to do with the politics of lending conditionality as structural adjustment programmes (SAPs) in Africa, Latin America and Asia were widely critiqued. In the process, the idea that basic infrastructure provision is a core anti-poverty activity lost its status for many decades (for an excellent overview of the different approaches to poverty see Wratten 1995). The resurgent interest in infrastructure and poverty across development circles (Pieterse and Hyman 2014) is now much more closely tied to resource debates through green infrastructure and financing and this offers, perhaps, one of the most exciting opportunities to fuse the climate mitigation and poverty reduction debates (see also Chap. 4 of this volume).

Throughout the 1990s and early 2000s both economic and infrastructure views were displaced as the leading edge of poverty studies, led by a group of social perspectives that drew inspiration from Nobel prize winner Sen (1999) and influential feminists like Moser (1998). Ideas of capabilities and assets provided the starting point for a livelihood approach to development. The focus is thus on harnessing the abilities and resources of the poor themselves and their relationships within a household and community. These micro-level relationships enable the poor to access resources, including natural assets, and provide the lens through which they relate to their wider neighbourhoods, the broader economy and natural systems. In the absence of adequate employment or social safety nets, the livelihood perspective has gained considerable attention as a means of reducing poverty by assisting the poor to build on their own assets, either through using or abusing the natural system. It also provided an accommodating intellectual frame for climate adaptation research, which flourished using the essential elements of the assets and livelihoods perspective (Bicknell et al. 2009; Frayne et al. 2012).

Livelihoods research has been criticised for being apolitical, for making the poor responsible for their own poverty and for ignoring the potential role of the state (Parnell and Robinson 2012). Two immediate intellectual alternatives to the livelihood work include: (1) a focus on the developmental state and; (2) foregrounding the long established ideas of environmental justice. The notion of environmental justice, or the mobilisation for social justice in environmental matters, refers to the wellbeing and rights of past and future generations. It is a movement that emphasises the overarching structures of inequality, for example in water rights, the location of waste and other hazards. Internationally, there is growing attention to the *environmental entitlement* movement that emphasises the rights of the poor to good quality and healthy neighbourhoods that are free of hazards and pollutants, a focus that sits well with climate mitigation projects. Opposition to degradation has, alongside the struggle to improve livelihood opportunities, become a new political vehicle of the poor, including in South Africa (Lawhon and Patel 2013). Environmental justice, as a social movement, should be enabling a politics of mitigation as well as adaptation, though the latter has dominated.

This brief overview suggests, improperly, that ideas evolve in a linear trend, in which one perspective gives way to another. This is not entirely true; differential understandings of poverty tend to co-exist, even while their influence rises and falls. In an acknowledgement of the value of the different traditions of poverty definition and measurement it is now common-place for a multiple definition of poverty to be asserted.

What has happened over the last decade or so is that, while poverty is almost always depicted as multi-dimensional, the city itself (how it is built, structured and managed) has slipped in priority as explanations of urban poverty have placed an increasing emphasis on human agency, livelihood strategies and community mobilisation (Rakodi and Lloyd-Jones 2002; Satterthwaite and Mitlin 2013). The focus on people-centred or bottom-up views of the last two decades are in no way wrong, but the human emphasis has meant that the structural and institutional role of 'the city' in shaping the experiences of poverty and the responses to poverty are now relatively poorly understood (Pieterse 2008). In the face of the boosterish calls to let the poor take more control of their lives, anti-poverty action that is imposed or top-down may be met with opposition or hostility. As a result, pro-poor actions which would be achieved through climate mitigation activities that work at the city scale (like enforced solar heating or improved public and non-motorised transport) can erroneously be presented as invasive, top-down or post-political technocratic urban management.

It has been suggested that the way poverty has been approached over the last two decades has resulted in failure to recognise the importance of the physical and organisational structure of cities and related impacts on the wellbeing of the poor, over multiple generations. The focus on people or agency, not structures and institutions appears to have distorted reform in urban planning, budget allocations and other mechanisms of urban transformation. A corrective is due that brings 'the city' back into public policy debates. Several authors within the broad poverty community have already shifted the overall orientation of the field. For example, a burgeoning literature is developing for the following themes: urban infrastructure and poverty (Silver 2014; Jaglin 2014; Pieterse and Hyman 2014); the importance not only of improved public transport for the poor, but a push for more systematic, large multi-modal transport interventions in poor cities (Brand and Davila 2011); a resurgent interest in what states do for the urban poor (Ballard 2013); and the enduring specialist areas of urban poverty that are bound up with sectoral elements of the built environment, such as sanitation and housing.

The field of poverty studies is shifting rapidly with the renewed emphasis on the role of states and a concerted effort to locate poverty and human vulnerability in larger systemic approaches to urban change, like that offered by resilience thinking. The potential of these changes to link better to climate science is easily seen in the scaling up of the disaster risk response from the early focus on individual and household preparedness, to more recent attention to comprehensive city and national disaster readiness and responses (Simon and Leck 2014). Similarly, biodiversity science has engaged the urban agenda, including on issues of poverty, and in the process is driving greater attention to the city scale and to issues of governance (Elmqvist et al. 2013).

Conclusion

This chapter has explored the suggestion that it would be appropriate for climate mitigation and urban poverty strategies to align more closely. Understanding the intellectual and political reasons why this has not hitherto happened clears the way for a more direct endeavour to link climate mitigation and poverty reduction as twinned developmental goals, as has been more successfully achieved in the climate adaptation agenda. For cities of the global south, a clearer integration of all aspects of climate action would aid already overburdened local governments. Better alignment of adaptation and mitigation in planning processes to the public good and the interests of the urban poor would further simplify the massive changes required in how cites are designed, built and operated.

Unfortunately, some structural barriers remain in the forums where the climate mitigation and poverty agendas should be jointly pushed. For example, the current structure of IPCC reporting, which only belatedly addressed any urban issues at all, continues to split mitigation and adaptation – thus reinforcing the north/mitigation, south/adaptation split. The almost complete exclusion of cities in the developing world from the C40 has begun to change; three more African cities were included in the organisation at the 2014 Johannesburg summit. But, until membership of mitigation lobbying and organising bodies like C40 include imperatives from large cities in poor countries, where one would have anticipated a clear mitigation policy to deal with transport and energy demands as well as building codes and forward planning in land use, a mitigation agenda with global traction will not be formulated.

The single biggest hope for inserting a mitigation agenda into the core of development thinking comes with the United Nations Sustainable Development Goals (SDGs) process now under debate. However, despite the shift to a more sustainable emphasis, climate is currently framed as a separate goal that is devoid of poverty related targets and it is likely to be linked to biodiversity through the process of slimming down the total number of SDGs. There is a parallel process for a standalone urban goal, which is currently not linked to climate debates.[6] Careful attention needs to be given to the targets and indicators, which will drive the overall SDG process, for these will influence international and national development priorities. For the new development approach, as articulated in the SDGs, to succeed realignment and reconciliation of the conceptions of both poverty and climate mitigation are needed.

References

ADB (2011) The middle of the pyramid: dynamics of the middle class in Africa, Chief Economist Complex, African Development Bank, Market Brief, April 2011

Anguelovski I, Carmin J (2011) Something borrowed, everything new: innovation and institutionalization in urban climate governance. Curr Opin Environ Sustain 3:169–175

Ballard R (2013) Geographies of development II: cash transfers and the reinvention of development for the poor. Prog Hum Geogr 37(6):811–821

Bank AD (2010) Focused action: priorities for addressing climate change in Asia and the Pacific. ADB, Manila

Berrisford S (2014) The challenge of urban planning law reform in African cities. In: Parnell S, Pieterse E (eds) Africa's urban revolution. Zed Books Ltd, London, pp 167–183

Bicknell J, Dodman D, Satterthwaite D (2009) Adapting cities to climate change: understanding and addressing the development challenges. Earthscan, London

Blanco H, Alberti M, Olshansky R, Chang S, Wheeler SM, Randolph J, Watson V (2009) Shaken, shrinking, hot, impoverished and informal: emerging research agendas in planning. Prog Plan 72(4):195–250

Brand P, Dávila JD (2011) Mobility innovation at the urban margins: Medellín's metrocables. City 15(6):647–661

Brenner N, Madden DJ, Wachsmuth D (2011) Assemblage urbanism and the challenges of critical urban theory. City 15(2):225–240

Bulkeley H, Betsill M (2013) Revisiting the urban politics of climate change. Environ Polit 22(1):136–154

Carmin J, Anguelovski I, Roberts D (2012) Urban climate adaptation in the global south: planning in an emerging policy domain. J Plan Educ Res 32(1):18–32

Cartwright A, Parnell S, Oelofse G, Ward S (eds) (2012) Climate change at the city scale: impacts, mitigation and adaptation in Cape Town. Routledge, Abingdon

Elmqvist T, Redman CL, Barthel S, Costanza R (eds) (2013) Urbanization, biodiversity and ecosystem services: challenges and opportunities: a global assessment. SpringerOpen, New York. doi:10.1007/978-94-007-7088-1_2

[6] UN-Habitat, UCLG, ICLEI, Cities Alliance, Metropolis, SDSN (18th September 2013), Why the World needs an Urban SDG? Accessed on 12 Feb 2014 from http://sustainabledevelopment.un.org/content/documents/2569130918-SDSN-Why-the-World-Needs-an-Urban-SDG.pdf. See also www.urbansdg.org

Frayne B, Moser CO, Ziervogel G (2012) Climate change, assets and food security in Southern African cities. Earthscan, Abingdon
Graham S, Marvin S (2001) Splintering urbanism: networked infrastructures, technological mobilities and the urban condition. Routledge, London
Hodson M, Marvin S (2010) Can cities shape socio-technical transitions and how would we know if they were? Res Policy 39(4):477–485
Hodson M, Marvin S (2013) Green cities: position paper, Unpublished. Mistra Urban Futures
Hunt A, Watkiss P (2011) Climate change impacts and adaptation in cities: a review of the literature. Clim Chang 104(1):13–49
Intergovernmental Panel on Climate Change (IPCC) (2012) Special report on managing the risks of extreme events and disasters to advance climate change adaptation (SREX). In: Field CB, Barros V, Stocker TF, Qin D, Dokken DJ, Ebi KL, Mastrandrea MD, Mach KJ, Plattner G-K, Allen SK, Tignor M, Midgley PM (eds) Cambridge University Press, Cambridge/New York. Available via http://ipcc-wg2.gov/SREX/images/uploads/SREX-All_FINAL.pdf. Accessed 5 Feb 2014
Jaglin S (2014) Regulating service delivery in southern cities: rethinking urban heterogeneity. In: Parnell S, Oldfield S (eds) A Routledge handbook of cities of the global south. Routledge, London, pp 343–446
Lawhon M, Patel Z (2013) Scalar politics and local sustainability: rethinking governance and justice in an era of political and environmental change. Environ Plan C 31(6):1048–1062
Lewis Y, Jooste M (2012) Opportunities and challenges in establishing a low-carbon zone in the Western Cape Province. In: Cartwright A, Parnell S, Oelofse G, Ward S (eds) Climate change at the city scale: impacts, mitigation and adaptation in Cape Town. Routledge, Abingdon, pp 99–121
Martine G, McGranahan G (eds) (2014) Urban growth in emerging economies: lessons from the BRICS. Routledge, London
Martine G, McGranahan G, Montgomery M, Fernandez-Castilla R (eds) (2008) The new global frontier: urbanization, poverty and the environment in the 21st century. Earthscan, London
Mitlin D, Satterthwaite D (2013) Urban poverty in the global south: scale and nature. Routledge, London
Moser CO (1998) The asset vulnerability framework: reassessing urban poverty reduction strategies. World Dev 26(1):1–19
Moser C, Satterthwaite D (2008) Towards pro-poor adaptation to climate change in urban centres of low- and middle-income countries, Human settlements discussion paper series. Human Settlements Group and Climate Change Group: International Institute for Environment and Development (IIED), London
Myers GA (2011) African cities: alternative visions of urban theory and practice. Zed Books Ltd, London
Napier M, Berrisford S, Kihato S, McGaffen R, Royson L (2013) Trading places: accessing land in African cities. African Minds, Somerset West
Parnell S, Pieterse E (2014) Africa's urban revolution. Zed Books Ltd, London
Parnell S, Robinson J (2012) (Re)theorizing cities from the global south: looking beyond neoliberalism. Urban Geogr 33(4):593–617
Parnell S, Simon D, Vogel C (2007) Global environmental change: conceptualizing the growing challenge for cities in poor countries. Area 39(3):357–369
Pelling M (2010) Adaptation to climate change: from resilience to transformation. Routledge, London
Pieterse E (2008) City futures: confronting the crisis of urban development. Zed Books Ltd, London
Pieterse E (2014) Filling the void: an agenda for tackling Africa's urbanisation. In: Parnell S, Pieterse E (eds) Africa's urban revolution. Zed Books Ltd, London, pp 200–220
Pieterse E, Hyman K (2014) Disjunctures between urban infrastructure, finance and affordability. In: Parnell S, Oldfield S (eds) A Routledge handbook of cities of the global south. Routledge, London, pp 191–205

Rakodi C, Lloyd-Jones T (eds) (2002) Urban livelihoods: a people-centred approach to reducing poverty. Earthscan, London

Revi A (2009) Climate change risk: an adaptation and mitigation agenda for Indian cities. In: Bicknell J, Dodman D, Satterthwaite D (eds) Adapting cities to climate change: understanding and addressing the development challenges. Earthscan, London/New York, pp 311–338

Rosenzweig C, Solecki WD, Hammer SA, Mehrotra S (eds) (2011) Climate change and cities: first assessment report of the Urban Climate Change Research Network. Cambridge University Press, Cambridge

Sánchez-Rodríguez R, Seto KC, Simon D, Solecki WD, Kraas F, Laumann G (2005) Science plan urbanization and global environmental change, IHPD report 15. International Human Dimensions Programme on Global Environmental Change, Bonn

Satterthwaite D, Mitlin D (2013) Reducing urban poverty in the global south. Routledge, London

Satterthwaite D, Huq S, Reid H, Pelling M, Romero Lankao P (2009) Adapting to climate change in urban areas: the possibilities and constraints in low- and middle-income nations. In: Bicknell J, Dodman D, Satterthwaite D (eds) Adapting cities to climate change: understanding and addressing the development challenges. Earthscan, London/New York, pp 1–47

Sen A (1999) Development as freedom. Oxford University Press, New York

Seto K, Güneralp B, Hutyra LR (2012) Global forecasts of urban expansion to 2030 and direct impacts on biodiversity and carbon pools. Proc Natl Acad Sci 109(40):16083–16088. doi:10.1073/pnas.1211658109

Silver J (2014) Locating urban retrofitting across three BRICS cities: exploring the retrofit landscapes of Sao Paulo, Mumbai and Cape Town. In: Urban retrofitting for sustainability: mapping the transition to 2050. Routledge, Abingdon

Simon D (2010) The challenges of global environmental change for urban Africa. Urban Forum 21(3):235–248

Simon D, Leck H (2013) Cities, human security and global environmental change. In: Sygna L, O'Brien K, Wolf J (eds) A changing environment for human security: transformative approaches to research, policy and action. Earthscan, London/New York, pp 170–180

Simon D, Leck H (2014) Urban dynamics and the challenges of global environmental change in the south. In: Parnell S, Oldfield S (eds) A Routledge handbook of cities of the global south. Routledge, London

Swilling M, Annecke E (2012) Just transitions: explorations of sustainability in an unfair world. UCT Press, Cape Town

Turok I (2013) Linking urbanization and development in Africa's economic revival. In: Parnell S, Pieterse E (eds) Africa's urban revolution. Zed Books Ltd, London, pp 60–81

Turok I, McGranahan G (2013) Urbanisation and economic growth: the arguments and evidence for Africa and Asia. Environ Urban 25(2):465–482

Tyler S, Moench M (2012) A framework for urban climate resilience. Clim Dev 4(4):311–326

UN-Habitat (2003) The challenge of slums: global report on human settlements. Earthscan, London

UN-Habitat (2009) Global report on human settlements 2009: planning sustainable cities. Earthscan, London

UN-Habitat (2014) The state of African cities, 2014: reimagining sustainable urban transitions. UN Habitat, Nairobi

United Nations (2012) World urbanization prospects: the 2011 revision, CD-ROM edition. United Nations, Department of Economic and Social Affairs, Population Division

Watson V (2009) The planned city sweeps the poor away…: urban planning and 21st century urbanisation. Prog Plan 72(3):151–193

Watson V (2013) African urban fantasies: dreams or nightmares? Environ Urban 0956247813513705

Watson V, Agbola B (2013) Who will plan Africa's cities? Counterpoints, Africa Research Institute, London

Winkler T (2009) For the equitable city yet to come 1. Plan Theory Pract 10:65–83

Winkler T (2011) On the liberal moral project of planning in South Africa. Urban Forum 22:135–148

Winkler T, Duminy J (2014) Planning to change the world? Questioning the normative ethics of planning theories, Planning Theory published online 19 Sept 2014. http://plt.sagepub.com/content/early/2014/09/15/1473095214551113. doi:10.1177/1473095214551113. Accessed 30 Sept 2014

Wratten E (1995) Conceptualizing urban poverty. Environ Urban 7(1):11–38

Ziervogel G, Parnell S (2012) South African coastal cities: governance responses to climate change adaptation. In: Cartwright A, Parnell S, Oelofse G, Ward S (eds) Climate change at the city scale: impacts, mitigation and adaptation in Cape Town. Routledge, Abingdon, pp 223–243

Chapter 12
The Way Forward: Climate Resilient Cities for Africa's Future

Sandra Fohlmeister, Stephan Pauleit, Adrien Coly, Hamidou Touré, and Kumelachew Yeshitela

Abstract Urbanisation and climate change are among the major contemporary challenges for sustainable development in Africa. On this background, the overall aim of the European Commission's 7th Framework Programme project CLUVA (**CL**imate Change and **U**rban **V**ulnerability in **A**frica) has been to develop innovative approaches to vulnerability analysis and for the enhancement of urban resilience. CLUVA was unique in that it combined a top-down perspective of climate change modelling with a bottom-up perspective of vulnerability assessment; quantitative approaches from engineering sciences and qualitative approaches of social sciences; a novel multi-risk modelling methodology; strategic approaches to urban and green infrastructure planning with neighbourhood perspectives to adaptation. This chapter synthesises important results of the CLUVA project by highlighting the critical factors for urban vulnerability that were identified by the African-European research team while working with the five case study cities of Dar es Salaam (Tanzania), Saint Louis (Senegal), Douala (Cameroon), Addis Ababa (Ethiopia) and Ouagadougou (Burkina Faso) from 2010 to 2013. It particularly explores the future path towards climate resilient cities from diverse perspectives, suggesting a set of building blocks for Africa's urban future.

S. Fohlmeister (✉) • S. Pauleit (✉)
Chair for Strategic Landscape Planning and Management, Technical University of Munich (TUM), Emil-Ramann-Str. 6, 85354 Freising, Germany
e-mail: fohlmeister@mytum.de; pauleit@wzw.tum.de

A. Coly
Department of Geography, University of Gaston Berger, B.P. 234, Saint Louis, Senegal
e-mail: adrien.coly@ugb.edu.sn

H. Touré
Department of Mathematics, University of Ouagadougou, UFR Sciences Exactes et Appliquées, 03 BP 7021, Ouagadougou 03, Burkina Faso
e-mail: toureh@univ-ouaga.bf

K. Yeshitela
Ethiopian Institute of Architecture, Building Construction and City Development (EiABC), Addis Ababa University, Addis Ababa, Ethiopia
e-mail: kumeyesh@googlemail.com

S. Pauleit et al. (eds.), *Urban Vulnerability and Climate Change in Africa*, Future City 4,
DOI 10.1007/978-3-319-03982-4_12

Keywords Africa • Climate change • Urban vulnerability • Climate resilient city • Multidisciplinary approach

Introduction

About 395 million people in Africa, corresponding to 38 % of the continent's population, resided in urban settlements in 2010. By 2050, this is expected to grow to approximately 1.3 billion people (UNDESA 2014). While the research community might debate the accuracy of forecast numbers related to African cities due to the lack of reliable census data (Potts 2012; McGranahan and Satterthwaite 2014), there is a clear consensus that the continent is challenged in a multi-faceted manner today.

Urban growth is galloping ahead, whilst economic and infrastructure development are still lagging behind. Urban informality is predominating in settings of under-regulation and under-investment, creating manifold benefits for political and economic entrepreneurship that seeks to perpetuate the status quo (UN-Habitat 2014). Local city governments and authorities fail in applying and enforcing adequate housing standards as well as the necessary provision of services, causing further negative impacts on human health and urban development.

A lack of livelihood options and overexploitation of land resources exacerbated by climate change-induced disasters result in rural exodus, leading to further pressure on urban areas by mass unemployment and the decline of green infrastructure as well as their corresponding vital functions for ecosystem services.

Nearly all African cities exceeding one million inhabitants are located in areas exposed to natural hazards, such as floods, droughts, desertification, heat waves or sea-level rise. And climate change is not only threatening African urban areas; it is already taking place: available data reports on climate-related hazards having caused damages of an average of 400 million USD per year in the African continent between 1990 and 2008, floods accounting for 45 % and droughts for 33 % (Chap. 1, citing Mendelsohn and Saher 2010).

In spite of an increasing number of policy reports (e.g. World Bank 2009; UN-Habitat 2011, 2014) and initiatives that address the issue of climate change adaptation of urban areas in Africa, research is still scarce to provide a scientific foundation. In particular, there is very little research on the interaction between processes of urbanisation and climate change in general, and in Africa in specific (see, for instance, the relevant chapters in IPCC's Fourth & Fifth Assessment Reports, Wilbanks et al. 2007; Revi et al. 2014; Niang et al. 2014, as well as Rosenzweig et al. 2011 to get a sense for the state of the art in science).

Instead, it seems more convenient to agree that there is a complex and hardly digestible "Spaghetti of Doom" (Chap. 5, citing López-Carresi et al. 2014: 5) on the menu of African local city governments. The question of how to disentangle it, remains mostly unanswered, though.

In the pursuit to create an improved evidence base for further action and strengthening of African urban policy in this regard, the European Commission funded research project CLUVA (**CL**imate Change and **U**rban **V**ulnerability in **A**frica) in the framework of its 7th Framework Programme (2010–2013)[1] undertook a systematic study of vulnerability of African cities to climate change with a multidisciplinary approach. It was unique in that it combined: a top-down perspective of climate change modelling with a bottom-up perspective of vulnerability assessment; quantitative approaches from engineering sciences and qualitative approaches of the social sciences; a novel multi-risk modelling methodology; strategic approaches to urban and green infrastructure planning with neighbourhood perspectives to adaptation. Moreover, CLUVA was conceptualised to foster cooperation and build up a strong network between European, East and West African research entities. It thus acknowledged that African cities are no stand-alone agents, but rather parcels of a shared context with many challenges of transboundary nature to address in a concerted manner (UN-Habitat 2014).

The present chapter synthesises important results of CLUVA, focusing on the paramount question of how to develop a future path towards climate resilient cities in Africa. It highlights the factors being considered as critical for the vulnerability of today's cities in Sub-Saharan Africa from the previous chapters of this volume, illustrates achievements of the project's research and capacity-building efforts and deduces a set of building blocks for Africa's urban future. Furthermore, the reader is offered an outlook on the desired way forward from the insider perspectives of researchers from the cities of Saint Louis (Senegal), Ouagadougou (Burkina Faso) and Addis Ababa (Ethiopia).

Urban Climate Resilience: Utopia or Vision?

Urban climate resilience is on the agenda of the wider climate change community. Numerous initiatives, which have emerged in recent years, are a proof of this; be it the Medellín Collaboration on Urban Resilience, ICLEI's "Resilient Cities" conferences, UN Habitat's City and Climate Change Initiative and Academy (CCCI/CCCA), UNISDR's Global Resilient Cities Campaign, the Rockefeller Foundation's Asian Cities and Climate Change Research Network (ACCCRN), or the Inter-American Development Bank's Emerging Sustainable Cities Initiative: the call for action is clear and immediate, while a postponement to future generations is "no longer tenable" (Simon 2014: vii).

Contemporary literature offers a variety of options for defining the term "climate resilient city". According to Twigg (2009: 8), a resilient city can be characterised by its capacity to "anticipate, minimise and absorb potential stresses or destructive forces through adaptation or resistance, manage or maintain certain basic functions and structures during disastrous events, and recover or bounce back after an event".

[1] Seventh Framework Programme, Grant agreement no. 265137: "CLimate change and Urban Vulnerability in Africa", 2010–2013, www.cluva.eu

Long-term changes of urban and climate systems, which exhibit both highly dynamic and fundamentally unpredictable behaviour, need to be considered. Therefore, resilience strategies should strengthen not only the ability of the urban system to bounce back after some disrupting climatic event, but enhance adaptation to these fundamentally changing conditions and even enable its transformation into an entirely new system if considered necessary (Pickett et al. 2004; Walker et al. 2004). It has been suggested that this adaptive and even transformative capacity of the urban system is first and foremost dependent on the ability of the social system at different levels of organisation – from households and communities to local and regional governments – to act in suitable ways (Walker et al. 2004; Satterthwaite 2013). With a focus on disaster reduction, UNISDR (2012) offers more details and insights into the must-have's of a resilient city with its "Ten Essentials for Making Cities Resilient" being in line with the five priorities of the Hyogo Framework for Action 2005–2015. Based on this set, a city may be considered "resilient", if it

- has a proper *organisation and coordination* in place to understand and reduce disaster risk, based on participation of citizen groups and civil society;
- earmarks a *budget* for disaster risk reduction;
- keeps data on hazards and vulnerabilities updated and uses *risk assessments* for urban development decisions;
- invests in and maintains *critical infrastructure* for risk reduction;
- checks the *safety* of schools and health facilities towards risks;
- applies and enforces building standards and *land use planning principles*;
- cares for *education and training* related to disaster risk reduction in schools and local communities;
- conserves *ecosystems* and natural buffers;
- provides for early warning systems and *emergency management* capacities;
- places needs of its *population at the centre* of reconstruction and response after a disaster.

Presenting principles and checklists is a quick step to take; achieving urban resilience under circumstances of pressing urbanisation, poverty and informality, is much more demanding. In the following, the reader is provided with a panoramic view on what emerges from the contributions to this book as critical factors and important building blocks on the way to urban resilience in Africa, trusting in its strength as guiding vision rather than considering it an intangible utopia.

Perspectives on the Development of Climate Resilient Cities: Critical Factors and Building Blocks

In 2010, the CLUVA project was launched with ambitious aims and a multidisciplinary African-European team, whilst the challenges ahead could only be roughly envisaged. The adopted approach served to explore future perspectives on the development of climate resilient cities in Africa, both from outsider and insider viewpoints. The outcome of this exploration matches well with the outlook done by UN-Habitat in its report on the state of African cities 2014: "At present, African cities are at risk. But the very concept of risk also implies choice." (UN-Habitat 2014: 30). CLUVA evidence indicates that 'business as usual' will not be enough. To opt for changing courses and embracing the developmental potential of urbanisation, whilst pro-actively providing infrastructure, policies and planning that support the transition to more inclusive, resilient and sustainable cities, will be a more promising choice (McGranahan and Satterthwaite 2014; UN-Habitat 2014).

The building blocks presented in the following subsections should be understood on this background; they are highlights of CLUVA achievements and perspectives on what is necessary for the establishment of climate resilient cities in Africa. As building blocks, they might be applied in a flexible manner, depending on each and every city's individual needs and aspirations for the path to resilience.

Building Block 1: Making Climate Information Relevant to Decision-Makers – From Climate Modelling to Multi-risk Assessment

Any strategy or measure towards creating climate resilient cities needs to be based on reliable information on how climate change will translate into hazards and how vulnerable the city is to these hazards. For Africa, the CLUVA project modelled climate change scenarios for the entire continent and down-scaled, for the first time, three of IPCC's scenarios to a scale of 8 km and even 1–2 km with two different approaches, hence a resolution at the scale where it becomes relevant to spatial urban planning (Chaps. 1 and 2). Temperature increases of at least 1 °C by 2050 were projected consistently for all five cities. As a consequence, the urban population would suffer from longer and more intense heat waves, in particular. Climate modelling suggests, for example, that the number of heat waves with a maximum length of five days could increase from three in the period 1950–1970 to up to 33 in Dar es Salaam and up to 40 in Addis Ababa in the period 2030–2050, whilst the frequency of longer lasting heat waves is also expected to rise (Chap. 4). Interestingly, according to these model outputs climate change would not lead to a significant increase in hazards for flooding, drought and desertification. This may be surprising given the fact that flooding incidences can be observed more and more in Africa (UNEP 2010: 166). However, the important message here is that an increase of probability for these disasters is less a consequence of climate change, but mostly caused by human factors such as the disorderly growth of urban areas into risk zones and the expansion of water impervious areas, being further exacerbated by lack of adequate drainage and poor sanitation, as well as overuse and exploitation of agricultural and forest land in surrounding rural areas.

Modelling resulted in a wealth of new information on the climate changes that lie ahead for Africa[2] (Chap. 2; Engelbrecht et al. 2011; CSIR and CMCC 2012a, b, 2013; Bellucci et al. 2012; Di Ruocco et al. 2012; Giugni et al. 2012b, 2013; Jørgensen et al. 2012; Nyed et al. 2013; Sibolla et al. 2012; Tonyé et al. 2012; Abo El Wafa 2013a, b; Buchta 2013a, b; Coly et al. 2013; De Paola et al. 2013a, b; De Risi 2013; De Risi et al. 2013a, b, c; Diop and Lo 2013; Kombe et al. 2013; Jalayer et al. 2013; Bucchignani et al. 2014; Weets et al. 2014).

A meaningful and challenging next step will be to insert this information into a multi-risk assessment that reflects the variety of interactions of different types of hazards and vulnerabilities to be encountered in specific urban settings. Here, a probabilistic framework for risk assessment offers new perspectives to come to terms with this complexity (Chap. 7; Garcia-Aristizabal and Marzocchi 2012; Garcia-Aristizabal et al. 2012, 2013, 2015). The framework does not only consider current risks, but also opens the road for taking into account the dynamics of both the climate and the urban system, including their uncertainties. Applying it may enable decision-makers to quantify losses via loss exceedance curves and to identify urban hotspots that need to be prioritised in risk management (De Risi et al. 2013c; Jalayer et al. 2014). Furthermore, the multi-risk assessment approach holds potential to serve as an integrating framework as it embeds the social context in a given area by considering socio-demographic characteristics (e.g. percentage of elderly people), risk perception, and social resilience (e.g. availability of resources).

Characterising the various hazards via climate change modelling, and tailoring methods, such as the identification of flood risk hotspots, to local needs, is one side of the coin for multi-risk assessment; exploring the vulnerabilities of the urban system is the other side. Addressing these vulnerabilities are further important building blocks towards climate resilient cities. Three dimensions of vulnerability were considered in the CLUVA project: the built environment and critical infrastructures, the life supporting system of urban ecosystems, and people. These will be discussed in turn in the following sections.

Building Block 2: Adapting Built Environment and Lifelines

The fact that the built environment in African cities is especially vulnerable to climate-related natural hazards such as flooding has manifold causes. In particular, the speed of the cities' growth in combination with limited administrative capacities makes well-planned urban development extremely difficult. In consequence, informal settlements prevail that are characterised by vulnerable building material

[2] Nevertheless, it should be kept in mind that any climate data-related forecasts rest on hypotheses being of anthropogenic offspring and are extremely difficult to predict. Furthermore, a variety of additional aspects with importance for city authorities, such as the economic evaluation of scenarios or the holistic survey on hazards and their possible impacts on the cities' peri-urban hinterlands, were beyond the scope of CLUVA's research. The findings should thus be interpreted on the background of these data limitations.

(e.g. mud, wood), inadequate standards of housing and services, as well as the common location in zones susceptible to flooding. According to estimates, 63 % of the urban population in Africa currently lives in such informal settlements and shanty towns (UN-Habitat 2009).

On this background, adaptation of the urban built environment and lifelines must start from the use of adequate approaches to vulnerability assessment being able to deal with limited availability and quality of local data.

In the framework of CLUVA, a mix of meso- and micro-scale methods has been adopted for executing vulnerability and risk assessments with the focus on flooding. The meso-scale methods have herein proven to be useful for the delineation of urban flooding risk hotspots, providing insights regarding the differences in exposure to risk, depending on the range of different residential types, for instance. To further explore a city's urban hotspots, micro-scale risk assessment methods can be applied to evaluate flooding risks in a smaller spatial extent, such as for portfolios of informal settlements on the neighbourhood level. Risk hotspot maps, such as the ones produced by CLUVA for Addis Ababa and Dar es Salaam (De Risi et al. 2013c; Nyed et al. 2013; Jalayer et al. 2014), are efficient screening tools for local urban planners to detect areas threatened by flooding risk, identify the need of immediate or long-term action and to make informed choices about locations for new buildings, the latter one probably being the most effective risk mitigation strategy for any city.

As findings from the case study surveys indicate, informal settlements, one-storey buildings and mixed residential areas will need most attention in the upcoming years. Urban planners will have to assume responsibility in foresighted planning and undertaking suitable measures to avoid the further uncontrolled sprawl of new informal settlements in flood-prone urban areas, to resettle endangered parts of the population, to take care of the enforcement of better housing standards and to upgrade informal settlements by using less vulnerable building materials. Guidelines containing low-cost measures and mitigation strategies for predominant housing categories, such as adobe houses (Ouagadougou), cement bricks (Dar es Salaam) and mud/wood houses (Addis Ababa) have been suggested by CLUVA for this purpose (De Risi et al. 2012). Furthermore, stormwater and road systems should be on top of the agendas of city authorities. While road planning needs to avoid that major corridors end up as "mouse-traps" for emergency action, stormwater systems have to be designed and managed in an appropriate manner (Esposito et al. 2012; Giugni et al. 2012a; Carozza et al. 2013). Above all, the responsible city authorities will have to intensify efforts to reduce further unfavourable land cover transformations as a precautionary measure.

Building Block 3: Promoting Urban Green Infrastructure

The importance of Urban Green Infrastructure (UGI) and its corresponding ecosystem services for urban sustainability and adaptation to climate change are increasingly acknowledged in academia and in practice (Gill et al. 2007;

Pauleit et al. 2011; Smit and Parnell 2012). In the African context, however, where extreme levels of poverty and the daily struggle for survival are affecting large portions of the human population, dealing with issues of green infrastructure is at risk of being considered a "luxury", and is still low on the political agenda (Chap. 4).

The CLUVA project employed a common method to assess for the first time the quantity and distribution of green spaces in the five case study cities. The mapping of urban morphology types (UMTs) and land cover surveys were applied for this purpose. This approach resulted in a database which is unique for African cities and provides sound evidence that urban ecosystems are an important way of mitigating the negative impacts of urbanisation and climate change (Cavan et al. 2012a, b, 2014; CLUVA 2013; Lindley et al. 2013; Pauleit et al. 2013; Kibassa 2014). Importantly, it allowed the establishment of the relationships between the form of urban areas, their green structure and ecosystem services.

Climate modelling at city level for Addis and Dar es Salaam and results from field measurements in Dar es Salaam showed the important role of green structure in moderating surface and air temperatures and thus in reducing heat stress for the urban population. Moreover, "provisioning services", e.g. food, fibres, medicine, wood for construction and fuel, continue to play a major role in African cities. As an example, 60 % of the eggs, 90 % of leafy vegetables and 70 % of milk consumed in Dar es Salaam were produced locally (several sources in Halloran and Magid 2013), but urban farming in its different forms is still by and large disrespected by decision-makers neglecting its paramount role for urban food security. However, not only farmland but green space in general is strongly under pressure from urban growth.

Numbers are more than alarming: for the cities of Addis Ababa and Dar es Salaam, CLUVA identified high loss rates of green structures over the last decade. For instance, in Dar es Salaam, 30 % of "green areas"[3] were lost between 2002 and 2008, and large trees have been cut down at an incredible rate of 11 % annually. For Addis Ababa, survey results hint to a similar sobering development: around 25 % of the agricultural land area of 2006 has disappeared by 2011. Both urban sprawl and the densification of central zones were identified as relevant drivers of these changes (Chap. 4).

Strategies should thus aim to protect and restore key green spaces, such as the river corridors in Dar es Salaam and Addis Ababa, promote urban agriculture and forestry through suitable policies as integral elements of the African city of the future, and maintain or re-establish adequate greening within already built areas. The latter may be particularly difficult to achieve. Household surveys in selected neighbourhoods of Dar es Salaam showed that residents appreciate their local green areas, but processes such as the extension of homes and appropriation of undeveloped plots of land by developers lead to overall losses (Lindley et al. 2013).

[3] Consisting of the UMTs *mixed forest, riverine, bushland, mangrove, marsh/swamp, parks and other open space*; see also Chap. 4, section "Development pressure".

While CLUVA managed to identify such challenges, further research will be needed to gain a deeper understanding of green infrastructure vulnerability as well as the different processes that cause the loss of urban green to design appropriate response strategies. More evidence is especially required on the UGI's ecosystem services to support green infrastructure planning at the strategic city- and city regional level. Interest and even adoption of this information for the development of the new Master plan for the city of Addis Ababa are a proof of the value of this approach.

Building Block 4: Preventive Action for Reducing Social Vulnerability

Today, many drivers significantly influence the vulnerability of communities facing natural hazards. The majority of them are directly connected to the rapid urbanisation, which the African continent is facing: deficient provision of social and technical infrastructure, the lack of water supply and storm water drainage, as well as the densification of informal settlements are common phenomena that especially intensify the social vulnerability of the poor (Chap. 6). Herein, vulnerability is not of static character, but evolves over time and thus needs careful attention, particularly because the increasing inequalities within African cities may cause a gradual decline of people's resilience. Therefore, a vital task of local city authorities lies in gaining an in-depth understanding of the corresponding drivers in order to mount up suitable preventive action.

In CLUVA, research related to social vulnerability was concentrated on the exposure, susceptibility and coping capacities of selected households and neighbourhoods of the case study cities Ouagadougou, Addis Ababa and Dar es Salaam. For assessing the vulnerability of households towards flooding events, CLUVA produced a novel framework that interlinks the four dimensions of asset, institutional, attitudinal and physical vulnerability, and thus facilitates accounting for contextual circumstances that are paramount for social vulnerability (Jean-Baptiste et al. 2011, 2013; John et al. 2012, 2014; Jean Baptiste and Kabisch 2013). Emerging from the framework's application on the case study of Bonde la Mpunga (Dar es Salaam) are the following insights:

First, it is the *urban planning sector and land use control* which deserve prominent attention in order to enable them to take up the responsibilities that they are currently unable to fulfil to the necessary extent. Be it the provision and maintenance of vital infrastructure, such as drainage systems, or the strict regulation, supervision and management of land cover changes to avoid further inappropriate densification of already crowded settlements, the tasks ahead are manifold. Tailored reforms and enforcement mechanisms should be initiated to expand the capacity of urban planning immediately.

Second, the vicious cycle of poverty plays a major role with regard to the asset dimension of social vulnerability. *Vulnerability assessment methods* as the ones employed by CLUVA produce valuable inputs for the detection and mapping of social vulnerability hotspots, which can orientate the further design of emergency relief actions accordingly. Besides, poverty reduction initiatives and public awareness building could be targeted to promote households' climate resilience in terms of improving income options and educational levels.

Third, a key opportunity to strengthen preventive action lies in the diversity of *social networks* (kin relationships, gender and religious groups, etc.), which remain an important source of mutual support and resilience on household level. These networks should be connected and coordinated in a more stringent manner for improving their effectiveness and to avoid negative side-effects, such as the re-direction of water flows to settlements by individualistic and short-sighted measures. Moreover, informal groups should be empowered by useful means (financial, material, human resources), whilst also strengthening formal institutions to become reliable partners for joint action. The recognition and support of these complex connections within African settlements are fundamental starting points for any preventive action to reduce social vulnerability.

In conclusion, the framework for social vulnerability assessment provides a suitable basis for identifying important "entry points" for reducing social vulnerability. While the framework was used for the study of selected neighbourhoods in the CLUVA project, it is suggested that it can be equally applied at higher levels up to entire cities. Moreover, it has the potential for integrating the study of social vulnerability with vulnerability assessments of the built environment and green infrastructure (physical dimension) as well as planning and governance related research (institutional dimension).

Building Block 5: Fostering Climate Change Governance

The recognition of cities as strategic actors for governing climate-related risks and enhancing resilience is no news to the international climate change community. Contrasting this image, reality proofs that cities find themselves bottlenecked with many barriers to the integration and institutionalisation of the climate risk agenda, which need to be better understood to cope with them (e.g. Bulkeley 2013).

Against this background, CLUVA examined the integration of climate change adaptation as a policy field at the interface of flood risk management within the multi-level governance settings of the two case study cities – Dar es Salaam and the much smaller Saint Louis – employing the concept of coproduction for shedding light on important drivers and barriers to municipal governance.

The comparison of the case study cities reveals important differences regarding their approaches to adaptation. While in Dar es Salaam adaptation and flood risk management do not have any distinct organisational homes at municipality level, public officials (municipal planners) substantially facilitate citizen participation in local development planning and flood risk management in Saint Louis. Interestingly, good agency performance turns out to result from the responsiveness to citizens and cross-sectoral cooperation rather than from the capacity of the public sector alone (Chap. 9). As for size and scale, findings indicate that smaller and simpler organisational structures, going hand-in-hand with committed leadership, ease active citizen input into service delivery and coproduction, and thus might be of advantage for a city's adaptation to climate change. This supports previous findings[4] and challenges the hypothesis of Rauken et al. (2014) that larger municipalities would have an advantage in integrating adaptation into governance due to higher capacity compared to smaller ones.

Meanwhile, evidence from CLUVA research on multi-level governance gives hope in that it provides new insights into factors hitherto considered as decisive for adaptation. One example is financial constraints encountered at municipality level, which did not prove to constitute a major hindrance for the adoption of incremental measures for adaptation. The case of Saint Louis also shows that a lack of political guidance and resources from higher-level government can be compensated to some degree by engaged municipal officials and a proper encouragement of citizens for coproduction. This reaffirms the importance of "local champions", the identification and support of whom may open windows to adaptation even when municipalities' power to act is severely constrained (Herslund et al. 2012, 2013; Vedeld et al. 2012, 2013).

What lies ahead as key task of national governments' agenda for successfully fostering climate change governance is a significant strengthening of institutions and coproduction through decentralising mandates and resources to city- and sub-city levels and a close cooperation across public and private actors. Importantly, decentralisation should not be misused as a 'universal buzzword' (Chap. 5 citing Ribot 1999), but indeed work to improve capabilities of the municipality level to coordinate and act by increased decisional and financial power. Governance strategies should be prioritised that foster a sound integration of disaster risk management, adaptation and development approaches; an improved land use planning, management and enforcement; an upgrade of urban drainage, waste and stormwater management as well as the adoption of urban development strategies that avoid the further expansion of settlements at the expense of agricultural land and other vegetated areas (Gasparini et al. 2013).

---

[4] Indeed, small cities and towns might offer several advantages when it comes to adaptation to climate change. From a socio-political perspective, short ways can ease the establishment of a political will to address climate change. Moreover, small cities' inhabitants are easier to mobilise for participatory planning than they might be in large informal settlements of big urban centres. From an economic perspective, small cities are nodes between rural produces and their markets and as such rest on a close relationship to their hinterlands, being in the position to be platforms for initiating a "climate-smart rural development" (see also Chap. 5, Box 5.1). Strategically, small cities and towns might thus play an important role to reduce the pressure on cities by stemming migration flows. More research will be meaningful to shed light on the different roles of places within African settlement hierarchies with concern to adaptation to climate change.

Building Block 6: Strengthening Strategic Urban Planning

How can strategic urban planning practically work in a setting, where the municipality is the main player in urban services and land use management, but simply not empowered enough to fulfil its responsibilities? And how can an African city identify its way forward to become more climate resilient?

In this context, it is worthwhile to reflect upon the particular understanding of "strategy-making" and "strategic planning" adopted by CLUVA. The research team of Herslund et al. defined it based on Healey (2009) to be "an interactive and collaborative process of identifying most important issues for a broad range of actors in the cities, a process of scoping and creating momentum for action, relying on both local and expert knowledge, and selecting and framing short and long term actions that resonate with stakeholders" (Chap. 10: section "Introduction").

The findings from the surveyed case studies, Addis Ababa and Dar es Salaam, demonstrate that there is only limited momentum among city stakeholders at the time being for citywide adaptation strategies. Too manifold are other tasks, and too scarce are the necessary resources at municipality levels. However, awareness and inherent interest exist among stakeholders to take immediate action in the form of short-term projects that address pressing matters and climate change at the same time. In the case of Dar es Salaam, integrated local projects in most vulnerable areas could serve as "kick-off measures" to a strategy-making process towards increasing the city's climate resilience. In Addis Ababa, reframing adaptation to the topic of integrated water management could pave the way to engage stakeholders in further long-term action (Chap. 10; Herslund et al. 2012, 2013; Jørgensen et al. 2014).

When selecting these 'door openers' it is important that they should bear the potential to deliver benefits going beyond those of resilience, – reduce poverty, improve health by better drainage systems, increase food security by urban agriculture-, to name a few (Chap. 10; UN-Habitat 2011; Simon 2013). Additionally, existing experience of local stakeholders with floods, droughts or other natural disasters may facilitate the immediate uptake of adaptation action.

This tailor-made, incremental *reframing of adaptation* to a city's own urgent matters does not only offer the opportunity to kick-start adaptation projects in practice, but also to identify local champions and suitable followers, create learning experiences and knowledge as well as to build up reliable networks for cross-sectoral collaboration with regard to long-term, transformative and city-wide strategic adaptation.

Bridging the 'Disconnects' Towards Realising Strategic Visions on Compact and Climate Resilient Cities

Several building blocks emerged from CLUVA evidence to be meaningful pieces on the way forward to climate resilient urban development in Sub-Saharan Africa. Likewise, numerous challenges are still ahead. To face them reveals the necessity of bridging several 'disconnects' in constructing tomorrow's African resilient city.

Urban Informality: Being Inclusive

Urban informality emerges as one of the most important, if not paramount, bottlenecks for adapting African cities to climate change. It may even be asked whether strategic action towards urban resilience is at all possible, considering the background of the continent's continued informal settlement growth (UN-Habitat 2009, 2014). Very likely, any further action will have to be "rethought from the slums" (UN-Habitat 2014: 43), engaging both formal and informal systems for developing new modes of participatory, inclusive governance (see also Chaps. 6 and 9). Planning theory approaches have hitherto predominantly sought to remove informal development instead of climbing the barriers of segregatory practices (UN-Habitat 2014); future efforts will thus have to find ways to embrace it. These new, participatory governance modalities may require time of transition being meaningful for testing, peer learning and mainstreaming. Technical and financial backstopping by national governments and the international community will be key.

Urban Transitions' Management: Being Supportive

Much has been said already about urban population growth on the African continent in recent years in this volume. Doubtlessly, the efficient and equitable accommodation of this many urban dwellers will be highly difficult. However, it seems that the real challenge is rather to "manage urban transitions than to control urban explosions" (McGranahan and Satterthwaite 2014: 5). What this calls for is a substantial commitment both from local governments as well as the international donor community in supporting African cities in this endeavour. In particular, three pillars are of utmost relevance for doing so: (i) to build a sound evidence base for urban policy action; (ii) to strengthen local governments' skills and resources for effective management of their cities and towns; and (iii) to foster urban stakeholder dialogue and exchange (UN-Habitat 2014).

The CLUVA project has deliberately sought to contribute to these three pillars with its research and capacity-building efforts. It focused to a great extent on the delivery of reliable tools and evidence to prepare the ground for action even beyond its particular case studies by providing: an open access climate database (see http://ict4eo.meraka.csir.co.za/cluva/); innovative downscaling techniques (Chaps. 1 and 2) and methods for vulnerability assessments of built infrastructure and lifelines (Chap. 3); methods for UMT-based evaluation of urban ecosystem services and green infrastructure planning (Chap. 4); frameworks to assess the social vulnerability on community- and household levels (Chap. 6); novel methodologies for multi-risk assessment (Chap. 7); approaches to the spatially explicit modelling and assessment of urban scenarios that support participatory approaches to strategic urban planning (Chap. 8) as well as methods of multi-level governance analysis (Chap. 9) and recommendations to integrate climate change adaptation into urban planning (Chap. 10).

Furthermore, the cooperation between African and European partners over a period of three years and the project's outcomes have contributed to local capacity development by building a critical mass of researchers in climate change issues on the continent, and also advanced the connections between local researchers and practitioners. PhD students benefitted from training sessions, summer schools and hands-on research (Simonis and Vahed 2011; CSIR 2013; Gasparini et al. 2013), and teaching modules on Master's level were developed for enriching university curricula by climate change-relevant contents in close cooperation with UN-Habitat's Cities and Climate Change Academy (CCCA) (UN-Habitat et al. 2011; Fohlmeister et al. 2013).

The valuable knowledge base generated in the CLUVA project will demand regular update and a sound management by responsible institutions for being efficient and useful in the long-term (Nyed and Herslund 2012).

An important and still open question, which needs both attention and further investigation, is how to get decision-makers of relevant "weight" to take action on climate change. Findings from CLUVA case study work show that it is particularly the relevant stakeholders, such as local leaders as well as higher-level politicians, who are difficult to engage and interest in matters of resilience. Without them, adaptation approaches will, however, likely remain in the range of incremental projects and not be able to turn into transformative action for the African city. Thus, the identification and fostering of suitable incentives will be of utmost importance. *City Climate Labs*[5] might be a possible option to host climate-related data and serve as *nuclei* for professional training and cross-sectoral exchange between political decision-makers, science, city authorities' staff, practitioners from urban planning and civil society groups, for instance. In such a way, the above-mentioned tools' application could be shared among all relevant parties, thus turning them into useful instruments for future African cities.

Reconnecting the Adaptation with the Mitigation Agenda

The need for strategies to climate change adaptation in African cities is clear, while mitigation is rarely considered of great relevance because per-capita emissions of greenhouse gases are low. Moreover, it has been argued that adaptation is connected to the issue of urban poverty as it is the urban poor who are particularly vulnerable to climate change (Moser and Satterthwaite 2008). However, Susan Parnell argues in her contribution to this book that "climate mitigation and adaptation must jointly form the foundation of a resurgent planning vision and practice" and "the treatment of climate change (mitigation and adaptation) and poverty

[5] The idea of City Climate Labs is inspired by the lead author's own field experience with UNIDO/UNEP's global network of National Cleaner Production Centres (NCPC). For further information, see http://www.unido.org/NCPC/. Accessed 31 Aug 2014

cannot be separated in any way, conceptually or operationally, in the reshaping of African cities" (Chap. 11: section "Introduction").

The growth of the African urban population alone provides a strong argument that mitigation is of great importance. While the urban population's per-capita emission rate is currently low, the desire for economic progress implies higher levels of consumption, and hence, future increases in per-capita greenhouse gas emissions. Moreover, analysis of urban dynamics, both past and future, via the scenario modelling approach applied in the CLUVA project to the cities of Dar es Salaam and Addis Ababa, showed the dramatic scale of urban expansion if 'business as usual' continues (Chap. 8; Abo El Wafa 2013a, b; Buchta 2013a, b). It will lead to inefficient settlement patterns that are poorly serviced by infrastructure and strongly dependent on motorised traffic; it will further destroy vegetated areas with important ecosystem services for urban livelihood and climate adaptation (Chap. 4). The loss rate of 30 % of Dar es Salaam's woody biomass between 2002 and 2008 makes it apparent that valuable potentials for carbon sequestration are presently endangered in African cities (Lindley et al. 2013). These developments will be difficult to correct if mitigation thinking is not being incorporated into urban planning already today.

Emphasising mitigation in cities of low and middle income countries would thus provide a necessary complement to prevailing people-centred or bottom-up approaches to adaptation and poverty reduction by refocusing on the overall physical and organisational dimensions of the city, which has important implications for poverty (e.g. in terms of equal access to vital infrastructures). To make this happen, as Parnell states, requires a fundamental reform of the urban planning sector in African cities.

Perspectives on the Way Forward to Climate Resilient Cities for Africa's Future: Local Viewpoints

Adaptation by Participation – The Saint Louis Perspective

Flooding and 'climate risk' in Saint Louis have not been sufficiently taken into account in urban planning to the point that the extent of the current situation gives the impression that local authorities are unable to respond to the challenges ahead. Only recently, the city authorities of Saint Louis have realised the impact of climate-related catastrophes on the development of the communities (Fig. 12.1).

The integration of risk regarding climate change in development planning exercises is therefore essential. However, at present this governance process is only in its 'infancy stage', and efforts are still focused on too many different single activities which are poorly coordinated (knowledge gathering, forecasting, planning, mitigation, adaptation, protection, recovery, etc.), thus not gaining the necessary "weight" for being efficient enough.

Fig. 12.1 Flooded street in Saint Louis, Senegal (Picture: Adrien Coly 2014)

Strengthening the participation of civil society in the governance of risk is a high priority for climate resilient African cities. All stakeholders should participate, from researchers to civil society, to local communities and local authorities. Political dialogues need to be established as a foundation for civil actions in which risk experiences are shared between the different actors, and local solutions and innovations are jointly produced. For the governance of risk, this participatory approach "facilitates the understanding of the public about the technical, organisational, social and political environment by citizens [. . .] This empowers them to participate effectively in the development of a common project, and the emerging public policies become operative, because of the adhesion of the greatest number" (Enda diapol 2009: 1) (Fig. 12.2).

In CLUVA, 'political dialogues' in combination with a systems analysis (LeFèvre 2003, 2011; Weets and Garcia-Aristizabal 2012) were used as an approach in order to investigate the underlying causes of risk events. This procedure helped to understand the failures leading to such events in a systematic manner. The approach can be tailored to the smallest entity of vulnerability that experiences these events, namely the household, neighbourhood or community. It clarifies the causes of flood events and makes transparent the degree of vulnerability with the ultimate aim of mobilising social networks in the management of crises, and to strengthen community dynamics on neighbourhood and kinship levels.

The political dialogue approach improves solutions for risk management and local adaptation by the provision of quality information, helping local officials to have a higher level of understanding through feedback from experience, a list of best practices, methods used by local authorities, and finally mobilising affected

Fig. 12.2 Women of a neighbourhood participating in a political dialogue session (Picture: Adrien Coly 2014)

communities for self-responsible action. For research, it supports the process of validation of knowledge for the communities that experience these phenomena. At the same time, the knowledge of scientists is confronted with, and enriched by local knowledge, which leads to the establishment of a common ontology on risks.

Usable Knowledge for Decision-Makers – The Ouagadougou Perspective

Ouagadougou City faces severe risks related to climate variability and change. The Sahel zone is most known for being hit by prolonged droughts, but in Ouagadougou, and indeed many other urban settlements of this region, flooding is a recurrent risk during the rainy seasons since more than a decade. The most disastrous event in the country history was the flooding of 1st September 2009, which revealed a number of weaknesses in connection with urban development and natural disaster management (Fig. 12.3).

The flooding of 1st September 2009 seriously impacted the country's economy. Damage and losses caused by the flood amounted nearly to USD 135 million and costs for building, reconstruction and repair to another USD 226 million (Ouedraogo 2011).

Floods are mainly caused by poor urban land development, the lack of an efficient response plan or its ignorance by stakeholders. Therefore, the origin and extent of the flooding damages are due to, among others, urban development in

Fig. 12.3 Flooding of September, 1st 2009 at the bridge to Ouagadougou University (Picture: Hamidou Touré 2009)

flood prone river valleys, the lack of sewage and rainwater drainage networks as well as to poor waste management and disposal.

Moreover, urban land use planning fails in involving various stakeholders in governance to the necessary extent. The individual roles and mandates of the stakeholders, notably the State, the municipality, the civil society and households, are not clearly specified in the local governance process. On top of this, non-compliance with good governance principles is widely persisting.

The CLUVA project has played a major role in the awareness building with regard to deficiencies in local governance of risks. Its results contribute especially to a better consideration of flood-related risks in public policies; additional studies will be essential to make Ouagadougou city more resilient in the future.

The present situation requires the identification of a set of strategies and concrete measures to reduce the vulnerability of natural and human systems in urban areas. It will be necessary to develop systems for forecasting, prevention and management of natural disasters as well as climate change impacts through predictive models and monitoring of extreme events.

The way forward towards more climate resilience of Ouagadougou calls for the improvement of information systems for being able to release real-time alerts to populations in case of natural disasters' risks; in particular, the collection system of meteorological and hydrological data should be optimised. Furthermore, we should continue to educate our citizens in the management of risks through more efficient local governance. Climate change issues have to be integrated into programmes and projects at national and local levels, as well as investments in infrastructure and other urban structures being earmarked in an appropriate manner (Fig. 12.4).

Fig. 12.4 Sewer clogged by waste in the city of Ouagadougou (Picture: Hamidou Touré 2011)

All this will have to be based on a close collaboration between scientists and decision-makers on local governance, as research must continue to produce some usable knowledge for policy makers on climate risks (Box 12.1).

Box 12.1: The CLUVA Project – Viewpoint on Challenges

CLUVA was the first European project in which our research team took part. Despite some shortcomings with respect to language barriers (French) and the still limited involvement of researchers from the South, we would judge the project as a success. Success is measured both in terms of the results we have achieved and the experience gained by our team in conducting multidisciplinary research involving decision-makers at local level.

One of the challenges for the success of this project was to conduct research activities on multiple sites in a multicultural context. CLUVA has made some achievements in this very difficult context. Another one was to produce scientific knowledge for decision-makers in local governance. *(Prof. Hamidou Touré, University of Ouagadougou, Burkina Faso)*

Institutionalising Flood Management – The Addis Ababa Perspective

Flooding is the most important climate related challenge the city of Addis Ababa is facing. Heavy flooding that resulted in the loss of property (houses, roads, electric and telephone lines) is becoming frequent during the rainy season. The increase in built-up structure has aggravated the problem of flooding (Fig. 12.5).

Recognising the severity of the flood problem in Addis Ababa, the Ethiopian government commissioned the Japan International Cooperation Agency to undertake a "Study on the Addis Ababa Flood Control Project" in 1998. The project developed a master plan, including both structural and non-structural measures, for flood control to mitigate flood damage in the city for the target year 2020. In addition, the project identified priority interventions and conducted a feasibility study for this set of selected measures. Although the study recommended the immediate implementation of the priority projects, neither of them has been implemented in the last 16 years.

The major problem with the implementation of the proposed projects and recommended measures was the absence of an institution for flood management in Addis Ababa. At the Construction and Housing Development Bureau a small unit is trying to manage the flood problems of the city. Absence of institutional arrangements for flood management was one of the major gaps identified during the course

Fig. 12.5 Informal settlements in flood prone urban areas: river bank of Bantiyiketu, Addis Ababa (Picture: Kumelachew Yeshitela 2012)

of the CLUVA project. To this end, CLUVA recommended the establishment of a full-fledged institution for integrated water/river management in Addis Ababa.

In order to make Addis Ababa climate resilient in the future, the local research team would recommend the following strategies and entry points:

- *Institutionalising climate change and flood management*: As the activities for dealing with climate change adaptation are fragmented in different sector offices, there must be an institutional set-up for coordinating all the activities of climate change adaptation. The Addis Ababa Environmental Protection Authority (AAEPA) should be strengthened as a regulatory authority responsible for regulating all environmental issues in the city. An organisation for implementing green space development and management and another one for the implementation of integrated water management should be established.

  Moreover, the collaboration of the federal government, city administration, city residents, academic and research institutions as well as non-governmental organisations is of paramount importance. The Ethiopian Institute of Architecture, Building Construction and City Development (EiABC) of Addis Ababa University is already working with the city administration in generating knowledge in the fields of urban planning, climate change adaptation, green infrastructure planning, and sustainable housing development. EiABC is seeking to enhance community-based adaptation by incorporating stakeholders, including into the research process, providing technical support, and disseminating research outputs to beneficiaries (Fig. 12.6).

Fig. 12.6 Participants of the CLUVA annual stakeholder workshop in Addis Ababa 2013 (Picture: Kumelachew Yeshitela 2013)

- *Mainstreaming climate change*: as the issue of climate change is a cross-sectoral one, it has to be mainstreamed into urban planning, waste management, housing development, transport, environmental and industrial development sectors of the city. At the Mayor's office, a sub-division that cares for the integration of activities related to climate change adaptation needs to be set-up.
- *Enhancing community-based adaptation*: Empowering communities (e.g. flood affected communities in the Southern part of the city) to adapt to climate change through collaborative research, capacity building, technical and financial support.
- *Green infrastructure planning based on ecosystem services*: since ecosystem services could be valued in economic terms, they are found to be important tools for the management of urban ecosystems.
- *Stakeholder participation*: Stakeholders should participate in the planning, design and implementation of development projects that have environmental and social impacts.
- *Capacity-building at all levels*: As the knowledge of climate change impacts, adaptation, mitigation, and disaster risk management are in their infancy stage in the African cities, the integration of these into school curricula at different levels allows development of the human capacity that is needed for managing the impacts of climate change.
- *Establishing networks*: For sharing knowledge, information exchange, capacity building and obtaining technical and financial support, Addis Ababa needs to establish networks with other cities, national and international organisations dealing with environmental management and climate change issues.

Conclusion

Eight climate scenarios using different modelling and downscaling techniques have been produced in the framework of the CLUVA project to better understand the future climate in the five selected case study cities Addis Ababa, Dar es Salaam, Douala, Ouagadougou and Saint Louis up to the year 2050. In parallel, spatial urban scenarios were developed which impressively indicate the enormous expansion of African cities and the concurrent loss of ecosystem services if current speed and pattern of low-density urban growth is to be continued in the coming decades. From these scenarios it becomes obvious that, although the risk of floods, droughts and desertification is expected to increase slightly, human factors such as the growth of urban populations, scarce development or disappearance of green areas, and unorganised informal settlement sprawl in and around the cities may be even more relevant to accelerate the consequences of climate change, and thus will need priority attention by science and decision-makers in the context of urbanisation and climate change.

> ...I spoke to [a colleague of mine] a few minutes ago. The wholesale destruction of old buildings to be replaced by high-rise buildings [in Dar es Salaam] is becoming obscene. At the moment those who are all out to attract the big investors hold the power...much power, but large scale protests are now frequent. How does one carry a voice of reason? (Statement by Tanzanian scientist, June 2014)

Indeed, how does one carry a voice of reason on the background of galloping urbanisation in today's African cities, with highly vulnerable, informal squatter settlements flourishing in flood-prone areas and escaping resettlement initiatives due to political patronage from up above? How to develop and implement compact urban development with adequate levels of green space, and climate resilient infrastructures? Local city governments are overwhelmed with many other tasks of the developmental agenda, whilst being understaffed with financial and human resources for doing so. At present, institutional set-ups at city level are simply too weak and fragmented to elaborate and implement the necessary strategies for dealing with disaster risks and climate related hazards in an integrated and foresighted manner. Small cities and towns and their rural hinterlands may hold the potential for climate-smart development but this remains utopia in the face of scarce decision-making power and resources.

The CLUVA project has encountered numerous challenges for today's African cities to become more climate resilient in the future. It has, however, also identified a set of entry points to do so, which encourages one to carry a voice of hope, at least. Moreover, CLUVA has catalysed valuable insights into better understanding the probable impacts of climate change for its partnering cities and brought forward innovative frameworks and tools for holistic vulnerability risk assessments, green infrastructure planning, multi-level governance analysis und urban spatial scenario modelling, to name a few.

Certainly, the CLUVA project also had its limitations. While it made important steps towards developing an integrated approach to vulnerability assessment, three years have been too short to see it through to its full implementation. There is further need for combining climate change dynamics and the dynamics of urbanisation into multi-risk and vulnerability frameworks in a foresighted approach. More work is due for identifying appropriate strategies to reduce the vulnerability of the built environment with its critical infrastructure, green infrastructure and society. Urban scenarios could gain additional value for decision-makers by accompanying economic research efforts. Further investigation is required to gain more in-depth evidence on the correlation of climate projections with the full set of cascading effects on the cities. For instance, natural disasters may produce a range of pressures on urban centres in complex contexts related to their respective hinterlands that remain unexplored. The fact that urbanisation and climate change processes involve the city, not only as stand-alone agent, but rather as nodes within the broader settlement hierarchy, points at the necessity of undertaking intensified research efforts in this regard.

A longer duration of CLUVA would have been beneficial to further develop the science-policy interface in order to enhance the transfer and implementation of the valuable data and new knowledge generated during the project into local authorities' daily work.

Innovative tools and strategies are also condemned to remain 'paper tigers' without appropriate responsiveness and application by local decision-makers, if further research, action and support to and by African cities are neglected. The continent's present situation in the 'cross-fire' of urbanisation and climate change is highly complex, and thus calls for a multi-faceted approach.

Top-down reform of the urban planning sector in the cities must be at the heart of such concerted action to enable uptake of innovative methods, such as sound assessments of physical, urban ecosystem and social vulnerabilities, and the efficient integration of their evidence into future urban development and responsible action. Furthermore, strengthened controlling and enforcement mechanisms are due from up above in order to avoid uncontrolled city development and political patronage that can block essential and possibly life-rescuing resettlement measures.

Planning must be strengthened, not only at the city-level, but further capacity needs to be developed for integrated planning and implementation at the city-regional level to establish a functioning system of urban settlements and enhance urban-rural relations. The scenario modelling approach for Addis Ababa and Dar es Salaam indicated that city areas will rapidly fill up, leaving no space for green structure, unless a more extensive plan is adopted to develop neighbouring towns into a polycentric urban pattern and thus relieve the pressure on the core cities. These goals call for an alignment with the need of securing farmland for food supply and reducing the flooding risk by developing approaches to integrated watershed management.

In parallel, existing initiatives on household and community levels should be supported by all means from bottom-up in order to utilise the capacity of local populations to mobilise kinship, religious and other informal networks for the purpose of increasing disaster preparedness.

In a nutshell: It is not bottom-up *or* top-down – it is action from *both directions* that is urgently needed to enable African cities to move forward to more climate resilience in the years to come.

It is not adaptation *or* mitigation – it is *both parts of the climate change agenda* being meaningful to advance necessary capacities and action on the continent.

Hence, it will have to be concerted action getting on board all entities, ranging from the wider research community, local city governments, and urban planning authorities to practitioners, the private sector, NGOs and civil society, if climate resilient cities shall become a reality in Africa.

And it is not the African cities and states alone who need to act. It is also in the relevant and decisive international fora, where all efforts must be joined in order to detect and eliminate barriers presently hindering African cities from their desired transformation towards improved climate resilience.

References

Abo El Wafa H (2013a) Urban settlement dynamics scenario modelling in Addis Ababa: background information. Available: http://www.cluva.eu/CLUVA_publications/CLUVA-Papers/USSDM_Addis Ababa_Background Information_JULY2013a.pdf. Accessed 1 Oct 2014

Abo El Wafa H (2013b) Urban settlement dynamics scenario modelling in Addis Ababa: technical user guide. Available: http://www.cluva.eu/CLUVA_publications/CLUVA-Papers/USSDM_Addis Ababa_Technical User Guide_JULY2013b.pdf. Accessed 1 Oct 2014

Bellucci A, Bucchignani E, Gualdi S, Mercogliano P, Montesarchio M, Scoccimarro E (2012) Data for global climate simulations available for downscaling. CLUVA deliverable D1.1. Available: http://www.cluva.eu/deliverables/CLUVA_D1.1.pdf

Bucchignani E, Garcia-Aristizabal A, Montesarchio M (2014) Climate-related extreme events with high resolution regional simulations: assessing the effects of climate change scenarios in Ouagadougou, Burkina Faso. In: Beer M, Au S-L, Hall JW (eds) Vulnerability, uncertainty, and risk: quantification, mitigation, and management. American Society of Civil Engineers (ASCE), Reston - VA, USA, pp 1351–1362. doi:10.1061/9780784413609.136

Buchta K (2013a) Urban spatial scenario design modelling (USSDM) in Dar es Salaam – background information. Available: http://www.cluva.eu/CLUVA_publications/CLUVA-Papers/USSDM_Dar es Salaam_Background Information_JULY2013.pdf. Accessed 1 Oct 2014

Buchta K (2013b) Urban spatial scenario design modelling (USSDM) in Dar es Salaam – technical user guide. Available: http://www.cluva.eu/CLUVA_publications/CLUVA-Papers/USSDM_Dar es Salaam_Technical User Guide_JULY2013.pdf. Accessed 1 Oct 2014

Bulkeley H (2013) Cities and climate change, critical introductions to urbanism and the city. Routledge, Oxon/New York

Carozza S, De Risi R, Jalayer F, De Paola F, Yonas N, Mbuya E, Feumba R (2013) Guidelines to decreasing physical urban vulnerability in the 3 considered cities. CLUVA deliverable D2.5. Available: http://www.cluva.eu/deliverables/CLUVA_D2.5.pdf. Accessed 1 Oct 2014

Cavan G, Lindley S, Roy M, Woldegerima T, Tenkir E, Yeshitela K, Kibassa D, Shemdoe R, Pauleit S, Renner F, Printz A, Ouédraogo Y (2012a) International evidence of the ecosystem services of urban green infrastructure in different climate zones. CLUVA deliverable D2.6. Available: http://www.cluva.eu/deliverables/CLUVA_D2.6.pdf. Accessed 1 Oct 2014

Cavan G, Lindley S, Yeshitela K, Nebebe A, Woldegerima T, Shemdoe R, Kibassa D, Pauleit S, Renner F, Printz A, Buchta K, Coly A, Sall F, Ndour NM, Ouédraogo Y, Samari BS, Sankara BT, Feumba RA, Ngapgue JN, Ngoumo MT, Tsalefac M, Tonyé E (2012b) Green infrastructure maps for selected case studies and a report with an urban green infrastructure mapping methodology adapted to African cities. CLUVA deliverable D2.7. Available: http://www.cluva.eu/deliverables/CLUVA_D2.7.pdf. Accessed 1 Oct 2014

Cavan G, Lindley S, Jalayer F, Pauleit S, Renner F, Gill S, Capuano P, Nebebe A, Woldegerima T, Kibassa D, Shemdoe R (2014) Urban morphological determinants of temperature regulating ecosystem services in two African cities. Ecol Indic 42:43–57

CLUVA (2013) Evaluation of the potential of urban ecosystem services. Final deliverable of Task 2.2 of the European Union CLUVA project [ENV.2010.2.1.5-1]. Available: http://www.cluva.eu/deliverables/CLUVA_D2.10.pdf. Accessed 1 Oct 2014

Coly A, Topa ME, Di Ruocco A (2013) Hazard scenario impacts for the six climate change models. CLUVA deliverable D5.6. Available: http://www.cluva.eu/deliverables/CLUVA_D5.6.pdf. Accessed 1 Oct 2014

CSIR (2013) Three PHD courses/summer schools. CLUVA deliverable D4.2. Available: http://www.cluva.eu/deliverables/CLUVA_D4.2.pdf. Accessed 1 Oct 2014

CSIR, CMCC (2012a) Regional climate change simulations available for the selected areas. CLUVA deliverable D1.5. Available: http://www.cluva.eu/deliverables/CLUVA_D1.5.pdf. Accessed 1 Oct 2014

CSIR, CMCC (2012b) Web climate service. CLUVA deliverable D1.6. Available: http://www.cluva.eu/deliverables/CLUVA_D1.6.pdf. Accessed 1 Oct 2014

CSIR, CMCC (2013) Climate maps and statistical indices for selected cities. CLUVA deliverable D1.7. Available: http://www.cluva.eu/deliverables/CLUVA_D1.7.pdf. Accessed 1 Oct 2014

De Paola F, Giugni M, Garcia A, Bucchignani E (2013a) Stationary vs. non-stationary of extreme rainfall in Dar es Salaam (Tanzania). IAHR Congress Tsinghua University Press, Beijing

De Paola F, Giugni M, Topa ME (2013b) Probability density function (Pdf) of daily rainfall heights by superstatistics of hydro-climatic fluctuations for African test cities. Wulfenia 20(5):106–126

De Risi R (2013) A probabilistic bi-scale framework for urban flood risk assessment. PhD dissertation, Department of Structures for Engineering and Architecture, University of Naples Federico II, Naples

De Risi R, Jalayer F, Iervolino I, Kyessi A, Mbuya E, Yeshitela K, Yonas N (2012) Guidelines for vulnerability assessment and reinforcement measures of adobe houses. CLUVA deliverable D2.4. Available: http://www.cluva.eu/deliverables/CLUVA_D2.4.pdf. Accessed 1 Oct 2014

De Risi R, Jalayer F, De Paola F, Iervolino I, Giugni M, Topa ME, Mbuya E, Kyessi A, Manfredi G, Gasparini P (2013a) Flood risk assessment for informal settlements. Nat Hazards 69(1):1003–1032. doi:10.1007/s11069-013-0749-0

De Risi R, Jalayer F, Iervolino I, Manfredi G, Carozza S (2013b) VISK: a GIS-compatible platform for micro-scale assessment of flooding risk in urban areas. In: Papadrakis M, Papadopoulos V, Plevris V (eds) COMPDYN, 4th ECCOMAS thematic conference on computational methods in structural dynamics and earthquake engineering, Kos Island, Greece, 12–14 June 2013

De Risi R, Jalayer F, De Paola F, Topa ME, Iervolino I, Giugni M, Yonas N, Nebebe A, Yeshitela K, Kibassa D, Shemdoe R, Cavan G, Lindley S, Printz A, Renner F (2013c) Identification of hotspots vulnerability of adobe houses, sewer systems and road networks. CLUVA deliverable D2.1. Available: http://www.cluva.eu/deliverables/CLUVA_D2.1.pdf. Accessed 1 Oct 2014

Di Ruocco A, Giugni M, Garcia-Aristizabal A, Sellerino M, Topa ME, Iavazzo P (2012) Software and methods for quantitative hazard scenarios. CLUVA deliverable D1.4. Available: http://www.cluva.eu/deliverables/CLUVA_D1.4.pdf. Accessed 1 Oct 2014

Diop I, Lo M (2013) An ontology design pattern of the multidisciplinary and complex field of climate change. ACSIJ Adv Comput Sci Int J 2(5). Available: http://www.cluva.eu/CLUVA_publications/CLUVA-Papers/ACSIJ-2013-2-5-258CLUVA.pdf. Accessed 1 Oct 2014

Enda diapol (2009) Promotion du dialogue citoyen et amélioration de la gouvernance locale. Communication lors première édition de l'Université des Acteurs Non Etatiques, 18-19-20 novembre, ex ENAM, Dakar, Sénégal, 3p. Available online: http://www.endadiapol.org/IMG/pdf/fiche_de_synthese_promotion_de_dialogue_citoyen_eetamelioratio_de_la_gouvernance_locale.pdf. Accessed 25 Sept 2014

Engelbrecht FA, Landman WA, Engelbrecht CJ, Landman S, Bopape MM, Roux B, McGregor JL, Thatcher M (2011) Multi-scale climate modelling over Southern Africa using a variable-resolution global model. Water SA 37(5). Available: http://dx.doi.org/10.4314/wsa.v37i5.2. Accessed 1 Oct 2014

Esposito S, Elefante L, Iervolino I (2012) Guidelines for reliability analysis of roadway network including procedures for emergency response management. CLUVA deliverable D2.3. Available: http://www.cluva.eu/deliverables/CLUVA_D2.3.pdf. Accessed 1 Oct 2014

Fohlmeister S, Pauleit S, Kombe W, Kyessi A, Coly A, Sarr C, Touré H, Yonkeu S, Tonyé E, Yeshitela K (2013) University curriculum(a) at master's level. CLUVA deliverable D4.3. Available: http://www.cluva.eu/deliverables/CLUVA_D4.3.pdf. Accessed 31 Aug 2014

Garcia-Aristizabal A, Marzocchi W (2012) Bayesian multi-risk model: demonstration for test city researchers. CLUVA deliverable D2.13. Available: http://www.cluva.eu/deliverables/CLUVA_D2.13.pdf. Accessed 15 Sept 2014

Garcia-Aristizabal A, Marzocchi W, Topa ME, Sellerino M, Di Ruocco A, Giugni M, De Paola F, Capuano P (2012) Harmonization of hazard indices and ranking of hazards. CLUVA deliverable D1.3. Available: http://www.cluva.eu/deliverables/CLUVA_D1.3.pdf. Accessed 15 Sept 2014

Garcia-Aristizabal A, Marzocchi W, Ambara G, Uhinga G (2013) Reports and map on multi-risk Bayesian scenarios on one selected city (Dar es Salaam, Tanzania). CLUVA deliverable D2.14. Available: http://www.cluva.eu/deliverables/CLUVA_D2.14.pdf. Accessed 15 Sept 2014

Garcia-Aristizabal A, Bucchignani E, Palazzi E, D'Onofrio D, Gasparini P, Marzocchi W (2015) Analysis of non-stationary climate-related extreme events considering climate-change scenarios: an application for multi-hazard assessment in the Dar es Salaam region, Tanzania. Nat Hazards 75:289–320. doi:10.1007/s11069-014-1324-z

Gasparini P, Di Ruocco A, Bruyas AM (eds) (2013) Climate change and vulnerability of African cities. Research briefs. CLUVA deliverable D4.5. Available: http://www.cluva.eu/deliverables/CLUVA_D4.5.pdf. Accessed 15 Sept 2014

Gill S, Handley J, Ennos R, Pauleit S (2007) Adapting cities for climate change: the role of the green infrastructure. J Built Environ 33(1):115–133

Giugni M, De Paola F, Topa ME (2012a) Guidelines on engineering design and management of storm water systems. CLUVA deliverable D2.2. Available: http://www.cluva.eu/deliverables/CLUVA_D2.2.pdf. Accessed 1 Oct 2014

Giugni M, Adamo P, Capuano P, De Paola F, Di Ruocco A, Giordano S, Iavazzo P, Sellerino M, Terracciano S, Topa ME (2012b) Hazard scenarios for test cities using available data. CLUVA deliverable D1.2. Available: http://www.cluva.eu/deliverables/CLUVA_D1.2.pdf. Accessed 1 Oct 2014

Giugni M, Capuano P, De Paola F, Di Ruocco A, Topa ME (2013) Probabilistic hazard and multihazard scenarios in GIS environment for the relevant hazards in selected cities. CLUVA deliverable D1.8. Available: http://www.cluva.eu/deliverables/CLUVA_D1.8.pdf. Accessed 1 Oct 2014

Halloran A, Magid J (2013) Planning the unplanned: incorporating agriculture as an urban land use into the Dar es Salaam master plan and beyond. Environ Urban 25:541–558

Healey P (2009) In search for the "strategic" in spatial strategy making. Plan Theory Pract 10(4):439–457

Herslund L, Mguni P, Lund DH, Souleymane G, Workneh A, Ouedraogo JB (2012) Exemplary policies, strategies and measures. CLUVA deliverable D3.6. Available: http://www.cluva.eu/deliverables/CLUVA_D3.6.pdf. Accessed 1 Sept 2014

Herslund L, Lund DH, Yeshitela K, Workalemahu L, Kombe W, Kyessi AG (2013) Strategic measures for the two cities. Most important strategies – recommendations. Recommendations for process and prioritisation of data needs. CLUVA deliverable D3.7 and D3.8. Available: http://www.cluva.eu/deliverables/CLUVA_D3.7.pdf. Accessed 1 Sept 2014

Jalayer F, De Risi R, Herslund L, Garcia-Aristizabal A, Carozza S, De Paola F, Yonas N, Nebebe A, Yeshitela K, Kibassa D, Shemdoe R, Jean-Baptiste N, Pauleit S, Lindley S (2013) Report describing the developed methods and its possible relevance for planning and decision-making. CLUVA deliverable D2.17. Available: http://www.cluva.eu/deliverables/CLUVA_D2.17.pdf. Accessed 1 Oct 2014

Jalayer F, De Risi R, De Paola F, Giugni M, Manfredi G, Gasparini P, Topa ME, Yonas N, Yeshitela K, Nebebe A, Cavan G, Lindley S, Printz A, Renner F (2014) Probabilistic GIS-based method for delineation urban flooding risk hotspots. Nat Hazards 73(2):975–1001

Jean-Baptiste N, Kabisch S (2013) Lessons learned and recommendations for assessing social vulnerability and adaptation strategies. CLUVA deliverable D2.12. Available: http://www.cluva.eu/deliverables/CLUVA_D2.12.pdf. Accessed 15 Sept 2014

Jean-Baptiste N, Kuhlicke C, Kunath A, Kabisch S, John R, Nguimbis JN, Fekade R, Herslund L, Lindley S, Kinda FB, Ouedraogo JB, Printz A (2011) Review and evaluation of existing vulnerability indicators in order to obtain an appropriate set of indicators for assessing climate related vulnerability. CLUVA deliverable D2.11. Available: http://www.cluva.eu/deliverables/CLUVA_D2.11.pdf. Accessed 15 Sept 2014

Jean-Baptiste N, Kabisch S, Kuhlicke C (2013) Urban vulnerability assessment in flood-prone areas in West and East Africa. In: Rauch S, Morrison GM, Schleicher N, Norra S (eds) Urban environment. Springer, Heidelberg/New York, pp 203–218

John R, Mayunga J, Kombe W, Fekade R, Ouedraogo JB, Tchangang R, Ndour NM, Ngom T, Gueye S, Sall F (2012) Preliminary findings on social vulnerability assessment in the selected cities. CLUVA deliverable D5.4. Available: http://www.cluva.eu/deliverables/CLUVA_D5.4.pdf. Accessed 1 Oct 2014

John R, Jean-Baptiste N, Kabisch S (2014) Vulnerability assessment of urban populations in Africa: the case of Dar es Salaam, Tanzania. In: Edgerton E, Romice O, Thwaites K (eds) Bridging the boundaries. Human experience in the natural and built environment and implications for research, policy, and practice, vol 5, Advances in people and environment studies. Hogrefe Publishing, Göttingen, pp 233–244

Jørgensen G, Herslund L, Sarr C, Nakouye N, Sine A, Workneh A, Workalemahu L, Bekele E (2012) Base line scenarios for urban development of selected case study areas. CLUVA deliverable D3.5. Available: http://www.cluva.eu/deliverables/CLUVA_D3.5.pdf. Accessed 1 Oct 2014

Jørgensen G, Herslund LB, Lund DH, Workneh A, Kombe W, Gueye S (2014) Climate change adaptation in urban planning in African cities: the CLUVA project. In: Gasparini P, Manfredi G, Asprone D (eds) Resilience and sustainability in relation to natural disasters: a challenge for future cities, SpringerBriefs in Earth sciences. Springer, Cham/Heidelberg/New York/Dordrecht/London, pp 25–37

Kibassa D (2014) Adaptation potential of green structures to urban heat island in urban morphological types of Dar es Salaam, Tanzania. PhD dissertation, Ardhi University, Tanzania

Kombe W, Kyessi A, Kassenga G, Shemdoe R, Yeshitela K, Workalemahu L, Workneh A, Coly A, Sall F, Touré H, Tonyé E (2013) Action plan to mitigate climate hazards. CLUVA deliverable D5.7. Available: http://www.cluva.eu/deliverables/CLUVA_D5.7.pdf. Accessed 1 Oct 2014

LeFèvre J (2003) Kinetic process graphs: building intuitive and parsimonious material stock-flow diagrams with modified bond graph notations. In: Proceedings of the 20th international conference of the system dynamics society, July 28–August 1, 2002, Palermo. http://www.systemdynamics.org/conferences/2002/proceed/PROCEED.pdf. Accessed 13 Sept 2014

LeFèvre J (2011) A tutorial on kinetic graphs to be used in the training sessions for the African researchers. CLUVA deliverable D2.15. Available: http://www.cluva.eu/deliverables/CLUVA_D2.15.pdf. Accessed 1 Oct 2014

Lindley S, Gill SE, Cavan G, Yeshitela K, Nebebe A, Woldegerima T, Shemdoe R, Kibassa D, Pauleit S, Renner F, Printz A, Buchta K, Abo El Wafa H, Coly A, Sall F, Ndour NM, Ouédraogo Y, Samari BS, Sankara BT, Feumba RA, Ambara G, Kandé L, Zogning MOM, Tonyé E, Pavlou A, Koome DK, Lyakurwa RJ, Garcia A (2013) A GIS based assessment of the urban green structure of selected case study areas and their ecosystem services. CLUVA Deliverable D2.8. Available: http://www.cluva.eu/deliverables/CLUVA_D2.8.pdf. Accessed 7 Sept 2014

López-Carresi A, Fordham M, Wisner B, Kelman I, Gaillard JC (2014) Introduction: who, what and why. In: López-Carresi A, Fordham M, Wisner B, Kelman I, Gaillard JC (eds) Disaster management: international lessons in risk reduction, response and recovery. Earthscan, London, pp 1–9

McGranahan G, Satterthwaite D (2014) Urbanisation concepts and trends. IIED working paper. IIED, London. Available: http://pubs.iied.org/pdfs/10709IIED.pdf. Accessed 25 Sept 2014

Mendelsohn R, Saher G (2010) The global impact of climate change on extreme events. Background paper for the report "Natural Hazards, UnNatural Disasters. The Economics of Effective Prevention" (2010). World Bank, Washington, DC

Moser C, Satterthwaite D (2008) Towards pro-poor adaptation to climate change in urban centres of low- and middle-income countries. Human settlements discussion paper series. Human Settlements Group and Climate Change Group: International Institute for Environment and Development (IIED), London

Niang I, Ruppel OC, Abdrabo MA, Essel A, Lennard C, Padgham J, Urquhart P (2014) Africa. In: Barros VR, Field CB, Dokken DJ, Mastrandrea MD, Mach KJ, Bilir TE, Chatterjee M, Ebi KL, Estrada YO, Genova RC, Girma B, Kissel ES, Levy AN, MacCracken S, Mastrandrea PR, White LL (eds) Climate change 2014: impacts, adaptation, and vulnerability. Part B: regional aspects. Fifth assessment report of the Intergovernmental Panel on Climate Change. Cambridge University Press, Cambridge/New York. Final draft available at: http://ipcc-wg2.gov/AR5/images/uploads/WGIIAR5-Chap22_FGDall.pdf. Accessed 4 Oct 2014

Nyed PK, Herslund L (2012) System of land use indicators for vulnerability to climate change. CLUVA deliverable D3.3. Available: http://www.cluva.eu/deliverables/CLUVA_D3.3.pdf. Accessed 13 Aug 2014

Nyed PK, Herslund L, Mangula A, Stysiak A (2013) Map of high-risk areas for selected case areas. Towards a holistic perspective on city-level vulnerability assessment. CLUVA deliverable D3.4. Available: http://www.cluva.eu/deliverables/CLUVA_D3.4.pdf. Accessed 1 Oct 2014

Ouedraogo S (2011) Approches d'intégration de l'adaptation aux changements climatiques dans les plans et projets locaux de développement: cas des plans communaux de développement au Burkina Faso, DGAT, ASDI, 37 p

Pauleit S, Liu L, Ahern J, Kazmierczak A (2011) Multifunctional green infrastructure planning to promote ecological services in the city. In: Niemelä J (ed) Handbook of urban ecology. Oxford University Press, Oxford, pp 272–285

Pauleit S, Buchta K, Abo El Wafa H, Renner F, Printz A, Kumelachew Y, Kibassa D, Shemdoe R, Kombe W (2013) Recommendations for green infrastructure planning in selected case study cities. CLUVA deliverable D2.9. Available: http://www.cluva.eu/deliverables/CLUVA_D2.9.pdf. Accessed 7 Sept 2014

Pickett STA, Cadenasso ML, Grove JM (2004) Resilient cities: meaning, models, and metaphor for integrating the ecological, socio-economic, and planning realms. Landsc Urban Plan 69:369–384

Potts D (2012) Whatever happened to Africa's rapid urbanisation? Africa Research Institute ARI Counterpoints. Available: http://africaresearchinstitute.org/files/counterpoints/docs/Whatever-happened-to-Africas-rapid-urbanisation-6PZXYPRMW7.pdf. Accessed 25 Sept 2014

Rauken T, Mydske PK, Winsvold M (2014) Mainstreaming climate change adaptation at the local level. Local Environ Int J Justice Sustain. doi:10.1080/13549839.2014.880412

Revi A, Satterthwaite D, Aragón-Durand F, Corfee-Morlot J, Kiunsi RBR, Pelling M, Roberts D, Solecki W (2014) Urban areas. Final draft. In: Field CB, Barros VR, Dokken DJ, Mach KJ, Mastrandrea MD, Bilir TE, Chatterjee M, Ebi KL, Estrada YO, Genova RC, Girma B, Kissel ES, Levy AN, MacCracken S, Mastrandrea PR, White LL (eds) Climate change 2014: impacts, adaptation, and vulnerability. Part A: Global and sectoral aspects. Fifth assessment report of the Intergovernmental Panel on Climate Change. Cambridge University Press, Cambridge/New York. Final draft available at: http://ipcc-wg2.gov/AR5/images/uploads/WGIIAR5-Chap8_FGDall.pdf. Accessed 4 Oct 2014

Ribot J (1999) Decentralization and participation in Sahelian forestry: legal instruments of central political-administrative control. Africa 69(1):23–43

Rosenzweig C, Solecki WD, Hammer SA, Mehrotra S (eds) (2011) Climate change and cities: first assessment report of the urban climate change research network. Cambridge University Press, Cambridge

Satterthwaite D (2013) The political underpinnings of cities' accumulated resilience to climate change. Environ Urban 25:381–391

Sibolla B, Vahed A, Engelbrecht F, Naidoo M, Mtsetfwa M, Simonis I (2012) A web-based data dissemination platform for climate-change risk and vulnerability. GISSA Ukubuzana general paper. Available: http://www.cluva.eu/CLUVA_publications/CLUVA-Papers/Bolelang-Sibolla.pdf. Accessed 1 Oct 2014

Simon D (2013) Climate and environmental change and the potential for greening African cities. Local Econ 28(2):203–217

Simon D (2014) New thinking on urban environmental change challenges. Int Dev Plann Rev 36(2):v–xi

Simonis I, Vahed A (2011) Training schedule and training programme for PHD students. Training needs, offered courses, requested courses, PHD courses/summer schools. CLUVA deliverable D4.1. Available: http://www.cluva.eu/deliverables/CLUVA_D4.1.pdf

Smit W, Parnell S (2012) Urban sustainability and human health: an African perspective. Curr Opin Environ Sustain 4:443–450

Tonyé E, Yango J, Tsalefac M, Ngosso AB, Nguimbis J, Ngohe Ekam PS, Moudiki C, Mgaba P, Tatietse T, Pancha Moluh PT, Giugni M, Capuano P, Topa ME, Kassenga G, Shemdoe R, Uhinga G, Kibassa D, Yeshitela K, Coly A, D'Almeida A, Diakhaté MM, Lo M, Sy BA, Ndour NM, Gueye S, Sall F, Sy AA, Touré H (2012) Report on climate related hazards in the selected cities. CLUVA deliverable D5.2. Available: http://www.cluva.eu/deliverables/CLUVA_D5.2.pdf

Twigg J (2009) Characteristics of a disaster-resilient community. A guidance note Version 2. Available via http://discovery.ucl.ac.uk/1346086/1/1346086.pdf. Accessed 31 Aug 2014

UN-Habitat (2009) Slums: levels and trends, 1990–2005. Monitoring the millennium development goals slum target. Available: http://www.unhabitat.org/downloads/docs/9179_33168_Slum_of_the_World_levels_and_trends.pdf. Accessed 1 Sept 2014

UN-Habitat (2011) Planning for climate change: a strategic, values-based approach for urban planners. UN-Habitat and Ecoplan International Inc., Nairobi

UN-Habitat (2014) The state of African cities 2014. Re-imagining sustainable urban transitions. United Nations Human Settlements Programme, Nairobi

UN-Habitat, TUM, IIED, UN University (eds) (2011) Global workshop on strengthening climate change in urban education – cities and climate change academy (CCCA). Final workshop report. Available: http://www.cluva.eu/CLUVA_publications/CLUVA-Papers/10205_1_594064.pdf. Accessed 1 Oct 2014

UNISDR (2012) Making cities resilient report 2012. My city is getting ready! A global snapshot of how local governments reduce disaster risk. UNISDR, Geneva. Available via http://www.unisdr.org/we/inform/publications/28240. Accessed 31 Aug 2014

United Nations, Department of Economic and Social Affairs, Population Division (UNDESA) (2014) World urbanization prospects: the 2014 revision. CD-ROM edition. United Nations, New York. Available online: http://esa.un.org/unpd/wup/. Accessed 15 Sept 2014

United Nations Environment Programme (UNEP) (2010) Africa Water Atlas. Division of Early Warning and Assessment (DEWA). United Nations Environment Programme, Nairobi

Vedeld T, Hellevik S, Kombe W, Kweka-Msale C, Coly A, Ndour NM, Guéye S, Klausen JE, Saglie I-L (2012) Report on planning system and government structure in 2 case cities. CLUVA deliverable D3.1. Available: http://www.cluva.eu/deliverables/CLUVA_D3.1.pdf. Accessed 15 Sept 2014

Vedeld T, Kombe W, Coly A, Ndour NM, Kweka-Msale C, Hellevik S (2013) Recommendation of how climate change can be better integrated in the planning and government system. CLUVA deliverable D3.2. Available: http://www.cluva.eu/deliverables/CLUVA_D3.2.pdf. Accessed 15 Sept 2014

Walker B, Holling CS, Carpenter SR, Kinzig A (2004) Resilience, adaptability and transformability in social–ecological systems. Ecol Soc 9(2):5 [online] URL: http://www.ecologyandsociety.org/vol9/iss2/art5. Accessed 13 Sept 2014

Weets G, Garcia-Aristizabal A (2012) A detailed description of the kinetic graph generic model to be instantiated by the test case researchers. CLUVA deliverable D2.16. Available: http://www.cluva.eu/deliverables/CLUVA_D2.16.pdf. Accessed 1 Oct 2014

Weets G, Coly A, Tonye E, Touré H, Yeshitela K, Kombe W (2014) Synthesis of the results obtained by the case studies. CLUVA deliverable D5.8. Available: http://www.cluva.eu/deliverables/CLUVA_D5.8.pdf. Accessed 1 Oct 2014

Wilbanks TJ, Romero Lankao P, Bao M, Berkhout F, Cairncross S, Ceron J-P, Kapshe M, Muir-Wood R, Zapata-Marti R (2007) Industry, settlement and society. In: Parry ML, Canziani OF, Palutikof JP, van der Linden PJ, Hanson CE (eds) Climate change 2007: impacts, adaptation and vulnerability. Fourth assessment report of the Intergovernmental Panel on Climate Change. Cambridge University Press, Cambridge, pp 357–390

World Bank (2009) Making development climate resilient: a World Bank strategy for sub-Saharan Africa. World Bank, Sustainable Development Department, Report no. 46947-AFR, Washington, DC

Index

S. Pauleit et al. (eds.), *Urban Vulnerability and Climate Change in Africa*, Future City 4,
DOI 10.1007/978-3-319-03982-4

Printed by Printforce, the Netherlands